Archimedes

New Studies in the History and Philosophy of Science and Technology

Volume 78

Archimedes has three fundamental goals: to further the integration of the histories of science and technology with one another; to investigate the technical, social and practical histories of specific developments in science and technology; and finally, where possible and desirable, to bring the histories of science and technology into closer contact with the philosophy of science.

The series is interested in receiving book proposals that treat the history of any of the sciences, ranging from biology through physics, all aspects of the history of technology, broadly construed, as well as historically-engaged philosophy of science or technology. Taken as a whole, Archimedes will be of interest to historians, philosophers, and scientists, as well as to those in business and industry who seek to understand how science and industry have come to be so strongly linked.

Submission / Instructions for Authors and Editors: The series editors aim to make a first decision within one month of submission. In case of a positive first decision the work will be provisionally contracted: the final decision about publication will depend upon the result of the anonymous peer-review of the complete manuscript. The series editors aim to have the work peer-reviewed within 3 months after submission of the complete manuscript.

The series editors discourage the submission of manuscripts that contain reprints of previously published material and of manuscripts that are below 150 printed pages (75,000 words). For inquiries and submission of proposals prospective authors can contact one of the editors:

Please find on the top right side of our webpage a link to our *Book Proposal Form.*

Paolo Galluzzi

Momento: Galilean Studies

Translated by Andrew Rodionov

Springer

Paolo Galluzzi
Istituto e Museo di Storia della Sc
Museo Galileo
Firenze, Italy

ISSN 1385-0180 ISSN 2215-0064 (electronic)
Archimedes
ISBN 978-3-032-35048-0 ISBN 978-3-032-35049-7 (eBook)
https://doi.org/10.1007/978-3-032-35049-7

Translation from the Italian language edition: "Momento. Studi galileiani." by Paolo Galluzzi, © Italian Research Institute 1979. Published by Ateneo e Bizzarri, Rome, on behalf of Italian Research Institute. All Rights Reserved.
Ateneo e Bizzarri, Rome, on behalf of Italian Research Institute. Momento. Studi galileiani. Italian. 1979. Italian Research Institute.

This Springer imprint is published by the registered company Springer Nature Switzerland AG
The registered company address is: Gewerbestrasse 11, 6330 Cham, Switzerland

Note on the Original Publication

Momento. Studi galileiani (Edizioni dell'Ateneo & Bizzarri, Rome 1979) was originally published in the series of the Lessico Intellettuale Europeo (LIE, today ILIESI), an Institute of the Italian Consiglio Nazionale delle Ricerche. ILIESI generously granted permission for the publication of the English edition of this work.

AUTHOR'S DEDICATION

In memory of Tullio Gregory (1929-2019), with gratitude.

Foreword

"A classic is a book that has never exhausted all it has to say to its readers."

This definition of 'classic' is by the Italian writer Italo Calvino who, in *Why Read the Classics?* (*Perché leggere i classici*, 1991; English translation 1999), provided a penetrating survey of what constitutes a 'classic', setting out fourteen distinct definitions of the notion.

Paolo Galluzzi's *Momento* is certainly a 'classic' of Galilean studies and, more broadly, of the history of scientific thought. Although published in 1979, the book continues to elicit innovative reflection, with each rereading offering novel prompts, insights, and perspectives.

The fruit of extraordinary analytical acuity and conceptual penetration, *Momento* shrewdly reconstructs the complex trajectory of a word to which different meanings were ascribed.

The book is conceived as both a lexical and a historical journey. From the opening pages, Galluzzi draws extensively on lexicographical and doctrinal sources—dictionaries, specialized lexica, biblical materials, and scholastic treatises and commentaries—and traces the evolution of *momentum* (and its Italian equivalent *momento*) as a technical term in Renaissance and early modern mechanics.

This careful, preliminary reconstruction of the term's history in scientific and philosophical contexts paves the way for a discussion of its role in Galileo's work. In that context, Galileo used the term to denote different concepts, interweaving statics, dynamical problems, and kinematic solutions. This generated a specific, distinctive semantic field with significant implications for understanding the evolution of Galilean thought.

Momento appeared in the study series of the *Lessico Intellettuale Europeo*—an important research center founded by the historian of philosophy Tullio Gregory, a leading figure in Italian culture across the twentieth and twenty-first centuries. The book was part of a project that sought, through lexical and terminological analysis, to clarify the conceptual repertoire that has shaped European philosophical and scientific thought since the Middle Ages.

Preceded by works on the lexicon of René Descartes, Marsilio Ficino, Giordano Bruno, Giambattista Vico, Baruch Spinoza, and other authors, Galluzzi's book addresses a fundamental problem in the history of science.

The Italian word *momento* and its Latin counterpart *momentum* are, indeed, key terms for understanding the development of Galilean mechanics and kinematics. As Galluzzi himself notes at the end of the book:

> an investigation into the Galilean "lexicon" – or rather, substantially, into a single term, although extremely significant – has been transformed, almost automatically, into a general essay on the genesis of the Galilean conception of movement, on its successive developments, on the theoretical elaboration, the "foundation", which accompanied it. If this was possible, it is because *momento* has proven to be, beyond initial expectations, a crucial term.

The variety of meanings of *momentum* was not peculiar to Galileo. Before him, the term had already seen conceptually significant use, dating back to its Greek antecedent *rhopé* (*ῥοπή*), attested even in translations of certain biblical passages (*Isaiah* 40:15; *Wisdom* 11:23) to denote the minimal addition of weight that upsets the balance between the two pans of a scale.

Within ancient mechanics, the term *rhopé* belongs to the Archimedean tradition: Archimedes uses the cognates *rhepein* and *isorropein* in *On the Equilibrium of Planes*, while *rhopé* appears in Eutocius's commentary on that work. After 1460, *rhopé* was often rendered as *momentum*. Beyond its static use within the Archimedean tradition, *momentum* was also employed in Aristotelian contexts to translate the concept of *κίνημα* (movement), as it occurs in two passages of Book VI of the *Physics* (232a 6–11; 240b 30–241a 6).

The in-depth examination of the development of these traditions in the Renaissance and early modern period introduces the analysis of *momentum/momento* in Galileo's writings. Galileo used the lemma in a wide range of different senses: with the general and common meaning of an instant of time; with that of "minimal quantity"; as denoting the natural inclination to motion produced by a body's weight and position; and even assigning it a role in the study of accelerated motion, employing it to designate the infinitesimal increments of velocity (*momenta velocitatis*).

The very question of the relation between weight and motion—that is, the relationship between the 'static' and 'dynamic' senses of the term *momento*—lies at the center of Galluzzi's analysis of Galileo's work.

It is a meticulous and incisive examination that considers the full range of Galileo's writings, from the early *Theoremata circa centrum gravitatis solidorum* to the *Discorsi e dimostrazioni matematiche intorno a due nuove scienze* (1638).

This study is of the utmost interest, as it sheds light on the uncertainties, vacillations, and even shortcomings that marked the evolution of Galilean thought. Accordingly, for anyone seeking to understand the trajectory of Galileo's research on motion, *Momento* is an indispensable resource. Due to its analytical and explanatory power, it deserves to be regarded as a veritable classic of Galilean studies.

It is much to Andrew Rodionov's credit that he undertook the formidable task of translating *Momento* into English.

The complexity of the undertaking makes his achievement all the more admirable and deserving of praise. *Momento* is not an easy text to read—still less to translate. It contains Greek and Latin passages alongside seventeenth-century Italian, the language of Galileo and many of his sources. This register can challenge even contemporary Italian readers, making the book intricate and exacting.

Rodionov surmounted these obstacles with distinction, producing a reliable translation of the entire work.

Thanks to his extraordinary tenacity, an international readership now has access to a landmark work in Galilean studies.

MICHELE CAMEROTA

Translator's Preface

When I first came across Paolo Galluzzi's *Momento* in September 2022, I had already spent several months attempting to decipher Galileo's use of the term *momento* in his *Discorsi e dimostrazioni matematiche intorno a due nuove scienze* (1638; *Discourses and Mathematical Demonstrations Relating to Two New Sciences*). I was particularly concerned with its curious uses contained in amendments by Vincenzo Viviani inserted into a scholium after *Theorem II* on naturally accelerated motion, additions generally included in later editions of the text (the English translation by Henry Crew and Alfonso de Salvio, 1914, being the most readily available today; see also the excellent translation by Stillman Drake, 1974; both translations are based upon the Edizione Nazionale version of *Le Opere di Galileo Galilei*, vol. VIII, edited by Antonio Favaro, 1898).

Throughout the work, *momento* is never explicitly defined; instead, it is introduced through a series of ostensive examples. Some uses are straightforward: within the context of an infinitesimal instant, it is easily understood to mean "momentary," and in explaining the lever and the balancing of weights, it is recognized as the "mechanical moment," that is, the product of distance and weight (in the static sense). Yet in the discussion of the dynamics of natural acceleration in free fall – later supplemented by Viviani's scholium – *momento* appears with an unusual function, serving as both a force and as something proportionate to speed, often in the sense of an incremental change (differential) of speed with respect to intervals of time. This understanding is foreign to modern physics, where "momentum" in classical mechanics denotes a quantity proportional to velocity, a quantity received from an impulse (i.e., the product of force and time), but not as a force alone, making the duality of Galileo's *momento* a source of confusion for modern readers and a frequently debated interpretation among Galilean scholars.

Turning to Galileo's earlier works, beginning with his *Le Mecaniche* (*On Mechanics*, ca. 1600), which contains his first definition and technical applications of *momento*, he employs the term predominantly within the context of statics. He also demonstrates, on several occasions, a quasi-dynamic sense of the term through hypothetical motions, that is, virtual displacements or virtual speeds of the respective weights of a system, even though those weights remain in equilibrium and at rest. Although no motion occurs, Galileo shows that if the weights were to move hypothetically, their virtual speeds would be proportional to their *momenti*, since the resulting speed is proportional to the weight's distance from the fulcrum.

From this, it was not immediately clear how the static and quasi-dynamic uses of *momento* in mechanical systems relate, if at all, to the actual, natural accelerated motions of free fall. Was *momento* intended as a dual concept, simultaneously a force and measure of speed, or did it maintain distinct functions, and if so, at which points in his reasoning should each interpretation be applied? When Galileo states that, at the bottom of its swing, the pendulum acquires a *momento* capable of lifting it to some height; when he equates the *momenta* of accelerated motion to the *momenta* of uniform motion; or when he relates the *momento* of a weight on an inclined plane to its speed of descent, is the term intended to denote speed, force, or both? A link between a purely static use and a purely dynamic one seems implied; yet in the absence of any explicit demonstration connecting the two, they remained semantically and contextually distinct.

Unable to follow Galileo's proofs coherently due to the ambiguity of *momento*, I turned to various publications on the subject for guidance, only to find that more recent scholarly discussions of the term respond to, or reference, Galluzzi's 1979 publication *Momento*. Realizing that Galluzzi's book was available only in Italian, I decided to obtain a copy and compile at least a rough translation of select portions for my own use.

By mid-November 2023, after translating the first several chapters, which proved to be among the most challenging parts of the text, it became increasingly clear that translating the entire work would be worthwhile. I was relieved to see that Galluzzi acknowledged the complexity of *momento*, the very difficulty that had motivated his book. As I continued reading, I became fascinated by his historical tracing of *moment-momentum*, including their origins, translations, and occasional mistranslations. This fascination led me to reach out to Professor Galluzzi, who expressed surprise that I had taken an interest in his "old book," to gauge his interest in publishing an English translation. Although initially apprehensive of the project's feasibility, he eventually gave his blessing for me to continue the translation and encouraged me to see it through to completion. Roughly twenty months later, this undertaking has reached fruition, and I am pleased to present a text that has given me, and I hope will give others, greater clarity in understanding Galileo's study of motion.

As we will see throughout *Momento*, the terms *moment-momentum* have a long history of mistranslation and misunderstanding, and have carried many different meanings. Since the present work is itself another translation, some errors of translation or interpretation may inevitably remain.

To at least partially mitigate these inevitable shortcomings, I have preserved key source passages in their original language, along with a translation contained within square brackets. In some instances, the original text utilized square brackets, which are now followed by an asterisk (*). Although my contributions to the cited passages are contained within brackets, this is not the case for the *Appendix* since Galluzzi relies heavily on brackets for passage citations to Galileo's texts. Instead, each section in the *Appendix* presents the original text in full, followed by its English translation.

References to works containing their original titles are preserved, but titles that were translated into Italian are likewise translated into English. For example, since Galluzzi presented the original title of Archimedes's *Ἐπιπέδων ἰσορροπιῶν* with the Italian title *Equilibrio dei piani*, I have opted for the English title, *Equilibrium of Planes*.

References to gravity and levity (e.g., *gravitatis* and *levitatis*) are respectively translated as "heaviness" and "lightness" to prevent any potential association with the modern, Newtonian sense of gravity. Thus, "heaviness" emphasizes the inherent quality of a heavy body to go downward rather than the effect of the gravitational force acting upon the mass of the body. The only exception to this rule is when the term is in direct reference to the gravitational force or to "specific gravity" as a measure of material density.

References to velocity (e.g., *velocità* and *velocitas*) are rendered as "speed," reflecting their original lack of directional meaning, in contrast to the modern sense of velocity.

These specific choices for "heaviness" and "speed" are influenced by Stillman Drake's translation of *Le Mecaniche* in his and I. E. Drabkin's 1960 publication of *On Motion and On Mechanics*; however, although Drake used "heaviness" for *gravità*, Drake still directly translated *centro di gravità* as "center of gravity." In contrast, I have chosen to use "center of heaviness" to maintain term consistency for the translation of *gravità*, which I have further applied to any pre-Galilean references to the same concept (e.g., *centrum gravitatis*).

Pre-Galilean uses of *momento* have been translated as "moment," while Galileo's uses of *momento* have been maintained to emphasize its distinction. *Momentum,* however, is preserved in both Galilean and pre-Galilean uses.

Some typographical errors inevitably escape review, and although this translation may also contain some of these errors, I made a few corrections to the original text. Most were minor and received no comment. There were a few instances, however, where the correction required additional explanation, as the literal meaning had changed significantly. For the most part – particularly with quoted Greek, Latin, and Italian passages – I relied directly on the text as Galluzzi transcribed it in his original manuscript. Only when I encountered significant difficulty in translation did I cross-reference the source material to verify and correct my rendering.

The most significant changes to the original text include (with bracket page numbers referencing the original page number of the manuscript):

1) Corrected several errors in the passage cited from Toledo, as indicated on page [111].
2) Corrections to the ratios or equations, as presented on pages [220], [279], and [342].
3) Restored missing portions of cited passages, as indicated on pages [278] and [371].
4) Correction to the assertion that "...Galileo was able to deduce from an incorrect principle (the increase in speed proportional to *time*) a correct conclusion: the distances traveled are the square of the times," in which the "incorrect" principle in the parenthetical should instead read "the increase in speed proportional to *distance*," as indicated on page [279].
5) Restored Fig. 5, which was originally depicted with a missing portion in its printing, now illustrated in full on page [208].
6) Edited Fig. 33 to include point *n* located at the intersection of lines *cg* and *hl*, now illustrated on page [338].

For change 1), I was unable to find a copy of Toledo's *Nunc quinto in lucem edita* of his *Commentaria*; instead, I was able to find the passage in the *Ab eodem authore emendata, nunc denuo in lucem edita*, p. 269, and in the *Nunc secundo in lucem edita*, c. 129*v*, both substantially providing the same. These versions were found on the Internet Archive website.

For changes 3) and 5), I relied upon the scanned digital copies of Carlo Manolessi's 1655 edition of *Opere di Galileo Galilei* stored on the United States Library of Congress website.

Other minor corrections to passages that were cited from the Edizione Nazionale edition of *Le Opere di Galileo Galileo* were recovered from the Internet Archive website.

Additional translator commentary to the corrections or elaborations is indicated in the page footnotes with daggers (†) or double-daggers (‡).

Although the *Index of Names* has been reproduced, several names have been translated into their recognized English spellings and differ significantly from the original forms. For example, I have translated "Guglielmo di Moerbeke" to its recognized English name, "William of Moerbeke." Accordingly, these names have been re-alphabetized with their respective English names and may not match the ordering of the original publication.

I would like to thank my family and friends for their interest and support in helping me complete this project. To Professor Paolo Galluzzi, who was surely patient with me and my numerous emails, and whose contributions to Galilean studies are nothing short of extraordinary, I owe my deepest thanks. To the staff at Museo Galileo, who graciously connected me with Professor Galluzzi, my thanks. To Professor Michele Camerota, whose continual support and urging to complete this project were, in no small way, inspirational and motivating. To the University of Nebraska Omaha, whose library resources and faculty helped inspire my pursuit of this endeavor. To my old friend and physics colleague, Zach Marquez, who helped review, edit, and format this translation. To my wife, Dr. Erica Schumann, whose invaluable guidance and assistance in navigating the many publications in Galilean studies were paramount. And, above all, to my daughter, Clara, who came into this world in the midst of this work and has been my *momento* for its completion.

ANDREW RODIONOV

Enimvero latinis *punctum* et *momentum* idem significabant: *momentum* autem est res quae movet; et cum *punctum*, tum *momentum* iisdem latinis quid indivisibile dicebatur [Indeed, *punctum* and *momentum* among the Latins signified the same: *momentum*, however, is the thing that moves; and both *punctum* and *momentum* were called by those same Latins something indivisible].
(G. B. Vico, *De antiquissima Italorum sapientia*, IV, § 2).

The two abstract elements, namely the opposites taken by themselves, in their separation, are called *moments* by him [Hegel]*, with imagery taken from the *moments of the lever*; and *moment* is sometimes also called the third term, that of synthesis.
(B. Croce, *Saggio sullo Hegel*, Bari 1913, pp. 15-16).

TABLE OF CONTENTS[†]

[†] Indicated page numbers refer to pages of the original publication, which are presented in the side margins throughout this text.

INTRODUCTION

When, at the end of 1974, in anticipation of the Second International Colloquium IX
of the Lessico Intellettuale Europeo [European Intellectual Lexicon (L.I.E.)], Professor Tullio Gregory, who is the Director of L.I.E., invited me to prepare a report on issues related to the Galilean Lexicon, of which I am overseeing the editing, I immediately thought of *momento*.

In fact, for some time I had been wondering about the exact meaning that the term took on in Galileo's work and the different functions it carried. Even then, I was fully aware that the countless studies dedicated to the mechanics and dynamics of the Pisan scientist, while unanimously emphasizing the centrality and importance of *momento* in Galileo, did not, however, serve to conveniently resolve such doubts and perplexities.

Having obtained approval for the project, I dedicated myself to the work, proposing to prepare a model of Galilean lexicon entries (*momento* | *momentum*) to which I could preface a short essay, in which the meaning of the term in Galileo's work is illustrated, and the ordering and articulation of the examples proposed in the entries are justified.

Having collected all the occurrences of *momento* | *momentum* in Galileo and made the first attempts at ordering them, insurmountable difficulties appeared to me. First, I couldn't always convincingly distinguish values of the term, often remaining perplexed in assigning it to this or that category. Furthermore, it was not clear to me whether I was faced with different meanings of the term that needed to be clearly distinguished, or whether they were different yet fundamentally connected uses. This last dilemma prefigured a general interpretative question, relating to the weight exercised by the "mechanical" foundation in the Galilean science of movement.

It was at this point that I decided to look for some help in the pre-Galilean history of *momento*, to understand the tradition from which Galileo had taken it and the function it had carried out in this tradition.

I thus began to systematically investigate general dictionaries, author lexicons, repertoires, etc., in search of precise indications, discovering, not without surprise, the absence of works that illustrate the history, albeit in small areas, of such an important and common term.

The scope of the search therefore had to broaden further and isolate three

X fundamental values of *momentum*, which seemed to me reasonably original, attempting to highlight, at least in broad terms, the history leading up to Galileo. At the same time, I was deepening the analysis of Galilean texts in which *momento / momentum* occurs, trying to reconstruct the complex evolution of Galilean reflections on movement. In this sense, the focus on *momento*, a key term for considering this sector of Galilean speculations, and the availability of valuable indications deduced from the previous history of the term, placed me in a privileged situation compared to other types of approach.

At this point, it was obvious the original plan had been surpassed. When, in January 1977, on the occasion of the Second International Colloquium promoted by the L.I.E., I read my report on *Momento* in Galileo (with an *Appendix*, which presented the complete index of occurrences, distinguished by meaning), I was fully aware that it was an intermediate stage of a much larger work, which would have required the space of an entire volume.

This short introduction serves to elucidate, not only the genesis of the work, but also the way in which it has taken shape: divided into two parts, both endowed with substantial autonomy, even though the first was conceived as a function of the second, to whose understanding it can bring benefit. In reality, as the title suggests, it is a truly "Galilean" work, which, by necessity, has been preceded by a series of essays on the pre-Galilean history of some aspects of the term and concept of *momento*.

Part One is not, nor does it pretend to be, a story of *momento*. Understood in this sense, it would be neither complete nor entirely plausible. In fact, only a few traditions have been subjected to careful investigation, even if they are of singular importance, and, with rare exceptions, the vicissitudes of the temporal *momento* and of the figurative and metaphorical uses have been neglected. In addition to presenting itself as an indispensable introduction to Part Two, it constitutes the only work available so far on the history of the term. It will thus be able to stimulate further research and constitute an initial orientation.

Considered from the point of view of a history of the technical meanings of *momento*, Part Two can also appear largely incomplete, at least in two respects. First, if Galileo's work marks an important stage in the current story, thus baptizing its entry into the modern scientific lexicon, this story nevertheless continues, as complex,, full of significant conceptual modifications, through the research of Huygens, Leibniz, Newton, Jacob Bernoulli, Euler, at least up to Lagrange's *Méchanique analitique*, just to name a few of the protagonists. Second, a precise investigation into the "influence" of the Galilean *momento*, conducted by carefully
XI examining the translations of his works, the echoes of his most famous pages in his disciples or by European scientists of the second half of the 17th century and the first of the 18th century, would greatly contribute to shedding light on significant aspects of the history of the term, as the few indications provided in this work demonstrate. The scope that the work had reached and the desire to keep it focused on clarifying the various functions that *momento* carried in Galileo's work led to the decision not to develop these research prospects, even though they are intriguing

and promising. I do not rule out attending to them in the future. Others, however, will be able to do so.

The main objective of this work is to focus on the problems on which the introduction of *momento* into Galileo's work depended, which was initially not recorded, and on the precise understanding of the very delicate functions that it carried. To achieve this objective, given the capital importance of the term and the concepts it expresses, it was necessary to proceed with an overall critical reconstruction of the Pisan scientist's reflections on static principles and the science of movement. The result of the investigation confirms that Galileo long tried to develop a general theory of mechanical and natural motions, in which any distinction between static and dynamic principles would effectively be eliminated. In pursuing this goal with extraordinary tenacity, he placed most of his hopes in *momento*. Prevented by insurmountable foundational difficulties, he was unable to implement this ambitious project to its end: in his final work, the *Discorsi* [*e dimostrazioni matematiche intorno a due nuove scienze* (*Discourses and Mathematical Demonstrations Relating to Two New Sciences*)], he presented "mechanical" questions and the science of natural motions as two substantially separate disciplines founded on different principles. The latter will appear as a field completely foreign to the contingent facts of the material world – heaviness and its "accidents" – rigorously geometric and obsessively dominated by the concept of time.

In this outcome, in which many have recognized the most revolutionary aspect of Galileo's entire contribution, his original plan seems to completely fade into the background. Indeed, it can be argued that those who follow Galileo's intellectual path only through the published works can hardly recognize that the unifying project, the idea of the "mechanical" foundation of natural motion, constitutes the general principle to keep in mind in understanding the full meaning of Galilean scientific construction. And it is precisely by emphasizing *momento*, carefully following the complex Galilean history, especially through the outlines of the theorems later published in the *Discorsi*, that we can perceive the extraordinary weight exerted by the unification project on the complex reflections of the Pisan scientist regarding movement. The history of the term, its clear initial connection with weight and distance, and the subsequent and increasingly close association with speed and time illustrate, in short, in an exemplary manner, the complex path, full of forced deviations and significant interruptions, traversed in Galileo's reflections over more than half a century.

The lesson that this work allows us to draw, on the level of general interpretations, appears to be of considerable significance. In fact, the impression of a chasm **XII**
between the "Pisan" dynamics of the *De motu* and the pages of *De motu locali* of the *Discorsi* seems to be subsiding. If the paths taken by Galileo in these two extreme poles of a long journey present notable differences, it is not difficult to recognize, when you look at them in transparency, the survival of different forms of the same design. Following the subtle game that Galileo plays with *momento*, unmasking its abuses and ambiguities, exposing the difficulties that led him to

resort to it, allows us to recognize a substantial continuity, even if not linear or obvious, in the investigations on movement, and clarifies the actual genesis of the position maintained in the *Discorsi*. This allows us to escape the blinding flash of easy and spectacular theories along the road to Damascus, of the striking intuitions that suddenly produce profound shifts in the orientation of the great men of science, making them authors of "revolutionary" changes, of abrupt paradigm shifts.

The story of *momento* in Galileo's work therefore also offers many useful ideas for the historian attentive to the "procedures" of scientific thought and to the cadences that mark its activity. It demonstrates that focusing attention on characteristic and central terms, following their movements precisely, and tracing from time to time the conceptual modifications or attempts to circumvent specific difficulties, is of great benefit in enriching the reconstruction of the actual genesis and progressive development of a scientific project with important nuances. Finally, it restores full meaning to an investigation that attempts to establish precise connections between the reflections delivered to the press and the first drafts, which often keep them much clearer and more transparent, allowing us to grasp, precisely through the terminological modifications, the author's intentions, the different paths attempted, the various methods adopted.

The work on *momento*, and on a series of fundamental terms with which it presents a complex system of connections (consider *impeto* [impetus], *grado* [degree], etc.), is also of considerable use as a significant, albeit incomplete, index of Galileo's sources. In the project of unification between static and dynamic principles, which we have emphasized, it is not difficult to recognize the Archimedean model, integrated and corrected with indications deduced from the *de ponderibus* tradition and, above all, from the pseudo-Aristotelian *Questions of Mechanics*, a text which appears increasingly decisive for understanding the genesis of many Galilean speculations.

It is from this static-mechanical tradition, absolutely "technical" and "rational", founded on not exclusively geometric approaches, that Galileo assumed *momento* – a sign that is certainly not "neutral", nor drawn from the lexicon of the practical arts – in which he had to recognize a term capable of instinctively suggesting an almost seamless transition between statics and dynamics. The choice of the term betrays, in short, the direction of thoughts.

These are not new sources, nor the only or most important scientific traditions
XIII present in Galileo. Others have repeatedly indicated their importance for the Pisan scientist. Through *momento*, however, it is possible to reconstruct precisely how he was able to bring together profitably the approaches of many of his contemporaries working with Archimedean *Equilibrium of Planes* and Pseudo-Aristotelian *Questions of Mechanics* into the same objective. And how, in a subsequent phase, and driven by precise needs, he enriched this singular graft with geometric analysis techniques of evidently terminist flavor.

In essence, the thesis, dear to Koyré, of a Galilean attitude towards the questions of motion in the purely Archimedean style, rigorously geometric and profoundly disinterested in *per causas* explanations, appears today in many ways not entirely satisfactory. As the story of *momento* demonstrates, the Galilean science of movement, in its long and contested process of definition, used, in addition to Archimedes, many other contributions and aids, including experimental ones, of considerable importance and significance. Furthermore, the Pisan scientist never stopped identifying and conveniently presenting the "physical" cause of natural motions. Even if the insurmountable foundational difficulties suggested that he avoid discussing the cause of motion, he must have remained intimately convinced to the end that it depended in some way on "heaviness" and its *momenti*.

If *momento* was able to offer Galileo many services, it was because the term historically presented – and the Pisan scientist was perfectly aware of this – an extraordinary range of different semantic values, even if they were all traceable, through mediations that are not always simple, to a common idea, connected to the image of the scale (weight × distance from the fulcrum along with inclination, as well as, again, the instant at which the equilibrium is broken). Like other important terms (consider *impeto* [impetus] or *talento* [ability]), the peculiar fate of *momento*, and the multiplicity of services that Galileo required of it, was favored by a certain flexibility that it presented. The Galilean science of movement, profoundly undermined by foundational difficulties and erected on principles that were anything but solid, did not need rigid, univocal, rigorously technical terms, but rather words and formulas capable of evoking different images and concepts as needed. Galileo, as we will see, placed considerable faith in the evocative abilities of *momento*.

Our work owes much to the benevolence of illustrious scholars and the collaboration of friends and colleagues. I would like to extend a very special thank you to Eugenio Garin, a master beyond the confines of the schoolroom, always ready to encourage, favor and guide those who dedicate themselves to new research, especially if it is not simple and obvious. The patience and understanding with which he listened to the exposition of the problems and difficulties that arose
during the work, the amount and importance of the specific suggestions and the **XIV**
architectonic indications that I received from him make a proportionate thank you impossible. Valuable advice and decisive suggestions came to me from Prof. Tullio Gregory, who enthusiastically accepted the first, timid hypothesis of a work on *momento* in Galileo, encouraging me to transform it into a general essay and helping me in every way to carry it out. I wish to express my gratitude to Prof. Aldo Duro for the valuable, highly competent assistance given to me in the more specifically lexical part and in the preparation of the *Appendix*. I am also indebted to Professor Marshall Clagett, of Princeton, for the exquisite generosity shown in allowing me to consult, before they appeared in print, notable parts of his fundamental research in *Archimedes in the Middle Ages* (vols. II and III). To many others, whom I promptly thank in the notes, I owe gratitude for suggestions, revisions and generous collaboration. I recognize in particular that I was singularly favored by the authoritative members of the Scientific Council of the L.I.E., to whom I explained several times the general outlines according to which the work was being carried

out, always receiving encouragement and suggestions. It is clear, however, that any errors or inaccuracies are exclusively my responsibility. Finally, allow me to recall the many courtesies received from colleagues and collaborators of the L.I.E., in the spirit of sharing and collaboration that characterizes this research body.

November 1978

ABBREVIATIONS

B.N.F. Biblioteca Nazionale Centrale di Firenze [National Central Library of Florence]. **XV**

CAG *Commentaria in Aristotelem Graeca*, Edita Consilio et Auctoritate Academiae Litterarum Regiae Borussicae [*Commentaries on Aristotle in Greek*, Published with the Council and Authority of the Royal Prussian Academy of Sciences].

Corp. Christ. Lat. *Corpus Christianorum, Series Latina* [*Christian Corpus, Latin Series*].

Corp. Christ. Med. *Corpus Christianorum, Continuatio Mediaevalis* [*Christian Corpus, Medieval Continuation*].

D.B.I. *Dizionario Biografico degli Italiani*, Istituto della Enciclopedia, Roma 1960- [*Biographical Dictionary of Italians*, Institute of the Encyclopedia, Rome 1960-].

D.S.B. *Dictionary of Scientific Biography*, Ch. Coulston Gillispie Editor in Chief, Ch. Scribner's Sons, New York 1970-1976, 14 vols.

G.G. *Le Opere di Galileo Galilei.* Edizione Nazionale, a cura di [*The Works of Galileo Galilei.* National Edition, edited by] A. Favaro, Barbèra, Firenze 1890-1909, 20 vols.

PG *Patrologiae Cursus Completus, Series Graeca* [*Complete Course of Patrology, Greek Series*].

PL *Patrologiae Cursus Completus, Series Latina* [*Complete Course of Patrology, Latin Series*].

Part One

"MOMENTO" IN TRADITION

CHAPTER I

DICTIONARIES AND LEXICONS

In the first part of our work especially, we will frequently emphasize the curious 3
nature of the relationships between *ῥοπή* [*rhopḗ*] and *momentum*, which emerge from the hesitations, errors, or sensibilities of the Latin translators and commentators of Greek works (or translated into Greek), fundamental to the formation of modern scientific and philosophical thought.

From this frequent comparison, which constrains those who intend to follow the Latin diffusion and the reception of certain discussions and themes that draw their origins from the Greek world, there emerges the unusual complexity of the two terms and the substantial overlap in the vast semantic range they present in their respective fields. Despite the strong similarities, they were not always perceived as corresponding terms. Many translators systematically avoided using *momentum* to render a semantic value of *ῥοπή* with which the Latin term was demonstrably endowed.

Naturally, behind the actions of the translators and commentators – especially when it appears homogeneous in certain chronological periods – specific reasons are hidden. They can only be perceived through a direct awareness of the history of meanings, that is, when it is possible to clearly distinguish the different semantic values and to identify, at least in broad terms, which ones were prevalent for one or the other term in the precise situation examined. Without this "historical" awareness of the specificity of the respective semantic developments, any attempt at comparison between the terms from different languages would be doomed to failure.

It will therefore be convenient, with the aid of available tools, to try to reconstruct at least the main stages of these "stories" – which will appear in many aspects quite different – both on the basis of the documentation offered by modern lexicons and dictionaries and by "reviewing" the testimonies recorded in the ancient lexicons and etymological repertoires. The latter serve, in fact, to give us the "state" of the term in certain chronological periods.

P. Galluzzi, *Momento: Galilean Studies*, Archimedes 78,
https://doi.org/10.1007/978-3-032-35049-7_1

1. Ῥοπή IN GREEK DICTIONARIES AND LEXICONS

4

For the entry of *ῥοπή* in Stephanus's[1] *Thesaurus Graecae Linguae*, it is not easy to recognize an ordering criterion. However, Stephanus provides some interesting indications, which are worth underlining. We will therefore attempt to schematically present the different meanings of *ῥοπή* documented by Stephanus (the etymology of *ῥοπή* from *ῥέπω*, "inclinare [to incline]", is not controversial):

1) Pressure exerted from top to bottom (Plutarch).
2) Moment (inclination) of the balance (Plutarch).
3) Generic inclination (physical and moral):
   a) internal of bodies downwards (Alexander of Aphrodisias).
   b) of man, in the sense of affection, attachment (Polybius).
   c) of the divine will (Patristic authors).
4) Weight:
   a) in a physical sense (Aristotle)
   b) phraseological use ("di molto, di nessun peso [of much, of no weight]"), late classical authors.
5) Moment of time, blinking (*ῥοπή ὀφθαλμοῦ*), only Christian examples (St. Basil, Gregory of Nyssa, John Chrysostom).
6) Cause of motion (from Aristotle to Chrysostom).
7) Critical moment (e.g., *ῥοπή βίου* = the moment of passing), decisive (classical examples).

It is not easy, on the basis of the articulation of the entry proposed by Stephanus, to form an idea, neither of the relationships between the different meanings, nor of their historical development. However, some findings are possible. First of all, the meaning of *ῥοπή* = "momento di tempo [moment of time]" has late evidence. That is, it does not appear to have been common in classical Greek. The same can be said – but this is not surprising – for the meaning of "inclinazione della volontà divina [inclination of divine will]".

In Liddell-Scott[2], the criterion followed is completely different. The entry is, in fact, organized into three sections, which collect the large semantic fields of *ῥοπή*. Within the sections, the variations attributable to the general meaning are documented. Here, again schematically, is the articulation of the entry in Liddell-Scott:

5 1) Moment (inclination) of the balance:
   a) In the physical sense (Aristotle, Theophrastus, Plutarch, Timaeus Locr.).

---

[1] *Thesaurus graecae linguae*, 8 vol., Akademische Druck-U. Verlagsanstalt, Graz 1954 (anast. reprint of the Parisian ed. of 1831-1865 [Firmin Didot]*), VII, pp. 2424-2426.

[2] H. G. Liddell-R. Scott, *A Greek-English Lexicon.* Revised and augmented by Sir H. Stuart Jones. With a supplement edited by E. A. Barber, At the Clarendon Press, Oxford 1973.

b) In a metaphorical sense:
   α) The act of weighing, of judging (Alcaeus, Aeschylus, Sophocles), the moment of judgment – *ῥοπή βίου* – (Sophocles, Menander).
   β) Collapse of the balance (in favor of a certain situation) and, therefore, "decisione [decision]" (classic examples).

2) Weight on the balance:
   a) Infinitesimal addition, sufficient to give motion to the balance (Herodas).
   b) In a metaphorical sense:
      α) Determinant weight (Plato, Sophocles, Euripides, Plutarch, Polybius).
      β) Decisive influence (Aristotle, Polybius, Aeschylus).

3) Critical moment, decisive moment; in general, moment of time (LXX version).

In the entry proposed by Liddell and Scott, we can glimpse the ordering criterion. *Ῥοπή* is considered a term deeply connected to the image of the balance. Starting from this basis, we try to identify its fundamental meanings (proper and metaphorical). Points 2 and 3 respectively present the documentation of *ῥοπή* = "peso [weight]" (or addition of weight) which determines the collapse of the balance (with all the metaphorical meanings connected to this value) and the *moment* (of time) in which the collapse occurs. Not everything in Liddell-Scott's articulation appears fully convincing, but it has the merit of striving to demonstrate that the considerable variety of values assumed by the term can be traced back to a common original meaning. As for the historical development, the entry suggests some interesting indications. First of all, we find confirmed what we already observed in Stephanus: the absence of the temporal meaning of *ῥοπή* in classical Greek. The value of "minimum weight", an "imperceptible addition", capable of making the arms of the balance tip from one side to the other, would also seem foreign to classical Greek. For both these values, the oldest documentation provided by Liddell-Scott in fact belongs to the LXX version[3].

A whole series of precious documentation is offered by G. W. H. Lampe's lexicon of patristic Greek[4]. The meanings assumed by the Greek term in the writings of the Fathers are collected by Lampe into seven groups.

1) Moment (inclination) of the balance (St. Basil).
2) Propensity, inclination: **6**
   a) constant, in the nature of God (Gregory of Nyssa)
   b) and in the angelic one
   c) unstable, and often oriented towards evil, in man (very numerous examples of many Fathers).
3) Movement:

[3] Cf., on this point, *infra*, Part One, Ch. II.

[4] *A Patristic Greek Lexicon*, edited by G. W. H. Lampe, D.D., At the Clarendon Press, Oxford 1961.

a) small motion (Chrysostom)
b) flow of water (Synesius)
c) change (Eusebius, Athenaeus)
d) act, sign also of God (Eusebius, Athenagoras).

4) Weight on the balance (Basil).
5) Determinant weight, decisive influence in a metaphorical sense (Chrysostom, Sozomen).
6) Divine assistance, indispensable for the success of human actions, therefore divine "grace" (numerous examples from many Fathers).
7) Decisive moment, critical moment; instant of time (Chrysostom, Palladius of Galatia, Clement, Melito of Sardis, Basil).

In the patristic lexicon, we find quite a few innovations compared to the state observed in classical Greek. In general, the renewed development of the metaphorical values of *ῥοπή* connected with the proper meaning of "weight" (or inclination of the balance) – of which practically no documentation remains – is evident. The clear prevalence of the use of the term in ethical-religious connotations ("unstable inclination" of man towards evil and "decisive intervention on the part of God", "grace", "inner illumination") – which is obviously not surprising in this genre of literature – is a fact worth emphasizing, as is the confirmation of the use of *ῥοπή* as an "instant" of time and, although still poorly attested, the emergence of the values (not new, but more clearly specified) of "change" and of "small movement".

As for the non-patristic documentation within the centuries of the Christian era, we can get a fairly precise idea of the semantic evolution of *ῥοπή* by consulting some famous onomastica and lexicons of authors, which record the state of the Greek language at different chronological levels.

The *Onomasticon* of Pollux (2nd century AD)[5] connects *ῥοπή* [inclination (n.)] to *ῥέπειν* [to incline (v.)], clearly showing the meaning of inclination produced by weight, also through a series of derivatives (*ἰσόρροπος* [of equal inclination (adj.)]
7 *Ἰσορροπία* [condition of equal inclination (n.)] etc.)[6]. In the *Onomasticon*, however, when defining other terms, *ῥοπή* is used also in the sense of "instant" of time[7] and of "minimum weight" that tips the balance[8].

---

[5] *Pollucis Onomasticon*, Edited and annotated by E. Bethe, vol. I-III, "Lexicographi Graeci", In aedibus B. G. Teubneri, Lipsiae 1900. The dating is obtained from the epistles prefaced to each of the 10 books, addressed to the emperor Commodus.

[6] *βαρύνειν, ῥέπειν (ῥοπή, βαρύτης βάρος, ἐλαφρότης, καταφέρειν ἀναφέρειν, ἄχθος ἐπαχθές, κοῦφον κουφότης, ὄγκος), ἰσόρροπον (ἰσορρόπως ἰσορροπία, (ἑτερόρροπον) ἑτερορρόπως ἑτερορροπία, ἀντίρροπον· βιαζόμενος δ' ἂν εἴποις καὶ ἀντίθετον...)* [to burden, to incline/tip (tipping-weight, heaviness, weight, lightness, to carry downward, to carry upward, burden, burdensome, light, lightness, bulk), equally balanced (in equilibrium, equilibrium, (tipping to one side) with unequal balance, unequal balance, counterbalancing; and if you were pressing hard you might also say counterpoised...)] (*ibid.*, IV, 172).

[7] *Ἐπὶ βραχεία ἦν ῥοπῆς* [it turned in the slightest inclination] (*ibid.*, III, 107, 29).

[8] In a series of expressions tending to define the absolute impartiality of the judge, Pollux inserts *ἀκριβεστέρος ζυγοῦ τὴν ῥοπὴν* [more precisely the tipping of the balance], to indicate how the judge can be compared to a balance of absolute precision, capable of doing justice up to the minimum *ῥοπή*. Here

Important documentation is offered by Hesychius's Alexandrian Lexicon (5th-6th century). For *ῥοπή*, Hesychius proposes the following values: *κλίσις* [slope], *νεῦμα* [*ῥάβδος*]* [nod [rod]], *δύναμις* [power], *βοήθεια* [aid][9].

Some sound familiar (*κλίσις* [*klī́sis*, alluding to "inclination"], *δύναμις* [dýnamis, alluding to "dynamics"]), while others require explanation. For *νεῦμα* and for *βοήθεια*, Lampe offers guidance. *Νεῦμα* is the "cenno di Dio [nod of God]", the imperceptible gesture that decides everything, the small effort (weight) that determines the outcome of a situation. We've seen that *ῥοπή* was used in this sense by the Greek Fathers. *Βοήθεια* refers to other meanings of *ῥοπή* recorded by Lampe (cf. section 6): "soccorso [assistance]", "intervento decisivo [decisive intervention]" from God (also in this case, by metaphor from the meaning of *ῥοπή* = addition of weight on one of the arms of the balance in equilibrium). In the Lexicon of Hesychius, the "patristic" values of *ῥοπή* are recorded. It seems more difficult to account for *ῥάβδος* = "verga [rod]", "bastone [stick]". So difficult that Schmidt, the editor of Hesychius's Lexicon, proposed its excision[10]. *Ῥάβδος* appears as a glosseme of the lemma preceding *ῥοπή*: *ῥόπαλον* = needle of the balance[11]. It may, therefore, be a repetition or a copyist's error. However, I would like to point out that – while admitting the thesis of the error and accepting the legitimacy of the expungement – in the codex of Hesychius, "bastone [stick]" and "ago della bilancia [needle of the balance]" were indicated among the values of *ῥοπή*. We will see shortly what peculiar developments this errorneous entry may have occasioned[12].

In the Lexicon of Photius (9th century)[13] for *ῥοπή*, we limit ourselves to **8**
indicating a single glosseme: *κλίσις*. We always find *κλίσις* in Suda[14], where however the meaning of "instant"[15] is also indicated. Furthermore, *ῥοπή*, as the "inclination" of the body, is contrasted with *ὁρμή*, which by contrast is the "inclination" (appetite) of the soul[16].

ῥοπή properly indicates "small weight", but in metaphor it reads as "la più trascurabile delle ragioni [the most negligible of reasons]" (*ibid.*, VIII, 10, 22).

[9] *Hesychii Alexandrini Lexicon*, Ed. M. Schmidt, Ienae 1861, 446.

[10] The square bracket that closes the term also in our quotation indicates, in fact, the editors's proposal for excision.

[11] *Ibid.*, 445: *ῥόπαλον, βακτηρία, ῥάβδος* [cudgel, support rod, staff]. Regarding *ῥόπαλον*, and its meaning of "needle" of the balance, other lexicons are rich in information, which often consider it a derivative of *ῥοπή*. See, for example, The *Etymologicum Gudianum* (sec. XI): *Ῥόπαλον. ξύλον, βάκλον... ἢ ἀπὸ τῇ ῥοπῇ ὅ ἐστι φορᾷ...* [Cudgel. a piece of wood, a stick... or from *ῥοπή*, which is to say from "to carry/bear"] (Ed. Fr. Guill. Sturzius, Lipsiae 1818, 493, 52). The etymology is reconfirmed by *Etymologicum Magnum* (sec. XII): *Ῥόπαλον. παρὰ τὸ εἰς ἕτερον μέρος ῥέπειν. ἢ παρὰ τὴν ῥοπὴν...* [Cudgel. from its inclining toward one end/side. Or from *ῥοπή*...] (*Etym. Magnum, seu verius Lexicon saepissime vocabulorum origines indagans*..., Recensuit et... instruxit Th. Gaisford ..., Oxonii 1848, p. 705, 23-24). *Ῥόπαλον* is explained essentially in the same way in Photius, in the Suda and in the lexicon of Zonaras.

[12] See *infra*, p. 17.

[13] *Photii Patriarchae Lexicon*, Recensuit, adnotationibus instruxit et prolegomena addidit S. A. Naber, Vol. I-II, E. J. Brill, Leidae 1864, p. 105.

[14] *Suidae Lexicon*, Edidit A. Adler, In aedibus B. G. Teubneri, Stutgardiae 1971 [reprint of the 1st edition, 1935]*, p. 231.

[15] *Ibid.*, p. 230.

[16] *Ibid.*

"Inclination", "change", "instant", "force" and "indivisible movement" are the glossemes that we find recorded in the Lexicon of Zonaras (12th century)[17]. However, no documentation relating to *ῥοπή* is found either in the *Etymologicum Gudianum* (11th century), or in the *Etymologicum Magnum* (12th century).

Lastly, in the testimonies preserved for us by the Greek-Latin lexicons of the humanistic age, the clear identification of two main meanings of *ῥοπή* emerges: "instant" (*ῥοπὴ ἐπὶ ὥρας* [inclination upon time], or simply *ῥοπή*) and "small weight" sufficient to tip the balance, and, therefore, also "small motion" (*ῥοπὴ ἐπὶ ζυγοῦ* [inclination upon the balance])[18].

In conclusion, it can be said that in the events of the semantic values of *ῥοπή* two main phenomena are observed: the notable evolution of the metaphorical values already present in classical Greek (connected to the image of the balance) towards new nuances of meaning – moral and theological – in the Christian writers; the progressive imposition, as the most common meaning, of the temporal value ("instant", "critical moment"), which we have seen practically absent in classical Greek.

2. "Momentum" in Latin Dictionaries and Lexicons

9 Before examining lexicons and historical dictionaries of the Latin language, it will be useful to briefly focus on the etymological question, which, moreover, is not controversial. In fact, there is complete unanimity on the derivation of *momentum* from *moveo* [to set in motion], at least ever since Isidore, as we will see shortly. Ernout-Meillet, in reiterating it, emphasizes that our term (like *momen*, to which he assimilates it) has an "abstract" sense (impulse, movement, change) and a "concrete" one (weight determines the movement and inclination of the balance). According to Ernout-Meillet, the different meanings depend on these two values: 1) moral

[17] *Ῥοπὴ. ἡ ἀπό τινος εἴς τι μετάκλισις, ἢ εἴς τι μεταβολή. ῥοπὴ ἐστι καὶ τὸ ἀκαιριαῖον τῆς συντομίας καὶ ταχυτῆτος. Ῥοπὴ καὶ ἡ δύναμις. Ῥοπὴ ἐστιν ἀδιάστατος κίνησις* [*Ῥοπὴ*. the inclining from something toward something, or a change toward something. *Ῥοπὴ* is also the instantaneous quality of brevity and swiftness. *Ῥοπὴ* is also force/power. *Ῥοπὴ* is indivisible motion] (*Zonarae Lexicon*, Edidit I. Tittmann, Lipsiae 1808, 1619).

[18] This is what emerges from the Greek-Latin lexicons printed in the 15th and 16th centuries and beyond which we will discuss later, regarding *momentum* (cf. *infra*, note 63). The impression, however, appears confirmed by a group of very important manuscript lexicons, which are preserved in the Biblioteca Laurenziana di Firenze [Laurentian Library in Florence]. I owe the report to the courtesy of Dr. Rosario Pintaudi. In *Pluteo* 57, 16 (15th-16th century), the Greek term is presented in two phrases: *ῥοπὴ ἐπὶ ζυγοῦ* [inclination upon the balance] and *ῥοπὴ ἐπὶ ὥρας* [inclination upon time] (c. 289r); the first is explained in terms of very rapid motion (i.e., instantaneity of time and motion): "subitus brachii motus ut librae inclinatio [sudden movement of the arm like an inclination of the scales]". In *Acq. e Doni 92* (c. 120*v*) they are *ῥοπή* (with indication of the "instant" value) and *ῥοπὴ ἐπὶ ζυγοῦ*, translated with *incrementum* [increment]. In the *Conventi Soppressi 587* (c. 178r) only *ῥοπή*, with the indication of the temporal value ("*quando* lanx infra vadit [*when* the measuring plate goes below]"). In the *Edili 219* (cc. 211*v*-212*r*) *ῥοπή* (temporal) and *ῥοπὴ ἐπὶ ζυγοῦ* (small weight, again indicated with *incrementum*). In the *Gaddi 15* (sec. XV, c. 237*v*) and in the *Conv. Soppr. 181* (written by Filelfo, c. 126*v*) *ῥοπή* (with temporal value).

(inclination, cause that determines a decision); 2) small weight (minimal division in general and, especially, of time; in this sense synonymous with *punctum* [point]); 3) increment (minimum weight addition)[19].

As for the great dictionaries of the Latin language, today we have a tool that simplifies our task quite a bit, at least until the 7th century AD. I intend to refer to the entry of *momentum* in the *Thesaurus linguae latinae*[20], rich in documentation, which presents a complex and interesting articulation. As we did for the main Greek lexicons, we will schematically give the structure of the entry edited by Wieland.

I) *Cause that determines motion*
- A) Weight, act of weighing:
  - 1) *own*
    - a) phrases that express the image of weighing (from Terence to Paulinus of Nola).
    - b) act of weighing (Vitruvius-Rufinus)
    - c) small amount of weight (Cicero-Boethius)
  - 2) *translate* (authority, dignity, importance)
    - a) particularly significant examples (Cicero-Boethius)
    - b) chosen phrases, distinguished by case: abl., gen., acc., with the verb *esse* (to be) (examples from all over Latinity).
- B) Impulse, cause of motion:
  - 1) *proper* (Ovid, Livy, Seneca)
  - 2) *translate*
    - a) cause (minimal) determining motion (Cicero-Eustathius)
    - b) impetus (Livy-Claudius Donatus).

II) *Movement (and, by metonymy, space and time)* **10**
- A) Act of moving:
  - 1) *own* (Cicero-Pseudo-Prosper)
  - 2) *translated*
    - a) change (Cicero-Origen)
    - b) motion of the soul (Ennodius-Origen).
- B) Reported place and distance:
  - 1) Space
    - a) short space (Quintilian)
    - b) specific space
      - α) small part of the sky (Chalcidius and Martianus Capella)
      - β) numerical unit (Chalcidius).
  - 2) Point (as minimal space)
    - a) *own* (Pliny)

[19] A. Ernout-A. Meillet, *Dictionnaire étymologique de la langue latine. Histoire des mots*, Vol. I-II, Librairie C. Klincksieck, Paris 1960 (4th ed.), II, p. 416.

[20] *Thesaurus linguae latinae*, in Aedibus B. G. Teubneri, Lipsiae 1933-, vol. VIII (1966), *momentum* (Wieland), col. 1391-1396.

b) *translate* (the individual parts of anything: Valerius Maximus, Pliny, Quintilian, Chalcidius, Boethius).
C) Referring to time (point, instant of time, but also of very short space or path):
1) Selected examples
a) time in general (small)
α) in a generic sense (Caesar-*Acta conciliorum*)
β) in a specific sense
1 - temporal part of the action (Quintilian-Hilary)
2 - measurement of syllables (Terentianus Maurus)
3 - in the legal sense, temporary possession (Symmachus-Gregory the Great)
b) part determined by tempo (Metrological Writers, Pseudo-Apuleius, Pseudo-Anatolius).
2) Selected phrases, different per case – abl., gen. – and according to the prepositions – *a*, *ab*, *ad*, *post* etc. – (examples from all over Latin, but few classics).

The arrangement of the examples implemented by Wieland responds to a precise criterion. He intentionally highlighted the close connection of *momentum* with *moveo* [to set in motion]. The two great distinctions into which the entry is articulated are specifically "weight" or disposition produced by the weight (which causes motion), and "motion". From these, Wieland believed he could derive the other values of *momentum*, spatial, temporal, phraseological, etc. The documentation provided in the entry is impressive and the articulation seems to hold up substantially, even if, in the rigid distinctions imposed by the need to create a consultable tool, many nuances of meaning are necessarily lost or flattened. However, IA1b does
11 not appear entirely satisfactory, since the examples provided do not present the necessary homogeneity. It would have been appropriate to assign a special section to the meaning of "inclination" (*proprie et translate* [literally and figuratively]) produced by weight, of which Wieland himself offers extensive documentation[21]. At a certain point it would come to be felt as the only genuine value of the term. Furthermore, the registration of the meaning of *momentum* as a precise linear unit of (the smallest) measurement, used in the measurement of fields[22], attested by

[21] The entry in the *Thesaurus* presents, in the same section, examples with very different meanings. First of all, the two occurrences of *momentum* in Vitruvius, in which our term has the meaning of "(variable) effectiveness of weight". Two further examples are documented from St. Augustine (*De civitate Dei*), excellent ones, especially the first, where *momentum* is felt as the natural inclination of the elements towards their own end. In the two passages of Augustine, the reference to the Aristotelian theory of the elements is explicit: "Nam velut amores corporum momenta sunt ponderum, sive deorsum gravitate sive sursum levitate nitantur... [For it is as if the affections of bodies are moments of weight, whether they lean down with heaviness or up with lightness...]" (*Sancti Aurelii Augustini De civitate Dei*, 2 vols., Typographi Brepols, Turnholti 1955, II, 348). For the clear distinction of the two meanings recorded in Vitruvius and Augustine, cf. my article *A proposito di un errore dei traduttori di Vitruvio nel '500*, in "Annali dell'Istituto e Museo di Storia della Scienza", anno I (1976), fasc. 2.

[22] "In secundo vero loco digito subscripserunt, in tertio stateram, id est semi unciam, in quarto quadrantem,

pseudo-Boethius, is missing.

The structure of Wieland's entry provides indications that allow some initial considerations on the prevalence, in the chronological context examined, of different *momentum* values. As a first indication, we will observe that the term, already widely used by classical authors, especially with abstract meanings, presents a series of "technicalizations", particularly evident in later authors (unit of measurement, legal sense, etc.). See, for example, points B and C of Section II, where the documentation is almost entirely late. It also seems that, in the meaning of an instant of time, it was not common in the "classics", in which – at least that's the impression suggested by the entry – the term is mainly used in the meaning of "movement" (usually small) and the "weight" that determines the movement. However, the metaphorical use of *momentum* in this last meaning (of no, of much, of little moment) appears very **12**
widespread. The relationship between *momentum* and the balance and its laws was not prevalent in "classical" authors, and was later more restricted. In fact, examples of *momentum* as "peso sulla bilancia [weight on the balance]" or as "movimento della bilancia [movement of the balance]", in the proper sense, are quite rare[23].

It will have been noticed that, in broad terms, the organization proposed by Wieland highlights as fundamental values of *momentum* the same meanings highlighted for *ῥοπή* in Liddell-Scott. The differences concern the already mentioned absence of a constant explicit reference of *momentum* to the balance and the lesser emphasis given by Wieland to the temporal meaning of *momentum* (probably due to the almost total absence of documentation in classical Latin, which, moreover, corresponds to the situation observed for *ῥοπή* in Greek). Exactly as we observed for the Greek term, even in the different values assumed by *momentum*, the idea of "movement" and "decisive action" always appears present.

Following the pattern already tested for *ῥοπή*, we will now try to trace the evolution of the meaning of *momentum* in Christian writers. Blaise's *Dictionnaire*

---

in V dragmam, in VI scripulum, in VII obolum, in VIII semiobolum…, in VIIII siliquam, in X punctum, in XI minutum, in XII momentum nominato posuerunt [In the second place they signed with a finger, in the third the stateres, which is half an ounce, in the fourth a quarter, in the 5th a drachm, in the 6th a scruple, in the 7th an obol, in the 8th a semi-obol…, in the 9th a siliqua, in the 10th a point, in the 11th a minute, in the 12th they placed a named momentum]". Further on, the relationships between the different quantities are specified more specifically "...in puncto autem duo minuta et dimidium et in minuto IIII *momenta* esse asseruerunt [further, they asserted that in one point there are two and a half minutes, and in one minute four *moments*]" (Pseudo-Boethius, *De institutione arithmetica libri duo. De institutione Musica libri quinque. Accedit Ars geometrica …*, Edidit G. Friedlein, In aedibus B. G. Teubneri, Lipsiae 1867, pp. 426 and 428). An echo of this meaning of *momentum*, with explicit identification of the source in Pseudo-Boethian *Institutio arithmetica*, is found in *Epitome de mensuris et ponderibus* of Georgius Agricola, which had no small influence and was published, among other things, together with *De Architectura* of Vitruvius in the famous Elzevirian edition of 1649: "Unciam Boethius geometrico more et eo quidem greco dividit in: Semiuncias duas, sive stateres duos…, Puncta 248, Minuta 576, *Momenta*, sive minutiae, 1152 [Boethius divides the ounce geometrically and indeed in the Greek manner into: two half-ounces, or two stateres... Points 248, Minutes 576, *Moments*, or minuteness, 1152]" (*Epitome in omnes Georgii Agricolae de mensuris et ponderibus libros per Guilielmum Philandrum Castilionum, in M. Vitruvii Pollionis Architectura libri decem*, Apud Ludovicum Elzevirum, Amstelodami 1649, p. 251).

[23] If I am not mistaken, only two "classics", but both by that peculiar and problematic writer Vitruvius (cf. *Thesaurus linguae latinae* cit., col. 1392).

for Christian authors offers scant documentation. In fact, it records only one meaning under *momentum*: process for the re-establishment of property stolen by violence (by Cassiodorus)[24].

Blaise does not appear entirely attentive on this point. Using the author's lexicons, where they are available, and the *Thesaurus* itself (which covers part of Christian Latinity), we can deduce that *momentum* is present, for example, in Tertullian[25], in Arnobius[26], in Augustine[27] and other patristic authors. The prevalence of the temporal meaning of *momentum* for these authors is very clear[28]. Blaise himself, in the *Lexicon Latinitatis Medii Aevi*, indicates three meanings for *momentum* with which it would have been used by later Christian authors: 1) small interval of time; 2) small determinate measure of time (fortieth part of an hour); 3) tongue of the balance[29]. Based on this documentation – and leaving aside the third
13 meaning, which we will focus on later – it seems we can therefore conclude that in the Latin lexicon of the Christian authors the temporal meaning of *momentum* ended up clearly imposing itself on the others.

If we compare the conclusions we reached by examining the reception of *ῥοπή* among the Greek Christian authors with the situation of *momentum* among the Latin Patristics, some macroscopic differences cannot be overlooked, which can only partly be attributed to the notable diversity of the instruments used in the two regions.

On a statistical level, the notable presence of *ῥοπή* corresponds to a negligible use of *momentum*. On the level of significance, it appears – and this is a most peculiar fact – that the widespread use of *momentum* and its metaphorical meanings to express ethical-religious concepts was not recorded. Unlike *ῥοπή*, *momentum* has never indicated either man's inclination towards evil (or God and the angels towards good), or God's omnipotent nod (in Christian authors, *nutus* [nod] is generally used to express this meaning).

The rarity of the use of *momentum* in late Latin and the prevalence of the temporal meaning is confirmed by Alexander Souter's *Glossary of Later Latin*, which is based on documentation up to 600 AD. According to Souter, the value of "instant" must be added to the technical-juridical one mentioned above[30].

A document of great importance, which reinforces the indications provided by medieval lexicons (Christian and otherwise) up to the 7th century, is the

---

[24] Albert Blaise, *Dictionnaire Latin-Français des auteurs chrétiens. Revu spécialement pour le vocabulaire théologique par H. Chirat*, Librairie des Méridiens, Paris 1954, p. 538.

[25] Cf. for example, *Adversus Hermogenem*, pp. 431 and 823 (*Tertulliani opera*, I-II, *Corp. Christ. Lat.*, I, Typographi Brepols, Turnholti 1954). See also G. Claesson, *Index Tertullianeus*, vol. 3, Études Augustiniennes, Paris 1974-1975, II, p. 966.

[26] Cf. L. Berkowitz, *Index Arnobianus*, G. Olms, Hildesheim 1967, p. 428.

[27] Cf. *Thesaurus linguae latinae* cit., col. 1392.

[28] *Ibid.*, especially col. 1394-1396.

[29] A. Blaise, *Lexicon latinitatis Medii Aevi, praesertim ad res ecclesiasticas investigandas pertinens*, *Corp. Christ. Med.*, Typographi Brepols, Turnholti 1975, pp. 596-597.

[30] *A Glossary of Later Latin to 600 A.D.*, Compiled by A. Souter…, At the Clarendon Press, Oxford 1949, p. 255. Souter, in fact, documents, beyond the legal meaning, only the expression *ad momenta*.

Etymologiarum Libri of Isidore. In the fifth book, he records a single meaning for *momentum*:

> *Momentum* est minimum atque angustissimum tempus a motu siderum dictum. Est enim extremitas horae in brevibus intervallis, cum aliquid sibi cedit atque succedit[31]. [*Momentum* is the smallest and narrowest unit of time from the motion of the stars. For it is the boundary of an hour in brief intervals, when one thing yields and another succeeds.]

Beneath the apparent simplicity of the definition lies a notable complexity that we should not overlook. In fact, Isidore emphasizes that the temporal meaning – which he presents as prevalent – derives from the "proper" meaning of movement (*a motu siderum dictum* [named from the motion of the stars]). In the second part of the quote, he highlights how *momentum* also indicates the ending of one thing and the beginning of another; that is, a "change". *Extremitas* [Extremity] must be read – of course – as a synonym of *terminus*, the final, indivisible component. The reception of this interpretation will be remarkable.

Returning to the dictionaries, Du Cange only indicates the temporal meaning **14**
(small space of time in general and specific measure: 1/40 of an hour)[32]. The documentation recorded in Arnaldi's *Lexicon Imperfectum* for the Italic Latin presents an enrichment of the definition of Isidore, revealing its influence:

> momentum a motu siderum celerrimo nuncupatur. Plurimi scriptores brevissimum illud temporis spatium quo palpebra moveri potest, momentum vocant[33]. [momentum is named from the swift movement of the stars. Many writers call that very brief interval of time by which the eyelid is moved, momentum.]

The second part of the entry recalls the *ῥοπὴ ὀφθαλμοῦ* [moment of the eye] (*ictus oculi* [blink of the eye]) previously observed[34].

Latham's *Word-List* confirms the exclusively temporal use of *momentum* (short duration and part of the hour), suggesting chronological terms between the 7th and 13th centuries[35]. The influence of the Isidorian *Origines* and the prevalence of the temporal sense appear evident in two definitions of the famous *Vocabulista* of Papias (11th century):

> Momentum minimum tempus a motu syderum dictum quasi movimentum.

---

[31] *Isidori Hispalensis Episcopi Etymologiarum sive Originum Libri XX*, Recognovit… W. M. Lindsay …, Vol. I (Lib. I-X)-II(XI-XX), Ex typographeo Clarendoniano, Oxonii 1911, I, Lib. V, ch. XXIX.

[32] Du Cange, *Glossarium ad scriptores mediae et infimae latinitatis* …, L. Favre, Niort 1885, V, p. 450.

[33] *Latinitatis italicae medii aevi … Lexicon Imperfectum*, care and study of F. Arnaldi, Vol. 3, Bruxelles 1939-1970, II (care and study of M. Turriani), p. 363.

[34] Cf. *supra*, p. 4.

[35] *Revised medieval Latin Word – List from British and Irish sources*, Prepared by R. E. Latham, Oxford University Press, London 1965, p. 303.

Momentum dictum a temporis brevitate ut in aliqua mora nullo modo producatur[36].

[Momentum, the smallest unit of time, named from the motion of the stars, as it were a setting into motion.

Momentum is named from the brevity of time, such that it is in no way prolonged.]

It is worth pointing out *quasi movimentum* [as it were a setting-into-motion], which betrays Papias's full awareness of the term's etymology (*a moveo* [from setting into motion]) and refers, by analogy, to some testimonies preserved in the *Corpus Glossariorum*:

Momentum quasi motamentum a motione[37]. [Momentum, as it were a setting-into-motion, from motion.]

It is difficult to escape the impression that in this case too the influence of the
15 aforementioned passage from Isidore is decisive. The bishop of Seville had in fact clearly emphasized that momentum occurs "*cum aliquid sibi cedit atque succedit* [when one thing yields and another succeeds]", that is, when there is a "change". In Aristotle's *Physics*, on the other hand, time was clearly defined as a *measure* of motion and change[38].

The relationship between the temporal meaning of *momentum* and the values of "movement", "change" appears even more explicit in a testimony by Huguccio of Pisa (13th century), documented in Blatt's *Novum Glossarium Mediae Latinitatis*:

momentum quasi movimentum, quia cito transitu moveatur per breve spacium temporis, unde et saepe ponitur pro mutacione et movimento[39]. [momentum, as it were a setting-into-motion, because it is moved by a swift transit through a brief interval of time, whence it is often used also for change and movement.]

The extensive documentation offered by Blatt confirms, however, that the by far prevailing meaning was temporal (small space of time and minimum part of the hour). We can deduce from Blatt that the phraseological value also remained in use (importance, authority, etc.: *nullius momenti* [of no moment], etc.)[40].

The *Corpus Glossariorum*, lastly, presents two recordings, in which it seems that *momentum* indicates the meaning of a small space[41].

---

[36] *Papias Vocabulista*, Impressum Venetiis per Philippum de Princ. Mantuanum, Anno domini 1496, die XIX aprilis. In the example of B.N.F. which I use (*Inc.* A, 4, 5) the page corresponds to the sheet 104*r* (modern pencil numbering). Of this edition of *Vocabulista* an anastatic reprint has recently been performed (Bottega di Erasmo, Torino 1966).

[37] *Corpus Glossariorum latinorum*, Edidit G. Goetz, vol. 7, In aedibus B. G. Teubneri, Lipsiae 1887-1901, IV (*Glossae Abavus*, sec. VIII-IX), p. 366, l. 3.

[38] Cf. *infra*, Part One, Ch. IV.

[39] *Novum Glossarium mediae latinitatis ab anno DCCC usque ad annum MCC*, Ejnar Munksgaard, Hafniae 1963. *Miles-Mozytia* (Presiding over the preparation of the text, F. Blatt), col. 743-746.

[40] *Ibid.*, col. 744.

[41] "Momento spatio [In the space of a moment]" (*Corpus glossariorum* cit., *Glossae Codicis Vaticani 3321*, IV, p. 119, 1.38) and "Momento spatio soliarius [Singular space of a moment]" (*vel solitarius quod est monachus*), *ibid.*, V, p. 466, 1.12 (*Excerpta ex glossis AA*).

In addition to this general and continued prevalence of the temporal meaning of the term, we find a different value indicated in the same used instruments. Given its specificity, we thought it appropriate to isolate it, subjecting it to a more accurate analysis. It is, in fact, an exclusively medieval meaning of *momentum*, absent in Roman Latinity, as well as in neo-Latin authors after the 13th century.

As we have already indicated, in examining the medieval Blaise, the meaning is that of "lingua [tongue]" (needle) of the balance. Alongside this meaning, we find an interesting term recorded for us, *momentana*, whose affinity with *momentum* is immediately evident[42]. *Momentana* is documented in the two Lexicons of Blaise[43] and in the *Glossary* of Souter[44], while *momentum* = needle of the balance is attested only in medieval Blaise[45]. To explain *momentana*, it is worth reading the description 16
given by Isidore in book XVI, Ch. XXV (*de ponderibus*), of the *Etymologiae*:

> Trutina est gemina ponderum lances aequali examine pendens, facta propter talenta et centenaria adpendenda; sicut *momentana* pro parva modicaque pecunia[46]. [The balance is a pair of platters hanging in equal measure, made for the hanging sums of money and hundred pound weights; just as the *momentana* is for small and moderate amounts of money.]

It is therefore a balance. Isidore is not explicit on the etymology of *momentana*. However, the indication "pro parva modicaque pecunia [for small and moderate amounts of money]", together with the name of this particular instrument, seems to indicate the derivation from *momentum* = minimum quantity of weight. That is, the *momentana* would be the steelyard of *momenta [moments]*. We will see, in the next chapter, on which authoritative tradition the derivation of *momentum* from *momentana* could be based.

Some notes recorded in the *Corpus Glossariorum* effectively contribute to clarifying, without the possibility of misunderstanding, the meaning of the term, and to reiterating its character as a balance for small weights. Firstly, *momentana* is described as:

> Momentana duabus lancis et uno in medio stilo aequaliter constat, est trutina modica pecunia[47]. [Momentana, consists of two platters and one middle stylus in equal agreement, a balance for a small amount of money.]

In two other *glossae*, we finally find *momentum*, clearly indicated as an "ago [needle or beam]" of a balance, the balance precisely being the *momentana*:

> Momentum, stilus in quo momenta[na]*, quia cito admodum inclinatur vocabulo supersit[48].

---

[42] For *momentana*, cf. also the entry of *Thesaurus linguae latinae* cit., VIII, col. 1390.
[43] Cf. the work cited in notes 24 and 29. The page is the same.
[44] *Glossary* cit., p. 255.
[45] *Lexicon latinitatis Medii Aevi* cit., p. 596.
[46] *Etymologiarum* cit., II, Lib. XVI, ch. XXV.
[47] *Corpus glossariorum* cit., V, p. 614, l. 43 (*Glossae Scaligeri*).
[48] *Ibid*., IV, p. 119, l. 8 (*Glossae Codicis Vaticani 3321*, sec. IX).

[Momentum, the stylus of the momenta[na], the term remains because it inclines quickly.]

And also

Momentum, stilus in quo momentana exequatur, nam ideo momentana vocabulum sumpsit quia cito et admodicum inclinatur si aequaliter non pensetur[49]. [Momentum, the stylus that follows the momentana, consequently, momentana takes the name because it quickly inclines if not equally weighed.]

In the two *glossae*, the reasons for the derivation of *momentana* from *momentum* are also suggested. In the first example, they are identified in the temporal meaning of *momentum* (if the weights are different it turns "in un momento [in a moment]"); in the second, reference is also made to the one with the least weight (it turns "in a moment" but also with very little weight). In the next chapter, we will see commentators predominantly assign the meaning of the needle of the balance to
17 the two occurrences of *momentum* in the *Vulgate*, at least until the 13th century. The *glossae*, which explain the etymology of *momentana*, do not enlighten us on the reasons for the identification of *momentum* with the needle of the balance. You can, however, try to guess them. It is evidently a derivation from the meaning *momentum* = minimum weight. The needle is in fact the part of the balance capable of indicating (with an inclination) even the smallest of weights. It was probably from this original meaning that *momentana* is derived[50].

It is possible however – let it be said with necessary caution – to indicate another reason, less logical and rational, for the attribution of *momentum* to the meaning of the needle of the balance. I am referring to *ῥάβδος* [rod], whose peculiar presence among the glossemes of *ῥοπή* in Hesychius's Lexicon we highlighted previously[51]. Let's assume that some scholar, perfectly aware of the relationship *ῥοπή-momentum*, used Hesychius's Lexicon, learning that *ῥοπή* also had the meaning of the needle of the balance. Nothing prevents us from thinking that he could have extended the same value to the corresponding Latin term.

We also find documentation of *momentana* and of *momentum* = "ago [needle]" in other dictionaries. Papias, for example, defines the *momentana* by repeating the words of Isidore[52]. Among the meanings of *momentum* recorded by Papias is also the *stilus* [stylus] of the *momentana* (with indication through a single analogy carrying a temporal meaning)[53]. *Momentana* and *momentum* are also recorded in Du Cange. The latter – on the testimony of *Glossae* – refers to the interpretation

---

[49] *Ibid.*, V, p. 311, l. 33 (*Glossarium Ampl. Secundum*, sec. IX).

[50] The derivation of *momentana* from *momentum* = small weight is confirmed in the aforementioned entry of *momentana* in the *Thesaurus linguae latinae* and, above all, the *Dictionnaire étymologique* cit. of Ernout and Meillet (p. 416).

[51] Cf. *supra*, p. 7.

[52] "Momentana duabus lancibus et uno in medio stilo librata aequaliter constat, dicta quod parva modicaque pecunia in ea solita sit appendi [The momentana contains two platters and one middle stylus balancing equally in agreement, it is said that small and moderate amounts of money are customarily attached to it]" (*Vocabulista*, edition cit., f. 104*r*).

[53] "Momentum stilus a quo momentana; quia cito admodum stili inclinatur sumpsit nomen [Momentum is the stylus from which the momentana takes its name, because the stylus inclines very quickly]" (*ibid.*).

of Graevius[54]. Latham assigns *momentana* to the 7th century, but does not record *momentum* = *stilus*[55]. In the *Lexicon imperfectum latinitatis italicae medii aevi*[56] neither *momentana*, nor *momentum* = *stilus* are documented. The same goes for the *Novum Glossarium Mediae Latinitatis*[57]. *Momentana* is registered, however, in the *Mediae Latinitatis Lexicon Minus*[58].

The use of *momentana* therefore appears documented between circa 5th-6th **18**
and 12th centuries. For *momentum*, in the specific meaning of "ago [needle]" of the *momentana*, it will be necessary to think of a more recent origin, since it is not recorded by Isidore, nor does it appear in the aforementioned Souter's Glossary. With this value, it remained in use at least until the 13th century. Naturally, the use of *momentum* = minimum weight has much more distant origins (see *Thesaurus*, IA1c)[59].

It therefore seems possible to argue, with reasonable foundation, that the semantic range and the use of *momentum* underwent a certain impoverishment in the centuries from the 5th to the 14th. In fact, the temporal meaning was enormously favored, and at the same time, some temporal expressions became widely used (*ad momentum* [at the moment], *per momenta* [for moments], *momento* [moment], etc.). We have already observed, however, that the awareness that *momentum*, although used as an instant of time, also indicated "action", "movement", "change", was never lost. It represented the smallest particles of time in their continuous flow, "ut in aliqua mora nullo modo producatur [so that in any delay it is by no means prolonged]" – as Papias's definition testifies aptly[60].

Although common in classical Latin, practically no evidence of momentum as a signifier of weight (inclination, the cause of motion) remains from the 5th to the 14th century. The idea of weight – or rather of small weight – is undoubtedly in the context of the "ago [needle]" of the balance. As we will see in the next chapter, many authors of the 12th and 13th centuries tried to explain *momentum* = *stilus*, precisely by observing that the *stilus* is made to tilt by the minimum of the weights (*momentum*)[61].

The "misfortune" of the term and its semantic impoverishment can justify the parsimonious use made of it by the medieval translators of the Greek works, such as Aristotle's *Physics* or Eutocius's *Commentary* on Archimedes's *Equilibrium of*

---

54 "Momentum, Stilus, quo momentaria inclinatur. Gloss. Is. Ubi Graevius legendum putat: *quo movente libra inclinatur* [Momentum, Stylus, that which briefly inclines. Gloss. Is. Where Graevius thinks it reads: *that which sets the scales in motion as to incline*]" (*Glossarium* cit., p. 450).

55 "Momentana, sort of balance 7c." (*Revised Medieval Word-List* cit., p. 303).

56 See *supra*, note 33.

57 See *supra*, note 39.

58 *Mediae latinitatis lexicon minus*, Composuit J. F. Niermeyer, E. J. Brill, Leiden 1954-1976, Fasc. 8 (1960), p. 701. The examples provided by Niermeyer concern Gregory of Tours and Isidore.

59 I mean to say, if from *momentum* = small weight derived *momentana* = balance for *momenta*, it was probably from *momentana* that the attribution to *momentum* of the meaning of the "ago [needle]" of the balance depended.

60 Cf. *supra*, p. 14.

61 Cf. *infra*, Part One, Ch. II.

Planes, in which ῥοπή appeared with the meaning of inclination produced by weight, that is, with a value not fully "filtered" by classical Latin of the Middle Ages[62].

The recovery of the "classical" meanings of the term was not easy, nor definitive, nor did it happen all at once. The Greco-Latin lexicons of the previously examined humanistic age sought to reconstruct the sensibility towards ῥοπή of the scholars of the 15th and 16th centuries, helping also to understand the meaning assigned to *momentum*. The term is still usually used to indicate ῥοπὴ ἐπὶ ὥρας [inclination upon time], while it is generally preferred *incrementum* [increment] (minimum addition
19 of weight, sufficient to make the balance tip) to make ῥοπὴ ἐπὶ ζυγοῦ [inclination upon the balance]. When *momentum* is used to translate this last expression, it is to indicate – and this is explicitly observed – the instantaneousness (the critical moment) of the collapse of the balance and, together, the rapidity of the movement, not the weight, that determines it: "momentum, *quando* lanx inclinat [momentum, *when* the platter inclines]"[63].

In any case, it was with the revival of interest in the classics and their language, with the rediscovery of the codices and with the printed editions, through which the classical works were widely circulated, that the conditions were created for a recovery of values, including linguistic ones.

The pursued imitation of the classics, of their elegant language – especially Ciceronian – together with the polemics against the acrid medieval Latin, could not help but lead the recovery of the genuine meaning of terms – such as *momentum* – disfigured and corrupted by "barbaric" encrustations.

Relative to our term, around the middle of the 16th century, we will see the first ripened fruits of a changed posture: among the Greek versions of the works of Aristotle curated by the humanists, such as Argyropulus[64], *momentum* appeared for the first time as the equivalent of ῥοπή, and in the precise meaning of inclination to motion produced by weight; a meaning documented in the classical authors, but among those, to cite an example, not a single occurrence is found in the works of Thomas[65]. We will verify the same thing – and in the same period of years – for the Archimedean *corpus* version[66]. It was not, then, a question of discovery or

---

[62] Cf. *infra*, Part One, Ch. III, paragraphs 1 and 5.

[63] Cf. *supra*, note 18. In addition to those mentioned above, the testimonies of some printed Greek-Latin lexicons can be added. See, for example, the *Lexicon Graeco-Latinum* published in Basel in 1554: "ῥοπή – momentum lancis huc aut illuc inclinatis ictu oculi [ῥοπή – the moment the platter inclines this way or that in a blink of an eye]" (col. 1250); cf. also *Lexicon Graeco-Latinum multis praeclaris additionibus locupletatum…*, Paris 1512: "ῥοπή – momentum quando lanx inclinat [ῥοπή – the moment when the platter inclines]" (p. 366, l. 21). In the latter Lexicon, *momentum*, together with *punctum*, was also proposed as the equivalent of στιγμή. Cf., lastly, the *Dictionarium graecum*, published in Venice by Aldo in 1497: "ῥοπή – momentum quando lanx infra vadit [the moment when the platter goes below]" (ch. not numbered).

[64] Cf. *infra*, Part One, Ch. III, paragraph V.

[65] This is what results from the precious *Index Thomisticus Sancti Thomae Aquinatis operum omnium indices et concordantiae…*, compiled by R. Busa S. J. with the IBM electronic automated system, … 1975, Sec. II, *Concordantia Prima*, vol. XIII, pp. 1112-1113. Thomas does not even employ *momentana*. In addition to the temporal meaning and metaphorical use, it is recorded by Thomas *momentum* = indivisible motion (cf. *infra*, Part One, Ch. IV).

[66] Cf. *infra*, Part One, Ch. III, paragraph 1. The now recovered full meaning of *momentum* and the

enrichment, but of a conscious revival of classical motifs, of a conscious return to the **20**
ancient, noble origins of a language corrupted and disfigured by a centuries-old habit of ignorance and barbarity. In the documentation offered by Perotti's *Cornucopiae*, who was a significant interpreter of this effort, there is no difficulty in recognizing this need. The clarification of the meaning of *momentum* was entrusted to examples taken from proven classic authors:

> Ab hoc [moveo]* derivatur ... motio et momentum, quod proprie significat exiguam quandam mutationem et pene indivisibilem, quod graeci atomom vocant (ut inquit Lucretius), illis corpusculis, quae infusis per fenestram radiis solis videmus. *Varro*. Momentum temporis uterque discessit, hoc apud Veteres etiam momen dicebatur. *Lucr*. Momine si a parvo possunt impulsa moveri. Hinc quodcunque minimum potest momentum appellari[67]. [From this [setting into motion] derives...motion and momentum, which properly signifies a certain slight change and almost indivisible, which the Greeks call an atom (as Lucretius says), those little corpuscles, which we see poured through the window by the rays of the sun. *Varro.*: a momentum of time each passing, that which the Ancients called impulse. *Lucr.*: A movement emergent from a small capable impulse. Hence any small and able thing is called momentum.]

Perotti "rediscovered" that the Latin term indicated any small quantity, in particular small movement, small body (weight), small space of time. And he deliberately ignored the meanings assumed by the term of the medieval period: neither *momentum* = *stilus* nor *momentana* were mentioned.

The documentation offered for our term by Nizolio is even more extensive. In the first volume of his *Thesaurus*, a true Ciceronian lexicon, which had enormous influence, there was apparently recorded, alongside the temporal one, the meaning of *momentum* as a dynamic inclination produced by weight, which remained essentially foreign to medieval Latin[68]. In the spirit of Nizolio, an epigone a few decades later was able to add to the Ciceronian Lexicon rich documentation – attesting to the vast range of meanings of *momentum* – obtained from the examination of innumerable other Latin authors, classified according to a very precise and eloquent hierarchy:

1) *Vetusti* [old], imperfect and inelegant, but proficient.
2) *Medii* [middle], of its kind, perfect and most elegant.
3) *Postremi* [late], imperfect and lacking.

The last of the *postremi* was Apuleius; beyond that nothing deserved to be saved. It goes without saying that not even the appendix to Nizolio's *Thesaurus*

---

complexity of its relationships with the different values of *ῥοπή*, especially as a term of statics, appear evident in the *Lexicon graecum et institutiones linguae graecae* published in Antwerp in 1572 (Plantin): "*ῥοπή* – lancis ad imum vergentis inclinatio atque momentum, pensio. Vitr. De statera usurpavit. Plut. *Ῥοπή*, translate, *νεῦμα* nutus, momentum, inclinatio, sive propensio, *Ῥοπή*, quod momentum affert, *ὁρμή*, impulsus, impetus [*ῥοπή* – the inclination and moment of a scale-pan tipping downward, a weighing. Vitruvius used it of the steelyard. Plutarch: *Ῥοπή*, translated as νεῦμα, a nod, moment, inclination, or propensity; *Ῥοπή* as that which brings a decisive weight; *ὁρμή*, impulse, impetus]. Paradoxically, the temporal sense of the term faded into the background!

[67] *Cornucopiae, sive linguae latinae commentarii, diligentissime recogniti atque ex archetypo emendati*... [of Niccolò Perotti]*, In aedibus Aldi et Andreae Asulani, Venice 1527, col. 936.

[68] Of the different editions of *In M. T. Ciceronem observationes utilissimae* one may consult the one printed by the heirs of Seb. Gryphius, Lugduni 1562, col. 1008. Among the countless prints of *Thesaurus Ciceronianus* you can consult the one from Basel of 1559 (Hervagius), col. 1285-1286.

21 mentioned the medieval meanings of *momentum*[69]. It cannot be admitted that they were unknown: not many years before, Papias's *Vocabulista* had been published, meeting with notable influence among humanists, in which, among the meanings of *momentum*, that of *stilus* of the *momentana* is registered[70].

It was probably on these elements and on the parallel revival of interest in the study of Greek – with the rediscovery of the multiple analogies between *ῥοπή* and *momentum* – that the progressive "emancipation" of *momentum* from the marginal role played in the Middle Ages and its progressive reinsertion – with dignity, soon, of first magnitude – into the modern philosophical and scientific lexicon. We will illustrate some important stages of this progressive acquisition in the following chapters, all of which are oriented towards Galileo. Here it is enough to observe how, without the work of removing the medieval encrustations and recovering past meanings, *momentum* would not have been of use to the Pisan scientist.

It remains to examine the story of the vernacular *moment*. A conclusive speech seems impossible without specific and patient research. However, it does not appear that the term – except for the temporal meaning and the phraseological use, which have been recorded since the authors of the 13th-century[71] – was widely used before Galileo. Some examples of the use of the vernacular term will be illustrated in the next chapters. We will see that these are almost always translations of Latin texts where *momentum* appears.

The investigation carried out on a group of 16th-century vernacular dictionaries gave completely negative results. I have never found *moment* reported, neither in the meaning of "inclination" produced by the weight, nor in that of "weight", nor in that of "movement"[72]. On the other hand, the exceptional nature of *moment* with these values seems to find confirmation in the first edition of the *Vocabolario* of the Academicians of Crusca, in which only the temporal meaning and the phraseological use are recorded under *moment* ("L'usiamo anche per Importanza, Cosa di momento [We also use it for Importance, Thing of the moment]")[73]. We will have to wait for
22 the second edition, from 1623, to see it appear – almost certainly on the basis of the examination of the Galilean *Discorso* of 1612 – in its mechanical meaning[74]. The documentation offered by modern dictionaries for these meanings indicates Galileo

[69] *Latinae linguae Thesaurus bipartitus. Prior quidem ex uno Cicerone, iam olim a Mario Nizolio collectus ... Posterior autem ex aliis quoque latini sermonis authoribus comportatus ..., Editionem procurante M. Ludovico Lucio ...*, Per Sebastianum Henricpetri, Basileae 1613, II, col. 3318.

[70] *Vocabulista* cit., c. 105*r*.

[71] Just refer to the entry of *momento* of Tommaseo-Bellini, *Dizionario della Lingua Italiana*, Torino-Napoli 1869, vol. III, Part I, pp. 336-337.

[72] I checked the following dictionaries: Adriano Politi (*Dittionario Toscano*, 1613), Fabricio Luna (*Vocabulario di cinquemila Vocabuli Toschi*, Napoli 1536), Francesco Alunno (*Ricchezze della lingua volgare*, Venezia 1512), Pierfrancesco Giambullari (*Origine della lingua fiorentina*, Firenze 1549), Orazio Toscanella (*Dittionario volgare et latino*, Venezia 1568), Luc'Antonio Bevilacqua (*Vocabulario vulgare et latino*, Venezia 1583), Giacomo Pergamini (*Memoriale della lingua*, Venezia 1602), Alberto Accarisio (*Vocabulario, Grammatica et Orthographia de la lingua volgare*, Cento 1543).

[73] *Vocabolario degli Accademici della Crusca ...*, Venezia 1612, p. 537.

[74] See *infra*, Part Two, Ch. II, p. 243. Cf. also my *A proposito di un errore dei traduttori di Vitruvio nel '500* cit.

as the oldest example of the use of *momento* [moment] in a technical sense. And the data is confirmed by the materials for the current entry of the *Grande Dizionario della Lingua Italiana* (UTET), the files of which I was kindly allowed to consult[75]. It seems that the same can be said for French, English, German, and Spanish[76].

[75] I take this opportunity to thank Prof. Bàrberi Squarotti, from whom I received such notable courtesy.
[76] Cf., for French, É. Littré, *Dictionnaire de la langue française*, Gallimard, Paris 1968, V, pp. 354-357. For English, cf. J. A. H. Murray, *A New English Dictionary on Historical Principles...*, Clarendon Press, 1908, VI, pp. 595-597. For German see J. Grimm and W. Grimm, *Deutsches Wörterbuch*, Hirzel 1887, VI, col. 2482. For Spanish, M. Alonso, *Enciclopedia del Idioma. Diccionario histórico y moderno de la lengua española*, vol. I-III, Aguilar, Madrid 1958, II, *momento*, p. 2871 (Alonso dates the non-temporal use of *momento* to the beginning of the 18th century); in the *Diccionario de Autoridades* of the Real Academia Española (3 vols., F. del Hierro, Madrid 1732, II, pp. 593-594), the use of *momento* "in statics" is cited, however, on the basis of a testimony of Tosca (*Compendio mathemático*, 1707-1715, IV, p. 4). In truth, the passage from Tosca is a literal translation of the definition of *momento* given by Galileo in the *Mecaniche* (cf. *infra*, Part Two, Ch. II, paragraph 1).

CHAPTER II

"MOMENTO" AS A MINIMUM OF WEIGHT: THE BIBLICAL TRADITION

1. THE TEXTS

In the Latin translations of two verses of the Old Testament, there is an 23
expression that the commentators' glosses have long emphasized: *momentum staterae* [momentum of the steelyard]. The two verses correspond to *Isaiah* XL, 15 and to *Wisdom* XI, 23†. They also appear close in terms of the meaning of the discussion that takes place. In fact, it has been argued that *Wisdom* depends on *Isaiah* on this point[1].

In the context of *Isaiah*, the unbridgeable distance that separates God from man is illustrated with effective metaphors. In the context of *Wisdom*, by contrast, it refers to a negligible element, an invisible fragment, when compared to God and his plans. Not only the individual, but entire nations, are nothing before the Creator. In *Wisdom*, in place of just man's disproportion to God, the insignificance of the entire created world in relation to the Supreme Being is emphasized[2].

The Hebrew text of *Isaiah* expresses, with a balance metaphor, the smallness of man before God. The term *šaḥaq* appears there, which, in the singular, is a *hapax* in the entire *corpus* of the Holy Scriptures. Today there is substantial agreement among Semitists that the term means "fine polvere [fine dust]", that which is broken into pieces, made fine, as dust. In the Hebrew Bible, however, the same term, in the plural, is used twenty times in the meaning of (fine) "nubi [clouds]". It was noted that, between the meaning of the term in the singular ("polvere [dust]") and

† The citation for this verse can vary between *Wisdom 11:22* and *11:23,* depending on which version of the Bible is used.

[1] This is the conclusion reached by M. Gilbert S. J. in his recent course on *La Philanthropie de Dieu. Exégèse de Sagesse XI, 15-XII, 27,* Pontifical Biblical Institute, Rome 1976 (handout for the 2nd Semester), p. 19. I take this opportunity to thank the learned Father Gilbert, from whom I received kind assistance as well as numerous and valuable indications.

[2] For questions relating to *Wisdom*, see C. Larcher, *Études sur le livre de la Sagesse*, "Études Bibliques", Paris 1959. For a recent commentary on *Isaiah*, cf. Claus Westermann, *Isaiah* 40-66. *A Commentary*, "The Old Testament Library", SCM Press Ltd., London 1969; for the commentary corresponding to our verse, cf. pp. 51-53.

P. Galluzzi, *Momento: Galilean Studies*, Archimedes 78,
https://doi.org/10.1007/978-3-032-35049-7_2

24 what it seems to have in the plural ("nubi [clouds]"), there is no insurmountable distance (think of the expression "una nube di polvere [a cloud of dust]"). According to others, *šaḥaq* would mean not "nubi [clouds]", but rather "il cielo [the sky]", or better, "ciò che è più remoto da noi [that which is most remote from us]", the "altissimo [highest]"[3].

Having clarified the meaning of the term, let us examine the entire expression: *ukšaḥaq mōznayim*. On the basis of what we have already established, it can be said that it corresponds to "la fine polvere della (o sulla) stadera [fine dust of (or on) the steelyard]", what remains on the platter of the balance, the solid "residuo [residue]", after several weighings. In the metaphor, this means that comparing man to God would be like trying to weigh a grain of fine dust with a steelyard. It will be appropriate to point out that the meaning of the metaphor remains substantially unchanged if the hypothesis *šaḥaq* = "nube [cloud]" is accepted: "come nube sulla bilancia [like a cloud on the balance]"[4].

The dust, then – this seems to be the genuine meaning – is not proportional to the balance and cannot be felt by the balance. In technical terms, this is equivalent to saying that such a small weight cannot move (give motion to) the balance.

Of both verses, we have, of course, the Greek text. It is worth remembering, however, that while the Greek text of *Isaiah* constitutes a translation, *Wisdom* was written directly in Greek. Let's see how the two passages appeared in the LXX [Septuagint] version:

εἰ πάντα τὰ ἔθνη ὡς σταγὼν ἀπὸ κάδου καὶ ὡς ῥοπὴ ζυγοῦ ἐλογίσθησαν, καὶ ὡς σίελος λογισθήσονται (*Is.* XL, 15)[5]. [truly the nations are as a drop from a bucket and are reckoned as the inclination of a balance, and shall be reckoned as spittle]

ὅτι ὡς ῥοπὴ ἐκ πλαστίγγων ὅλος ὁ κόσμος ἐναντίον σου καὶ ὡς ῥανὶς δρόσου ὀρθρινὴ κατελθοῦσα ἐπὶ γῆν (*Wis.* XI, 23)[6,†]. [because the whole world before thee is as the inclination of a balance and as a drop of

[3] Not knowing the Hebrew language, for the part of the work relating to the analysis of the Hebrew text and its meaning, I depend on the information contained in the only work known to me which focuses on the different values of *šaḥaq*: D. Winton Thomas, "*A drop of a bucket?*" *Some observations on the Hebrew text of Isaiah 40, 15*, in *In memoriam Paul Kahle*, Herausgegeben von Matthew Black und Georg Fohrer, Töpelmann, Berlin 1968, pp. 214-222 (cf. especially pp. 214-218). Other precious suggestions come to me from the amiable courtesy of Prof. J. L. Teicher, who explained to me his thoughts on the Hebrew text of the verse in a long letter, of which I will report the essential points later. I also take this opportunity to thank Dr. Pietro de Marco, from whose broad biblical expertise I have received valuable advice.

[4] Cf. D. Winton Thomas, "*A drop of a bucket?*" cit., p. 216.

[5] *Isaias*, Edidit J. Ziegler... ("*Septuaginta, Vetus Testamentum Graecum*", auctoritate Academiae Litterarum Gottingensis editum, vol. XIV, Vandenhoeck und Ruprecht, Göttingen 1967, p. 269). The critical edition does not present a significant variant relative to the expression *ῥοπὴ ζυγοῦ*.

[6] *Sapientia Salomonis*, edidit J. Ziegler, "*Septuaginta*" cit., vol. XII, 1, Vandenhoeck und Ruprecht, Göttingen 1962. The critical edition does not register a significant variant by the expression *ῥοπὴ ἐκ πλαστίγγων*.

[†] The citation for this verse can vary between *Wisdom 11:22* and *11:23*, depending on which version of the Bible is used. In this particular case, see above note 6.

morning dew descending upon the earth]

In regard to *Is.* XL, 15, the expression *šaḥaq mōznayim* is translated *ῥοπὴ ζυγοῦ* [inclination of the balance]. In truth, this raises some doubts. That is, one has the impression that the version does not convey the exact spirit of the original text. There is, in fact, some difficulty in admitting that *ῥοπή* can mean "polvere [dust]", that which does not weigh on the balance. This meaning of *ῥοπή* is not recorded in the lexicons, with the exception of our two contexts and in one occurrence in Herondas (3rd century BC)[7], which is very difficult to decipher. *Ῥοπή* is often used in a meaning only apparently similar to that of *šaḥaq* in the Hebrew text: "piccolo peso che determina il tratto della bilancia in equilibrio [small weight that causes a balance in equilibrium to tip]". It indicates an infinitesimal amount (among other things, of any material) that is enough to add to one of the two platters to break equilibrium. In general, *ῥοπή* seems closely connected to an idea of movement. The values of the term, when referred to the balance, can be grouped around three meanings whose interdependence is evident: 25

1) The minimum weight sufficient to tip the balance (perhaps even a minimum unit of weight; see the example of Herondas).
2) The tipping, itself, of the balance (that is, its movement).
3) The instant in which the tipping occurs (critical moment)[8].

Ῥοπὴ ζυγοῦ [inclination of the balance], is also (or will become) a "technical" expression, the meaning of which will be, depending on the context, one of the three indicated above. We saw this value in some of the manuscripts and printed Greco-Latin lexicons examined in the previous chapter[9]. An authoritative confirmation that this was the value of the expression is offered by a commentator on the Greek version of the verse. This refers to the *Paraphrasis* of St. Cyril of Alexandria on *Is.* XL, 15. Cyril takes care to explain the meaning of the metaphors used in the verse, showing how they successfully indicate the negligible state of man compared to God:

τί δ'ἂν εἴη καὶ ζυγοῦ ῥοπὴ πρὸς ἰσότητα φέρουσα τὸ ἡττῆσθαι δοκοῦν[10]. [but what could the inclination of the balance be that brings what seemed defeated toward equality?]

This passage leaves no doubt about the meaning of *ῥοπὴ ζυγοῦ*. The expression in fact indicates a negligible weight that is enough to restore equilibrium to the 26

[7] "*Οὐδ' ὅσον ῥοπὴν ... ψεῦδος*" (Herondas, *Mimes*, 7, 33): "a lie no bigger than a *ῥοπή*". Except for this example, which is not easy to decipher, neither the Stephanus nor the Liddell-Scott record the use of *ῥοπή* in the meaning of negligible, imperceptible weight (cf. Ch. I). In the example of Herondas, *ῥοπή* could simply indicate the smallest unit of weight.
[8] These are the three main meanings in which the entry *ῥοπή* is articulated in the *Greek-English Lexicon* cit., by Liddell-Scott. They are also recorded in the *Patristic Greek Lexicon* cit., by G. W. H. Lampe (cf. *supra*, Ch. I).
[9] Cf. *supra*, Ch. I, notes 18 and 63.
[10] *S. Cyrilli Alexandrini, Commentaria in Isaiam, PG*, vol. LXX (cura et studio J. Auberti, Parisiis 1859), p. 811.

balance (of equal arms) in which one of the two weights is insensibly smaller. *Ῥοπὴ* therefore means a minimal addition of weight and, at the same time, a small movement.

If our impression is not wrong, we can say that the Greek translator of *Isaiah* introduced a significant change. In fact, he transforms "polvere [dust]", which is incapable of conferring movement, into a "minima aggiunta di peso [minimal addition of weight]" that is sufficient to tip the balance. The difference may appear subtle, but it is not without implications. The very image preserved for us by the biblical metaphor underwent a transformation: man – as was stated in the Greek version – although minimal and petty, is "pesabile [weighable]", that is, he has a positive reality. The theological implications, as well as the problem of whether it was an accidental inadequate translation or an intentional interpretation, remain outside the scope of this work. It will therefore be sufficient to record the suggestive hypothesis communicated to me by Prof. Teicher. He believes it cannot be ruled out that the Greek translator understood *šaḥaq* not in the sense of "polvere [dust]", but in that of "il cielo [the sky]". Since for him "the sky" was nothing other than the motion of the celestial spheres, he thought it was the motion of the balance and rendered it as *ῥοπὴ ζυγοῦ*. According to this explanation, this would be a true translation error[11].

Obviously, the case in *Wisdom* appears different. The considerations made for *Isaiah* would in fact apply to *Wis.* XI, 23 only if the derivation of the verse from the Hebrew text of *Is.* XL, 15 was admitted. The fact is that the two texts in Greek appeared to have the same meaning and were perceived as such by the Latin translators. Only later, when we returned to the Hebrew text of *Isaiah*, did we realize the difference.

As is known, between the 2nd and 3rd centuries AD, a *corpus* of Latin translations of the Old and New Testaments was formed, performed on the Greek text of the LXX version, without consulting the Hebrew text. This *corpus* of translations, which is best called *Vetus Latina*, but essentially resolves itself into a series of national traditions (*Italian*, *Spanish*, etc.), appeared outdated and was completely put aside following the great translation performed on the Masoretic text by St. Jerome at
27 the turn of the 4th-5th century[12]. This new version, known as the *Vulgate*, long

[11] Prof. Teicher, in the kind letter I mentioned above, observes that "There is an obvious discrepancy between the meaning of the Hebrew phrase and the translations (Latin and Greek), far from being unique. The reason for this discrepancy is perhaps due to the fact that in Hebrew the entry *šaḥaq* has two meanings: in *Is.* 40, 15 it means "dust", and, in other places, "the sky". It seems that the Greek translator of the Bible understood the second meaning of the term. Since for him the "heavens" are nothing other than the motion of the celestial spheres, he thought that it was the "motion" of the balance in *Is.* 40, 15, especially because, in this way, the Hebrew phrase appeared to correspond exactly to the metaphorical meaning of *ῥοπὴ ζυγοῦ*". It is worth remembering the etymology of *momentum* proposed by Isidore, already previously emphasized: "*a motu siderum dictum* [by the motion of the stars]" (v. *supra*, Ch. I, p. 13).

[12] The biblical texts of the *Vetus Latina* can still be read in large part today in the valuable edition edited by the French Benedictine Pierre Sabatier in the 18th century: *Bibliorum Sacrorum Latinae versiones antiquae, seu Vetus Italica et caeterae quaecunque in codicibus mss. et antiquorum libris reperiri potuerunt. Quae cum Vulgata latina et cum textu graeco comparantur*..., Apud Franciscum Didot,

remained as the official text of the Church. Some books remained extraneous to Jerome's version, including *Wisdom*. For those, the *Vetus* is still used. We therefore have a dual situation for our two texts. In fact, we have *Isaiah* both in the *Vetus* version and in the *Vulgate*, while we only have the *Vetus* version of *Wisdom*. Let's see how our two verses are rendered there.

Vetus Latina	*Vulgata*
Si omnes gentes sicut stilla situlae et quasi *momentum staterae* reputatae sunt et quasi saliva reputatae sunt (*Isaia* XL, 15).[13] [As if all the nations were as a drop in a bucket and regarded as a *momentum of a steelyard*, regarded as spittle.]	Ecce gentes quasi stilla situlae et quasi *momentum staterae* reputatae sunt. Ecce insulae quasi pulvis exiguus (*Isaia* XL, 15).[14] [Behold, nations are as a drop in a bucket and as a *momentum of a steelyard*. Behold, the islands are regarded as petty dust.]

Vetus Latina
(resumption of *Vulgata*)

Quoniam tamquam *momentum staterae* sic ante te est orbis terrarum et tamquam gutta roris antelucani quae descendit in terram (*Wisdom* XI, 23).[15] [For as the *momentum of a steelyard*, so is the world before you, just as a drop of morning dew that descends to the earth.]

We are faced with a single choice, at least regarding the two expressions that interest us. In the three versions, they are always in fact rendered with *momentum staterae* [momentum of the steelyard]. We know that the *Vetus Latina* was conducted exclusively on the Greek text. The translators therefore strove to render ῥοπὴ ζυγοῦ [inclination of the balance] without the possibility of verifying the different value recorded in the Hebrew text. In fact, the choice of *momentum staterae* appears just. We have in fact shown, on the basis of lexicons and dictionaries, that *momentum* also had this value of "infinitesimal quantity", both before and after the chronological **28**
context to which *Vetus Latina*'s version is assigned[16]. And, exactly like ῥοπή, it

Parisiis 1739-41 and (2nd ed.) 1751, 3 vols. in 6 tomes. I used the 1751 edition. A complete renovation of the *Vetus Latina* began a few years ago, again under the care of the Benedictines: [*Vetus Latina*]* *Die Reste der altlateinischen Bibel nach Petrus Sabatier neu gesammelt und herausgegeben von der Erzabtei Beuron*..., Herder, Freiburg 1963.

[13] *Bibliorum Sacrorum... seu Vetus Italica* cit., Tomus II, Pars II, p. 581; the italics are ours. In *Notae ad versionem antiquam*, Sabatier referred to the paraphrase of some Fathers to this place in *Isaiah*. Since they will be illustrated later, we avoid mentioning them here.

[14] *Biblia Sacra iuxta Latinam Vulgatam Versionem ad codicum fidem..., cura et studio monachorum Abbatiae Pontificiae Sancti Hieronymi in Urbe, Ordinis Sancti Benedicti Edita. Liber Isaiae ex interpretatione Sancti Hieronymi*..., Typis Poliglottis Vaticanis, Romae 1969, p. 154; the italics are ours.

[15] *Biblia Sacra iuxta Latinam Vulgatam Versionem* cit., *...Sapientia Salomonis. Liber Hiesu Filii Sirach*..., Typis Poliglottis Vaticanis, Romae 1964, p. 65; the italics are ours.

[16] See *supra*, Ch. I, p. 9.

was used in the specific meaning of a small *decisive* quantity, capable, that is, of determining a situation. As indicated by the derivation from *moveo* [to set into motion], the term suggested action, movement. In short, *momentum* corresponded to *punctum* [point], *atomus* [atom], that is, it did not refer to any particular matter, but had the implicit idea of the minimum *effective* quantity. In the (metaphorical or scientific) context of the balance, *momentum* presented the same three closely linked meanings that we observed in *ῥοπή*: "minimum weight" (minimum addition of weight), sufficient to tip a balance of equal arms; "movement" of the balance produced by a minimum weight; "instant" in which the tip occurs. In the Latin versions – even in that of Jerome, which also returned to the Hebrew text – the unification of the two verses is carried out. The formal differences preserved from the Greek text are also suppressed. Later, we will see that, particularly through the reversion to the original text of *Isaiah*, many will show that they do not consider the proposal of *momentum* for *šaḥaq* correct and adequate, while they will confirm it for *ῥοπή* in *Wisdom*. The review of the different opinions of translators and commentators on the two verses serves greatly to illuminate the history of this semantic value of *momentum*, often through the explicit declarations of what our term *cannot* mean.

Before examining, with respect to *Isaiah* XL, 15 only, the attempts to reform the *Vulgate* carried out in the 1500s, it will be worth emphasizing that Tertullian, whose activity is much closer to the *Vetus Latina* than to the version of St. Jerome, showed an awareness of the exact meaning of the Hebrew text and avoided using *momentum* to render *šaḥaq*. In fact, in an almost literal paraphrase of our verse he writes:

> Nationes vero nunquam penes Deum, nisi stillicidium de situla et pulvis ex aera...[17]. [The nations are but, in the hands of God, only as a drop in a pitcher and as bronze dust...]

And, elsewhere,

> Semel in stillam situlae et in pulverem aereae et in salivam nationibus deputatis a Deo...[18]. [All at once, the nations are reduced to drops in a pitcher, to bronze dust, and to spittle, in comparison to God]

The absence of *momentum* and the substitute use of *pulvis* [dust] would seem to admit access (not necessarily of Tertullian) to the Hebrew text, on which the awareness of the improper use of *momentum* must have depended.

29 As we anticipated, it was especially in the 16th century that there was a general revival of interest in the Hebrew text. Many returned, animated by the need – so characteristic of Renaissance culture – both to subject ancient texts to refined philological analyses (in order to extract their genuine meaning), and to adapt the medieval versions of fundamental works, restoring an elegance of classical Latin to the harsh speech of the dark ages.

---

[17] *De praescriptione haereticorum*, in *Tertulliani Opera* cit., I, p. 194.

[18] *De fuga in persecutione*, *ibid.*, II, p. 1138 (II, 6).

Among the 16th century attempts to revise the *Vulgate* on the basis of the Hebrew text, Santi Pagnini's version, which included *Isaiah* and *Wisdom*, had particular importance and had great influence. Here are the solutions he proposed.

> Ecce gentes sicut stilla situlae et sicut *pulvis tenuis staterae* reputate sunt. Ecce insulas tamquam pulverem proiiciet[19] (*Is.* XL, 15). [Behold, the nations are just as drops in a pitcher and regarded as *fine dust of the steelyard.* Behold, the islands are as cast dust.]

> Quoniam tamquam *momentum staterae,* sic et ante te orbis terrarum et tam quam gutta roris antelucani, qui descendit in terram[20] (*Wis.* XI, 23). [For as the *momentum of a steelyard,* so also is the world before you, and so much as a drop of the morning dew that descends to the earth.]

It will be observed that Pagnini reacted differently in the two cases. His version of the *Wisdom* verse in fact repeats, almost literally, the *Vetus Latina.* As for *Isaiah,* he instead proposed a different solution. The *momentum staterae* of the *Vetus* and of St. Jerome are replaced by *pulvis tenuis staterae* [tenuous dust of the steelyard]. We have already previously pointed out that this was the correct translation of the Hebrew text of *Isaiah.* In restoring it, Pagnini quite clearly distinguished the two contexts that were unified within the *Vulgate.* In *Wisdom* he left *momentum staterae*, to indicate that it was the "tratto della bilancia [tipping of the balance]", an expression that he felt was adequate to indicate *ῥοπὴ ζυγοῦ* [inclination of the balance]. In *Isaiah*, however, he skipped the Greek version – which had *ῥοπή*, as used in *Wisdom* – and corrected, so to speak, the error of the Hebrew text. What is interesting is the fact that Pagnini shows that he conceives *momentum staterae* as *necessarily* connected to an idea of movement: the inclination or weight that causes the steelyard to tip.

The correction introduced by Pagnini is confirmed in Francisco Foreiro's famous and rich commentary on *Isaiah*, in which he presents a "nova Hebraico versio [new Hebrew version]":

> Ecce gentes quasi gutta austri et *quasi levitas bilancium* reputate sunt, ecce insulas quasi tenue operiet[21] (*Is.* XL, 15). [Behold, the nations are like a falling drop and *are light on the balance*; behold, the islands are as if they were fine.]

Thus, not even Foreiro thought he could confirm the *momentum staterae* of the 30
Vulgate. The introduction of *levitas* indicates that he also felt the inadequacy of *momentum staterae* to express *ukšaḥaq mōznayim*. In fact, he explicitly states in a note that he attaches to the *momentum* used by Jerome:

> nomen *šaḥaq* notat levitatem, et sumitur hoc loco pro tenuissimo pulvere in lance incidente, qui *nullius*

---

[19] *Biblia Sacra ex Sancti Pagnini tralatione, sed ad hebraicae linguae amussim novissime ita recognita et scholiis illustrata, ut plane nova editio videri possit*, Apud Hugonem a Porta, Lugduni 1542, c. 155*v*; the italics are ours.

[20] *Ibid.*, c. 139*r*; the italics are ours.

[21] *Iesachiae Prophetae, Vetus et Nova Hebraico Versio, cum commentario in quo utriusque ratio redditur, vulgatus interpres a plurimorum calumniis vindicatur… F. Francisco Forerio… auctore*, Ex officina Iordani Zileti, Venetiis 1563, c. 145*v*; the italics are ours.

est ponderis. Non raro autem pro nube accipitur; *Deut*. 33. Ps. 89, pro aethere videtur accipi[22]. [the word *šaḥaq* denotes lightness, and it is taken to mean the finest dust falling on a scale-platter, which *is of no weight at all*. Not infrequently, however, it is taken to mean cloud; *Deut*. 33. Ps. 89, it seems to be taken to mean ether.]

Unfortunately, this note does not tell us what meaning is attributed to *momentum staterae* by Foreiro. It states, however, what it *cannot* mean: an entity of negligible weight and, therefore, not capable of tipping the balance. It can, then, be stated that Foreiro also introduced a modification in the *Vulgate* version, probably because he felt it was arbitrarily introduced, alongside the concept of minimum weight, that of the ability to tip the steelyard.

We find confirmation of a general dissatisfaction in relation to *Isaiah* XL, 15 in the *Vulgate* version, and, at the same time, of the distinction introduced between the two verses unified by the Greek and Latin translators, both in the version by Sebastiano Castiglione[23] and in the "dalla hebraica verità in lingua toscana [Hebrew Truth in the Tuscan Language]" by Antonio Brucioli:

Ecco le genti, come gocciola di secchia e come *polvere di bilancia* sono reputate, ecco gittare le isole come spazzatura[24] (*Is*. XL, 15). [Behold, the nations are regarded as drops in a bucket and like *dust of the balance*; behold, he casts the islands like debris.]

Perché è come *momento di stadera*, cosi e davanti a te l'universo de la terra, et come gocciola di rugiada avanti al giorno, che scende in terra[25] (*Wis*. XI, 23). [Because it is like a *moment of a steelyard*, so is the universe of the earth before you, and like a drop of morning dew that descends to the earth.]

Even Brucioli, then, in whose version it is easy to recognize the influence of Pagnini, considered the use of *momentum* arbitrary in the verse of *Isaiah*. As he clarified in the commentary on this passage, the genuine meaning of the Hebrew text was that all men, compared to divine omnipotence, "sono come niente et di nessun conto et pregio [are like nothing and of no account or value]"[26]. *Momentum* could then not serve this purpose.

2. Glosses and Commentary

31

[22] *Ibid*., c. 139*r*; the italics are ours.
[23] Castiglione translated the *Is*. verse as "*proque staterae granulo* [and for a granule of the steelyard]", while for *Wis*. it confirmed "staterae momentum" (*Biblia interprete S. C.*, Joan. Oporinum, Basileae 1554).
[24] *La Biblia, quale contiene i Sacri Libri del Vecchio Testamento tradotti da la hebraica verita in lingua toscana per Antonio Brucioli…*, Venetia 1538, c. 345*v*.
[25] *Ibid*., c. 310*r*.
[26] *Commento di Antonio Brucioli in tutti i Sacrosanti Libri del Vecchio e Nuovo Testamento…*, Tomi V, in Venetia 1546, III, c. 41*r*.

In the glosses and in the commentary on our two verses, we see the reasons emerge for the uncertainty and contrast noted in the versions. From these invaluable documents, we obtain very important indications for the history of *momentum*.

Let's start with the brief indications contained in the so-called *Glossa Interlinearis* which, according to prevailing opinion, is attributed to Anselm of Laon (12th century), *magister divinitatis*:

Is. XL, 15 (*momentum staterae*): Quod velut ex nihilo in utramque partem declinatur[27]. [That which, as if out of nothing, inclines to either side.]

Wis. XI, 23 (*momentum*): lingua[28]. [tongue.]

The *Glossa Interlinearis* for *Isaiah* XL, 15 clearly expresses the idea of *momentum* as a small weight that causes the balance to tip, or as "ciò che un niente basta a far inclinare [a mere trifle that is enough to cause a tilt]". This second meaning seems to be suggested by the *gloss* on *momentum*: *Lingua* (*staterae*) [tongue (of the steelyard)] is precisely that which can be moved (displaced from the equilibrium position) with a minimum weight.

The identification of *lingua staterae* [tongue of the steelyard] with *momentum* is also suggested in the so-called *Glossa Ordinaria*, today attributed in scholarly tradition to Anselm. The passage that interests us is a commentary on the verse of *Wisdom*, in which reference is made to *Is*. XL, 15 in a similar context:

Vult enim intelligi quod mens et ratio et sensus domini per quem omnia facta sunt et sine quo factum est nihil, ille sit de quo dicitur Verbo domini celi firmati sunt etc. *Omnes gentes quae non cognoverunt creatorem suum* ad comparationem Dei quasi stilla situlae et quasi *momentum staterae quod leviter declinatur*[29]. [For he wants it to be understood that the mind and reason and sense of the Lord by whom all things were made and without whom nothing was made, is that of whom it is said 'By the word of the Lord were the heavens established', etc. *All the nations that have not acknowledged their creator* are, in comparison to God, as a drop in the pitcher *and as the momentum of a steelyard that gently declines.*]

The agreement of *quod* [that which] with *momentum* indicates that our term is considered the subject of the last proposition; *momentum* therefore indicates "ciò che viene fatto abbassare dolcemente [that which is made to lower gently]". It seems there can be no doubt that the commentator considers *momentum* as the "ago [needle]" of the steelyard. It's a meaning that we have seen assigned to our term in **32**
medieval lexicons and glossaries[30]. It will be necessary to immediately emphasize that the deviation of this attribution from the meaning of the value of *effective* "piccolissimo peso [very small weight]", conferred by the first Latin translators, is

[27] *Biblia Sacra, cum glosis ordinaris et interlinearibus et Nicolai de Lyra Postilla…*, vol. I-VI, Paganino de' Paganini, Venetiis 1495, IV, c. 733*v*.

[28] *Ibid*., III, c. 617*r*.

[29] Cf. *PL*, Tomus CXIII (Parisiis 1852), from which we take the quotation (I, col. 1175), the *glossa ordinaria* is published under the name of Walafridus Strabo (9th century), who was then considered its author (*Walafridi Strabi… opera omnia ex editione Duacensi*, vol. I-II). The commentary of the *Glossa ordinaria* on *Is*. XL, 15 does not present interesting ideas for our purposes.

[30] Cf. *supra*, Ch. I, pp. 16-18.

more apparent than it actually is. *Leviter* [gently] indicates, in fact, that it is a very slow motion, caused by an imperceptible weight. It remains to be noted that, even in this interpretation, *momentum* is felt to be closely linked to the idea of movement; better yet, it is invoked precisely as a term capable of indicating the maximum disposition to motion. It should not, of course, be forgotten that the image of the needle of the balance adequately expresses the metaphorical sense of the "peso decisivo [decisive weight]" that determines a situation. Obviously, it is primarily this metaphorical value that is recorded in the two verses.

An explicit explanation of *momentum*, in the meaning illustrated above, is found in De Lyra's famous *Postillae* for *Is*. XL, 15. Referring to the *momentum staterae* of the *Vulgate*, he in fact specifies:

> Id est lingula staterae quae valde de facili ad utramque partem movetur de se nullam habens stabilitatem[31]. [This is the tongue of the steelyard that moves very easily towards either side, having no stability in itself.]

In De Lyra's *Postilla*, the idea of *momentum* = *lingua staterae* and its natural inclination to motion are emphasized with absolute clarity. The deviation from the Hebrew text was, at this point, truly notable.

The same interpretation is reiterated in the *Postillae* for *Isaiah* by Hugh of Saint-Cher. Here are his observations on our verse:

> sicut stathera defacili et quasi per nihilo in utramque partem inclinatur; sic nihil sunt gentes coram eo; et ita defacili potest eas convertere ad dominum quemcunque voluerit, sicut movetur stathera, et sicut *lingula quae est in medio stathera* convertitur ubi vult librans[32]. [like a sensitive steelyard and how, by nothing to either side, it inclines; in this way, the nations are viewed as nothing; and in such a way they can be easily turned and rotated according to the Lord, just as the steelyard moves, and just as the *tongue in the middle of the steelyard* turns as it wills when balancing.]

The meaning of "lingua mobile [mobile tongue]" of the steelyard is also assigned
momentum in the commentary of Saint Thomas *in Isaiah*[33]. This interpretation
33 returns, with an abundance of details, in the substantial commentary by Robert
Holcot *in Sapientiam* [*in Wisdom*]. Holcot, illustrating our verse, believed he could
recognize a universal meaning in it: nothing can oppose God. He added that this

---

[31] *Biblia Sacra cum glosis* cit., IV, c. 713*v*.

[32] *Biblia Sacra cum Postilla Hugonis Cardinalis…*, vol. I-VII, Per Johannem de Amerbach, Basileae 1498-1502, IV, c. not numbered, col. a; the italics are ours. In column b, Hugh returned to *momentum staterae*, emphasizing how this expression was used to indicate "instabilitas vitae humanae. Momentum enim statherae de facili ad utramque partem inclinatur [the instability of human life. For the moment of the steelyard is inclined easily to either side]".

[33] "Ecce gentes quasi stilla, parva gutta, *situlae*, cadens de situla, quae secundum nos est vas aquaticum sitientibus aptum, quae pro nihilo reputatur, *momentum* quod inter brachia staterae est, quod de facili inter utrumque inclinatur [Behold, the nations are like drops, little specks, of a *pitcher*, falling from the pitcher, which according to us is a water vessel suitable for those who are thirsty, which is regarded as nothing, the *momentum* that is between the arms of the steelyard, that is easily tilted between the two]" (*In Isaiam, Sancti Thomae Aquinatis Doctoris Angelici…*, *Opera Omnia…*, Typis Fiaccadori, Parmae 1863, I, p. 528).

idea was expressed by comparing the Creator with two things incapable of resisting:

Primum est momentum staterae. id est lingua staterae quando pendicula sunt aequalia et in eis aequalia pondera apposita ceteris paribus stat erecta perpendicularis. sed quantuncunque modicum gravitatis apponatur, statim recedit a situ suo et facit motum sine quacunque resistentia. Isto modo omnis creatura est in manu dei, quam si deus voluerit annihilare nulla resistentia potest esse. Et hoc est quod dicit quomodo tamquam momentum staterae[34]. [First is the momentum of the steelyard. This is the tongue of the steelyard that stands perpendicularly upright when the hanger arms are equal, equal weights are placed upon them, and all else is equal. But whenever a small bit of heaviness is placed, immediately it withdraws from its own position and makes movement without any resistance. In this manner, all of creation is in the hands of God, which, if God should wish to annihilate, nothing can resist it. And this is the reason for the saying "just as the momentum of the steelyard".]

Once again we see the meaning of *momentum* indicated in "lingua della stadera [the tongue of the steelyard]". Holcot, however, specified that the term expressed the inability to offer resistance, to defend one's position in equilibrium, against the smallest of weights. *Momentum* is that which "facit motum sine quacunque resistentia [makes the movement without any resistance]". It is the epitome of instability; in other words, that whose fate is determined by something else. In the commentary of Holcot, which had numerous printed editions at the end of the 15th century, the idea of the inclination of the balance, of its "tratto [tipping]", was very clear. The small weight remained, however, in the background, as the "cause" of the inclination of *momentum*. The *Commentarius* of Denis the Carthusian must also be included in this family of interpretation[35].

It can, therefore, be concluded that in the Latin commentaries of the Old Testament, up to the 12th century, the interpretation of *momentum* as the "ago [needle]" (*stilus* [stylus], *lingua* [tongue]) of the balance appears prevalent. By the 13th century, the situation started to change. Alongside the occurrences, such as those mentioned above, in which the interpretative tradition is followed, new "readings" appear, which tend to restore the meaning presumably assigned by the first translators of *momentum*.

A decisive step in this direction was taken by Albertus Magnus, whose *Postilla in Isaiam* presents notable interest and originality:

Momentum indivisibile temporis est. Ita in ponderibus minimum pondus aut quod nullum pondus notabile est, momentum vocatur. Et est sensus quod tota natura humana notabile pondus non habet in statera tam immensae Creaturae[36]. [Momentum is the indivisible of time. Likewise, in weights, the smallest weight, or that which has no notable weight, is called momentum. And the sense is that the

---

[34] *Opus praeclarissimi eximii Doctoris Roberti Holkot… super Sapientiam Salomonis*, Impressum anno Incarnationis Domine 1489 [Aput J. Otmar, Reutlingen]*.

[35] "Quoniam tamquam momentum, id est stilus seu lingua staterae…[Because just as *momentum*, this is the stylus or tongue of the steelyard…]". He also emphasized the close analogy with *Is.* XL, 15 (*Doctoris Ecstatici D. Dionysii Cartusiani in Sacram Scripturam commentaria, cura et labore monachorum Sacri Ordinis Cartusiensis…*, Typis Cartusiae Sanctae Mariae de Pratis, Monstrolii 1898, Tome VII, *In…Sapientiam*, p. 522).

[36] *Alberti Magni…Postilla super Isaiam*, Primum edidit Ferdinandus Siepmann…, In aedibus Aschendorff, Monasterii Westfalorum 1952, p. 415.

entire human nature lacks notable weight on the steelyard against the immeasurable Being.]

34 According to Albertus, *momentum* would therefore originally mean "instant" or minimum time. Thus, by analogy, it would also indicate "minimum", "indivisible" in general, and then, also minimum weight. Albertus seems to restore the genuine meaning of the Hebrew text to the verse. *Momentum* is in fact considered by him as a minimal, negligible quantity of weight, incapable of breaking the equilibrium of the balance. Albertus tries, in these lines, to correct the impression that the *momentum* is what determines the tipping of the balance. That is, he retains the term, but proposes a distorted interpretation of it[37].

The abandonment of the interpretation of *momentum* as a needle of a balance is also found in the *Commentarius in Sapientiam* by St. Bonaventure of Bagnoregio. Although in a different way from Albertus, he proposes the equivalence of the two infinitesimal meanings of the term (*moment* of weight, *moment* of time):

tanquam momentum staterae, id est illud modicum, quo statera huc vel illuc inclinatur; inde dicitur momentum a movendo, vel quia, sicut faciliter et quasi in momento movetur huc vel illuc, sic divina virtute mundus ad nutum Dei movetur. Unde Isaia quadragesimo...[38]. [*just like the momentum of the steelyard*, this is that small amount, toward which the steelyard inclines this way or that way; henceforth called momentum from 'moving', or rather, because it moves easily and as if in a moment this way or that way, thus by divine power the world moves according to the nod of God. Hence Isaiah forty...]

Momentum is a modest amount of weight, but sufficient, however, to tip the balance. It is no coincidence that St. Bonaventure recalled, without doubt drawing from Isidore, *a movendo* [from setting into motion] etymology of our term[39]. And he derived the other meaning from it, that of "instant": almost instantly (*in momento* [at this moment]) the small weight causes the inclination of the balance.

A further turning point is recorded in the 16th-century commentators of *Isaiah*. As we have previously observed, regarding the versions, it was mostly determined by a general revival of interest in the Hebrew text.

The awareness of the original text shines through very clearly and is frequently "exhibited" in Leone Castro's *Commentarius*. Castro, analyzing the verse, emphasized how the mercy of God is exalted there,

35 quia qui venit quaerere perierat, nec stillicidium illud tanquam vile contempsit et quo per stateram significatum erat, Deum hominem in iustitia condidisse, allevavit *momentum staterae* et sputo

---

[37] Later, however, Albertus returned to the expression, showing that he understood it to mean "minimum weight". In fact, he stated that "momentum staterae conversationis significat rectitudinem [momentum of the steelyard signifies the correctness/uprightness of the exchange]" (*ibid.*). Honesty in business can be symbolized – this is the meaning of the beautiful expression – by the *moment of the steelyard*, that is, from the maximum possible approximation to an exact weighing.

[38] He warns, a few lines later, "Et notandum quod dicit creaturam esse tamquam momentum etc., respectu Dei quantum ad facilitatem agendi in ipsam [And it should be noted that he says that creation is like momentum, etc., with respect to God regarding the ease of acting upon it]" (*Doctoris Seraphici S. Bonaventurae...Opera omnia*, Ex typographia Collegii S. Bonaventurae, Quaracchi 1893, p. 183).

[39] *Etymologiarum* cit., V, XXIX.

substantiam boni corporis dedit...[40]. [because he who came to seek had perished, he did not despise those drops as something vile, and as indicated by the steelyard, God formed man in righteousness, lifting *the momentum of the steelyard* and gave spittle to a good body...]

The commentator, after having expressed the smallness of man with *momentum*, focused on *šaḥaq*.

šaḥaq dictionem haebraeam Hebraei putant significare pulverem et ita vertunt pulverem staterae. Hiero. et 70 viri putant significare momentum et ita vertere sicut *momentum staterae*. Significat etiam caelum[41]. [the Hebrews think that the Hebrew word *šaḥaq* means dust and so translate it as dust of the steelyard. Hieronymus (i.e., St. Jerome) and 70 men[†] hence think it means *momentum* and so translated it as *momentum of the steelyard*. It also means sky.]

Castro showed that he knew the meaning of the Hebrew term. He also knew that elsewhere it referred to "il cielo [the sky]". Highlighting the contrast between the opinion of the Jews and the choice of *momentum* made by St. Jerome, he felt the need to emphasize the significant difference: *šaḥaq* in fact indicates a negligible weight, while the *momentum*, even if minimal, is what breaks the equilibrium condition. Even Hector Pinto, who was responsible for another 16th-century commentary *in Isaiam*, showed that he conceived *momentum* as a weight that moves[42].

Two other commentators of the 16th-century took a position against the meaning of the expression in *Isaiah* suggested by the Greek version of LXX and the Latin ones (*Vetus* and *Vulgate*) of the 16th century. Taddeo of Perugia, "Augustinian hermit", declared that he was using the Hebrew text. Commenting on *Is*. XL, 15, he emphasized how petty and inconsistent human creatures appeared in the presence of God, an emphasis which invited the reader to a correct interpretation of our verse, which precisely expressed this meaning:

Per stilla situlae sic intelligas ut ipsae creaturae a Deo comparatae sint velut parva guttula, quae aliquando decidit et prodit ex aliquo vase aqua repleto et velut aliquid minimum et nullius fere ponderis

---

[40] *Commentaria in Esaiam Prophetam ex Sacris Scriptoribus Graecis et Latinis confecta... authore Leone Castro...*, Mathias Gastius, Salmanticae 1570, pp. 640-641; the italics are ours.

[41] *Ibid*., p. 641; the italics are ours.

[†] The "Septuagint (LXX)" comes from the Latin phrase *Vetus Testamentum ex versione Septuaginta Interpretum*, which translates as "The Old Testament from the version of the Seventy Translators". Thus, the "70 men" is in reference to those translators.

[42] "Nemo Deum docuit est enim eius sapientia summa et sempiterna nec indiget alicuius hominis consilio et adiumento. Imo omnes gentes cum eo comparatae sunt gutta situlae et momentum staterae. Situla est vas accomodatum ad hauriendam aquam ex quo aliquando excidit quaedam guttula cui genus humanum comparatur. Comparatur etiam momento staterae quoniam levissimo pondere in alteram partem inclinatur, quidam pulvis facit bilancem in alteram partem pendere [No one has taught God, for His wisdom is supreme and eternal, and does not need the counsel or assistance of any man. Indeed, all nations compared to Him are like drops in a pitcher and the momentum of a steelyard. A pitcher is a vessel suitable for drawing water, from which sometimes a certain drop falls, to which humanity is compared. It is also compared to the moment of the steelyard, since with the slightest weight the other side inclines, a sort of dust that makes the balance hang to the other side]" (*F. Hectoris Pinti Lusitani, Ordinis D. Hieronymi, in Esaiam Prophetam Commentaria...*, In aedibus Viduae et Haeredum Ioan. Stelsii, Antuerpiae 1572, p. 253).

positum in altera parte staterae[43]. [By pitcher drops, understand that creatures themselves are in comparison to God as small drops, which sometimes fall and come from filled water vessels and are as something miniscule and bear no weight placed on the other side of the steelyard.]

Thus Taddeo also restored the correct meaning recorded in the Hebrew text to
36 the verse. Mentioning the insensitivity of the steelyard to a weight too small, he
avoided, not by chance, using *momentum*, which he evidently felt was inadequate
for the purpose.

Even more explicitly, Juan Maldonado emphasized the difference between the Hebrew text and the Latin version in his commentary *In Isaiah*:

Et quasi momentum. Id est quasi minimus scrupus staterae, ita illae apud Deum minimum habent ponderis. H[ebraicum]* *šaḥaq*, quasi pulvisculus, id est, sicut minimus pulvis, si in bilance sit non deprimit eam, ita omnes gentes nihil deum movere possunt[44]. [*And like momentum*. This is like the minuscule pebble of the steelyard, in such a way that they have minuscule weight before God. The Hebrew *šaḥaq*, like fine dust, that is, just as minuscule dust, if on the balance it does not depress, in such a way that every nation can do nothing to move God.]

It should be noticed how, in explaining the *momentum staterae* in the sense of a minimum weight (*scrupus* [pebble]) detectable by the scale, Maldonado followed it with the Hebrew term, emphatically highlighting its meaning as “polvere che non pesa in bilancia [dust that does not weigh on the balance]”.

It can therefore be stated that *momentum* was generally viewed as inadequate in expressing the unnoticeable presence of dust on the steelyard platter. *Momentum* instead appeared highly commensurate to *ῥοπή*: as the Greek term indeed expressed consistently the idea of movement, of inclination. This is why, in the commentators of *Wisdom* – differently from what was observed for *Isaiah* – we find a substantial uniformity. Everyone considers *momentum staterae* to mean “infinitesimal weight”, capable of tipping the balance. This interpretation could be extended to *Is*. XL, 15 – and it was – on the basis of the *Vetus* and the *Vulgate*, which assimilated the two verses, even more decisively than in the Greek version of LXX. However, when the translators regained access to the Hebrew text, they could not escape the fact that the meaning in *Isaiah* was different. They therefore proposed the correction of *momentum* with *pulvis* [dust], or they specified in commentary the meaning of the Hebrew term, incorrectly rendered by the translators of the *Vetus Latina* and by Jerome. Thus, two texts that were closely connected, perhaps from the origin, ended up being commented and annotated in substantially different ways. And *momentum* appeared as an adequate term only in *Wisdom*, that is, in a context in which it was beyond question that it was a minimal weight, but capable of unbalancing.

It is no coincidence that, in Jesuit Jean Lorin’s extensive review of the main opinions of the commentators of *Wis*. XI, 23, a characteristic consensus emerged on the value of the expression *momentum staterae*. None of the opinions recorded

[43] *F. Thaddaei Perusinii Heremitae Augustiniani...Explanatio in Esaiam Prophetam duobus tomis comprehensa…*, Apud Vicentium Columbarium, Perusiae 1598, I, p. 734.

[44] *Joannis Maldonati S. I. Commentarii in praecipuos Sacrae Scripturae libros Veteris Testamenti*, vol. I-II, Sumpt. Sebastiani Cramoisy, Parisiis 1643, I, p. 288.

by Lorin can in fact be traced back to the one that we ultimately saw prevail among the interpreters of *Isaiah*.

Firstly, the French Jesuit proposed his own interpretation of the verse:

> Nam aeque omnia Dei levissimo nutu regi ac dirigi et commigrare ut in lances omni ex parte et **37**
> accurate, exacteque perfectas et collibratas vel minimus scrupus atomusve illapsus, alterutram commutat ad momenta ac deprimit...[45]. [For all things are equally governed, directed, and moved by the slightest nod of God, as on scales perfect and accurately balanced in every part, even the smallest scruple or atom falling in changes and depresses either side at the critical moment.]

In his *explanatio* [explanation], the two values of *momentum* appear unified: "movement" and "instant" in which the movement occurs. Furthermore, it was clear that Lorin considered a minimum *momentum* of weight, placed on one of the platters of the balance in equilibrium, sufficient to cause its collapse. The meaning was therefore that of minimal addition of weight. It should be recalled that, in the humanist lexicon manuscripts cited in the previous chapter, we have often found *ῥοπὴ ζυγοῦ* [inclination of the balance] indicated with *incrementum* [increment][46], whose value and derivation undoubtedly now appear clearer to us.

Further on, Lorin explains the opinion of the commentators on the verse:

> *Momentum staterae ῥοπὴ ἐκ πλαστίγγων* Momentum ex lancibus. Glossa Interlin. stateram exponit linguam. Hugo lingulam staterae, cuius, inquit, rectitudine indicatur aequalitas ponderum, e facillime movetur: quae similitudo indicat, ut ait Carthusianus, continuam rerum a Deo dependentiam. Cantacuzenus affert illam aliam rationem, quam prius diximus, ab atomo, qui satis sit ad lancem alteram deprimendam. Id vocat Bonaventura modicum, quo statera huc aut illuc inclinatur: et ait dici momentum a movendo. Innuit tamen etiam aliam expositionem de staterae lingua quae facile movetur ad motum lancium. Arbitrium nostrum, cum inter bonum et malum agitatur, in neutram partem pronum, et praeceps, ac velut aequipondium habens...
>
> Graecum vocabulum *πλάστιγξ* est lanx e iugo, seu librili, suspensa, cui pondus imponitur: *ῥοπή* propensio, devergentia, inclinatio, nutus, momentum lancis ad imum vergentis[47].
>
> [*Momentum staterae ῥοπὴ ἐκ πλαστίγγων* Momentum of the scales. Interlinear Glossary explaining the steelyard tongue. Hugh's tongue of the steelyard, which, he says, is indicated by the straightness of the equality of weights, from which it is most easily moved: which the likeness indicates, as the Carthusian says, the continuous dependence of things on God. Kantakouzenos gives that other reason, which we mentioned before, from the atom, which is sufficient to depress the other plate. Bonaventure calls it the measure by which the steelyard is tilted this way or that, and says that 'moment' is so called from 'moving'. However, it also hints at another explanation of the tongue of the steelyard which is easily moved to the motion of the plate. Our decision, when it is agitated between good and evil, inclined to neither side, and precipitous, and having, as it were, equivalent weight...
>
> The Greek word *πλάστιγξ* is a plate of the beam, or swing, that is suspended, to which weight is placed: *ῥοπή* propensity, divergence, inclination, nod, the moment of the plate approaching the bottom.]

The awareness in encountering an expression that indicates the occurrence of a movement, produced by weight, shines through in the effort to define the meaning

---

[45] *Ioannis Lorini Avenionensis S. J. Commentarii in Sapientiam*..., Sumptibus Horatii Cardon, Lugduni 1607, p. 408.

[46] Cf. *supra*, Ch. I, note 18.

[47] *Ioannis Lorini* cit., p. 409.

of *ῥοπή* through numerous Latin equivalents. They all point to an idea of dynamism.

It can therefore be concluded that different values emerge, or its meaning is illustrated, from the versions and commentary of the two verses in *Isaiah* and *Wisdom* in which *momentum* is used. In general, it can be said that *momentum* was perceived as synonymous with *punctum* [point] and *atomus* [atom][48]. In the
38 context of the balance, to which the metaphor alludes in both verses, *momentum* was conceived as an indivisible minimum of weight. In the following chapters, we will see *momentum* take on, in addition to the meaning of "minimum time", the value of "indivisible of motion and space". What we are interested in emphasizing here is that, in this entire family of meanings, attributable to the general value of "small quantity", *momentum* always takes on the explicit value of the *decisive* minimum; minimum "that moves", "that which moves", "over which one moves", or that "measure that which moves".

The expression *momento della stadera* [moment of the steelyard] (or of the balance) in the indivisible meaning of weight became common. We have seen that, at a certain point in medieval Latin, *momentana* indicated specifically a precision balance that is capable of recording *momenta* [moments][49]. *Momenta* [moments] or *puncta* [points], in turn, were also called the small notches marked on the shaft of the steelyard[50]: the indices of the smallest differences in weight.

The ideas of the *momento della bilancia* [moment of the balance] and *della bilancia del momento* [of the balance of moment] are therefore closely linked. In the vernacularization of the *Tratto della coscienza* of St. Bernard, we see *moment* appear, in the meaning of minimum weight index, marked in the body of the balance:

> La qual cosa fanno i santi propriamente, quando, abbagliati dall'opere e dalle maraviglie di Dio sopra li nascosi misteri che intendere non si possono, comprendono che non sanno nulla, ma pesano tutte le opere loro a uno sottile punto e momento di stadera[51]. [Which is what the saints do exactly when, dazzled by the works and wonders of God over the hidden mysteries that cannot be understood, they understand that they know nothing, yet weigh all their works to a fine point and moment of the steelyard.]

In one of Pietro Bembo's letters – and Bembo must have known the Scriptures and biblical tradition very well[52] – we instead see the *bilancia del momento* [moment

---

[48] See Vico's observation, which was precisely reminiscent of the analogies between *momentum* and *punctum*: "Enimvero latinis *punctum* et *momentum* idem significabant, momentum autem est res, quae movet; et cum punctum, tum momentum, iisdem latinis quid indivisibile dicebatur [Indeed, in Latin, *point* and *momentum* meant the same thing; however, momentum is something that moves; and both point and momentum were used by the same Latin speakers to denote what was indivisible]" (*De antiquissima italorum sapientia*, IV, paragr. 2, in *Le orazioni inaugurali; il* De italorum sapientia *e le polemiche*, edited by G. Gentile and F. Nicolini, Laterza, Bari 1968 [rist. Anast. of the 1914 edition]*).

[49] Cf. *supra*, Ch. I, pp. 16-18.

[50] For *momento*, in this sense, cf. the example of St. Bernard cit. further on. For *punctum*, cf. the passage by Vitruvius cit. *infra*, Part One, Ch. III, p. 64.

[51] Vernacularization of *Tratto della coscienza* of St. Bernard, Pomba, Torino 1830, p. 33. The vernacularization almost certainly dates back to the 15th century.

[52] For his familiarity with one of the leading biblical scholars of the time (Jacopo Sadoleto), cf. H. Hurter S. J., *Nomenclator Literarius Theologiae Catholicae…*, vol. I-IV[2], Libraria Academica Wagneriana

of the balance] described:

Voi avete mandato un peso da troppo gran ducati che i miei non sono, che son piccioli. Ma da vero la bilancia è un bilancion, che per la grandezza non si move per ogni poca differenzia; però pigliatene un'altra, che sia come quella che già mi mandaste, anzi più tosto vorrei che fosse ancora assai più 39
picciola che quella. Che così sentono bene ogni variazione e momento[53]. [You have sent too large a burden of ducats – they are not mine; mine are small. But truly, the balance is a large balance, which in terms of size does not budge due to any small difference; so take another one, which is like the one you already sent me, or rather I would like it to be even much smaller than that. For this way they feel every variation and moment well.]

The influence of the biblical metaphor of *momento della bilancia* [moment of the balance] alone deserves long and detailed work. It will be enough for us to have emphasized that *momentum*, in the meaning of minimum quantity of weight with which it appears in the two biblical verses, ended up also establishing itself as a technical term in statics.

Henry Monantheuil, a French scientist and philologist to whom we will return shortly[54], translated the pseudo-Aristotelian *Questioni Meccaniche* [Questions of Mechanics] at the end of the 16th century. In his *Tractatus de puncto*, which circulated in connection with the translation, he provided more than one example of the "technification" of the biblical expression that had occurred. Firstly, he specified *momentum*, as an "indivisible in weight", as being equivalent to an "instant" in time and to a "point" on a line:

Nam *τὸ νῦν*, seu instans in tempore, unitas in numeris, momentum in ponderibus et alia multa partibus omnino carent, sicque ab his punctum non est distinctum...[55]. Solum enim punctum in Geometria partitionis est expers, ut unitas quidem, sed in Arithmetica *nunc* et *momentum* in Physica...[56]. [For *τὸ νῦν*, or instant in time, unity in numbers, moment in weights, and many other things entirely lacking in division, so that the point is not distinct from these… Only the point in Geometry is lacking of division, as a unity indeed, but in Arithmetic it is *now*, and *momentum* in Physics…]

Later, wanting to illustrate the utter instability of a perfect mechanical system, he used *momentum* to indicate, as in the Latin versions of *Isaiah* and *Wisdom*, that a minimum weight was sufficient to break the equilibrium:

Sphaericum enim corpus, quantum ars tornatilis exquisito fabricari potest, plano insistens horizonti parallelo, ipsum ita tangit ut uno puncto sensibili tangere videatur, unde est ut in eo perpetuo moveatur, nequaquam ipsius puncti contactu praepeditum: vel undiquaque tota sui dividia [*sic*]* parte nutet et propendat in ipsum planum, tamquam consistens in aequilibrio. Qua ratione fit ut levissimo momento huc illuc impellatur et ad quamlibet partem, addito ponderis, vel flatus, vel exigui pulsus unico grano, statim moveatur...[57]. [For a spherical body, as far as the art of lathing can perform, standing parallel to

---

Oeniponte, 1903-1913, II (1906), col. 1510-1511.

[53] *Opere di Pietro Bembo*, vol. I-XII, Milano 1808-1810, vol. IX (1810), *Lettere*, p. 361.

[54] Cf. *infra*, Part One, Ch. III, p. 75.

[55] *Aristotelis Mechanica Graeca, emendata, latina facta et commentariis illustrata ab Henrico Monantholio…*, *Accedit eodem authore Tractatus de puncto*, Ex bibliopolo Commelianiano, Lugduni 1600, p. 12.

[56] *Ibid.*, p. 13.

[57] *Ibid.*, p. 5.

the horizon, touches in such a way that it seems to touch one perceptible point, whence it is perpetually in motion in it, in no way hindered by the contact of that point: or on each side the whole divided [*sic*]* part of itself nods and leans towards the plane itself, as if standing in equilibrium. For this reason, it happens that with the slightest moment it is pushed this way or that way, and to any part, by the addition of a weight, or a blow, or a slight pulse with a single grain, it is immediately moved…]

We will see later that Galileo knew and used *momentum* with the meaning of
40 infinitesimal addition of weight, negligible in itself, but capable of producing the tipping effect of the balance. He used the term in this sense in *Mecaniche*[58] It is difficult to establish whether Galileo was aware of its origin. Nothing allows us to affirm (or deny) that he had the two verses in mind and that he had reflected on them. The only clue – albeit vague – could be the fact that *Wis*. XI, 23 closely follows the famous verse which states that God arranged everything in *numero, pondere et mensura* [number, weight and measure]. A theme that was certainly very familiar to Galileo, and which he certainly reread and reused in a "Platonic" framework[59].

Even though we are forced, then, to leave the question unresolved, a characteristic occurrence must not escape our notice: two years after the death of the Pisan scientist, Giovanni Battista Odierna, in publishing Galileo's *Bilancetta*, of which the Sicilian scientist presented it as the reborn work of Archimedes, did not hesitate to characterize, right from the title, the dear and precise Galilean instrument, the *Stadera del momento*[60].

[58] Cf. *infra*, Part Two, Ch. II, pp. 208-209.

[59] Here is the verse of *Liber Sapientiae*: "Sed et sine his uno spiritu poterant occidi persecutionem passi ab ipsis factis suis et dispersi per spiritum virtutis suae: sed omnia in mensura, et numero et pondere disposuisti [But even without these they could have been slain by one spirit, suffered persecution by their own deeds, and dispersed by the spirit of their virtue: but you arranged everything in measure, both in number and weight]" (XI, 21). Plato insists on the theme of creation inspired by a "geometrical" intelligence in the *Timaeus*.

[60] *Archimede redivivo con la Stadera del Momento, del Dottor Don Gio. Battista Hodierna della città di Ragusa. Dove non solamente si insegna il modo di scoprir le frode nella falsificazione dell'oro e dell'argento, ma si notifica l'uso delli pesi e delle misure civili presso diverse nazioni del mondo…* [Archimedes revived with the steelyard of moment, by Doctor Don Gio. Battista Odierna of the city of Ragusa. Where not only is the way to detect fraud in the counterfeiting of gold and silver taught, but the use of civil weights and measures is notified in various nations of the world…], For Decio Cirillo, Palermo 1644.

CHAPTER III

THE SHIFTS OF "MOMENTUM": FROM NATURAL INCLINATION TO TECHNICAL TERMS IN STATICS

1. THE ARCHIMEDEAN TRADITION: THE "EQUILIBRIUM OF PLANES" AND EUTOCIUS'S "COMMENTARY"

It is in large part thanks to the research of Marshall Clagett that today we begin to 41
have satisfactory and reliable information regarding the diffusion of Archimedes's
texts in the Middle Ages[1], a diffusion that the progress of studies shows to be
increasingly less sporadic and negligible. The Syracusan scientist's writings were, 42
in fact, translated into Latin and circulated widely. And when, in the 16th century, the
return-to-Archimedes phenomenon took on macroscopic proportions, the medieval

[1] Clagett rightly and emphatically insisted on the need, before attempting an overall assessment, to reconstruct the various aspects of Archimedes's permeation into the Middle Ages, through not only translations, but also compilations, more or less faithful uses, etc. In the first of the valuable volumes that the North American historian has dedicated to medieval Archimedes, already published some years ago (*Archimedes in the Middle Ages*, Vol. I. *The Arabic-Latin Tradition*, University of Wisconsin Press, Madison 1964; hereinafter cit. *Archimedes*, Vol. I), he focused attention on the Arabic-Latin tradition, that is, on the permeation of Archimedean texts and themes drawn from Arab sources into medieval culture. And in fact the circulation of Archimedes and his works depended mainly on the Arabic versions until beyond the middle of the 13th century, that is, before Moerbeke translated the entire corpus of the Syracusan scientist's writings into Latin from the original Greek. The second volume, recently released, is entirely dedicated to the publication of the Flemish Dominican version: *Archimedes in the Middle Ages*, Vol. II. *The Translations from the Greek by William of Moerbeke*, Part I (*Introduction*) - II (*Texts*) - III (*Variant readings* etc.), in 2 vols., The American Philosophical Society, Philadelphia 1976 (hereinafter cit. *Archimedes*, Vol. II). Among Clagett's works, in addition to *Archimedes*, see, for the Latin tradition, the very useful essay *The use of the Moerbeke translations of Archimedes in the works of Johannes de Muris* ("Isis", vol. 43, part. 3, n. 133, 1952, pp. 236-242). Significant publications and references to Archimedes's influence in the Middle Ages and the Renaissance can be found in other essays by Marshall Clagett: *Leonardo da Vinci and the medieval Archimedes* ("Physis", XI, 1969, fasc. 1-4, pp. 100-151); *Mechanics* in the *Leonardo da Vinci* entry of the *D.S.B.*, VIII (1973), pp. 215-234; *The works of Francesco Maurolico* ("Physis", XVI, 1974, fasc. 2, pp. 149-198); *The life and Works of Giovanni Fontana* ("Annals of the Institute and Museum of the History of Science of Florence", I, 1976, fasc. I, pp. 5-27). See also by Clagett, naturally, *La scienza della Meccanica nel Medioevo* (Italian translation by Libero Sosio, Feltrinelli, Milano 1972), especially the first two chapters.

P. Galluzzi, *Momento: Galilean Studies*, Archimedes 78,
https://doi.org/10.1007/978-3-032-35049-7_3

translations played a decisive role, offering a useful starting point for attempts at revision and improvement[2], when the text was not dependable for the first printings[3].

Of particular interest for our specific research is the influence of two of the writings that make up the Archimedean *corpus*: Archimedes's *On the Equilibrium of Planes*[4] and Eutocius's *Commentary* on the first of the two books into which the Syracusan scientist's writing was divided[5]. As is well known, the *Equilibrium of Planes* presented a series of geometric demonstrations that determine the center of heaviness of numerous two-dimensional planar figures, which Archimedes treats
43 as having weight (*τὸ βάρος* [the burden])[6]. Furthermore, in the first book, there is a geometric demonstration of the law of the lever. According to this law, there is a balance of power and resistance when an inverse proportionality relationship is given between applied weights and the length of the arms (in this case, the two weights are described by Archimedes as *ἰσόρροποι* [of equal moment], that is, having the same *ῥοπή*, even if they are different in weight). In other words, it shows an

---

[2] For the influence and versions of the various writings making up the Archimedean *corpus*, which did not have common events, see the first chapter of *Archimedes*, Vol. I, pp. 1-14. For the Greek Archimedes codex, see the *Prolegomena* of Heiberg in *Archimedis Opera Omnia cum Commentariis Eutocii* (vol. 3, in Aedibus G. B. Teubneri, Stutgardiae 1972, I, pp. VIII-XCVIII [anastatic reproduction of the 1915 edition]). Finally, also important is the *Introduction* (chap. II, *Manuscripts and principal editions*) by T. L. Heath in the English edition of Archimedes's works (*The works of Archimedes*, Cambridge 1897-1912, pp. XXIII-XXXVIII).

[3] This is the case, as we will see later, of the 1543 Venetian edition of some of Archimedes's works (*Opera Archimedis Syracusani Philosophi et Mathematici Ingeniosissimi, per Nicolaum Tartaleam Brixianum…*, Per Venturinum Ruffinellum sumptu et requisitione Nicolai de Tartaleis Brixiani, Venetiis 1543 *mense aprili*) in which Tartaglia proposed Archimedes's text according to Moerbeke's version. Luca Gaurico also used the translation of the Flemish Dominican for the edition of *Dimensio circuli* and the *De quadratura parabolae* (*Tetragonismus id est circuli quadratura, per Campanum, Archimedem Syracusanum atque Boetium mathematicae perspicacissimos adinventa*, Per Ioan. Bapt. Sessa, Venetiis 1503).

[4] Of the *Equilibrium of Planes*, no translations from Arabic were made, unlike what happened for other Archimedean writings. The first Latin version was therefore from the Moerbeke Greek. However, many of the Archimedean concepts that are presented in the *Equilibrium* have been circulating since the 12th century through two treatises belonging to the *corpus de ponderibus*: the anonymous *De Canonio*, translated from Greek, and the famous *Liber Karastonis* of Thābit ibn Qurra (9th century), translated from Arabic by Gerard of Cremona (12th century). These two treatises, evidently influenced by the Archimedean theory of centers of heaviness, addressed the questions of statics – the laws of balance – with an approach fully attributable to that of the Syracusan scientist: the geometric demonstration of equilibrium and the attribution of "weight" to geometric quantities. Cf. *The medieval science of weights* (*Scientia de ponderibus*), Edited by E. A. Moody and M. Clagett, The University of Wisconsin Press, Madison 1960, pp. 55-86 (*De canonio*, edited and with *Introduction* by Moody) and 77-117 (*Liber Karastonis*, edited and with *Introduction* by Marshall Clagett). For the *Liber Karastonis*, see the edition of the Arabic text edited by K. Jaouiche (*Le livre du Qarasṭūn de Thābit ibn Qurra*, Brill, Leiden 1976).

[5] Not even in Eutocius's *Commentary* of the *Equilibrium of Planes* is it known to have any Latin translations prior to that of Moerbeke.

[6] Archimedes's work lacks any declaration of the terms used (center of heaviness, balance, weight, etc.), which leads us to believe that such definitions would have been found in a previous lost writing. On this problem, see the elegant and still usable essay by Giovanni Vailati, *Del concetto di centro di gravità nella Statica di Archimede*, "Atti della R. Accademia delle Scienze di Torino" (vol. XXXII, Adunanza del 9 maggio 1887, pp. 500 ff.), reprinted in *Scritti di G. Vailati*, Leipzig-Firenze 1911, pp. 79-90; cf. also *Le Opere di Archimede* edited by A. Frajese, UTET, Torino 1974, pp. 389-395 and the editor's notes on the Italian translation of the *Equilibrium of Planes* (*ibid.*, pp. 397-440).

explicit awareness of a fundamental principle of statics: the same weight, applied to varying distances from the center of heaviness of a mechanical system, exerts a greater force at a greater distance[7].

The *Commentary* composed by Eutocius (5th century) is valuable for its numerous preserved historical reports regarding the Greek mathematicians and their activity[8]. Of the *Commentary*, what is of particular interest to us is the opening passage of the first book, in which the contrasting opinions of Plato and Aristotle on the problem of positive lightness are presented. Namely, Plato excluded that the "light" had its own "inclination" to move (*ῥοπή*), while Aristotle and Ptolemy assigned this specific "inclination" to both the "heavy" and to the "light". In the **44**
Latin reception of these two writings, which usually circulated as closely linked, there are elements capable of illuminating several aspects of the history of our term. In both, the Greek text records numerous occurrences of *ῥοπή* (the term, however, has meanings that are only seemingly similar to each other) which we have often seen indicated in dictionaries as corresponding to the Latin *momentum*. However, medieval translators of these texts did not employ *momentum*, and it will be necessary to wait for Iacopo da Cremona [Iacopo da San Cassiano] (c. mid-15th century) to see our term become the protagonist.

Before examining this important tradition, it will be appropriate to remember that the *Equilibrium of Planes* and Eutocius's *Commentary* were texts very familiar to Galileo from the first years of his activity. For example, *De motu* was drafted with the Archimedean model of the *Equilibrium of Planes* explicitly in mind, to which the *Theoremata circa centrum gravitatis solidorum* and the *Bilancetta*[9] must

---

[7] The fundamental demonstration is presented by Archimedes in Propositions VI and VII of the *Equilibrium of Planes* (*Le Opere di Archimede* cit., pp. 403-408, especially notes 7 and 8). Proposition VI aims to demonstrate that "Le grandezze commensurabili sono in equilibrio se sospese a distanze inversamente proportionali ai pesi [Commensurable quantities are in equilibrium if suspended at distances inversely proportional to the weights]" (*ibid.*, p. 403). The objective is achieved by using proportions to reduce the case of unequal weights, hanging at inversely proportional distances from the fulcrum of the balance, in the case of the component parts of the unequal weights distributed along a balance of equal arms. In this way, the assumed mechanical system is demonstrated equivalent to that of the balance of equal arms (to which identical weights are applied), which, by the First Postulate of the Archimedean writing, is in equilibrium (*ibid.*, p. 397). In Proposition VII, the validity of this decomposition is extended to the case of incommensurable quantities (*ibid.*, p. 406, note 8). The fundamental assumption advanced in these pages by Archimedes is that the equilibrium in a mechanical system, such as the balance, is not disturbed if a weight applied at a given point is replaced by two weights, each equal to half the first, applied at equal distances from the first weight. For a rigorous analysis of the meaning of the Archimedean demonstration, cf. Giovanni Vailati, *La dimostrazione del principio della leva data da Archimede nel libro primo dell'equilibrio delle figure piane*, "Atti del Congresso Internazionale di Scienze Storiche", Roma 1903, vol. XII (Acc. dei Lincei, Roma 1904), reprinted in *Scritti di G. Vailati* cit. pp. 497-502. For an evaluation of the relationship between the demonstration of the lever principle contained in the pseudo-Euclidean *Book on the Balance* and the Archimedean one, see M. Clagett, *La scienza della meccanica*, trad. it. cit., pp. 27 ff. and 43 ff.

[8] For information on Eutocius and the importance of his *Commentaries*, cf. the entry *Eutocius of Ascalon* by Ivor Bulmer-Thomas, in *D.S.B.*, IV (1971), pp. 488-491, which also provides bibliographical information.

[9] For an evaluation of the influence exercised by the *Equilibrium of Planes* and Eutocius's *Commentary* in Galileo's early works, see *infra*, Part Two, Ch. I.

be reconnected in an even more explicit manner.

The first known Latin translation of the *Equilibrium of Planes* and of the entire *corpus* of Archimedes's writings[10] is the one preserved in the Ottobonian Vatican Latin Codex 1850, uncovered by Rose in 1884[11]. It was based on the original Greek[12]. Some of the writings translated in the Vatican Codex bear the date 1269[13]. Although there is no indication of the author of the translation, Rose, Heiberg, and also Clagett (who first published the entire version) all agree on its attribution to
45 William of Moerbeke[14], the Flemish Dominican whose valuable translation work was used by Thomas[15]. According to Heiberg's opinion – recently confirmed by Clagett – the Vatican Codex could even be an original manuscript of Moerbeke[16]. Besides the *Equilibrium of Planes*, the Vatican Codex also contains the translation from Greek Eutocius's *Commentary*, of which some significant occurrences of *ῥοπή* are recorded in the first book. The text in question is the one that opens the *Commentary*. It will be worth quoting it in full.

Τὴν ῥοπήν, ὦ γενναιότατε Πέτρε, κοινὸν εἶναι γένος βαρύτητος καὶ κουφότητος ᾿Αριστοτέλης τε λέγει καὶ Πτολεμαῖος τούτῳ ἀκολουθῶν· ὁ δέ γε παρὰ Πλάτωνι Τίμαιος πᾶσαν ῥοπὴν ἀπὸ βαρύτητος λέγει γίνεσθαι· τὴν γὰρ κουφότητα στέρησιν νομίζει. ὧν ἔξεστι τὰς δόξας τοῖς φιλομαθέσιν ἀναλέγεσθαι ἐκ τε τοῦ Περὶ ῥοπῶν βιβλίου τῷ Πτολεμαίῳ συγγεγραμμένου καὶ ἐκ τῶν ᾿Αριστοτέλους φυσικῶν πραγματειῶν καὶ ἐκ τοῦ Πλάτωνος Τιμαίου καὶ τῶν ταῦτα ὑπομνηματισάντων. ὁ δὲ ᾿Αρχιμήδης ἐν τούτῳ τῷ βιβλίῳ κέντρον ῥοπῆς ἐπιπέδου σχήματος νομίζει, ἀφ᾿ οὗ ἀρτώμενον παράλληλον μένει τῷ ὁρίζοντι, δύο δὲ ἢ

[10] With the exception of the *Arenario*, the *Problema dei buoi* and, naturally, the *Metodo*, found, as is shown, only in 1906 in Constantinople by Heiberg and immediately published by himself, in collaboration with Zeuthen, in the "Bibliotheca Mathematica", s. III, vol. VII (1906-1907), pp. 321-363 (*Eine neue Schrift des Archimedes*).

[11] Rose announced the discovery of the Vatican Codex in a short essay with a suggestive title, *Archimedes im Jahre 1269*. "Deutsche Litteraturzeitung", 5 (1884), n. 6, pp. 210-213. For a description of the Vatican Codex and for the *inscriptions* of various Archimedean writings that contain it, cf. M. Clagett, *The use of the Moerbeke translations* cit., pp. 240-241, and, also by Clagett, the *Archimedes*, Vol. II, which gives the first complete printed edition. Heiberg publishes from the Vatican Codex the Flemish Dominican versions of the *De lineis spiralibus* and the *Dimensio circuli*, as well as, parts of the *De sphaera et cylindro* (*Neuen Studien zu Archimedes*, "Zeitschrift für Mathematik und Physik", 34, 1890, *Suppl.*, pp. 1-84).

[12] For the two Greek codices used by the translator, cf. M. Clagett, *Archimedes*, Vol. II, I, pp. 54-55.

[13] The versions of the Archimedean writings dated in the Vatican Codex are the *De spera et cylindro*, 25 Sept. 1269 (Cod. Vat., c. 33*v*V; *Archimedes*, Vol. II, I, p. 219); the *Eutokii Ascalonitae rememoratio in secundum Archimedis de spera et cylindro*, 10 Kal. Nov. 1269 (Cod. Vat., c. 44*v*P; *Archimedes*, Vol. II, I, p. 285); the *De conoydalibus et sperydalibus*, Idibus Nov. 1269 (Cod. Vat., c. 52*v*D; *Archimedes*, Vol. II, I, p. 337); and the *Eutokii Ascalonitae rememoratio in secundum de equerepentibus*, 21 Nov. 1269 (Cod. Vat., c. 55*r*W; *Archimedes*, Vol. II, I, p. 355).

[14] Rose was the first to suggest this attribution (art. cit.). Heiberg reached similar conclusions based on a terminological analysis and evaluation of translation techniques (*Neuen Studien* cit.). As for Clagett, he declares himself convinced that the translation was carried out by Moerbeke and promptly explains his arguments in his *Archimedes*, Vol. II (*Introduction*, I, pp. 3-73).

[15] There is vast literature on Moerbeke. For information on his impressive activity, see the classic and important work by M. Grabmann, *Guglielmo di Moerbeke O.P., il traduttore delle opere di Aristotele* (Roma 1946) and the recent entry *William of Moerbeke* by Minio Paluello, in *D.S.B.*, IX (1974), pp. 434-440, which also provides numerous bibliographical references, as well as recording all the translations attributed to him. Also useful is the first chapter of *Archimedes*, Vol. II by Clagett, *William of Moerbeke as a translator of Archimedes* (I, pp. 3-28).

[16] Cf. *Neuen Studien* cit.; M. Clagett, *Archimedes*, Vol. II, I, pp. 60 ff.

πλειόνων ἐπιπέδων κέντρον ῥοπῆς ἤτοι βάρους, ἀφ' οὗ ἀρτώμενος ὁ ζυγός παράλληλός ἐστι τῷ ὁρίζοντι[17]. [The *ῥοπήν*, O most noble Peter, as Aristotle and Ptolemy both say, belongs to the common category of heaviness and lightness, with Ptolemy following Aristotle in this. However, Plato’s Timaeus, on the other hand, says that every *ῥοπὴν* arises from heaviness, for he considers lightness to be a deprivation. Of these opinions, those who love knowledge can review from Ptolemy’s book on *ῥοπῶν*, from Aristotle’s works on physics, and from Plato’s Timaeus, as well as from those who have made notes on these. Archimedes, in this book, considers the center of *ῥοπῆς* to be a point of a planar figure from which, when suspended, it remains parallel to the horizon. For two or more planes, he considers the center of *ῥοπῆς*, namely the center of burden, from which the suspended balance remains parallel to the horizon.]

Moerbeke must have found himself in some difficulty when faced with a term like *ῥοπή* (which, in this context, indicates the inclination to move naturally with a certain speed, proportional to one’s weight or lightness)[18], for which he evidently felt he could not propose in Latin a perfectly adequate equivalent. So, he decided to transliterate. In the Vatican Codex, in fact, the title of Archimedes’s writing, *’Επιπέδων ἰσορροπιῶν ἢ κέντρα βαρῶν ἐπιπέδων*, is translated *De centris gravium vel de planis eque repentibus*[19]. And, in the *versio* [version] of Eutocius’s *Commentary* on the first book of *Equilibrium*, we see this choice confirmed:

Euthochii Ascalonite rememoracio in primum librum Archymenidis equerepentibus. **46**

Ropen, i.e. reptionem, o generosissime Petre, genus esse commune gravitatis et levitatis, Aristotiles et Ptolemeus ipsum sequens dicunt. Qui autem apud Platonem Timeus omnem ropen a gravitate dicit fieri, levitatem enim putat privationem. Quorum opiniones amantibus addiscere licet legere ex libro de ropis a Ptolemeo conscripto et ex naturalibus tractatibus Aristotelis, et ex Timeo Platonis et ex exponentibus hec.

Archimedes autem in hoc libro centrum ropes plane figure putat a quo suspensa manet equidistans orizonti, duorum autem aut plurium planorum centrum ropes aut gravitatis a quo suspensa libra est equedistans orizonti[20].

[Eutocius of Ascalon commentary in the first book of Archimedes’s “equal repentibus (i.e., *Equilibrium of Planes*)”.

Ropen, i.e., reptionem, O most generous Peter, is a common type of heaviness and lightness, as Aristotle and Ptolemy declare. But Plato’s Timaeus says that all ropen is produced from heaviness, and lightness to be a privation. For those who love knowledge, read from the book on ropis by Ptolemy and from the natural treatises of Aristotle, from Plato’s Timaeus, and from those who teach upon it.

However, Archimedes in his book considers a center of “ropes” of a plane figure remaining suspended equidistant from the horizon, moreover the center of “ropes” or heaviness of two or more planes, from which the suspended scales remain equidistant to the horizon.]

We are faced with an interesting statement. The transliteration of *ῥοπή* with which Eutocius’s passage opens in Moerbeke’s version is accompanied by a clarification: “idest reptionem”. Of *reptio* [to creep or crawl slowly], I confess that I have not been able to find evidence in the lexicons of classical Latin, and

[17] *Archimedes opera omnia*, III, 264, 1-15.

[18] It is, essentially, the same meaning assumed by *ῥοπή* in the fourth book of Aristotle’s *Physics*, as we will see later (cf. *infra*, paragr. 5).

[19] Ottobonian Vatican Latin Codex 1850, c. 17*r*A; see now the Latin text in *Archimedes*, Vol. II, I, p. 116.

[20] Cod. Vat., c. 53*r*A; *Archimedes*, Vol. II, I, p. 340.

only one – but the term does not have this meaning – in the medieval lexicons[21].
47 Nonetheless, this meaning of this peculiar clarification, made by introducing a newly coined term, must not have been difficult for the translator to understand. The Flemish Dominican began with the analogy *ῥέπω-repo*. This is demonstrated, without a shadow of a doubt, by his translation of *ἰσόρροπος* [equal moment] with *equerepens* [equal inclination]. And he tried to extend the analogy to the level of nouns, proposing *reptio* for *ῥοπή*, without evaluating with due diligence the significant "conceptual" difference that separates the two contexts (on the one hand, it is, in fact, an "inclination" produced by weight × distance, and on the other, an "inclination" proportional only to weight or lightness)[22]. *Repo* and *reptio* probably

---

[21] Neither Forcellini, Du Cange, Latham, nor the *Corpus Glossarierum* contain the headword reptio, which seems to be completely absent in classical and medieval Latin. In truth, consulting the Latin lexicons cited in Chapter I, one finds evidence of reptio in the *Glossary of Later Latin to 600 A.D.* by A. Souter (p. 351): "crawling (Hil. *in psalm.* 68.28)". The term also appears in other modern Latin dictionaries (such as Georges), but always on the basis of the only occurrence in the commentary on the Psalms of St. Hilary (4th century). For the context of St. Hilary of Poitiers, cf. *PL*, Tome IX, *Sancti Hilarii…opera omnia…*, Tome I, col. 487: "In his enim quae in mari repunt (cum tamen in aquatilibus natatus magis sit ex natura quam reptio), ea quae in inferis degunt docentur [For in those that crawl in the sea (however, swimming in water is more by nature than crawling), they are taught about what dwells in the underworld]." In the text of St. Hilary, reptio indicates crawling; it therefore has a very distant meaning from that indicated at the end in this passage by Moerbeke. It would seem, therefore, that the Flemish Dominican was actually the first to use reptio with this meaning. Furthermore, Moerbeke used reptio as the equivalent of *ῥοπή* also in Aristotle's version of *On the Heavens*, both to indicate Ptolemy's *Περὶ ῥοπῶν* and to express the tendency of all bodies to move with a speed proportional to their weight or lightness: "Ptolomaeus autem Mathematicus in libro *de reptionibus* contrariam Aristotelis [opinionem]* habens [a proposito della *gravitas elementorum in proprio loco*]* conatur astruere etiam ipse quod in sui ipsorum regione, neque aqua, neque aer habet gravitatem...Nam si naturalis reptio est appetitus (*ἡ φυσικὴ ῥοπὴ ἔφεσις ἐστὶ*) proprii loci, sortita ipsum non adeo unquam appeteret neque repent utique in ipso existentia sicut neque saturatum appetit alimentum [But Ptolemy the Mathematician, in his book *de Reptionibus*, having a contrary [opinion] to Aristotle [regarding *the heaviness of the elements in their proper place*], tries to establish that in their own region, neither water nor land has heaviness… For if natural crawling is a natural desire (natural ῥοπή is a tendency) for its own place, having obtained itself, it never so much desires, nor abruptly seeks, certainly not in its very existence, as neither does a fully satisfied entity desire food]" (from *Simplicii… Commentaria in quatuor libros de Coelo. Noviter fere de integro interpretata, ac cum fidissimis codicibus graecis recens collata.* Apud Hieronymum Scotum, Venetiis 1548, Lib. IV, pp. 710-711; the italics are ours). It is probably on the model of Moerbeke's translation that the authors depend, who, generally referring to the text of Eutocius, presented Ptolemy's lost work, *Περὶ ῥοπῶν*, in Latin with the title *de reptionibus*. This is the case, for example, of Jacopo Mazzoni who, in an effort to show what advantages Plato derived from adopting the methods of mathematicians, cited the example of Archimedes and Ptolemy. Precisely because they were mathematicians, they avoided falling into the error committed by Aristotle of considering the elements heavy in their own place: "secunda est suppositio ex eodem Archimedis et libello Ptolomei de reptionibus, elementa in proprio loco actu nullam habere gravitatem, quicquid secus dixerit Aristoteles lib. quarto de Caelo [the second supposition is from the same work of Archimedes and Ptolemy's book *de reptionibus*, claiming that the elements in their proper place do not actually have any heaviness, contrary to what Aristotle says in the fourth book of *On the Heavens*]" (*In universam Platonis et Aristotelis philosophiam praeludia, sive de comparatione Platonis et Aristotelis.* Apud G. Guerilium, Venetiis 1597, p. 190; the italics are ours). Mazzoni immediately followed by quoting the passage from Simplicius's commentary that we gave above.

[22] Clagett states in the *Commentary* to this text that Moerbeke resorted quite often to the technique of transliteration (*Archimedes*, Vol. II, II, pp. 561-562). It is important to observe that the *equerepens* adapted by Moerbeke to make *ἰσόρροπος* did not adequately clarify that the planes, in this case, exerted

appeared to Moerbeke as adequately capable of expressing the different nuances of the equivalent Greek terms: not only the inclination to move the arms of the balance upwards or downwards, or the forces capable of producing this phenomenon, but also the "sudden" unfolding of the operation [the author's claim here appears to align with *repens* [sudden, unexpected], which also pairs well with *repo* [creep, crawl] in the context of the balance, giving the impression that the tipping of the balance appears suddenly while its motion also appears slow]. In the medieval and humanistic lexicons and glossaries, we have seen this dual connotation assigned to momentum: force, impulse that causes one of the arms of the balance to incline and, together (or, better, consequently), rapid, abrupt, instantaneous movement[23].

It may seem strange that William of Moerbeke did not think of expressing this complex meaning with our term, preferring to transliterate. In his translation, in fact, *ῥοπή* is constantly rendered as *rope*, considered as a noun of the Greek third declension. *Rope* is also used to make the expression *κέντρον ῥοπῆς* (*centrum ropes*) [center of inclination]; while *centrum gravitatis* [center of heaviness][24] is used corresponding to *κέντρον βάρους* [center of burden][25]. An explanation of the peculiar solution adopted by Moerbeke can be provided on the basis of the conclusions we reached in Chapter I, examining the semantic evolution of *momentum*. In fact, we have seen that, in the Middle Ages, the "classical" meaning of *momentum* was, so to speak, temporarily "lost" (or underwent, in any case, considerable attenuation): the inclination to motion (or movement) produced by (and proportional to) weight (or lightness)[26]. As for the technical meaning of the
term (inclination produced by the compound effect of weight × distance), it was **48**
practically absent even in the "classical" tradition, with the sole exception of two occurrences in Vitruvius (see *infra*, paragr. 2). Moerbeke probably did not feel,

---

an equal force not by virtue of their identical "weight", but because they were equipped with the same *ῥοπή* (weight × distance).

[23] Cf. *supra*, Ch. I, note 18 and 63.

[24] Cod. Vat., c. 53*r*C; *Archimedes*, Vol. II, II, p. 340.

[25] Cf. *Archimedis Opera omnia* cit., III, 264, 13-14.

[26] To be sure, at least one medieval author, Robert Grosseteste, appears, in one case, to be aware of the *ῥοπή-momentum* analogy. The Bishop of Lincoln, in fact, in *Commentum in Aristotelis Ethicam* (cf. E. Franceschini, *Roberto Grossatesta e le sue traduzioni latine*, "Atti del Reale Istituto Veneto di Scienze, Lettere ed Arti", XIIIC, 2nd part [1933-34]*, p. 110), added the following note to an occurrence of *ῥοπή* of the Greek text, with the meaning of "moral inclination": In greco habetur *ropin*: est autem *rope* subitus motus brachii libre ex parte qua pondus est maius; quemadmodum enim appensa in iugis que maioris ponderis sunt utiliora sunt, sub hac metaphora loquitur hic, et posset quoque hoc nomen rope convenienter hic transferri per momentum, vel motum, vel inclinationem, secundum praedictam metaphoram… [In Greek we have *ropin*: but *rope* is the sudden movement of the arm to the side where the weight is greater; just as things hanging on yokes that are of greater weight are more useful, he speaks under this metaphor here, as this name rope could also be fittingly be transferred here as *momentum*, or *motion*, or *inclination*, according to the aforementioned metaphor…]". As Clagett pointed out (*Archimedes*, Vol. II, II, pp. 561-562) from whose exquisite courtesy I had received the notification of Grosseteste's note, even before the printing of the second volume of *Archimedes*, the Bishop of Lincoln, however, never used *momentum* to render the occurrences of *ῥοπή* in the *Nicomachean* text that he gave it in the Latin translation (cf. "Aristoteles Latinus", XXVI, 1-3, *Ethica Nicomachea. Translatio Roberti Grosseteste*…, Edidit R. A. Gauthier, E. J. Brill, Leiden-Desclée de Brouwer, Bruxelles 1972). Grosseteste translated *ῥοπή* as *incrementum*, *auxilium* and *inclinatio*.

therefore, *momentum* as an adequate term to successfully express this dual meaning of *ῥοπή*.

Although, apart from the Vatican, only a handful of codices containing the translations of the Archimedean *corpus* attributed to Moerbeke are known – and no one has recorded them in full[27]– they had considerable influence, direct and
49 indirect. Leonardo da Vinci[28], Luca Gaurico[29] and Tartaglia[30] knew them and used them. According to Heiberg[31], Iacopo da Cremona[32], who was responsible for the second Latin translation of the Archimedean *corpus* around 1450[33], kept in mind the previous *versio* of the Flemish Dominican. The influence of this second translation was undoubtedly much greater. In fact, it constituted the basis for the *editio princeps* of Archimedes's works[34]. Iacopo da Cremona, unlike Moerbeke,

---

[27] In addition to the Ottobonian Vatican Codex, there is the Cod. Vaticano Reg. Lat. 1253, which only retains *On Spirals*; the Codex 9119 of Bibl. Nac. of Madrid (XV-XVI centuries), in which, compared to the Ottobonian Codex, *On the Sphere and Cylinder*, with Eutocius's *Commentary* is also preserved (cf. Archimedes, Vol. II, I, pp. 54-73). For codexes that contain a compilation obtained by combining a proposition of the *Measurement of a Circle* with thirteen propositions of *On Spirals*, cf. also Clagett, *The use of the Moerbeke translations* cit., pp. 237 ff.

[28] For Leonardo's use of Moerbeke's version, cf. W. Schmidt, *Zur Textgeschichte der "Ochumena" des Archimedes*, "Bibliotheca Mathematica", 3 Folge, vol. 3 (1902), pp. 176-179. Heiberg had already put forward the hypothesis that in one of Leonardo's references to Archimedean manuscripts ("Borges ti farà avere archimede del vescovo di padova [Borges will get you Archimedes from the Bishop of Padua…]", Ms.L., f. 2*r*) should be referred to the same Ottobonian Lat. Cod. 1850, then evidently in the hands of Bishop Pietro Barozzi in Padua (*Archimedis opera omnia* cit., III, p. LXVII). For interpretations of Leonardo's piece, cf. the massive update of Carlo Pedretti of the classical work of Richter (*The Literary Works of Leonardo da Vinci*, London 1883; 2nd ed., with notable additions, Toronto 1939; 3rd ed., London 1970, vol. 2); *The Literary Works of Leonardo da Vinci… Commentary*, 2 vol., Phaidon, London 1977, II, pp. 332-333. Clagett, in the aforementioned essay, *Leonardo da Vinci and the medieval Archimedes*, authoritatively confirms Heiberg's hypothesis, according to which Leonardo probably made use of Moerbeke's *versio*. He claims, however, to have proof that the codex in Barozzi's hand was not the Ottobonian but the 9119 of Bibl. Nac. of Madrid (pp. 102-103). For Clagett's judgements and reasons on this problem, cf. also pp. 126 ff. and 151.

[29] Gaurico used Moerbeke's version for the above-mentioned edition of *Tetragonismus* (see note 3). According to Clagett, the codex used by Gaurico was not the Ottobonian, but the Madrilenian 9119 (cf. *Leonardo da Vinci and the medieval Archimedes* cit., p. 103).

[30] The *Opera Archimedis Syracusani* cit., published in 1543 by Tartaglia resumed the text of the two works present in the previous edition by Luca Gaurico. Tartaglia also added *On* the *Equilibrium of Planes* (without Eutocius's *Commentary*) and the first book of *On Floating Bodies*, in Moerbeke's versions. It seems that the scientist from Brescia also used the Madrid Codex (cf. M. Clagett, *Archimedes*, Vol. I, p. 13; see also P. L. Rose, *The Italian Renaissance of Mathematics* cit., *infra*, note 42, pp. 152-153).

[31] *Neue Studien* cit., p. 93; Clagett expresses the same opinion (*Archimedes*, Vol. II, I, p. 12).

[32] Regarding the attribution to Iacopo da Cremona, despite the lack of codexes that clearly indicate him as the author of the version, there is substantial agreement among historians on this matter. M. Clagett (*ibid.*, p. 12), however, highlights the recording ("Interprete Francisco Cereo de Burgo Sanctii Sepulchri"), present in the codex of the Bibl. Nat. of Paris (Nouv. Acquis. Lat. 1538), which contains the compilation we mentioned in note 27. For a more in-depth discussion of this problem, we will have to wait for the forthcoming Vol. III of Clagett's *Archimedes*.

[33] Cf. Heiberg, *Neue Studien* cit., p. 93.

[34] *Archimedis Syracusani Philosophi ac Geometrae excellentissimi Opera quae extant omnia, latinitate iam olim donata, nuncque primum in lucem edita…*, Per Ioannem Hervagium, Basileae 1544. The edition was edited by Thomas Geschauff (Venatorius). For the codex containing the version of Iacopo da Cremona used by Venatorius and for the *emendationes* written by Regiomontano, cf. Heiberg, *Archimedes opera omnia* cit., III, pp. LXVII-LXIX.

abandoned the transliteration of ῥοπή and attempted to render the Greek term with a dual technique. On the one hand, in fact, he introduced *momentum* to translate the ῥοπή of Eutocius's *Commentary* (inclination to move proportional to weight or lightness). On the other hand, he resorted to *pondus* [weight] (and *gravitas* [heaviness]) to render *ἰσόρροπος* [equal moment] and *κέντρον ῥοπῆς* [center of inclination] respectively with *aequepondus* [equally weighted] and *centrum ponderis* [center of weight] (or *centrum gravitatis* [center of heaviness]), thus showing that he perceived the difference in the meaning of ῥοπή in the two contexts and that he did not feel *momentum* as a suitable term to express the inclination produced by weight × distance. Thus the original title, *Ἐπιπέδων ἰσορροπιῶν*, was translated by Iacopo da Cremona into *Planorum aequeponderantium inventa, vel centra gravitatis planorum*[35]. In the translation of the initial passage of *in libros de aequeponderibus* in Eutocius's *Commentary*, we can find confirmation of the understanding reached by Iacopo. In fact, he renders ῥοπή with *momentum*, and yet carefully avoids using our term to translate *κέντρον ῥοπῆς* [center of inclination] (where ῥοπή no longer indicates the inclination produced by simple weight).

Momentum ipsum, o generosissime Petre, commune gravitatis et levitatis esse genus, et Aristoteles **50**
asserit et Ptholomeus eum sequutus. Timeus vero apud Platonem *momentum* omne dicit a gravitate produci. Nam existimat levitatem privationem quandam esse. Quorum opiniones licet disciplinae studiosis legere et Ptholomei libro quem *de momentis* conscripsit et ex naturalibus negotiis Aristotelis et ex Timeo Platonis et ex his qui illos exposuerunt. Archimedes vero in hoc *centrum ponderis* figurae planae existimat id ex quo suspensa manet equedistans orizonti, duorum vero vel plurium planorum *centrum ponderis, hoc est gravitatis*, a quo libra suspensa stat orizonti aequedistans[36]. [*Momentum* itself, O most generous Peter, is said to be a common kind of heaviness and lightness, as affirmed by Aristotle, and as followed by Ptolemy. But Timaeus in Plato says that every *momentum* is produced by heaviness. For he thinks that lightness is a kind of privation. Whose opinions may be read by those who study the discipline, and in the book of Ptolemy, which he wrote *de momentis*, and from the dealings of nature of Aristotle, and from the Timaeus of Plato, and from those who expounded them. Archimedes, however, thinks that the *center of weight* of a plane figure is that from which it remains suspended equidistant from the horizon, whereas the *center of weight* of two or more planes, *is that heaviness*, from which the scales are suspended equidistant from the horizon.]

Another translation of Eutocius's *Commentary* on the *Equilibrium of Planes*, directly from the original Greek in his possession[37], was conducted by Giorgio Valla and was included in *De expetendis*[38]. The translation is limited to a part of the initial passage cited above and is rather unrestricted. Valla uses *momentum* to translate ῥοπή, following Iacopo da Cremona in this decision, whose *versio* he was

[35] I quote from the Basel edition of 1544, p. 125. Not even Iacopo's *aequeponderantium* successfully conveyed the Archimedean concept of *ἰσορροπία*.

[36] *Ibid*., p. 57; the italics are ours.

[37] The Codex of Valla provided the model for the Laurentian Codex (*Plut.* 28, 4), on which Heiberg exemplified the Syracusan scientist's edition of the *Opera omnia*. For information on this very important and ancient codex, on which the Parisian Codexes 2360 and 2361 (B and C) also depend, cf. Heiberg's *Prolegomena* in *Opera omnia* cit., vol. III, pp. XI ff. The Valla Codex has been lost.

[38] *G. V. Placentini…*, *De expetendis et fugiendis rebus…*, In Aedibus Aldi Romani, Venetiis 1501 (without page numbering).

likely aware of, as suggested by the remarkable terminological similarities.

> Momentum commune genus gravitatis et levitatis Aristoteles et secutus in hac parte Aristotelem Ptolomaeus. Plato quoque ante Aristotelem et Timaeus esse dixerunt. Ptolomaeus certe in libro, cuius est index de momento et Aristoteles in naturalibus indagationibus. Archimedes autem, in libro qui aequalium momentorum inscribitur, centrum momenti planae figurae putat, a quo elevatus manet finienti parallelus, duorum, aut plurium planorum momenti centrum, aut gravitatis; unde lanx elevata parallelus est finienti[39]. [Momentum is a common concept of heaviness and lightness as taught by Aristotle and Ptolemy who followed Aristotle in this regard. Plato, prior to Aristotle and Timaeus, also declared the same. Ptolemy certainly did in a book, of which there is an index of moment, as well as Aristotle in his natural investigations. However, Archimedes, in a book titled On Equilibrium of Moments, takes the center of moments of a plane figure to be that from which, when raised, it remains parallel to the horizon; and for two or more planes, he takes the center of moments — or of heaviness — to be that from which the raised scale-pan remains parallel to the horizon.]

Valla's version is very curious. First of all, the reader is not warned in the slightest that the passage is from Eutocius, or that it is a translation. Furthermore, the interpretation is so incorrect as to radically distort the meaning of Eutocius's text: the claim that Plato would have supported the same thesis as Aristotle and Ptolemy! And it must be excluded that these errors depended on deficiencies in the Codex, given that Valla owned and used the archetype of the same Florentine Codex on which Heiberg based the edition of the works of the Syracusan scientist[40]. Valla's version
51 covers, however, an undoubted interest in the terms used. The title of *Equilibrium of Planes* was, in fact, rendered with *liber aequalium momentorum* [book of equal moments], thus extending *ῥοπή-momentum* relationship to the compound *ἰσόρροπος* [equal moment]. Valla remained faithful to this criterion, even in the translation of *κέντρον ῥοπῆς* [center of moment] with *centrum momenti* [center of moments]. This undoubtedly ended up creating a certain mystification. The use of the same term diluted or shrouded the significant difference between the two contexts. In this sense, Valla's version did not contribute to a perfect understanding of the meaning of these passages. What we are interested in pointing out is that the Piacentine considered *momentum* a term suitable to express both the inclination to move with a speed proportional to the weight, and the ability to counterbalance that depends not only on the weight, but also on the distance from the fulcrum. It should be noted that, a few lines beyond the passage cited above, Giorgio Valla used *repo* to indicate the inclination of one of the two arms of the balance[41].

The compilation of *Equilibrium of Planes* carried out by Francesco Maurolico, dated 1548, is also considered to be based on the Greek text[42]. The original has been

---

[39] Cf. *Liber XII, Caput II* (*Ut Dionysodorus*).

[40] Cf. *supra*, note 37.

[41] "Non dubium putat quin aequilibrae sint a, b, c partes sibi, neque altera magis altera *rhepet* in finientem [no doubt that parts a, b, c are in equilibrium with each other, and neither one *creeps* more than the other towards the surface]" (*ibid*).

[42] Francesco Maurolico compiled the four books *de momentis aequalibus* between 1547 and 1548, between Castelbuono and Palermo. For Maurolico, see the essay of M. Clagett, *The Works of Francesco Maurolico* cit.; also very useful is the recent work of P. L. Rose, who has effectively included Maurolico in the ranks of the most authoritative protagonists of the "renaissance" of mathematics, which occurred

lost, as well as the first printing from Messina in 1670, to which Giovanni Alfonso Borelli, among others, contributed[43]. It remained incomplete, and subsequently was dispersed owing to the events that followed the Messinese anti-Spanish insurrection of those years[44]. We therefore only retain the edition published in Palermo in 1685[45].
It demonstrates that Maurolico gave a loose paraphrase of the Archimedean writing **52**
– here and there integrating it with his own propositions – in which he had not even respected the original distribution of the subject matter. The Messina scientist articulated it, in fact, into four books[46]. For the title of Archimedes's writing, he proposed the same solution suggested by Valla: *Archimedes de momentis aequalibus ex traditione Francisci Maurolici*. The scientist from Messina knew Valla's *De expetendis*[47], from which it cannot be excluded that he derived the use of *momentum*.

---

in Italy, through the recovery of classical sources, in the 15th and 16th centuries (*The Italian Renaissance of Mathematics. Studies on Humanists and Mathematicians from Petrarch to Galileo.* Travaux d'Humanisme et Renaissance CXLV, Librairie Droz, Genève 1975, Ch. 8, pp. 159-184).

[43] The events of the first Messina print, of its fortunate rediscovery, and, finally, of the new Palermo impression of 1685, are illustrated in the letters of Cyllenio Hesperio, Francisco Arias, and Petro Carlo de Balsamo (the latter two of the Society of Jesus) prefaced to the *Admirandi Archimedis Syracusani monumenta omnia mathematica quae extant, ex traditione D. Francisci Maurolici…*, Apud D. Cyllenium Hesperium, Panormi 1685 (first unnumbered 6 chapters).

[44] The first impression, over which, as Arias reports in the aforementioned letter, Borelli presided, remained incomplete due to the forced and sudden abandonment of the city to which the Messina scientist was driven by informers who denounced his anti-Spanish sentiments, in 1672. Arias, among other things, fully exonerated Borelli, not many years later (the letter is dated June 1681), from the accusations of conspiracy ("cum exitiali Superum fato, publicae rei exitio Borellus, quamvis vitae integer, omnisque sceleris purus, paulo ante illos rumores exulare nonnullorum delationibus cogitur. [With the destructive fate of the Superiors, the destruction of the public affairs of Borellus, though blameless in life and entirely free from all crime, is compelled to go into exile a little before those rumors by the reports of some]").

[45] *Admiranda Archimedis Syracusani monumenta* cit. They are included "ex traditione Maurolyci", the *De circuli dimensione*, the *De Sphaera et cylindro*, the *De aequeponderantibus, sive de momentis aequalibus libri IV, De quadratura parabolae, De lineis spiralibus, De conoidibus et sphaeroidibus libri II.* It was then given "textus purus Archimedis" of the *Arenarius* and of the *De iis quae vehuntur in aqua libri II.* Lastly, the volume closed with a series of *Exotica, sive externa e variis auctoribus asserta.*

[46] We keep a record of Maurolico regarding the meaning of his Archimedean translation-compilation and its division into four books. The intentions by which the work was inspired are indicated: "Neque me paenitebit unquam de momentis aequalibus libellos quatuor scripsisse: cuius materiae inventio et laus Archimedi nostro debetur. Ego tamen multa copiosius super ea re demonstravi. In primo quidem de momentorum proportione; in secundo de centris triangulorum et planarum figurarum; in tertio de proportione parabolas. In his Archimedes succinte nimium se praestitit, si modo quod extat opus integrum est, pondus enim et momentum, cum sint magnitudinum notandae species, erant multo latius tractandae. In quarto demum libello totam mihi laudem vendicare non erubescam, nam de centris solidorum quod ab Archimede praetermissum magnopere admiror, disserui… [Nor will it ever displease me to have written four little books on equal moments, the invention and praise of which belong to our Archimedes. However, I have demonstrated many things more fully on this matter. In the first, concerning the proportion of moments; in the second, concerning the centers of triangles and plane figures; in the third, concerning the proportion of parabolas. In his work, Archimedes presented himself too briefly, provided that the existing work is complete; for weight and momentum, since they are species of noteworthy magnitudes, required much broader treatment. In the fourth book, I will not hesitate to claim all the praise for myself, for I have discussed what I greatly admire as omitted by Archimedes, namely, the centers of solids…]" (cit. F. Napoli, *Scritti inediti di Francesco Maurolico*, in "Bullettino di Bibliografia e di Storia delle Scienze Matematiche e Fisiche", IX, 1876, p. 32).

[47] Cf. *supra*, note 38. Maurolico's in-depth knowledge of Giorgio Valla and his work as a translator

In the *Diffinitiones* that open the *Liber Primus* we find a definition of momentum recorded and some of its indicated fundamental characteristics:

> VIII. Momentum est vis ponderis a spatio quopiam contra pendentis. [Momentum is the force of weight from a certain distance against a hanging object.]
>
> IX. Unde ponderum aequalium momenta possunt esse inaequalia; et e contrario continget momentorum aequalium pondera esse inaequali. [Hence, equal weights can have unequal moments, and conversely, it will happen that equal moments can have unequal weights.]
>
> X. Aequalia enim momenta sunt gravium aequaponderantium sive aequependentium[48]. [Equal moments belong to heavy bodies that are equal-weighing or equal-hanging]

In this passage, for which it is not easy to find the equivalent in Archimedes's text, Maurolico showed that he had achieved full awareness of the meaning of *momentum*. With this term, he indicated, in fact with extreme precision, not a generic inclination to descend or ascend, but rather the variation of the effect of the same weight, according to the point of application. *Momentum*, clearly distinct from *pondus*, had exactly the meaning in which we will find it used in Galileo's
53 *Mecaniche*. We will see shortly that the transition, which we take for granted today, was neither easy nor automatic. Unfortunately, Eutocius's *Commentary* was not included in the compilation of Maurolico. It is therefore impossible to establish how he would have responded when faced with the occurrences of *ῥοπή* in the initial passage of the manuscript.

It can be stated with some certainty that Maurolico's *De momentis aequalibus* remained practically unknown in the 16th century, due to the lack of codices and the absence of prints. However, the existence of this compilation and the translation of the Greek title of the Archimedean treatise proposed by Maurolico were known through a list of his manuscript works published as an appendix to some of his very successful and widespread works. For example, the *Cosmographia* published in 1543 in Venice[49] indicated, among the works already completed by Maurolico, the translation of the *Archimedis de momentis aequalibus*, *sive de aequiponderantibus libellus ex traditione Eutotii Ascalonitae*[50]. Furthermore, the famous *Index lucubrationum Maurolyci*, which saw the light for the first time in 1558[51] in partial edition, presented *De momentis aequalibus libri quatuor. In*

---

and compiler is evident from some handwritten pages of the Villacanese Codex, in which the Messina mathematician records news of the works and translations of the Piacentine (cf. M. Clagett, *The Works of Francesco Maurolico* cit., p. 150; cf. also p. 153).

48 *Admiranda Archimedis Syracusani monumenta* cit., p. 86.

49 *Cosmographia Francisci Maurolyci… in tres dialogos distincta…*, Apud haeredes L. A. Juntae, Venetiis 1543.

50 Naturally, it was a different version from the one we focused our attention on. And it is not easy to say whether it was the version of the text of the *Equilibrium of Planes* with the Eutocius's *Commentary*, or only of the latter. That it is a different translation is attested both by the "ex traditione Eutotii Ascalonitae" and, above all, by the fact that the compilation cited above was carried out in the years 1547-48, that is, subsequently. I confess that I am unaware of whether any trace remains of this translation of Maurolico. Clagett makes no mention of it and seems to confuse it with the compilation of 1547-48, to which he refers (*The works of F. Maurolico* cit., p. 176). Maurolico's reference to the Archimedean version is found in the dedicatory letter to Pietro Bembo (dated 1540) with which the *Cosmographia* opens.

51 In the *Theodosii Sphaericorum elementorum libri III*, Petrus Spira, Messanae 1558.

quorum postremo de centris solidorum ab Archimede omissis agitur. Et de centro solidi paraboles[52] among the completed works. In this second case, it was the writing published in 1685[53].

Proof of the widespread knowledge of Maurolico's *Index lucubrationum* is provided to us by Commandino, who declared, in the introduction to his *De centro gravitatis*[54], that he had postponed the publication of his own treatise, after reading that the learned Maurolico had finished one on the same subject[55]. Commandino had certainly deduced the information from the edition of the spherical elements of 54
Theodosius, published by Maurolico in 1558[56].

The *Editio Princeps* of *quae extant omnia* [all the extant works] of Archimedes, published in Basel by Venatorius in 1544 and including Eutocius's *Commentaries*[57], presented a Latin version based on the version of Iacopo da Cremona, accepted entirely by the publisher, who limited his own intervention with a few marginal corrections[58]. The title changed slightly (*Aequeponderalium Archimedis*, instead of *de aequeponderibus* by Iacopo da Cremona[59]), but the choice of *momentum* to translate *ῥοπή* remained unchanged. *Κέντρον ῥοπῆς* [center of inclination], still on the model of Iacopo, was rendered with *centrum ponderis* [center of weight]. The Basel edition is of considerable importance for us, both because it allowed a considerable diffusion of the Archimedean *corpus*, and because we know that Galileo owned a copy and used it in his first years of activity, as demonstrated by some of his youthful annotations on a copy of the Basel press[60].

---

52 The indication can be found on the title page.

53 Cf. *supra*, note 43.

54 *Federici Commandini Urbinatis Liber de centro gravitatis solidorum…*, Ex officina Alexandri Benacii, Bononiae 1565.

55 "Cum autem ad hoc scribendum aggressus essem, allatus est ad me liber Francisci Maurolici Messanensis, in quo vir ille doctissimus, et in iis disciplinis exercitatissimus affirmabat se de centro gravitatis corporum solidorum conscripsisse. Cum hoc intellexissem, sustinui me paulisper, tacitusque expectavi, dum opus clarissimi viri, quem semper honoris caussa nomino, in lucem proferretur: mihi enim exploratissimum erat Franciscum Maurolicum multo doctius et exquisitius hoc disciplinarum genus scriptis suis traditurum. Sed cum id tardius fieret, hoc est, ut ego interpretor, diligentius mihi diutius hac scriptione non supersedendum esse duxi, praesertim cum iam libri Archimedis de iis quae vehuntur in aqua opera mea illustrati typis excudendi essent… [But when I had begun to write this, there was brought to me the book of Francis Maurolicus of Messana, in which that man, the most learned, and the most trained in those disciplines, affirmed that he had written about the center of heaviness of solid bodies. When I understood this, I held my ground for a while, and waited in silence, while the work of a most famous man, whom I always name out of honor, was brought to light: for I was most certain Francis Maurolicus would deliver this type of discipline in his writings much more learnedly and meticulously. But when that was done more slowly, that is, as I interpret, I thought more carefully that I should not dwell on this writing any longer, especially since the books of Archimedes on things that float on water were already to be printed in my illustrated works…]" (*ibid.*, ff. n.n.).

56 Cf. *supra*, note 51.

57 Cf. *supra*, note 34.

58 Cf. Heiberg, *Archimedes opera omnia* cit., III, pp. LXVII ff.

59 For a comparison between the *incipit* of Archimedean works in the translations of Moerbeke and Iacopo da Cremona, cf. M. Clagett, *The use of the Moerbeke translations* cit., pp. 240-241.

60 Galileo's annotations were in the hands of Antonio Santini in 1671, as can be seen from a letter from him to Vincenzo Viviani (B.N.F., Ms. Galil. cc. 188-194; the letter is dated 27 September 1671), in which he sent a copy to the faithful disciple of the Pisan scientist. It seems that the annotations concerned

A year before the Venatorius edition, in 1543, some works by Archimedes had seen the light of day, including the *Equilibrium of Planes*, edited by Nicolò Tartaglia[61]. Although Tartaglia ambiguously suggested to the reader that it was an original translation from the Greek text preserved in a codex of his property, reduced
55 to very poor conditions[62], he actually published Moerbeke's version with very few changes[63]. Tartaglia's edition, however, did not include Eutocius's *Commentaries*. On the other hand, in Moerbeke's version of the *Equilibrium of Planes*, there were no examples of the translation of *ῥοπή* with *rope*. The transliteration was in fact adopted only at the level of verbal forms: *ῥέπω* = *repo*. Furthermore, *κέντρον ῥοπῆς* [center of inclination] was always rendered by the Flemish Dominican as *centrum gravitatis* [center of heaviness], instead of the *centrum ropes* used in the *Commentary* version. Finally, in Tartaglia's version, *repere* is characteristically alternated with *inclinare*, as shown in *Petitio III*.

> Si gravitatibus *aequaliter inclinantibus*, ab aliquibus longitudinibus ad alteram gravitatem apponatur, non aequaliter inclinare, sed inclinare ad gravitatem illam cui additum est. Similiter autem et si ab altera gravitatum auferatur aliquid *non aequaliter inclinare sed repere* ad gravitatem a qua non ablatum est[64]. [If, for heavy things *equally inclined*, a heavy thing is placed away from one length toward the other, it does not equally incline, but inclines toward the heavy thing that was added. Likewise, and yet, if from the other, the heavy thing is diminished to some extent, it does *not equally incline but creeps* toward the heavy thing away from that which is taken.]

Tartaglia's edition ensured a significant dissemination of Moerbeke's version, even if the incorrectness of the text and the use of constructions and terms that would have already been perceived by the mid-sixteenth century reader as archaic and devoid of elegance, generally making them prefer the *versio* of Iacopo da Cremona, offered by the Basel press.

To complete the picture, it remains to be noted that an anonymous translation of the *Equilibrium of Planes*, along with Eutocius's *Commentary*, is preserved in the Magliabechiano collection of the National Library of Florence[65]. This document is

---

only *De sphaera et cylindro*. From the quotations of the text, which accompany the annotations, we can deduce that Galileo used the Geschauff edition. Viviani transcribed Galileo's marginal notes on his copy of the Basel edition, in the corresponding places (now in B.N.F., Notes, 102^1). They were also published by A. Favaro, *Opere di Galileo Galilei*, Edizione Nazionale, Barbèra, Firenze 1890-1909, 20 vol. (reprints: 1929-1939; 1964-1966; 1968), I, pp. 230-242; I used the reprint of 1968, from here onwards cit. *G.G.*

[61] *Opera Archimedis Syracusani* cit.; on the weight that the Archimedean tradition exerted on Tartaglia, see P. L. Rose, *The Italian Renaissance of Mathematics* cit., Ch. 7, pp. 151 ff.

[62] In the dedicatory letter to Ricardo Venfort Britannico (c. 2*r*), Tartaglia declared that he had translated it into Latin "fracti et qui vix legi poterant quidam libri manu graeca scripti illius celeberrimi Philosophi Archimedis [some of the books written in the Greek hand of that famous philosopher Archimedes were broken and could hardly be read]".

[63] For the Codex used by Tartaglia, see *supra*, note 30.

[64] *Ibid.*, c. 3*v*; the italics are ours.

[65] B.N.F., Magliab. Cod., cl. XI, 50. The *incipit* of the *Equilibrium* version is as follows: "Archimedis planorum aequeponderantium sive gravium planorum centra. Petimus pondera aequalia a distantiis aequalibus aequeponderare [Archimedean centers of planes of equal weight or heaviness. We seek to equally balance at equal distances with equal weights]" (c. 3*r*). In the same codex, there is a

of great interest, given that it offers us a new translation variant for *ῥοπή* and for *κέντρον ῥοπῆς* [center of inclination]:

Propensionem, o generosissime Petre, commune esse genus gravitatis atque levitatis et Aristoteles et qui eum sequutus est Ptolemaeus asserit. Thimaeus autem apud Platonem, *propensionem* omnem, inquit, a gravitate fieri, arbitratur enim levitatem esse privationem. Quorum opiniones et ex Ptolomaei quem de *propensionibus* conscripsit libro et ex Aristotelis negotiis naturalibus et ex Thimaeo Platonis et ex eis qui haec commentati sunt licet disciplinarum studiosis eos legere. Archimedes vero in hoc libro *centrum propensionis* figurae planae putat id a quo suspensa manet aequedistans horizonti, duorum vero vel plurium planorum *centrum propensionis sive ponderis*, a quo suspensa libra est aequedistans **56**
horizonti[66]. [*Propensity*, O most generous Peter, is a common property of both heaviness and lightness as asserted by both Aristotle and Ptolemy who followed him. However, Timaeus, by Plato, asserts that all *propensity* arises from heaviness, for he considers lightness to be a privation. These opinions, from Ptolemy who wrote a book on *propensity*, from Aristotle's work on nature, from Plato's Timaeus, and from those commentators, are able to be read by the disciplined student. Indeed, Archimedes in his book considers that a flat figure's *center of propensity* is that from which, when suspended, it remains equidistant from the horizon, and that, for two or more planes, the *center of propensity or weight* is that from which, suspended, the scale remains equidistant from the horizon.]

The Magliabechian translation closely follows that of Iacopo da Cremona, from which it differs only in the choice of *propensio* [propensity] instead of *momentum* and that of *centrum propensionis* [center of propensity] (*sive ponderis* [or weight], in one case) instead of *centrum ponderis*.

Of less interest, from the point of view of our work, focused on Galileo, is Giuseppe Averani's version, preserved in a Riccardiana Codex[67]. It was in fact

second, a different translation of the *Equilibrium of Planes*, which recalls Moerbeke's version. *Centrum gravitatis* and *aequeponderare* are widely used in it. The title is *Archimedis Syracusani planorum aequeponderantium, sive centra gravitatis planorum*: inc. "Centrum gravitatis planae magnitudinis est punctum a quo suspensa magnitudo manet aequidistans horizonti [The center of heaviness of a planar magnitude is the point from which a suspended magnitude remains equidistant to the horizon]" (c. 1*r*., old num.). The Catalog assigns the codex to the 17th century. In the annotations contained in the last pages of the Magliabechian Codex, I seem to recognize the hand of Viviani.

66 *Ibid*., c. 3*r*; the italics are ours.

67 Riccardian Codex 107, indicated by M. Clagett (*Archimedes*, Vol. I, p. 14). It contains the version of all the works of Archimedes known at the time. See Averani's version of the opening passage of the *Eutocius in primum Archimedis aequeponderantium*: "Momentum, o strenuissime Petre, commune genus gravitatis et levitatis, et Aristoteles dicit et hunc sequens Ptolemaeus. Timaeus vero apud Platonem omne momentum a gravitate fieri asserit. Levitatem enim privationem putat, ut licet discendi cupidis opinionis ipsorum recognoscere et ex libro de Momentis a Ptolomaeo conscripto et ex Aristotelis physicis exercitationibus, et ex Platonis Timaeo, et ex iis qui haec commentati sunt. Archimedes autem centrum gravitatis figurae planae in hoc libro existimat ex quo ea supensa parallela manet horizonti. Duorum vero aut plurium planorum centrum gravitatis vel momenti ex quo iugum seu libra suspensa parallela manet horizonti. [Momentum, O vigorous Peter, is a common property of heaviness and lightness, as both Aristotle asserts and Ptolemy, following him. However, Timaeus asserts in Plato that all momentum arises from heaviness. For he considers lightness to be privation, as those eager to learn can recognize from their opinions, both from the book on Momentum written by Ptolemy and from Aristotle's physics exercises, and from Plato's Timaeus, and from those who have commented on these matters. But Archimedes, in his book, determines the center of heaviness of a plane figure to be that which, when suspended, remains parallel to the horizon. But for two or more planes, the center of heaviness or moment is that from which a yoke or scale, when suspended, remains parallel to the horizon]" (c. 1582). In the choices relating to *ῥοπή* and *κέντρον ῥοπῆς*, Averani followed Iacopo da Cremona.

performed in the 18th century.

We can, therefore, draw the first conclusions. The contrast between Aristotle and Ptolemy on the one hand, and Plato on the other, relating to the problem of positive or absolute lightness, expressed by Eutocius's passage, relates to a discussion that – as we will see – was central in the 16th century. In *De motu*, the young Galileo dedicated quite a bit of space to this problem, showing that he significantly fluctuated in his choice of position to adopt. On the other hand, Galileo knew very well the very texts we have been quoting[68]. Therefore, he must have been familiar with the meaning of *momento* [moment] which, in the context of Eutocius, means "inclination", "propensity", "tendency" to move up or down according to the lightness or heaviness of the body. It is worth remembering that, on the occasion of Galileo's first definition of *momento*, it was not by chance that he attempted to explain its meaning, emphasizing its analogy with *propensity*[69].

Furthermore, with the exception of Maurolico, none of the translators or compilers examined above used *momentum* to express a precise value of *ῥοπή*:
57 inclination produced by the compounded effect of weight and distance from a fulcrum of a mechanical system (in other words, the "technical" meaning in statics terminology that *momentum* will take on with Galileo).

Finally, the various translations we have examined show the non-automatic relationship between *ῥοπή* and *momentum*. Indeed, prior to the middle of the 15th century, with Iacopo da Cremona, then with Valla and with Maurolico, it does not seem that the legitimacy of the translation with *momentum* was established. And perhaps never definitively, as seems to be demonstrated by the choice of *propensio* made by the anonymous Magliabechian translator, almost certainly active at the beginning of the 17th century.

The use of *momentum* in the 15th and 16th centuries, to express both the inclination produced by weight or lightness, and the ability to counterbalance determined by weight × distance, is documented, of course, even outside of the translations of the *Equilibrium of Planes* and Eutocius's *Commentary*. The theme, for example, presented in the opening passage of the *Commentary* – that is to say, of the natural inclinations to the motion of heavy and light bodies – is, in fact, often addressed in contexts that do not explicitly refer to Eutocius's passage, even if it appears evident that they depend upon it.

This is the case that we observe in a passage from *Liber XX* of Giorgio Valla's *De expetendis*. In Chapter XII, where, among other things, the problem of the *locus* [location] occupied by the body is discussed, we find a significant occurrence of *momentum*:

> Corpus est quod trinis distinguitur intervallis, longitudine, latitudine, crassitudine aut tumor sensui obiectus quantum in ipso est, aut locum occupans. Plato autem quod neque grave neque leve suapte

---

[68] Galileo's attitude towards the positive lightness and the consequent "naturalness" of the sursum [upward] motion, will be discussed exhaustively in the First Chapter of Part Two (paragr. 3).

[69] *G.G.*, II, *Le Mecaniche*, p. 159: "Momento è la propensione di andare al basso… [Momento is the propensity to go downwards…]". We will return to the definition used in *Mecaniche* in the Second Chapter of Part Two.

natura est cum suum tenet locum; at si in alieno sit, tum *vim habere ad nutandum, ex nutu porro momentum fieri, vel in gravitatem, vel in levitatem*. Aristoteles simpliciter terram esse gravissimam, tam aquam, ut levissimum ignem et aerem alias aliter. Stoici ex elementis quatuor bina levia ignem et aerem, reliqua duo, aquam et terram gravia. Leve siquidem est quod natura suapte in suo non nutat medio si ab eo absit. quod autem ad medium grave, grave non est medium[70]. [A body is that which is distinguished by three dimensions: length, breadth, and thickness; or a volume as perceived to the senses to the extent that it is in itself, or occupying space. However, Plato reasoned that things are neither heavy nor light in their own nature when within their own proper place; and yet, if somewhere else, then *it has a nodding force, and from the nod momentum is further produced, towards either heaviness or lightness*. Aristotle simply reasoned that earth is the heaviest, as is water, while fire is the lightest, with air variously in between. The Stoics reasoned that, of the four elements, both fire and air are light, and the remaining two, water and earth, are heavy. For that which by its own nature does not nod in its own middle, when away from it, is light. But what is heavy toward the middle is not itself the middle.]

As we have seen, Giorgio Valla knew Eutocius's text very well; it is not surprising, therefore, that he reports the contrasting positions of Aristotle and Plato[71], nor that he uses *momentum* in the sense of *nutus* [nod], "appetite," "tendency," or the "natural inclination" of heavy and light bodies to move downward or upward. In fact, our passage indicates a precise relationship between *nutus* and *momentum*. *Momentum* is in fact generated by *nutus*. It would be, namely, the dynamic effect.

In Francesco Maurolico's *De quantitate sermo noster*[72], we find some occur- **58**
rences of *momentum* clearly attributable to Archimedes's *Equilibrium of Planes* and the laws of balance established there. In fact, the Messinese abbot, in addition to *body* and *weight*, also introduces *momentum* among the *potentiae* [power] of bodies:

aequalia pondera a diversis spaciis non aequaliter, et inaequalia aequaliter ponderant; nam pondus a longiori spacio appensum ponderosius est, ut patet in statera. Erit ergo tertia quaedam potentia, sive tertia magnitudinis differentia, diversa a corpore, diversa a pondere, quam *momentum* vocant. Corpus igitur acquirit pondus a quantitate et qualitate; pondus autem momentum suscipit a spacio ad quod appenditur. Unde quando spacia sunt ponderibus reciproca, momenta sunt aequalia: ut Archimedes in libro aequalium momentorum demonstravit[73]. [equal weights at different spaces do not equally weigh, and unequal weights equally weigh; because the weight hung at the longer distance is heavier, as is evident on the steelyard. Therefore, there is a certain third power, or a third difference of magnitude, different from the body, different from weight, which is called *momentum*. Therefore, the body acquires weight from quantity and quality; but the weight receives *momentum* from the space at which it is hung. Hence when space is reciprocal to weight, the *moments* are equal: just as Archimedes demonstrated equal moments in his book.]

Here, *momentum* indicated the variation in effectiveness of the same weight, according to different applications. On the balance – again, according to the Archimedean teaching of the *Equilibrium of Planes* – the *moment* of the weight

---

[70] *De expetendis* cit., *Liber XX, Caput XII* (*De corporibus*); the italics are ours.

[71] However, remember that previously (*Liber XIII, Cap. II*) Valla had translated Eutocius's *Commentary*, profoundly distorting its meaning (cf. *supra*, p. 50).

[72] Dated "Catania 18 June 1554", it was published in *Maurolyci Abbatis Prologi, sive Sermones, quidam de divisione artium, de quantitate, de proportione*, Edidit G. Bellifemine…, Ex typographia Mezzina, Melphicti 1968.

[73] *Ibid*., p. 46; the italics are ours.

increases with the distance from the fulcrum. Maurolico therefore establishes, on a purely Archimedean basis, the difference between *momentum* and *pondus*[74].

With Federico Commandino, we return to an author who was very familiar to Galileo, in the first years of his career[75]. The scientist and scholar from Urbino carried out a truly significant restoration of classical texts on mathematics, geometry
59 and mechanics[76]. As for Archimedes, he presented several works with many improvements in 1558, translating directly from the Greek text[77]. The case is different for *de iis quae vehuntur in aqua*, published in Urbino in 1568. Lacking the original, Commandino limited himself to correcting, on the basis of conjectures, the previous, very corrupt edition of Tartaglia, which reproduced the version of Moerbeke[78]. In the anthology of Archimedean writings published in 1558 by Commandino, both the *Equilibrium of Planes* and Eutocius's *Commentary* were missing[79]. We therefore do not have precise references to establish how the Urbino scholar would have reacted when faced with the occurrences of *ῥοπή* recorded in those texts. We can, however, form a fairly precise idea of his orientation by examining the *Liber de centro gravitatis solidorum*, in which Commandino proposed to extend the demonstrative methods used by Archimedes for flat figures to solid figures. It was published after a few years of delay – as Commandino informs us – for fear that his efforts would be rendered useless by the announced

---

[74] Maurolico, loyal to Archimedes's rigorously static interpretation, distinguished *momentum* from *impetus*. The latter constituted the fourth *potentia*. "Est ultra corpus, pondus, momentum, quarta quaedam potentia, quae impetus aut vis appellari potest, a tribus praedictis omnino diversa: de qua quaerit Aristoteles in mechanicis quaestionibus. Quod enim quis impetu aut vi incussi mallei aut securis impactae, sive cudendo, sive secando facit, nulla quamvis magis pondere aut momento fieri potest [Beyond body, weight, and momentum, there is a fourth power, which may be called impetus or force, quite different from the three aforesaid: of which Aristotle inquires in the mechanical questions. For what a man does by the impulse or violence of a hammer or an ax, or by hammering or cutting, nothing can be done by greater weight or moment]" (*ibid.*, p. 47).

[75] The *Theoremata circa centrum gravitatis solidorum* was composed by Galileo in 1585, inspired by the young scientist's dissatisfaction with Commandino's *Liber de centro gravitatis solidorum*; cf. *G.G.*, II, *Avvertimento*, pp. 181-182.

[76] Beyond the *Liber de centro gravitatis* and the *Opera nonnulla* of Archimedes, he published, in valuable editions from a philological point of view, *Apollonii Pergaei conicorum libri quattuor una cum Pappi Alexandrini lemmatibus et commentariis Eutocii Ascalonitae* (Bononiae 1556), the *Planisphaerium* (Aldo, Venetiis 1558) and the *Liber de Analemmate* of Ptolemy (Paolo Manuzio, Romae 1562), the *De superficierum divisionibus Liber Machometo Bagdedino ascriptus* (in collaboration with John Dee, Concordia, Pesaro 1570), the *De gli Elementi d'Euclide Libri Quindici* (Frisolino, Urbino 1575), *Heronis Alexandrini spiritalium liber* (Urbini 1575), the *Mathematicae Collectiones* of Pappus of Alexandria (Concordia, Pesaro 1588). On Commandino's contribution to the "renaissance" of mathematics, cf. P. L. Rose, *The italian renaissance of mathematics* cit., Ch. 9, pp. 185-221.

[77] *Archimedis opera nonnulla a Federico Commandino Urbinate nuper in latinum conversa et Commentariis illustrata*, Apud Paulum Manutium, Venetiis 1558. According to Clagett, Commandino's translation takes into account that of Moerbeke (*Archimedes*, Vol. I, p. 13).

[78] The Greek Codex used by Moerbeke for his version was lost, and the Flemish Dominican translation of *De iis quae vehuntur in aqua* remained the text until 1906 when the Greek original of the Archimedean writing was rediscovered by Heiberg in the famous Codex of Constantinople, which also contained *The Method* (cf. *supra*, note 10).

[79] In the 1558 edition, the following Archimedean writings were included: *Measurement of the Circle*, *On Spirals*, *Quadrature of the Parabola*, and *The Sand Reckoner*.

publication of the four books of *Archimedis de momentis aequalibus*, in which Maurolico promised the exposition *de centris solidorum ab Archimede omissis*[80]. Commandino's work opened with a definition of the center of heaviness, behind which one easily recognizes Pappus of Alexandria's model definition, of whose *Collectiones* Commandino would later become editor:

Dicimus autem centrum gravitatis uniuscuiusque corporis punctum quoddam intra positum, a quo, si grave appensum mente concipiatur, dum fertur quiescit; et servat eam quam in principio habebat positionem; neque in ipsa latione circumvertitur[81]. [We call the center of heaviness of any body a certain point located within it, from which, if a heavy thing is conceived to hang in the mind, while it is carried, it remains at rest and preserves the position it had at the beginning, nor does it revolve in its motion.]

Immediately afterwards, the Urbino scholar proposed a different way of defining **60**
the center of heaviness of solids:

Centrum gravitatis uniuscuiusque solidae figurae est punctum illud intra positum, circa quod undique partes *aequalium momentorum* consistunt. si enim per tale centrum ducatur planum figuram quomodocumque secans semper in partes aequeponderantes semper dividet[82]. [The center of heaviness of any solid figure is that point placed within, about which the parts of *equal moments* rest on each side. For if a plane is drawn through such a center, cutting the figure in any way whatsoever, it will always divide it into equally weighted parts.]

Commandino's translation, like the previous ones by Valla and Maurolico, was particularly respectful of Archimedes's Greek text: *ἰσόρροπος* means precisely "di eguale momento [of equal moment]". The center of heaviness was thus defined, according to the same indications of Archimedes – who had not given an explicit definition – as the fulcrum of the ideal balance constituted by the solid, the point of equilibrium of all its "inclinations" and *moments*. This concept is more explicitly expressed in a passage from Pappus's *Collectiones Mathematicae*, again in the Commandino version:

Si intelligatur planum ABCD productum, secabit utique super impositum corpus in duas partes *aequalium momentorum*, quae circa planum, veluti circa punctum suspensionis in libra inter se aequeponderabunt[83]. [If we imagine the plane ABCD extended, it will certainly cut the body placed on

[80] Cf. the *Index lucubrationum Maurolyci*, in M. Clagett, *The Works of Francesco Maurolico* cit., p. 182. Commandino's declaration is found in the *Liber de centro gravitatis solidorum* cit., *Epistula* dedicated to Alessandro Farnese (c. n. num.).

[81] *Liber de centro gravitatis solidorum* cit., p. 1. See the version, practically identical, of the definition of center of heaviness by Pappus of Alexandria, in the edition prepared by Commandino himself: "Dicamus autem centrum gravitatis uniuscuiusque corporis esse punctum quoddam intra positum a quo si grave dependens mente concipiatur, dum fertur quiescit et servat eam quam in principio habebat positionem, neque in ipsa latione circumvertitur [But let us say that the center of heaviness of any body is a certain point located within it, from which, if a suspended weight is conceived in the mind, while it is being carried, it remains at rest and preserves the position it had at the beginning, nor does it revolve in the course of the carrying itself]" (*Pappi Alexandrini Mathematicae Collectiones a Federico Commandino Urbinate in Latinum conversae et commentariis illustratae*, Apud Hieronymum Concordiam, Pisauri 1588, *Liber VIII*, p. 306).

[82] *Liber de centro gravitatis solidorum* cit., pp. 1-2; the italics are ours.

[83] *Mathematicae Collectiones* cit., p. 306; the italics are ours.

it into two parts of *equal moments*, which, with respect to the plane, as if about the point of suspension on the scales, will equally balance each other.]

In these contexts, *momentum* appears clearly distinct from *pondus* [weight]. Our term indicates, in fact, the inclination to descend, which depends on two factors: the weight and the distance from the fulcrum. However, despite the fact that the term has already acquired the precise technical meaning of a load variable, there is no other occurrence of *momentum* in the works and versions of Commandino outside of the definition of the center of heaviness.

The work of Federico Commandino and the investigations of Guidobaldo dal Monte are closely linked. The latter showed his appreciation of the Urbino definition of the center of heaviness, including it in his *Paraphrasis* of the *Equilibrium of Planes*. In correctly stating its provenance, he took the opportunity to state that it was a description rather than a definition:

Centrum gravitatis uniuscuiusque solidae figurae est punctum illud intra positum, circa quod undique partes *aequalium momentorum* consistunt.... Hanc postremam definitionem seu potius descriptionem tradidit Federicus Commandinus in libro de centro gravitatis solidorum[84]. [The center of heaviness of any solid figure is that point located within it, around which the parts of *equal moments* balance on all sides.... This last definition, or rather description, was given by Federico Commandino in his book on the center of heaviness of solids.]

61 As in Commandino, and also in Guidobaldo, the use of *momentum* is limited exclusively to the definition of the center of heaviness of solids. As we have already said, the meaning that *momentum* presents in the definition is different from that of an inclination to move downwards or upwards as determined by heaviness or lightness, which we have seen recorded in some versions of the opening passage of Eutocius's *Commentary* (corresponding to *ῥοπή*). However, as it appears clearly in another passage of the *Paraphrasis* on the *Equilibrium of planes*, it does not seem that Guidobaldo was fully aware of using a term that had two distinct meanings:

dicamus quoniam planum ACD, quatenus est corpori coniunctum, horizonti aequidistans permanere debet; si seorsum a corpore illud intelligamus, ut si ADC ex eius centro gravitatis G suspendatur, tunc quocunque modo reperiatur, hoc est sive horizonti aequidistans, sive minus, idipsum permansurum nihilominus intelligere possumus, partesque undique *aequalium momentorum* consistentes. Neque enim Aristoteles gravibus dumtaxat, sed etiam levibus *momenta* tribuit; idipsum (ut Eutocius in horum librorum commentariis refert) Ptolaemeo quoque placuit, ut habetur in libro (a nobis tamen desiderato) quem *de momentis* scripsit... et si mente concipiatur eadem figura levis cuiuspiam esse, tunc si detineatur in G, partes undique *aequalium momentorum* consistent, essetque G (ut ita dicam) centrum levitatis[85]. [let us say that the plane ACD, as far as it is connected to the body, must remain equidistant to the horizon; if understood separately from the body, as if ADC were suspended from its center of heaviness G, then, however it is found, whether equidistant to the horizon or less, we can nevertheless understand

[84] *Guidiubaldi e Marchionibus Montis in duos Archimedis Aequeponderantium libros Paraphrasis, scholiis illustrata*, Apud Hieronymum Concordiam, Pisauri 1588, p. 9; the italics are ours. Also for a detailed illustration of Guidobaldo's mathematical and mechanical work, one can refer to the fundamental work of P. L. Rose, *The italian renaissance of mathematics* cit., Ch. 10, pp. 222 ff., who devotes considerable attention to the Archimedean revival promoted by Dal Monte.

[85] *Ibid*., p. 16; the italics are ours.

that it will remain the same, and the parts balanced on every side with *equal moments.* For Aristotle not only attributes *moments* to heavy bodies but also to light ones; the same thing pleased Ptolemy (as Eutocius reports in the commentaries on these books), as is found in the book (desired by us, however) that he wrote *on moments*... and if it is conceived in the mind that the same figure is light, then if it is detained at G, the parts balanced on all sides with *equal moments*, and G would be (so to speak) the center of lightness.]

Here, the Marquis dal Monte took to the defense of the Archimedean method consisting in proposing demonstrations of *centro gravitatis* [center of heaviness] with purely geometric procedures applied to flat figures. Aristotle, in fact, as Eutocius attests, maintained that there is a *moment* of both heavy and light things. It is therefore possible to establish the center of heaviness of a figure independently of its heaviness. In this passage, Guidobaldo demonstrated that he had not perfectly understood the differences between *momentum* as equivalent to the *ῥοπή* of Eutocius's *Commentary* (proportional only to weight or lightness) and *momentum* as corresponding to the *ἰσορροπία* of the *Equilibrium of Planes* of Archimedes, or in the meaning which the term had assumed in the Commandinian definition of the center of heaviness (i.e., proportional to weight and distance).

Aside from Maurolico, whose unpublished writings did not circulate, none of the authors we have dealt with so far consciously used *momentum* in the meaning of variable effectiveness of the same weight on the balance, except in the definition of the center of heaviness proposed by Commandino. It seems that the concept of "inclinazione prodotta dal peso [inclination produced by weight]"
(or by lightness), expressed in Eutocius's passage, was the only one these authors **62**
considered legitimate to indicate with our term. Guidobaldo, for example – like Commandino – having to express the variable effect of the same weight at different distances from the fulcrum of a mechanical system, never used *momentum.* He preferred to use the old scholastic terminology of *virtus* [strength], *vis* [force], *potentia* [power], *magis* or *minus gravare* (*maior* or *minor gravitas*) [greater or lesser heaviness]:

Aristoteles enim (gratia exempli) querens cur vecte magna movemus pondera? causam esse ait longitudinem vectis maiorem ad partem potentiae; et recte quidem; cum ex principio ab ipso constituto manifestum sit, ea quae sunt in longiori a centro distantia *maiorem* quoque habere *virtutem*. Archimedes vero ulterius adhuc progredi voluit, hoc admisso, nempe quod est in longiori distantia *maiorem vim* habere, quam id quod est in breviori, inquirere etiam voluit quanta sit vis eius, quod est in longiori distantia, ad id quod est in breviori[86]. [For Aristotle (for example) inquires: why is it that large weights are moved with a lever? The reason is said that the greater length of the lever is towards the power side; and rightly so; because, from the principle established by him, it is manifest that those things which are at a longer distance from the center also have a *greater strength.* But Archimedes wished to proceed yet further, this having been admitted, namely that that which is at a longer distance has a *greater force* than that which is at the shorter, he wished also to inquire how great the force of that is, which is at the longer distance, in relation to that which is at the shorter.]

Guidobaldo – like Commandino and many others – therefore encountered some difficulty in isolating a term or expression capable of expressing, without ambiguity,

[86] *Ibid.*, p. 4; the italics are ours.

the concept of "momento del peso [moment of weight]" – which depends, in addition to the weight, on other circumstances – in order to appropriately distinguish it from the generic concept of "inclination". Thus both, in illustrating what a few years later would be defined by Galileo as the law of *moments* (the *moments* of two equal weights, at unequal distances from the fulcrum of the balance, are proportional to the distances from the point of suspension), will limit themselves to following the unfortunate traditional method consisting in distinguishing the (constant) *pondus* [weight] from its variable effectiveness on the balance (generally indicated with *gravitas* [heaviness]). As Guidobaldo will say in the *Mecanicorum liber*, "maiorem gravitatem habebit pondus in E quam pondus in F [the weight at E will have a greater heaviness than the weight at F]". And he will counteract the increase in *gravitas* with "naturalis potentia ponderis [natural weight power]"[87].

In general, therefore – as we will also see emerging from the subsequent chapters – the legitimacy of using *momentum* to indicate the variable effectiveness of weight on the balance (namely, the technicalization of the term) must have appeared much less automatic and obvious in the 1500s than we might be led to think today.

[87] *Guidi Ubaldi e Marchionibus Montis Mecanicorum Liber, in quo haec continentur, De Libra. De Vecte. De trochlea. De axe in peritrocheo. De cochlea*, Apud Hier. Concordiam, Pisauri 1577, p. 20. The vernacular translation of the mechanical works of Guidobaldo performed by Filippo Pigafetta (*Le Mechaniche dell'Illustriss. Sig. Guido Ubaldo de' Marchesi del Monte, tradotte in volgare da F. P...*, Appresso Francesco de' Franceschi Sanese, in Venetia 1581). The corresponding passage is translated by Pigafetta: "gravezza maggiore haverà il peso posto in E che il peso posto in F [the weight placed in E will have greater heaviness than the weight placed in F]", while "naturalis potentia ponderis" is rendered with "la naturale possanza del peso [the natural might of the weight]" (p. 22*r*). Pigafetta's vernacular is valuable for the translator's numerous notes in which – as we read on the title page – "si dichiarano i vocaboli et luoghi più difficili [the most difficult words and passages are declared]". In reality, Pigafetta, as he himself explained in one of his first notes, posed the problem of the technicalization of the vernacular language, in which he considered it extremely difficult to deal with specialized topics: "Conciò sia che in questo trattato della Bilancia et ne gli altri ancora l'Autore usi alcune parole, le quali non si sono potute trasportare commodamente in volgare, per non essere esse anco state accettate in questa lingua, né intese da ognuno, io le ho lasciate così latine.... Trutina è quella cosa, che sostiene tutta la Bilancia, la quale Trutina piglia il Perno, overo l'Assetto et nomasi in questi paesi Gioa, altrove Gioiola, overo l'orecchie della Bilancia, et in altre contrade Scocca... [Since, in this treatise on the Balance and in others, the Author uses some words which could not be easily translated into the vernacular because they had not yet been accepted in this language or understood by everyone, I left the Latin as is... Trutina is that thing, which supports the whole Balance, in which the Trutina grasps the Pin, or the Fulcrum, and is named in these regions "Yoke", elsewhere "Little Yoke", or the "ears" of the Balance (i.e., a load-bearing loop, such as a lifting lug, lifting eye, or pad eye), and in other places "Frame" (Structure or Skeleton)...]" (*ibid.*, p. 2).

2. VITRUVIUS

Liber X of Vitruvius's *De Architectura*[88] is entirely dedicated to the examination 63
of construction machinery and contains quite a bit of information on mechanical
knowledge – Vitruvius calls it *principia machinarum* [principles of machinery] –
essential for allowing an architect to conceive tools suitable for lifting and moving
large weights. The success of this work and the influence it exerted are enormous.
The Renaissance, above all, made it a canonical text and the vernacular translations
were countless[89]. Alberti[90], Leonardo[91], Raphael[92], Francesco di Giorgio[93], to 64
mention just a few names, had it constantly in mind.

It is primarily in the initial part of *Liber X* that Vitruvius provides some general notions of mechanics. Drawing directly from Greek sources, he often took care to indicate the corresponding Greek terms for the Latin ones he used. Consequently, he also provided important contributions to linguistic foundations[94].

Among the topics dealt with in this first part of *Liber X*, the attempt to explain why a weight on a steelyard, as its point of suspension moves away from the fulcrum, gains – while remaining itself equal – the ability to balance increasingly larger

---

[88] *Vitruvii de Architectura Libri Decem*, Edidit F. Krohn, in aedibus B. G. Teubneri, Lipsiae 1912.

[89] The Latin *Editio Princeps* was released in Rome in 1486. The first published vernacular translation was that of Cesare Cesariano, *M. L. Vitruvio Pollione de Architectura Libri decem, traducti de Latino in Vulgare, affigurati, comentati*, Gotardus de Ponte, Como 1521. Subsequently, in 16th century, the edition edited by Luzio Durantino, *Di Architettura dal vero esemplare Latino in vulgare… con le figure…*, in the houses of Joane Antonio and Piero Fratello da Sabio, Venetia 1524 (Durantino literally takes up Cesariano's text). Only the first five books were included in the version of *Architettura col suo comento e figure. Vitruvio in volgar lingua raportato per G. B. Caporali*, Perugia 1536. Lastly, *I dieci libri dell'architettura di M. Vitruvio tradutti et commentati da Monsignor Barbaro…*, Venezia 1556 (1st ed.). For the influence of Vitruvius's text in Humanism and the Renaissance, cf. V. P. Zoubov, *Vitruve et ses commentateurs du XVI*^{e} *siècle*, in *La Science au seizième siècle*. Colloque International de Royaumont, 1957, Paris 1960, pp. 67-90; Lucia D. Ciapponi, *Vitruvio nel Primo Umanesimo*, "Italia medioevale e umanistica", III, 1960, pp. 59-99; and F. Buerger, *Vitruv und die Renaissance*, "Repertorium für Kunstwissenschaft", XXXII (1909), pp. 199-218. Cf. also J. Pellati, *Vitruvio nel medioevo e nel Rinascimento*, in "Bollettino del R. Istituto di Archeologia e Storia dell'Arte", 1932; and Carol Herselle Krinsky, *Introduction* to the anastatic reprint of Cesariano's Vitruvius (Wilhelm Fink Verlag, München 1969), pp. 5-28. See, lastly, J. Schlosser, *La letteratura artistica*, trad. it., 2nd ed., Firenze 1953, pp. 251 ff. and Giorgio Tabarroni, *Vitruvio nella Storia della scienza e della tecnica*, "Atti dell'Accademia dell'Istituto di Bologna", Classe di Scienze Morali, *Memorie*, vol. LXVI (1971-1972), pp. 1-37 (extract).

[90] For Alberti, cf. R. Krautheimer, *Alberti and Vitruvius*, in *Acts of the 20th International Congress of the History of Art*, Princeton 1963.

[91] Leonardo certainly knew the *Architectura*, which he cites in Ms. *L* (c. 53*v*) and in *G* (c. 96*r*) of the Istituto di Francia (cf. Richter, *The literary works* cit., II, p. 375 and Pedretti, *Commentary* cit., II, p. 374).

[92] Cf. *Vitruvio e Raffaello. Il* De Architectura *di Vitruvio nella traduzione inedita di Fabio Calvo Ravennate*, edited by V. Fontana and P. Morachiello, Officina Edizioni, Roma 1975.

[93] For the influence exercised by Vitruvius on Francesco di Giorgio, it is enough to remember that the famous architect had a vernacular version of the *De architectura*. It is preserved today at the B.N.F. (II, I, 141). Cf. the *Introduzione* of C. Maltese to the *Trattati di Architettura, Ingegneria e Arte Militare*, I-II, Il Polifilo, Milano 1967, primarily pp. XVII ff. Cf. also in my *A proposito di un errore dei traduttori di Vitruvio* cit., pp. 71-88.

[94] For the different opinions of Baldi, Alberti and Philandrier on the difficulties of the Vitruvian text, cf. in my *A proposito di un errore dei traduttori* cit., pp. 71-72.

counterweights, is of particular importance. In this context, he used *momentum* twice:

> Id autem ex trutinis, quae staterae dicuntur, licet considerare cum enim ansa propius caput unde lancula pendet, ibi ut centrum est conlocata et aequipondium in alteram partem scapi, per puncta vagando quo longius aut etiam ad extremum perducitur, paulo et inpari pondere amplissimam pensionem parem perficit per scapi librationem examinatio. ita longius ab centro recedens imbecillior aequi pondii brevitas maiorem vim ponderis *momento* deducens sine vehementia molliter ab imo sursum versum egredi cogit, quemadmodum etiam navis onerariae maximae gubernator ansam gubernaculi tenens… una manu *momento* per centrum pressionibus ratione artis agitans, versat eam amplissimis et immanibus mercis et penus ponderibus oneratam[95]. [This, however, can be considered from the balances, which are called steelyards, for when the handle is near the top from which the platters hang, placed there as a center, and the counterweight on the other side of the shaft, which wanders from point to point, the further it is brought, even to the extremity, a small and inferior weight weighs a provided large one performed by the leveling test of the shaft. Thus, as it recedes further from the center, the smallness of the weak counterweight, drawing with its *moment* the greater strength of weight, compels it without violence to rise gently from the bottom upward, just as the pilot of a large cargo ship holding the handle of its rudder… one hand with *moment* with respect to the center pressing skillfully, turns the massive vessel while burdened with abundant and enormous goods and weighty provisions.]

We are in the presence of the only two occurrences of *momentum* in the entire *De Architectura*[96] and placed in a context with a meaning that is not entirely clear. However, as Wieland's authoritative reading also confirms, *momentum*, in both cases, indicates precisely the variable counterbalancing capacity that the same weight takes on according to the distance from the fulcrum of the point of application[97]. The
65 relationship between *pondus* and *momentum* in the steelyard appears very precise in the first occurrence. The aequipondium (the counterweight), moved to a greater distance, can balance the greater "weight", i.e., *moment*, by virtue of its own superior "moment". The *pondus* remains constant, while its ability to counterbalance, its *momentum*, grows *per puncta vagando* [through roaming points] (*puncta* indicate here the notches marked on the axis of the steelyard[98]), as Vitruvius eloquently states. Even in the second occurrence, it seems that a similar meaning should be assigned to *momentum*. That is, it is not the – insignificant – weight that determines the steering of the very heavy ship, but rather the very large *moment* that that small weight exerts *per centrum* [through the center].

The use of *moment* in this meaning, so exquisitely technical, and the absence of any reference to the corresponding Greek term would seem to indicate that Vitruvius is aware of using a word that is in use and easy to understand. It is strange, however,

---

[95] *Vitruvii de architectura…*, Edidit F. Krohn cit., pp. 233-234; the italics are ours.

[96] Cf. *Index Vitruvianus*, Confecit H. Nohl. Sumptibus et typis B. G. Teubneri 1876, p. 80.

[97] *Thesaurus Linguae Latinae* cit., VIII (1966), *Momentum*, col. 1392. The entry is edited by Wieland, which places both Vitruvian occurrences of the term in the section indicating the technical meaning of *momentum* (*in ipso librando*).

[98] Barbaro speaks of "segni [marks]" and "croci [crosses]" ("perché i segni e le croci nel fusto segnati sono più larghi [because the marks and crosses in the marked shaft are wider]": *I dieci libri dell'Architettura* cit., p. 455). Philandrier repeats *punctum*, which accompanies *denticulum*, defining it as "certi ponderis index [index of a certain weight]" (*Gulielmi Philandri Castilioni Galli Civis… in decem Libros M. Vitruvii Pollionis de Architectura annotationes*, Apud A. Dossena, Romae 1544, p. 346).

that, although he still refers on several occasions to the concept of varying weight according to the point of application, he no longer uses *momentum*[99].

In addition to numerous printed editions of the Latin text of *De Architectura*, numerous vernacularizations were published in the 16th century[100]. The translators, in general, misunderstood the meaning of the above passage. What seems particularly important to us is that it was precisely the inability to understand the meaning of *momentum* that betrayed them.

The first printed vernacular edition of *De Architectura* saw the light in Como in 1521. Magnificently "depicted", the translation was, not by chance, the work of a pupil of Bramante, Cesare Cesariano[101]. The aforementioned passage from Vitruvius was rendered thus by Cesariano:

> lequipondio nelaltra parte del scapo per li ponti vagando, quanto più lontano o vero a lo extremo fu
> perducto anchora cum aequal peso un poco più amplissimo perfice la aequale pensione per la libratione 66
> del scapo et examinatione più longa dal centro retirandose, cossì la brevitate del aequipondio più imbecilla
> *in momento* del peso deducendo magiore forza constringe senza vehementia andare mollemente dal
> basso verso a lalto; per qual modo anchora il gubernatore de la maxima nave oneraria, tenendo l'ansa del
> Gubernaculo… cum una mane *in momento* per la ratione del centro cum le pressione de larte agitando
> volge quella onerata de maximi pesi de mercantia et da victuali[102]. [the counterweight on the other
> side of the shaft, by wandering through points, brought however far or even brought to the extremity,
> retreating further from the center than a more abundant weight achieves the equal weighing according to
> the leveling of the shaft and the tongue of the balance, thus the smallness of the counterweight, more
> weak, *by the moment* of weight leads the greater force, compelling without vehemence to go gently from
> the bottom toward the top; in like manner, the helmsman of a large cargo ship, holding the handle of
> the Rudder…with a hand *in moment* with respect to the center, together with the pressing of skillful
> steering, he turns it, burdened of great weights of goods and provisions.]

The translator understood, at least in the first occurrence, the technical value of *momentum*, while offering a version of the not simple passage of *De Architectura* which presents notable ambiguities. Even in the second context, it seems to me that we can affirm that Cesariano perceived the exact meaning of the Vitruvian *moment*, yet again manifesting an obvious confusion in the face of the cumbersome construction of the Latin text, which shines through from the lack of clarity of the version.

There is a curious fact to add. It is known that, due to disagreements with the financiers of the expensive publishing initiative, Cesariano suddenly abandoned the

---

[99] Typically he resorts to the *magis* (*minus*) *premere*. Elsewhere, speaking of two different weights placed on the balance, he states: "cum autem impares sunt eorum virtutes et unum plus valendo premit alterum, loco traiecto fit una pars iugi longior, quae imbecilliori auxiliatur iumento [But when the virtues of the two are unequal, and one being stronger presses the other, one part of the yoke becomes longer when it has been crossed, which helps the weaker beast]" (*De Architectura* cit., p. 251). It will be appropriate to remember that, apart from the two by Vitruvius, there is no classical evidence of *momentum* in this precise sense.

[100] Cf. note 89. For a bibliography of the editions of *De Architectura*, cf. B. Ebhardt, *Die zehn Bücher der Architektur des Vitruv und ihre Herausgeber seit 1484*, Berlin s.d.

[101] On Cesariano, cf. the *Introduction* of Carol Herselle Krinsky of the anastatic reprint of the original edition of 1521, pp. 15-28 (with bibliographical information and index of place and person names).

[102] *Ibid*., c. CLXVIII*r*; the italics are ours.

venture when it was practically completed[103]. The commentary on the translation thus remained stuck approximately halfway through *Liber IX*, and two "experts" (Benedetto Giovio and Bono Mauro) were called in to complete it, who, rather than commenting, limited themselves to paraphrasing. Now, the two good commentators, in annotating the aforementioned passage of Cesariano's version, did not understand the technical meaning of the expression *in momento* and believed that it was to be understood "in a moment"[104].

No less serious, and moreover without mitigating circumstances, was the repetition of the error by another Vitruvian translator in the 16th century: Daniele Barbaro, the famous Patriarch of Aquileia. Barbaro's version – enriched by an extensive commentary in which he drew heavily from the pseudo-Aristotelian *Questions of Mechanics* – was in fact widely disseminated: within a few years, three editions were printed[105]. Daniele Barbaro's misunderstanding of Vitruvius's passage was complete:

> **67** et questo si può considerare dalle stadere, perché quando la orecchia è vicina al capo, dove pende la lance, nel qual luogo ella è come centro, e che il marco, o romano, detto equipondio, nell'altra parte del fusto vagando per li segni, quanto è più lontano condotto, se ben fusse all'estremo del fusto, ancho con men pari peso agguaglia il peso che è dall'altra parte, se bene è grandissimo, e questo adiviene per lo bilancio del fusto, e perché la leva è lontana dal centro. Et con la picciolezza del marco più debile levando *in un momento* maggior forza di peso senza vehemenza dolcemente constrigne dal basso al disopra levarsi. Ancho sì come il nocchiero d'una gran nave da carico, tenendo l'ansa del temone... *in un momento* con una mano per la ragione del centro calcando artificiosamente volge la nave carica di pesi grandissimi, di merci e d'altre cose necessarie[106]. [and this can be considered from the steelyard, because when the ear (i.e., lifting lug) is close to the head, wherein the platter hangs and serves as its center, and that the "mark", or "Roman", also called a counterweight, wanders through the points on the other side of the beam, to wherever it is brought, even if it were to the extreme of the beam, still with a lesser weight it matches the weight on the other side, even if the weight is very large, and this occurs according to the balancing of the beam, and because the lever is distant from the center. And with the smallness of the weaker mark (i.e., counterweight), it lifts *in a moment* a greater force of weight, without vehemence, gently driving it to rise from below to above. Just as the helmsman of a large cargo ship, holding the handle of the rudder... *in a moment* with a hand with respect to the center, skillfully

---

[103] Cf. *ibid.*, p. 10 of the *Introduction.* At p. 154*v* of the Como edition, Cesariano's resignation was noted and readers were informed that from that point on, the authors of the commentary were Benedetto Giovio and Bono Mauro.

[104] *Ibid.*: "Con questa medesima ratione dice l'authore che il Gubernatore delle maxime nave onerarie: cioè il nochero de la grande nave de mercantia le quale vano a vela serza remi *in uno momento* volge la nave... [with this same reason, the author says that the Helmsman of the largest cargo ships: that is, the pilot of the large merchant ship, which goes by sail without oars, turns the ship *in a moment*...]".

[105] Venice 1556, 1567, 1584 (the last is the same as the previous). For the use of *Questions of Mechanics* in Barbaro's commentary, cf. P. L. Rose, S. Drake, *The pseudo-aristotelian "Questions of Mechanics" in Renaissance Culture*, in "Studies in the Renaissance", XVIII (1971), p. 81. About Daniele Barbaro, see the entry in *D.B.I.*, VI (1964), pp. 89-95 (edited by G. Alberigo).

[106] *I dieci libri dell'Architettura* cit. (1556), p. 260; the italics are ours. In Barbaro's commentary on these two passages there is no mention whatsoever of the instantaneousness of the operation, suggested by translating *moment* as "in a moment". To explain the meaning of Vitruvius's passages, Barbaro refers, for the steelyard, to the twenty-fifth pseudo-Aristotelian *Question*, while, for the steering of the ship through the rudder, he refers to the twentieth.

pressing, he turns the ship loaded with very large weights, goods and other necessary things.]

The Patriarch of Aquileia's mistake is evident. He confused the ablative of *momentum* in the technical sense with the ablative adverbial *moment*. And he then introduced into Vitruvius's text the idea of a sudden, instantaneous action which is completely foreign to it. This error, certainly not comparable to the ambiguities of Cesariano, seems to me to demonstrate the difficulty that a 16th-century reader, even an educated one – like Daniele Barbaro – would find when faced with a term – such as *momentum* – with multiple semantic meanings. It also demonstrates – and this may appear even more significant for our investigation – that the use of *moment* to indicate the variation in the effectiveness of the *pondus* in mechanical systems was anything but common. Aside from his failure to perceive the technical meaning of *momentum* in the aforementioned text of Vitruvius, in the extensive commentary in *Liber X* of the *Architettura*, the Patriarch of Aquileia never used the term *moment* to express the variable capacity that the same weight presents in counterbalancing according to the distance from the fulcrum of the point of suspension. In general, illustrating this phenomenon, he asserts that the weight "calca maggiormente [presses more]"[107]. However, Barbaro knew and took the time to indicate the natural tendency of heavy bodies to move, each towards its own place, in the sense with which we have seen it used in the translations of the passage from Eutocius's *Commentary*:

Io dico che per peso non solo s'intende quella gravità che hanno le cose ponderose e grandi, ma **68**
ancho quel *momento, e quella inclinatione* naturale di andar ciascuna al suo proprio luogo: e quando artificiosamente si costrigne una cosa grave a salire, e che la natura più presto che dar il vacuo consente, che gli elementi oltra la loro inclinatione o ascendino, o discendino, certamente questa è una gran virtù o forza: e questo costrignere gli elementi è con somma solertia dall'arte stato ritrovato[108]. [I say that by weight we mean not only that heaviness that weighty and large things have, but also that *moment, and that natural inclination* to go each to its own place: and when a heavy thing is skillfully forced to rise, and that nature, rather than permitting a vacuum, more quickly allows the elements, beyond their inclination, either to ascend or to descend, this certainly is a great virtue or strength: and this forcing of the elements has been discovered with utmost skill through craft.]

In this context, weight and *moment* appeared so closely linked that they seemed to be interchangeable terms. For as much weight, so too is its *moment*. The "static" use of *moment*, on the model of the balance, presupposes the breaking of the identity

[107] *Ibid.*, p. 260. Naturally, Barbaro, an attentive reader of the *Questions of Mechanics*, knew well the laws of *moments* of the balance. In fact, he gave an effective explanation in the commentary on the aforementioned passage of the translation: the steelyard "è come una sola bilancia, perché ha un centro solo e due raggi, ma noi mutando il peso mutiamo il centro, perché il marco non calca egualmente essendo più vicino o più lontano dal centro, imperoché, quando pesamo alcuna cosa, quanto più il centro, dove è l'orecchia, è vicino al peso, tanto più si leva, perché la linea, cioè il fusto, che è dal centro al marco si fa maggiore. [it is like a single balance, because it has only one center and two beams, but by changing the weight we change the center, because the mark (i.e., counterweight) does not weigh equally being closer or further from the center, because, when we weigh something, the more the center, where the ear (i.e., the lifting lug) is, is close to the weight, the more it rises, because the line, that is, the beam, measured from the center to the mark becomes greater]" (*ibid*).

[108] *Ibid.*, p. 255; the italics are ours.

relationship with weight. The weight remains constant, but its *moments* vary. In Barbaro, as in Commandino and Guidobaldo[109], this step is not taken: each *weight* corresponds to a specific *moment*. These texts also demonstrate, therefore, that *moment* is predominantly used to indicate a natural inclination, produced solely by weight or lightness. The general context that our term seems to evoke appears to be the Aristotelian theory of the elements and, associated with it, that of natural places[110].

3. The "Corpus de Ponderibus"

Further confirmation of the uncommonness of the use of *momentum* in the meaning of variable effectiveness of weight in mechanical systems is provided by the complete absence of our term in the group of writings – assignable to the 13th century – that constitute the so-called *de ponderibus* tradition[111]. The research of
69 Duhem[112] and Vailati[113], between the last century and ours, integrated more recently, especially by Maier[114], Moody[115] and Clagett[116], has contributed to highlighting – not without some exaggeration – the influence exerted on late medieval and

---

[109] With the exception of the definition of center of heaviness (cf. *supra*, p. 60).

[110] For a more detailed analysis of the misunderstandings regarding this passage among later translators of Vitruvius and also in other languages, I would like to refer you again to my *A proposito di un errore dei traduttori di Vitruvio* cit.

[111] For writings of *de ponderibus*, cf. *The medieval science of weights* cit.

[112] Of Pierre Duhem, cf. primarily, *Les origines de la statique*, 2 vol., Paris 1905-1906 and the *Études sur Léonard de Vinci*, Première et Seconde Série, 2 vol., Paris 1906-1909 (especially Ch. VII). It goes without saying – as it is well known – that Duhem saw in the notable achievements of the medieval authors of the *scientia de ponderibus* one of the most convincing confirmations in favor of the thesis of a medieval origin of modern science.

[113] Of Vailati's work it is worth mentioning the fine essay of 1897, in which, evaluating the *Tractatus de ponderibus* attributed to Jordanus, he emphasized the implicit use of the principle of "virtual speeds", bringing it closer to the pseudo-Aristotelian *Questions of Mechanics* (*Il principio dei lavori virtuali da Aristotele a Erone d'Alessandria*, "Atti della R. Accademia delle Scienze di Torino", vol. XXXII, meeting of 13 June 1897; reprinted in *Scritti* cit., pp. 91-106).

[114] Cf. A. Maier, *Die Vorlaufer Galileis im 14. Jahrhundert*, Roma 1949.

[115] Of Ernest A. Moody, as well as introductions to some of the writings *de ponderibus* published in *The medieval science of weights* cit. (*Liber Euclidis de ponderoso et levi, Liber de canonio, Elementa Jordani, Liber de ponderibus, Jordani de ratione ponderis, Bradwardini de proportionibus velocitatum, A fourteenth century commentary on Proposition One of the* Liber de ponderibus), cf. *Galileo and Avempace: The dynamics of the Leaning Tower "experiment"*, "Journal of the History of Ideas", XII (April-June 1951), 2-3, pp. 163-193 and 375-422; and *Laws of Motion in Medieval Physics*, "Scientific Monthly", LXXII, 1 (January 1951), pp. 18-23. Both essays republished in *Studies in Medieval Philosophy, Science and Logic. Collected papers 1933-1969*, University of California Press, Berkeley-Los Angeles-London 1975 (respectively, pp. 203-286 and 189-201).

[116] Of Marshall Clagett, as well as introductions to some of the writings of *de ponderibus* published in *The medieval science of weights* cit. (*Liber Archimedis de insidentibus in humidum, Liber Karastonis, Tractatus Blasii de ponderibus, A fragment on the Roman balance attributed to Euclid, A variant form of proposition Eight of the Elementa Jordani*), cf. *La scienza della meccanica nel medioevo*, trad. it. cit., and *Some general aspects of Medieval Physics*, "Isis", 39 (May 1948).

Renaissance culture by this *corpus* of writings on statics, which is commonly, albeit with multiple, legitimate doubts, attributed to the name of Jordanus Nemorarius[117].

In the *de ponderibus* treatises, numerous questions of statics are finely analyzed, for which solutions are proposed that will be largely accepted in the 16th and 17th centuries. As Duhem, Moody, Clagett and others have demonstrated, the manuscript influence[118] of these works was significant: Leonardo da Vinci[119],
Nicolò Tartaglia[120], Alessandro Piccolomini[121], Benedetti[122] and many others knew **70**
and used the *Elementa Jordani*, the *De ratione ponderis* or the *Liber de ponderibus*. The *Tractatus Blasii de ponderibus* also exerted considerable influence[123].

In the writings *de ponderibus* attributed to Jordanus, among other things, the *pondus* was precisely distinguished by the variable effectiveness that it exerts according to different inclinations. The latter was indicated with an expression destined to have enormous success: *gravitas secundum situm* [heaviness according to location]. See this passage in the *Elementa Jordani*:

> Secundum situm gravius esse, quando in eodem situ minus obliquus est descensus. Obliquiorem autem descensum, in eadem quantitate minus capere de directo[124]. [Greater heaviness according to location occurs when the descent at the same position is less oblique. However, a more oblique descent,

[117] On the attribution to Jordanus, on the significance of "Nemorarius", on his identity and on the unlikelihood that he can be confused with the General of the Dominican Order, Jordanus of Saxony, cf. *The medieval science of weights* cit., *General Introduction*, pp. 121-125; cf. also M. Clagett, *La scienza della meccanica*, trad. it. cit., pp. 92 ff. Worth noting is the fine entry *Jordanus de Nemore* in *D.S.B.*, VII (1973), pp. 171-179 (edited by E. Grant).

[118] The only writings of *corpus de ponderibus* that had an edition in 1500 were the *Liber de ponderibus*, published by Petrus Apianus (*Liber Jordani de ponderibus Propositiones XIII et earundem demonstrationes*, Norimberga 1533) and the *De ratione ponderis* edited by Nicolò Tartaglia (*Jordani opusculum de ponderositate. Nicolai Tartaleae studio correctum, novisque figuris auctum*, Apud Curtium Troianum, Venetiis 1565).

[119] For Leonardo's knowledge of the text *de ponderibus*, cf. P. Duhem, *Études sur Léonard de Vinci*, Première Série cit., Ch. VII (*La scientia de ponderibus et Léonard de Vinci*), pp. 257-316. Useful information can also be found in R. Marcolongo, *Memorie sulla Geometria e la meccanica di Leonardo da Vinci*, Napoli 1937. Also worth consulting Uzielli, *Ricerche su Leonardo da Vinci*, Serie II, Salviucci, Roma 1884, and R. Caverni, *Storia del metodo sperimentale in Italia*, 6 vol., Firenze 1891-1900, vol. IV, Ch. I. See, lastly, the section *Mechanics* by M. Clagett of the entry *Leonardo da Vinci*, in *D.S.B.*, VIII (1973), pp. 215-234.

[120] The aforementioned edition of *De ratione ponderis* by Curzio Troiano (1565) edited by Tartaglia was based on a codex belonging to the same Brescian scientist (cf. M. Clagett, *La scienza della meccanica*, trad. it. cit., p. 122). Tartaglia, too, repeated many of Jordanus's propositions in his *Quesiti*, without disclosing the source (*Quesiti e invenzioni diverse*..., Venezia 1546, pp. 81 ff.).

[121] Alessandro Piccolomini's knowledge of the *de ponderibus* writings is especially evident in the *In mechanicas quaestiones Aristotelis Paraphrasis paulo quidem plenior*..., Apud Antonium Bladum, Romae 1547.

[122] The Venetian scientist shows notable familiarity with Jordanus's texts. In the *Diversarum speculationum mathematicarum et physicarum liber* (Torino 1585), he harshly criticized Jordanus and Tartaglia, noting how both started from the incorrect assumption that the directions of the forces exerted by the weights were at right angles to the axis of the balance, that is, parallel to each other.

[123] Cf. *The medieval science of weights* cit., pp. 231 ff.

[124] *Ibid.*, p. 128.

in the same quantity, encompasses less of the direct (descent).]

For the author of the *Elementa*, the *gravitas secundum situm* [heaviness according to location] of the same weight along different inclinations was measured by the ratio between the length of the oblique path and the length of the corresponding vertical descent. In the case of the balance, the "gravità di posizione [positional heaviness]" of the weight was instead determined, in relation to that of the counterweight, by the ratio between the segments intercepted on the arcs described by the ends of the arms, were the balance to be set in motion[125]. That is, Jordanus observed that the greater *gravitas secundum situm* [heaviness according to location] of the same weight essentially depended on the tendency to move with a greater speed than the counterweight (determined by the position) in the case of a displacement of the mechanical system. In this demonstrative process, the method of the so-called "virtual speeds", set forth in the pseudo-Aristotelian *Questions of Mechanics*, was foreshadowed, instead of the strictly static one adopted by Archimedes.[126]
71 Nonetheless, in the writings that go by the name of Jordanus, the "positional heaviness", at least in the schema of the balance, is often made to depend not on the "virtual speed", but on the "horizontal distance" from the fulcrum of the system[127].

No less interesting are the considerations on the inclined plane contained in the *Liber de ratione ponderis*, which introduces quite a few corrections to the *Elementa*[128]. See, for example, the *Propositio X*, where it talks about the *virtutes descendendi* [strengths of descending] of bodies along different inclined planes and establishes the rules of their equivalence:

Si per diversarum obliquitatum vias duo pondera descendant, fueritque declinationum et ponderum una proportio eodem ordine sumpta, una erit utriusque virtus in descendendo[129]. [If two weights descend by paths of different inclinations, and if the inclinations and weights are taken in one proportion in the

---

[125] For a discussion of the method adopted by Jordanus, cf. M. Clagett, *La scienza della meccanica*, trad. it. cit., pp. 94 ff.

[126] See *ibid*., pp. 98 ff. The use of virtual speeds is observed in the proof of *Propositio VIII* of the *Elementa* ("Si fuerint brachia libre proportionalia ponderibus appensorum, ita ut in breviori gravius appendatur, eque gravia erunt secundum situm appensa [If the arms of the scale are proportional to the suspended weights, so that the heavier weight hangs on the shorter arm, they will be equally heavy according to their position when suspended]"; *The medieval science of weights* cit., p. 138). The *Propositio VIII* of the *Elementa* is presented unchanged in the *De ratione ponderis*, as *Propositio VI* (cf. *ibid*., p. 182).

[127] Cf. for example, the *Propositio V* of *Liber de ponderibus*: "Si fuerint brachia inequalia, equalibus ponderibus appensis, ex parte longiori fiet motus [If the arms are unequal, and equal weights are attached, the movement will take place on the longer side]". However, in the commentary on the Proposition, the dynamic attitude re-emerges: "Probatur sic: Ex parte longioris describitur circulus maior et sic patet, per suppositionem tertiam, quod pondus secundum situm est gravius [It is proved as follows: on the side of the longer, a larger circle is described, and thus it is clear, by the third supposition, that the weight is heavier according to its position]" (*The medieval science of weights* cit., pp. 156-157).

[128] On the relationship between the *Elementa* and the *De ratione ponderis*, cf. *The medieval science of weights* cit., pp. 169-173.

[129] *The medieval science of weights*, cit., p. 190.

same order, the power of both in descending will be the same.]

This concept, which Galileo, at a certain point, indicated with the expression "momento del discendere [moment of descent]", is of considerable importance, given that it seems to open the way to the possibility of a derivation of dynamics from the same principles of statics[130].

Without dwelling on the *de ponderibus* tradition, already well analyzed by Clagett, Moody and other scholars, we wish to point out that, as far as I know, in the group of texts that constitute it – chronologically assignable to the 13th century – not a single example of momentum is found neither in the generic meaning of inclination produced by the weight, nor in the more specific meaning of the variable of the effectiveness of the weight (weight $\times$ distance, or weight $\times$ "virtual" speed). *Momentum* therefore remained foreign to this group of texts, nor was it used, as we will soon see, by those later authors who attached themselves to this tradition.

The influence of the expression that occurs most frequently in the *corpus de ponderibus* to indicate the varying effectiveness of the weight, *gravitas secundum situm* [heaviness according to location], was, on the other hand, enormous. It is present, for example, in Leonardo, in whose manuscripts there are numerous references to Jordanus and citations of works attributed to him. See, for example, **72**
this passage from the *Codice Atlantico*, in which he promptly translates *Propositio VIII* of the *Elementa*:

> Se saranno le braccia della bilancia proporzionali alli pesi attaccati, in tal modo che al più corto braccio sia il peso maggiore, saranno egualmente gravi secondo il sito[131]. [If the arms of the balance are proportional to the attached weights, such that the greater weight is on the shorter arm, they will be equally heavy according to their position.]

Gravitas secundum situm [heaviness according to location] also occurs, naturally, in Tartaglia and with very high frequency. The engineer from Brescia was, in fact, extremely familiar with Jordanus's texts. Of the *De ratione ponderis*, for example, he possessed the codex on which Curzio Troiano's print was based in 1565[132]. In the *Quesiti et inventioni diverse*, we find the two principles of Jordanus mentioned

---

[130] Cf. Part Two.

[131] *Il Codice Atlantico di Leonardo da Vinci*, Hoepli, Milano 1894-1904, c. 154va. For the *Propositio VIII* of *Elementa*, cf. *supra*, note 126. An evident derivation from Propositions IX and X of *De ratione ponderis* shows another note contained in the Codice Atlantico: "La equalità della declinazione osserva le equalità de' pesi. Se le proporzioni de' pesi e dell'obliqua, dove si posano, saranno equali, ma converse, essi pesi resteranno equali in gravità e in moto [The equality of the declination observes the equality of the weights. If the proportions of the weights and of the oblique, where they are placed, are equal, but converse, the weights will remain equal in heaviness and in motion]" (p. 354*v*; cit. from R. Marcolongo, *op. cit.*, p. 173; for the Propositiones IX and X of Jordanus, cf. *The medieval science of weights* cit., pp. 188-189).

[132] *Jordani Opusculum de ponderositate* cit. Cf. *The medieval science of weights* cit., p. 10. It is also worth consulting, for the many notes contained therein, the essay by the deserving editor of Tartaglia, Arnaldo Masotti, *Nicolò Tartaglia e i suoi* "Quesiti", in *Quarto Centenario della morte di Nicolò Tartaglia*, Proceedings of the History of Mathematics Conference (30-31 May 1959), edited by A. Masotti, in the Ateneo di Brescia, Supplement to the Commentaries (1960), pp. 33 ff.

above included in the *Diffinitioni*: *gravitas secundum situm* [heaviness according to location] and *virtus descendendi* [strength of descending]:

Quesito XV Diffinitione XII
[Question XV Definition XII]

Un corpo si dice essere più grave di un'altro secondo la specie, quando che la sostantia material di quello è più ponderosa della sostantia material dell'altro, come che è il piombo del ferro, e altri simili. [A body is said to be heavier than another according to its species, when the material substance of the one is heavier than the material substance of the other, such as lead to iron, and the like.]

Quesito XVI Diffinitione XIII
[Question XVI Definition XIII]

Un corpo si dice essere più grave, over men grave d'uno altro, secondo il luoco, over sito, quando che la qualità del luoco dove che lui se reposa e giace, lo fa essere più grave dell'altro anchor che fussero simplicemente egualmente gravi[133]. [A body is said to be heavier, or less heavy, than another, according to the location, or position, when the quality of the location where it rests and lies causes it to be heavier than the other even if they were simply equally heavy.]

73 In the whole of Tartaglia's works, there is not, as far as I know, a single occurrence of *moment*. Conversely, beyond that of Leonardo and Tartaglia, *gravitas secundum situm* [heaviness according to location] is present in many authors[134].

In the *corpus de ponderibus*, it is worth noting another significant expression, *nutum facere* [to give a nod], which is recorded both in the *De ratione ponderis*[135], and in the *Tractatus Blasii*[136]. We will see later that *nutus* [nod] was also frequently used to translate *ῥοπή*[137]. If this Latin *corpus* originally depended – as has been supposed – on the now lost Greek sources[138], it cannot be ruled out that *ῥοπή* stood in the Greek texts at the points corresponding to where the Latin *nutus* [nod] appears[139]. See in the *De ratione ponderis*:

Cum fuerint appensorum pondera equalia non *faciet nutum* in equilibri appendiculorum inequalitas.

---

133 *Quesiti et inventioni diverse de Nicolò Tartaglia, di nuovo ristampati…*, Appresso l'autore, 1554 (reproduction facsimile edited and with introduction by A. Masotti, in the University of Brescia, Supplement to the Commentaries, 1959), p. 84.
134 It is, for example, common in Cardano, in Scaliger, in Benedetti.
135 Cf. *The medieval science of weights* cit., p. 180.
136 *Ibid.*, p. 244.
137 Cf., for example, the next paragraph.
138 Cf. the *General Introduction* of *The medieval Science of Weights* cit. According to C. Thurot (*Recherches historiques sur le principe d'Archimède*, in "Revue archéologique", 19, 1869, p. 117), The *De ratione ponderis* would even be the translation of Ptolemy's lost *περὶ ῥοπῶν*. But, as Clagett points out, there is no evidence to support this hypothesis (*La scienza della meccanica*, trad. it. cit., p. 94). Regarding the influence exerted on the *scientia de ponderibus* by the pseudo-Aristotelian *Questions of Mechanics*, cf. M. Clagett, *ibid.*, pp. 104-105; and G. Vailati, *Il principio dei lavori virtuali* cit., in *Scritti* cit., pp. 99-103.
139 Another hypothesis is that the Greek term was *ὁρμή* (usually translated into Latin as *impetus*).

> Si brachia libre fuerint inequalia, equalibus appensis, ex parte longiore *nutum faciet*[140].
>
> [When the attached weights are equal, the inequality of the arms will not *give a nod.*
> If the balance arms are unequal, with equal attachments, it will *give a nod* on the longer side.]

Here *nutum facere* [to give a nod] clearly indicates not force, but rather its exertion (i.e., a movement) at a specific point, in a specific direction. In Alessandro Piccolomini and Galileo, we will find exactly equivalent expressions, respectively with "momento" and "impeto" in place of *nutus*[141].

4. The Pseudo-Aristotelian "Questions of Mechanics"

Similar results are achieved by examining the Latin and vernacular influences of **74**
the *Questions of Mechanics*[142]. The writing which, in the 16th and 17th centuries, was generally considered the work of Aristotle, despite there being no shortage of contrary opinions, is certainly attributable to another author, perhaps to Strato (3rd century B.C.), as some have suggested[143]. However, it also had enormous diffusion, especially after the publication of the Greek text, edited by Bessarion (1497)[144], particularly in the 16th century, quickly becoming a school manual. Galileo himself used it in Padua for some courses and wrote a Commentary, which has not survived[145]. As is known, the *Questions of Mechanics* presented the method of "virtual speeds" in a more explicit manner than the *de ponderibus* texts to demonstrate the laws of equilibrium for the balance. That is, it emphasized that the variable effectiveness of the same weight in mechanical systems depended not so much on the variation of the distance from the fulcrum of its point of application, but rather on the different speeds with which it would move, in the event of a breakdown of the equilibrium[146]. In this sense, the *Questions of Mechanics* presented a different

---

[140] *The medieval science of weights* cit., p. 180; the italics are ours.

[141] "Cum deprimat ac momentum faciat, loco ponderis existimandum est [when it presses down and produces momentum, it is to be considered instead of weight]" (*In mechanicas quaestiones... paraphrasis* cit., pp. XXIII-XXIV). "Se nella medesima circonferenza fusse applicata forza animata, la quale avesse momento di far impeto per tutti i versi... [If an animated force were applied to the same circumference, which had the opportunity to make impeto in all directions...]" (*G.G.*, II, p. 168).

[142] *Μηχανικὰ Προβλήματα.* The edition I used is the Teubnerian one of O. Apelt, *Aristotelis quae feruntur. De plantis. De mirabilibus auscultationibus. Mechanica. De lineis insecabilibus. Ventorum situs et Nomina de Melisso Xenophane Gorgia...*, In aedibus B. G. Teubner, Lipsia 1888.

[143] On the attribution to Strato and on the possible dependence of the *Questions* on the environment and studies of the Lyceum, cf. the excellent essay by P. L. Rose and S. Drake, *The pseudo-aristotelian Questions of Mechanics in Renaissance Culture* cit., pp. 65-104 (primarily pp. 72-73). Cf. also M. Clagett, *Greek science in antiquity*, New York 1963, pp. 93-94.

[144] *Aristotelis Opera Graece cum Theophrasti Historia plantarum et Alexandri Aphrodisiensis Problematis*, Apud Aldum Manutium, Venetiis 1495-1498.

[145] Cf. P. L. Rose, S. Drake, *The pseudo-aristotelian Questions of Mechanics* cit., pp. 94-96. For the course held in Padua by Galileo, in 1598, on *Questioni Meccaniche*, see Document VII in A. Favaro, *Galileo Galilei e lo Studio di Padova*, 3 vol., Editrice Antenore, Padova 1966, I, p. 115.

[146] Cf. M. Clagett, *La scienza della meccanica*, trad. it. cit., pp. 25-27. Truly valuable for its clarity

approach from Archimedes's rigorously static one. And it hinted at the possibility of a dynamic extrapolation of the same static principles. In fact, most of the thirty-five problems for which solutions were proposed in the *Questions of Mechanics* pertained precisely to questions of movement[147]. The number of problems dedicated to statics was also notable, especially to the laws of equilibrium for the balance.

75 It does not appear that Latin translations of the pseudo-Aristotelian writing were made in the Middle Ages[148]. The influence of the text became notable in the 15th century and later largely significant with the first printings and translations in the 16th century[149] (the first, Latin, is due to Fausto). There is no mechanical author of any importance from the 16th century in whom one does not find a direct or indirect reference to some of the "problems" analyzed in the pseudo-Aristotelian writing, each of which became a real *topos*.

The diffusion of the *Questions of Mechanics* was mainly contributed by the translation by Leonicus Thomaeus, published in 1525 and subsequently reprinted numerous times, almost always with Greek text on the front and without figures[150]. It is a very faithful translation, even if based on a Greek text that is anything but certain[151]. Thomaeus also prepared an array of notes for the translation[152]. Galileo owned a copy of the 1525 edition of Thomaeus's version[153]. A second translation was carried out at the end of the 16th century by a pupil of Ramus, Henri de Monantheuil[154].

---

remains the essay by G. Vailati, *Il principio dei lavori virtuali da Aristotele a Erone* cit., in *Scritti* cit., especially pp. 91-98.

147 A list of the *Questions* discussed can be found in P. L. Rose, S. Drake, *The pseudo-aristotelian Questions of Mechanics* cit., pp. 71-72.

148 There are no codices that record medieval Latin translations of the *Questions of Mechanics*. However, the doubt is legitimate, as Haskins pointed out (*Studies in medieval science*, 2nd ed., Cambridge Mass. 1927, pp. 316-317), highlighting the presence, in *De arte venandi cum avibus*, of one of the *Questions*, and Clagett (*La scienza della meccanica*, trad. it. cit., pp. 91-92), which records the reporting of a *Repertorium super Mechanica Aristotelis*, in the *Chartularium* of the Bolognese Studio (dated 18 August 1413).

149 We have already mentioned the Aldine edition. As for the versions, the first Latin one was performed by the humanist Vittore Fausto (*Aristotelis Mechanica Victoris Fausti industria in pristinum habitum restituta, ac latinitate donata*, In Aedibus Iodoci Badii, Parisiis 1517), very rare today. Of Fausto's version I was only able to consult the exemplar manuscript copy of the print (Codice Vaticano Urbinate Latino 1321; copied by hand in the 16th century). On the version of Fausto, cf. P. L. Rose, S. Drake, *The pseudo-aristotelian Questions of Mechanics* cit., pp. 77-78, and the excellent work of D. De Bellis, *Niccolò Leonico Tomeo interprete di Aristotele naturalista*, in "Physis", XVII (1975), fasc. 1-2, p. 90.

150 *Conversio mechanicarum questionum Aristotelis…*, Venetiis 1525. For the countless editions of Aristotle's works of the 16th century in which the *Questions of Mechanics* appear in the Thomaeus version, cf. E. Cranz, *A bibliography of Aristotle editions 1501-1600*. Verlag Valentin Koerner, Baden-Baden 1971, p. 218.

151 On the nature of the translation carried out by Thomaeus, on the codex used and, in general, on his contribution as a translator of Aristotle's naturalistic writings, cf. P. L. Rose, S. Drake, *The pseudo-aristotelian Questions of Mechanics* cit., pp. 79-80 and D. De Bellis, *Niccolò Leonico Tomeo* cit., especially pp. 90-92.

152 Cf. De Bellis, *Niccolò Leonico Tomeo* cit., p. 90.

153 Cf. A. Favaro, *La libreria di Galileo Galilei*, "Bullettino di Bibliografia e Storia delle Scienze Matematiche e Fisiche", XIX (1886), p. 243.

154 *Aristotelis Mechanica graeca, emendata, latina facta et commentariis illustrata ab Henrico*

The Greek text of the *Questions of Mechanics* has numerous occurrences of *ῥοπή* and different forms of *ῥέπω*. The first interesting context is found in the opening passage of the *Questions*, which states that the gifted man competent in 76
mechanical science is capable of craftily conquering nature. Let's see how the piece was rendered within the contexts of the Thomaeus and the Monantheuil versions.

PSEUDO-ARISTOTLE	THOMAEUS	MONANTHEUIL
καὶ τὰ ῥοπὴν ἔχοντα μικρὰν κινεῖ βάρη μεγάλα[155] [and small things, having inclination, move large weights]	Huiusmodi autem sunt in quibus et minora superent maiora, *et quaecunque momentum parvum habent, magna movent pondera*[156]. [Of such, however, are those in which the smaller overcome the greater, *and that which has a small momentum moves large weights.*]	Ut in iis quae, cum minora sint, superant maiora; *et parvum momentum cum habeant, ingentia dimovent pondera*[157]. [Just as in those where, although they are smaller, they overcome the greater, *and having a small momentum, move huge weights.*]

In this case, both translators agreed to translate *ῥοπή* as *momentum*. The concept implied in the passage and subsequently developed upon is that a small weight can overcome even enormous resistances, when its "natural" *moment* is craftily increased. This increase occurs – as the *Questions* suggest – in the case that the small weight operates with greater speed (or – which is the same thing – is applied at a greater distance from the fulcrum of the system). *Ῥοπή* here indicates the "natural inclination" produced by the weight (and proportional to it). We have previously observed – and will observe better in the following paragraph – that the translation of this meaning of the Greek term with *momentum* was a common phenomenon at the beginning of the 16th century.

However, in the rest of the version, both translators used terms other than *momentum* to render the multiple occurrences of *ῥοπή*. See *Question VIII*, in which it tries to clarify why circular things move more easily.

PSEUDO-ARISTOTLE	THOMAEUS	MONANTHEUIL
ῥέπει [to turn]	Ad haec quo *nutat* pondus eo motor movet. Cum igitur ad rectum super plano circuli fuerit diameter, planum secundum	Praeterea quo corpus *vergit* eo motor impellit. Quum igitur diameter circuli recta insistit plano, circulo in puncto planum

Monantholio..., Apud Hieremiam Perier, Parisiis 1599 (2nd ed., 1600, cit. *supra*, Ch. II, note 55).

155 847^{a}, 11-14.

156 For the Thomaeus translation, I used the *Aristotelis opera omnia quae extant grece et latine*, I-II, Typis Regiis, Lutetiae Parisiorum 1619; vol. I: *Aristotelis Stagiritae Quaestiones Mechanicae Nicolao Leoniceno interprete*, p. 1183; the italics are ours. Fausto had also resorted to *momentum* to render *ῥοπή* in this context: "Id genus autem ea sunt in quibus et minora plus maioris possunt ac parvi momenti res movent pondera immensa [this kind of thing, however, is one in which smaller things can move larger ones, and small moments can move immense weights]" (Vat. Urbinate Lat. 1321, c. 3*r*).

157 *Mechanica graeca* cit., p. 8; the italics are ours.

ῥέπον [turning]

ἐπὶ τὸ ἐναντίον τῆς ῥοπῆς [against the inclination]

ῥοπή [inclination]

ῥοπή [inclination]

ῥοπή[158] [inclination]

punctum contingente circulo aequale utrinque pondus disterminat diameter. Cum autem movetur, plus illico, ad quod movetur, ceu inde *nutans*, ab impellente facilius in ante movetur. Quo enim unumquodque vergit movetur ex facili. Siquidem difficulter *ad contrarium nutus sui* movetur modum… Celerius enim ab aequali moventur potentia maiores circuli, moventque onera, quoniam circuli maioris angulus ad minoris angulum circuli *nutum* habet quendam: e sicuti diameter ad diametrum, ita maior circulus ad minorem. Infiniti autem sunt minores. Si autem ad alterum *nutum* habet circulus, similiter est bene mobilis. Et aliam sane habet *inclinationem* circulus, et ea quae a circulo moventur[159].

attingente aequaliter utrimque diameter pondus disterminat. 77
Quando vero movetur statim ad id movetur, veluti eo *repens* motu facilius impellente in anteriorem partem. Eo enim quo vergit unumquodque facilius movetur. Quandoquidem si *in contrarium quam quo vergat* moveatur difficulter movebitur… Celerius enim ab aequali vi maiores moventur, et pondera movent, quia maioris circuli angulus *nutum* quendam habet ad minoris angulum. Et est ut diameter ad diametrum sic omnis maior circulus ad minorem. Infiniti autem sunt minores. Si vero etiam circulus *nutum* habet ad alterum. Similiter vero facile mobilis alium *nutum* habet circulus et quae a circulo moventur[160].

[To these things, the weight *nods* in the direction in which the mover moves. Therefore, when the diameter is at right angles to the plane of the circle, the plane passing through the point touching the circle, on both sides, determines the weight equally on either side of the diameter. But when it moves, immediately, towards which it is moved, as if *nodding* from there, it moves more readily forward from the impeller. For indeed, everything moves easily toward where it inclines. Indeed, with difficulty does it move in *the opposite direction to its own inclination*… For larger circles move more swiftly from an equal power, and they move burdens, because the angle of a larger circle has a certain *nod* compared to the angle of a smaller circle: just as diameter is to diameter, so is the

[Moreover, where the body *turns*, the mover impels it. Therefore, when the diameter of the circle lies straight on the plane, the circle, at the point where the plane touches it, equally determines the weight on both sides of the diameter. But when it moves, it immediately moves towards that point, as if *swaying* in that easy motion, propelled more easily towards the front. For indeed, everything moves more easily toward where it inclines. Since if it were to move in the *opposite direction to where it turns*, it would move with difficulty… For larger ones move more swiftly from an equal force, and they move weights, because the angle of a larger circle has a certain *nod* to the angle of a smaller one. And just as diameter is to diameter, so is every larger circle to a smaller one. But there

[158] 852^{b}, 12-15.

[159] *Quaestiones Mechanicae… Leoniceno interprete* cit., p. 1189; the italics are ours. Fausto had resorted, again, to *momentum*: "Siquidem et quae contra momentum suum propellunt… [Indeed, also those things which propel against their own momentum…]" (Vat. Urb. Lat. 1321, c. 9*v*).

larger circle to the smaller. But there are infinitely smaller ones. If, however, the circle has an inclination towards another *nod*, it is similarly easily movable. Indeed, the circle has another *inclination*, and so do the things moved by the circle.]

are infinitely smaller ones. If, however, the circle also has a *nod* towards another, similarly, the circle is easily movable, and those things moved by the circle also have another *inclination*.]

Nutus [nod] thus took the place of *momentum* to express *ῥοπή*. Other terms also appear: both Thomaeus and Monantheuil use *nutare* [to nod] and *vergere* [to turn] to translate *ῥέπω*. Monantheuil only once utilizes *repens* for *ῥέπον*, recalling the translation of Archimedes's *Equilibrium of Planes* as suggested by Moerbeke (*de plani aequerepentibus*)[161]. Again, in one case, Thomaeus alternates – probably to avoid excessive repetitions – *inclinatio* [inclination] with *nutus*.

It remains to be clarified why, after the initial use of *momentum*, the term no longer occurs even once either in the Thomaeus translation or in that of Monantheuil, who certainly knew the predecessor's version[162]. Giving a convincing answer is **78**
not easy. It will be noted, however, that if it is true that the occurrences of *ῥοπή* in the recently cited passage all indicate an "inclination" to move produced by the weight, nevertheless of such "inclination", instead of the direct proportionality with the weight (i.e., the "intensity", as in the opening passage of the *Questions of Mechanics*), only the "direction" (*deorsum*) is considered and emphasized. The difference is subtle, but undeniable. And it cannot be ruled out that the translators, being aware, decided to avoid any possible ambiguity, rendering the occurrences of *ῥοπή* in the two passages with different terms. This means that, both in Thomaeus and in Monantheuil, *momentum* appeared to be a suitable term to express the specific "inclination" or *natural* "motive force" that heavy bodies possess by virtue of their own weight. It is, essentially, the same meaning that *ῥοπή* presented in a famous context of Aristotelian *Physics*, of which we will focus on shortly[163].

If the Latin translation of Thomaeus had enormous influence, no less was the influence exerted by Alessandro Piccolomini's famous *Paraphrasis* on the *Questions of Mechanics*[164], and by the vernacularization of the *Paraphrasis*, carried out in

---

160 *Mechanica graeca* cit., p. 105; the italics are ours.

161 Cf. *supra*, pp. 46-47.

162 In fact, he explicitly states it in the comment to *Caput IX* (*Mechanica graeca* cit., p. 106). Instead, Fausto usually translated *ῥοπή* with *momentum*, but his version circulated very little.

163 Cf. *infra*, paragr. 5.

164 *In mechanicas quaestiones Aristotelis paraphrasis paulo quidem plenior* cit. On Piccolomini and his activities in Padua, cf. F. V. Cerreta, *Alessandro Piccolomini*, Siena 1960 (especially pp. 19-48). For an evaluation of Piccolomini's scientific contribution, see Rufus Suter, *The scientific work of Alessandro Piccolomini*, "Isis", LX (1969), pp. 210-222. For the Paduan environment and the mechanical interests inspired by the tenth book of Vitruvius's *De architectura*, from which Piccolomini's paraphrase of the *Questions of Mechanics* drew inspiration, cf. P. L. Rose and S. Drake, *The pseudo-Aristotelian Questions of Mechanics* cit., pp. 81-85. A second edition of the *Paraphrasis*, without any significant modifications, was published by Curzio Troiano in Venice in 1565.

1582 by Oreste Biringucci[165]. Although it was a paraphrase and not a translation, Piccolomini's Latin text closely followed the treatment of the pseudo-Aristotelian writing, respecting its division into 35 questions. Let's see how he responded in the two contexts of the *Questions* examined above.

In the first, he preserved the precise distinction of the Greek text between *ῥοπή* and *βάρος* [burden], rendered by Thomaeus with the binomial *momentum-pondus*:
79 "ea autem sunt, tum quibus maiora a minoribus superantur, tum etiam quae et si modicae virtutis sint, graviora tamen commovent [and these are things, both by which greater ones are overcome by smaller ones, and also those which, even if of moderate strength, still move heavier things]"[166]. Piccolomini therefore preferred the more generic *virtus* of *momentum* to render *ῥοπή*. The vernacular translator, in turn, rendered *virtus* with "forza [force]", literally translating the rest of the passage[167].

In the second, the Archbishop of Patras essentially followed the model of the Thomaeus, whose translation he knew[168], using *nutus* [nod] and alternating the verb forms *nutare* [to nod], *inclinare* [to incline], *vergere* [to turn]:

Huc accedit quod quo *nutat* pondus eo facilius motor movet. movens autem quae rotunda sunt *nutu ponderis* iuvatur semper; nam cum ad perpendiculum super planum circuli dimetiens fuerit, planum in puncto fere contingente circulo; aequale utrinque pondus a dimetiente relinquetur. Tunc ergo, quamvis a modica admodum potentia, huiusmodi orbiculatum corpus, quia in puncto fere planum tangit, cui innititur, moveri poterit; cum primum tamen deiecta linea dimetiente, ex illo puncto, impulsum fuerit, illico, quocumque deverget superior dimetientis pars, eodem necessario sese *inclinaverit* contiguae partis pondus, atque ad hanc contiguae alterius partis; et sic deinceps. ita quod succedentibus semper partibus, ad eas quae praecesserint, *nutum facientibus*, nullo negocio inceptus motus a minori etiam potentia quam illa fuit quae a primo impulerit, perseveranter perpetuabitur. Quo enim unumquodque *suo nutu vergit*, eo facillime motum recipit: quemadmodum, vice versa, difficulter *ad contrariam sui nutus partem* moventur omnia[169]. [To this it is added that the mover moves more easily toward that in which the weight *nods*. In moving round things, the *nod of the weight* always assists; for when the diameter of the circle is perpendicular above the plane, it leaves equal weight on both sides. Therefore, although by a very small power, such a circular body, as it touches at a point at which supports it, is able to move; for as soon as the previously drawn diameter line, from that point, is impacted upon, immediately, the

[165] *Parafrasi di Monsignor Alessandro Piccolomini Arcivescovo di Patras sopra le Mecaniche d'Aristotele, tradotta da Oreste Vannocci Biringucci, Gentiluomo Senese*, by Francesco Zanetti, in Rome 1582. Biringucci added his own *Introduction* and a final chapter, dedicated to the examination of the screw, left out in the pseudo-Aristotelian treatise. On Biringucci's version and on the dependence of the final chapter on Guidobaldo dal Monte's *Liber Mechanicorum*, cf. P. L. Rose, S. Drake, *The pseudo-Aristotelian Questions of Mechanics* cit. p. 85. On Oreste Vannocci Biringucci, who must obviously be distinguished from Vannoccio Biringucci, author of *De la Pirotechnia*, cf. the entry of the *D.B.I.*, X (1968), pp. 624-625 (edited by C. Perina), to which we refer for bibliographical information.
[166] *In mechanicas quaestiones… paraphrasis* cit., c. VI*r*.
[167] "Con questa dunque, o scientia od arte meccanica restiamo superiori in quelle cose nelle quali dalla natura saremmo sopraffatti, cioè nel superare le cose maggiori col mezo delle minori, et in quelle che, essendo di poca forza e gravezza, non dimeno muoveno cose più gravi [Therefore, with this science or mechanical art we remain superior in those things in which we would be overwhelmed by nature, that is, in overcoming greater things by means of smaller ones, and in those which, being of little strength and weight, nevertheless move more heavy things]" (*Parafrasi* cit., p. 15).
[168] In fact, he mentions it in the *Proemio* (cf. P. L. Rose, S. Drake, *The pseudo-aristotelian Questions of Mechanics* cit., pp. 83-84).
[169] *In mechanicas quaestiones... paraphrasis* cit., p. XXXr-v; the italics are ours.

upper portion turns away, and to the same purpose, it necessarily *inclines* the adjacent weighted portion towards the other part; and so forth. In such a way, with the parts always succeeding, toward those which have preceded, *producing the nod*, the motion is perseveringly perpetuated without difficulty by a lesser power than that which initially set it in motion. For things most easily receive motion in the direction that *it turns in its nod*; just as, conversely, all things move with difficulty *in the opposite side to its nod.*]

Biringucci, in vernacularizing this passage, rendered *nutus* [nod] with "inclinatione [inclination]", *nutus ponderis* [nod of the weight] (the only significant terminological innovation introduced by Piccolomini) with "inclinatione del peso [inclination of the weight]", *ad contrariam sui nutus partem* [to the opposite side of its nod] with "verso la parte contraria all'intrinseca sua inclinatione [towards the part contrary to its intrinsic inclination]", translating *nutare* [to nod] and *vergere* [to turn] with "inclinare [incline]". Furthermore, in one case he introduced "impeto [impetus]"[170].

Another translation of the *Questions of Mechanics* which deserves attention **80**
is that "di greco in volgare idioma [of Greek into the vernacular]", performed by Antonio Guarino, Superintendent of the fortifications of Alfonso II d'Este, Duke of Modena[171]. Guarino followed the version with the *Dechiarationi* of the most difficult places of the pseudo-Aristotelian *Questions*[172]. In Guarino's version, which

[170] "A questo si aggiunge che verso quella parte il motore muove più facilmente alla quale il peso è inclinato e pronto ad andare, ma che muove cose rotonde dal inclinatione del peso loro è sempre aiutato; perché quando il diametro del circolo sarà sopra alla superficie piana a piombo, sarà il piano tocco dal circolo quasi in un punto, resterà dall'una parte e dall'altra del diametro peso eguale. Allhora da forza molto piccola potrà esser mosso questo corpo rotondo, perché tocca il piano quasi in un punto sopra del quale si posa, che spinto sarà tal corpo in qual si voglia banda, subito, rimosso il diametro di quel punto, la sua parte superiore s'inchinarà alla banda medesima e seco il peso della parte contigua, appresso della quale seguirà il peso dell'altra parte contigua, e così di mano in mano, talché, succedendo sempre le parti *et aggiugnendo gravezza* alle precedenti *o vogliamo dir impeto*, con pochissima fatica il principiato moto, ancor con minor potenza di quella che prima lo concitò, continuamente si perpetuarà, perché qual sia cosa facilissimamente verso quella banda riceve il moto verso la quale è pronta ad andare per propria inclinatione; come per contrario difficilmente si muoveno tutte le cose verso la parte contraria all'intrinseca sua inclinatione [To this it is added that the mover moves more easily toward the side to which the weight is inclined and prepared to go, and yet in moving round things, they are always aided by the inclination of their weight; because when the diameter of the circle is plumbed above the flat surface, the plane touched by the circle at almost a single point, the same weight will remain on both sides of the diameter. Then this round body can be moved by a very small force, because it touches the plane almost at a point on which it rests, so that when this body is pushed in any direction, immediately, once the diameter of that point is removed, its upper part it will bow to the same side and with it the weight of the contiguous part, after which will follow the weight of the other contiguous part, and so forth in succession, so that, with the parts always succeeding *and adding heaviness, or we say impetus*, to the previous ones, with very little effort, the initiated motion, even with less power than that which first stimulated it, will continuously perpetuate itself, because things readily receive motion towards that side which it is prepared to go by its own inclination; just as, on the contrary, all things move with difficulty towards the side contrary to their intrinsic inclination]" (*Parafrasi* cit., pp. 55-56); the italics are ours.

[171] *Le Mechanice d'Aristotile, trasportate di greco in volgare idioma…*, Printed by Andrea Gadaldino, in Modona 1573. This is a very rare print. I use the Vatican Library copy, which presents a handwritten numbering of the papers. On Guarino's version, cf. P. L. Rose, S. Drake, *The pseudo-Aristotelian Questions of Mechanics* cit., p. 97.

[172] *Dechiaratione sopra le Mechanice d'Aristotile volgarizzate*. Printed by Andrea Gadaldino, in Modona 1574 (I use the copy of the Bibl. Vaticana, with handwritten numbering of the folios).

never uses *moment*, there is no shortage of interesting aspects. Let's see how he rendered the opening passage of the *Questions of Mechanics*.

> Per arte noi vinciamo le cose, che da natura siamo superati. Di questa sorte sono quelle nelle quali le cose minori vincono le maggiori, et quelle che avendo *poco moto e tempo* muovono pesi grandi[173]. [By craft we overcome things, that which surpasses us in nature. Of this sort are those in which smaller things overcome larger ones, and those which, having *little motion and time*, move large weights.]

It is easy to recognize that the good translator found himself considerably confused by this passage of the Greek text, especially in relation to ῥοπή. He then resorted to the available aids, i.e., the Latin version of Thomaeus, in which he verified the use of *momentum*. The confusion, at this point, must have even increased, probably as a consequence of his unawareness of the meaning that *momentum* takes on in this context. Not considering himself authorized to transliterate, Guarino therefore decided to use an equivalent expression and this – based on the idea he had of *momentum* – he believed to be "motion and time". In essence, the Modenese engineer evidently considered *momentum* (and ῥοπή) to have two fundamental values, which were those, as we have seen, most widely attested by tradition: "small movement" and "instant of time". Motivated by this awareness, he therefore did not hesitate to translate ῥοπή by proposing both meanings together. Naturally, the meaning underwent a radical distortion. In fact, small weights can only move large
81 ones with *much* motion and *much* time, when, and under the conditions that, they are equipped with great "virtual" speed and carry out the operation in not so rapid times[174].

The Modenese translator followed Thomaeus in the second of the passages of the *Questions of Mechanics*, of which we have focused our attention, using "inclination" to render the different occurrences of ῥοπή[175]. Not even in the *Dichiarazioni*, where Guarino shows that he has perfectly understood why the same weight can balance different counterweights in the steelyard[176], is there any use of *moment*.

---

[173] *Le Mechanice* cit., c. 3*r*; the italics are ours.

[174] See the preamble of *Mecaniche*, in which Galileo warns against those who claim to deceive nature by promising to multiply force with machines, without losing speed and time (*G.G.*, II, pp. 155-156). It should be noted that Guarino was much more successful in translating ῥοπή in another context, which, not by chance, Thomaeus had utilized *onus* instead of *momentum*: "Se adunque sarà posto il tratto nel β, il β sarà ove è lo ε e il γ dove è il ζ, di modo che la linea che dividea la bilanza in due parti, prima era il $\delta\mu$ della sua perpendiculare che postovi il tratto sarà il $\delta\beta$... [Therefore, if the piece is placed at β, β will be where the ε is and γ where ζ is, so that the line that divided the balance into two parts, previously the $\delta\mu$ of its perpendicular, once the feature is placed, the line will become $\delta\beta$...]" (*Le Mechanice* cit., c. 5*r*). The "tratto [piece]" perfectly conveyed the meaning of ῥοπή, not a simple "weight", but a weight capable of tipping the balance.

[175] "Più facile è da essa mosso dal motore all'innanzi, essendo che ciascuna cosa è facile ad esser mossa al luogo al quale essa inclina, sì come al contrario della sua inclinatione è difficile [It is easier for it to be moved forward by the mover, since each thing is easy to be moved to the place to which it inclines, just as, contrary to its inclination, it is difficult]" (*Le Mechanice* cit., c. 7*r*).

[176] Here is how Guarino explained this principle: "il marco della statera il quale di peso è sempre il medesimo e lieva tanto maggiore peso quanto viene tirato più a dietro, di modo che la linea dal centro sopra il quale egli si muove sia fatta più lunga, e ciò fa egli sempre proportionalmente per le noti et segni delle libre o pesi, come si vede. Però dice l'authore che quella ragione e proportione è servata

At this point, it is worth observing that in the Greek text of *Questions of Mechanics*, alongside the concept of "inclination" produced by weight (*ῥοπή*) and proportional to it, the mechanical principle was illustrated on several occasions – as we have already confirmed – according to which a small force or weight is able to balance or overcome a greater force or weight, if equipped with a superior "virtual†" speed. This concept, which today we define as a "static" moment, was never explicitly expressed in the Greek text by *ῥοπή*. To clarify this, the author of the *Questions* always resorted to complex periphrases, demonstrating that he did not have a term capable of adequately indicating it. In fact, here is how, in the *Questions*, the effect of distance (and, therefore, speed) in the lever was explained:

ὅ οὖν τὸ κινούμενον βάρος πρὸς τὸ κινοῦν, τὸ μῆκος πρὸς τὸ μῆκος ἀντιπέπονθεν. αἰεὶ δὲ ὅσῳ ἂν μεῖζον ἀφεστήκῃ τοῦ ὑπομοχλίου, ῥᾶον κινήσει ... ὥστε ἀπὸ τῆς αὐτῆς ἰσχύος πλέον μεταστήσεται τὸ κινοῦν τὸ πλεῖον τοῦ ὑπομοχλίου ἀπέχον[177]. [so the moving weight is opposed to the mover as length to length. But always, the further away from the fulcrum, the easier it will move… thus, from the same force, the mover will displace more when it is farther from the fulcrum.]

Thomaeus followed the Greek text to the letter, repeating the image of the same **82**
force that "muove di più [moves more]" (that is, faster) the greater its distance from the fulcrum:

quod igitur motum pondus ad movens longitudo patitur ad longitudinem semper autem quanto ab hypomochlio distabit, tanto facilius movebit... quare ab eadem potentia plus separabitur movens illud quod plus ab hypomochlio distabit[178]. [so, a moved weight is affected by the mover as length is to length, but always the farther it is from the fulcrum, the easier it will move… wherefore, from the same power, the mover that is farther from the fulcrum will be displaced more.]

"Facilius movebit [move more easily]" successfully rendered the *ῥᾶον κινήσει* [move easily] of the Greek text. The solution offered by Monantheuil was essentially

---

dalla maggior linea verso la minore [the mark (i.e., the counterweight) of the steelyard is always the same in weight and lifts a greater weight the further it is pulled away, so that the line from the center over which it moves is made longer, and this it always does proportionally according to the notes and signs of the pounds or weights, as can be seen. However, the author says that that cause and proportion is served by the major line towards the minor]" (*Dechiarationi* cit., c. 15*v*).

† The text originally indicated "visuale [visual]," which is believed to be a typographical error intended to be "virtuale [virtual]."

[177] 850^{b}, 1-3, 5-7. And, further on, the author once again resorted to periphrasis to express the concept: *θᾶττον δὲ κινοῦται καὶ πλέον απὸ τῆς αὐτῆς ἰσχύος αἱ τῶν μειζόνων κύκλον ἢ αἱ τῶν ἐλαττόνων* [Those of the larger circles move faster and more from the same force than those of the smaller ones] (852^{b}, 15-16).

[178] *Quaestiones Mechanicae…Leoniceno interprete* cit., p. 1187. The idea of "muovere più velocemente [moving faster]" is rigorously adhered to again by Thomaeus in other contexts (see the one cited in the previous note): "Celerius autem et plus ab eodem vi lineae maiorum sunt circulorum ab eadem vi quam quae minorum. Ab eadem vi plus transfertur id extremum, quod longius e centro distat [But the lines of the larger circles are more rapid and more from the same force than those of the smaller ones. That end, which is farther from the center, is moved more by the same force]" (*ibid.*, p. 1191). The solution proposed by Fausto in this case was not very different: "Citius atque etiam plus pari de vi feruntur quae maiores circulus ducunt [Those which trace larger circles are carried faster and also with more than their equal share of force]" (Vat. Urb. Lat. 1321, c. 11*r*).

no different, as he limited himself to replacing Thomaeus's *separare* [to separate] with a more clear *mutare* [to change][179].

The analysis of the concept, as proposed by Piccolomini's *Paraphrasis*, is more complex. The Archbishop of Patras, free from obligations of literal compliance with the text, lingered at length on the principle of virtual speeds, which he rightly considered central:

> Quo magis linea a centro distat, *eo celerius* quia maiorem describit circulum *ab eadem vi movetur* atque ob id facilius; necesse est quod quamvis exigua sit vis, et quasi exiguum pondus, tamen quia longitudine iuvatur lineae maius pondus moveat, quam illa sit; quod enim est pondus motum ad ipsum movens, illud versa vice longitudo patitur a longitudine. hoc est in quo pondus excellit pondus, longitudo vincitur a longitudine. pondus enim motum, quia maius est, *magis moveret ex sui natura* quam virtus longiorem a centro deprimit lineam; propterea, cum ab hac longitudine, non solum *eius infirmitas* compensetur, sed etiam iuvetur magis, nil mirum erit si pondus illud *longe minus movere* possit; atque ob id eiusdem potentiae eo *facilius movebit* movens quo magis a fulcimento distat[180]. [The further the line is from the center, *the faster* because it describes a larger circle while *moved by the same force* and therefore more easily; it is necessary that, even though the force is small, and as if the weight were small weight, nevertheless because it is helped by the length of the line, it moves a greater weight than it is; for that which the motion of the weight is to that of the mover, so too does the respective length undergo the other. This is where weight excels over weight, length is overcome by length. For the weight in motion, because it is greater, moves *more from its own nature* than the force which depresses by having a line longer line from the center; therefore, since from this length, not only is *its weakness* compensated, but it is also more assisted, it will not be surprising if that weight can *move much less*; and for this reason, the mover will move more easily, the more it is distant from the fulcrum, of the same power.]

Piccolomini has therefore clearly distinguished, on the one hand, the thrust of the greater weight, which "ex sui natura [by its own nature]" would be superior,
83 and on the other hand, the compensation of the *infirmitas* [weakness] of the lesser *vis* [force] or *virtue* [strength], determined by the greater distance from the fulcrum and, therefore, by the more intense speed. However, not even Piccolomini is able to express the difference between the "natural" inclination and the "mechanical" or "artificial" one with simple and precise terminology. The same goes for the vernacularization of Guarino[181] and for Biringucci, who strictly followed the Latin

---

[179] "Quoniam radius maior maiorem describit circulum. Itaque ab eadem vi plus mutabitur movens illud, quod plus distat a pressione [Since a larger radius describes a larger circle. Therefore, by the same force, moving that which is more distant from the pressure will be more changed]" (*Mechanica graeca* cit., p. 58). In the other context (see previous note) Monantheuil literally took up Thomaeus's solution: "At et celerius movetur et plus ab eadem vi lineae maiorum circulorum quam minorum. Ab eadem enim vi extremum quod longius est a centro plus transfertur [But it moves more quickly and with more force along the lines of the larger circles than the smaller ones. For by the same force the end which is farther from the center is moved more]" (*ibid.*, p. 118).

[180] *In mechanicas quaestiones…paraphrasis* cit., cc. XXIII*v*-XXIV*r*; the italics are ours. In the second context cited above. (cf. notes 178 and 179), Piccolomini adopted the same solution: "Quoniam igitur, quo longiora exeunt a centro lineae eo celerius ac facilius ab eadem vi moventur ac circumagantur, cum maiorem describant ambitum [Because, therefore, the farther the lines go from the center, the more quickly and easily are they moved and revolved by the same force, since they describe a larger circle]" (*ibid.*, c. XXXV*v*).

[181] "Oltre di questo più velocemente da uno istesso peso è mossa la maggiore delle linee che vengono dal centro; sono però tre le cose che sono intorno alla leva, l'hypomochlio, che è il sparto e centro, e le due cose gravi, quella che muove e quella che è mossa; quello però che la cosa grave mossa compatisce

text. At least in one case, however, the translator of the *Paraphrasis* allowed himself an interesting liberty, mentioning the different "potentia [power]" that the same weight can exert[182].

In Piccolomini's *Paraphrasis*, however, *momentum* appears, albeit only once. The occasion was provided by the description of the lever and the exposition of the reasons that allow it to be traced back to the same principles as the balance.

> Vectis autem partes, sicut etiam librae sunt lineae a centro exeuntes ac circulum ex motu ponderum describentes et quemadmodum in libra duo debent imponi pondera, in singulis lancibus quidem alterum; sic etiam in vecte duo sunt pondera, alterum quod elevari debet, alterum vero vis ipsa movens; quae, cum deprimat ac momentum faciat, loco ponderis existimanda est[183]. [The parts of the lever, just like the scales, are lines emanating from the center and describe a circle by the motion of the weights, and just like two weights placed in the scales, in each of their respective platters; so also there are two weights on the lever, one that is raised, while the other is the moving force; which, when depressed down and *makes momentum*, should be considered in place of weight.]

The lever, to which Piccolomini refers here, is a balance in which one of the weights is replaced by a *vis movens* [moving force] which pushes downwards and exerts *momentum*, exactly as a weight would. The *momentum facere* [to make momentum] of this context refers to the *nutum facere* [to make a nod] previously observed and suggests an equivalence. However, in this case, the idea of movement, of displacement, appears more explicit. Here *momentum* has the value not of a generic inclination produced by the weight, but of "force" whose intensity does not depend exclusively on the weight, but also on the point of application (or how **84**
fast the weight would act if the system were moving). That is, it is the "technical" meaning, the same that the term presents in the two contexts of Vitruvius examined previously. The Archbishop of Patras, using *momentum* to indicate this concept, at

---

da quella che muove, la longhezza compatisce dalla longhezza, peroché sempre quanto serà più remota dall'hypomochlio muoverà più facilmente, e la causa è l'antedetta, cioè che la più remota dal centro descrive maggiore circolo, onde la cosa che muove, con l'istessa forza moverà maggiormente come più serà remota dall'hypomochlio [Besides this, the greater of the lines coming from the center is moved more quickly by the same weight; however, there are three things that are around the lever, the fulcrum, which is the pivot and center, and the two heavy objects, the one that moves and the one that is moved; however, what the moved heavy object bears from the mover, length bears from length; therefore, the more remote it is from the fulcrum, the more easily it will move, and the reason is the aforementioned one, namely that the one more remote from the center describes a larger circle, so that the moving object will move more with the same force as it is more remote from the fulcrum]" (*Le Mechanice* cit., cc. 5*v*-6*r*). Not even Guarino introduced significant innovations in the second of the contexts cited. "Peroché più tosto muovono e con possanza maggiore sono mosse da una medesima forza le linee delli maggiori circoli che quelle delli minori [Because the lines of the larger circles move faster and are moved with greater power by the same force than those of the smaller ones]" (*ibid.*, c. 7*v*).

[182] "Tanto più velocemente e, in conseguenza, più facilmente dalla medesima forza è mossa… [The faster and, consequently, the more easily it is moved by the same force…]" (*Parafrasi* cit., p. 45). Furthermore, Biringucci rendered *infirmitas* with "debolezza [weakness]" and "nil mirum erit si pondus illud longe minus movere" with "che meraviglia dunque sel maggior peso havrà molto minor potentia? [what marvel then if greater weight will have much less power?]" (*ibid.*, p. 46). In the second context cit., Biringucci literally kept Piccolomini's Latin text: "perciò più facilmente e più presto si potranno muovere [therefore they will be able to move more easily and more quickly]" (p. 64).

[183] *In mechanicas quaestiones... paraphrasis* cit., c. XXII*r*-*v*; the italics are ours.

least once, confirms that, in the mid-sixteenth century, the process of technicalizing *momentum* as a mechanical term had begun. Nevertheless, quite a few hesitations remain, which limit the use of the term in this precise meaning to very few cases.

Biringucci, in translating into vernacular, did not adhere strictly to Piccolomini's Latin text and rendered the expression "cum deprimat ac momentum faciat loco ponderis existimanda est [when it depresses and creates momentum, it is considered as a weight]" with "deprimendo e aggravando, si considera come peso [depressing and aggravating, it is considered as a weight]"[184]. *Facere momentum* [to make momentum] meant something different from "aggravare [aggravating]" and was much more precise; but evidently Biringucci did not consider it legitimate to transfer the Latin expression directly into the vernacular. As we have already seen, "fare momento [making momento]" recalls the "fare impeto [making impeto]" used by Galileo[185].

Even the examination of the versions and compilations of the *Questions of Mechanics*, therefore, leads to the conclusion – as confirmed by the investigation conducted in the preceding paragraphs – that the use of *momentum*, both in the meaning of "inclination" produced by weight and as a "technical" term in mechanics (weight × distance, or weight × speed), was anything but common in the 15th and 16th centuries. If the value of *momentum* as a variable effectiveness of weight now seems fixed, nevertheless, the use of the term with this specific connotation remains exceptional. To express this fundamental concept, in fact, other more generic, non-technical terms such as *nutus*, *inclinatio*, *vis*, *potentia*, *virtus* etc. are preferred to *momentum*, or recourse is made to phrases that are not always perspicuous and crystal clear.

On the other hand, this widespread hesitation to use *momentum* in discussions of statics to indicate the variation of the driving force generated by the same weight appears to be a general phenomenon characteristic of all authors, who, under the influence of the Archimedean tradition, the *de ponderibus* texts, Vitruvius's *De architectura* or the *Questions of Mechanics*, dealt, in the 15th and 16th centuries, with the problems of statics, focusing in particular on the laws of balance.

If we are not mistaken, only in two passages written in the *recto* of folio 1 of
85 the *Codice sul volo degli uccelli*, on a page assignable to 1505, Leonardo used our term. In both contexts, *momento* was explicitly referring to the balance.

Quanto il polo della bilancia sarà di minor grossezza, tanto la sua ventilazione fia di minor *momento*.

Quanto la bilancia è più lunga, tanto è di minor *momento*, perché il polo, a tanta lunghezza, viene a essere più sottile che alla bilancia corta, essendo all'una e all'altra bilancia il polo d'una medesima

---

[184] "E sì come nella libra si deven por due pesi, uno per ciascuna bilancia, cosi sono due pesi nella lieva, uno che si deve inalzare, l'altro è la forza movente, la quale, deprimendo e aggravando, si considera come peso [And just as in the scales two weights must be placed, one on each of the two platters, thus there are two weights on the lever, one which must be raised, the other is the moving force, which, depressing and aggravating, is considered as a weight". (*Parafrasi* cit., p. 45).

[185] Cf. *supra*, paragr. 3, note 141.

grossezza[186].

[However much less the pole (tongue) of the scale extends, the less the *moment* will fluctuate. However much longer the balance, the lesser its *moment*, because the pole (tongue), at such length, is smaller relative to a small balance, the pole (tongue) being of the same size on both balances.]

Let us first try to correctly interpret these passages, the meaning of which is anything but crystal clear. "Polo" corresponds here to the fulcrum, or "lingua [tongue]", of the balance. "Ventilazione" indicates the fluctuation of the arms when the balance condition is altered. Leonardo means that the smaller the tongue of the balance is, the more easily (i.e., with the application of a smaller force) it will lose the equilibrium condition. "Di minor momento", means that to cause the collapse of a balance with a small pole it is sufficient to apply a smaller weight.

There is more. Leonardo was about to write *movimento* [movement], as he had already written *mov*, when his hand stopped with sudden regret. He deleted the *v* with a transversal stroke, completing the word with *momento*. This helps us understand the meaning that Leonardo assigned to this term: *moment* identified with the movement produced by a weight applied in a certain way to the balance; the terms also appeared interchangeable.

The second part of the passage established another condition – partly dependent on the first – capable of determining the greater sensitivity of the balance: the longer its arms are, the more precise the instrument will be (a smaller *moment*, i.e., a smaller weight, will be enough to remove it from equilibrium).

It is difficult to escape the impression that Leonardo here is under the influence of the first of the pseudo-Aristotelian *Questions of Mechanics*[187]. In this passage of the *Codice sul volo*, like in the Question, an effort is made to clarify why balances
with longer arms are more "certain",[188] in other words, more sensitive, with lesser **86**
moment. Against this evident analogy lies the difficulty found in admitting that

[186] *Il Codice sul volo degli uccelli nella Biblioteca Reale di Torino.* Diplomatic and critical transcription by A. Marinoni. National edition of Leonardo da Vinci's drawings and manuscripts, edited by the Vinciana National Commission, Giunti-Barbèra, Firenze 1976, pp. 26-27; the italics are ours. I owe the indication to the courtesy of Prof. Marinoni.

[187] In the Thomaeus version, the first of the *Questions* was introduced in this way: "quamnam ab causam exactiores minoribus maiores sint librae [for what reason are greater scales more exact than smaller ones]" (*Quaestiones Mechanicae… Leoniceno interprete* cit., p. 1184).

[188] In Daniele Barbaro's commentary on the vernacularization of Vitruvius there is an explicit reference to the first Question: "Quando adunque questi raggi sono eguali di grandezza e di peso, la teste loro, essendo la bilancia sospesa, non piegano una più dell'altra, ma sono egualmente distanti dal piano; ma quando se le dà peso da uno de' capi, forza è che trabocchi la bilancia, et più presto traboccherà et *con minor peso* quanto il raggio sarà maggiore et il peso più lontano dal centro; per la sopradetta ragione però dicesi nelle Meccaniche che le bilancie che hanno i fusti maggiori sono più certe, cioè più presto e *con minor peso* bilanciano et più certo dimostrano il peso, perciò che per ogni lieve aggiunta si muovono et in egual, o minore, spatio di tempo fanno maggior spacio di luogo [Therefore, when these rays are equal in size and weight, their ends, the balance being suspended, do not bend one more than the other, but are equally distant from the plane; but when weight is given to it at one end, its power overcomes the balance, and it will overcome more quickly and *with less weight* the greater the radius and the further the weight is from the center; for the aforementioned reason, however, it is said in the Mechanics that the balances which have larger shafts are more certain, that is, they balance more quickly and *with less weight* and demonstrate the weight more certainly, because for every slight addition they move and in

Leonardo could have known the pseudo-Aristotelian writing: until 1517, as far as we know, it could, in fact, only be read in the Greek text, that is, in a language that was not accessible to him[189]. Leonardo probably depended here on some Latin or vernacular text in which the *Question* was referring. Given the rarity of the use of *moment*, in this sense, at the beginning of the 16th century, it would be extremely interesting to be able to identify Leonardo's source, from which he most likely drew our term. It would not otherwise be possible to explain why in the other numerous pages, preceding and subsequent to this one of the *Codice sul volo*, dedicated by Leonardo to statics and the laws of balance, no other example of *moment* is found[190].

Tartaglia also dedicated some pages of his *Quesiti et inventioni diverse* to the attempt to establish "per che causa le maggior libre, over bilance, sono più diligente delle minore [why the greater scales, or balances, are more sensitive than lesser ones]"[191]. But here, like in the rest of his work, the Brescian scientist does not use *moment.*

The rare presence of *moment*, in its "technical" meaning, is confirmed by the almost complete absence of the term by authors and in writings on statics and mechanics between the 15th and 16th centuries. It is not found in Nicholas of Cusa's *Idiota*[192], nor in Alberti's *De re aedificatoria*[193]. It appeared, as we have seen, only on one occasion with this meaning in Valla's *De expetendis*[194]. It was
87 entirely absent in that group of technical treatises, characteristic of the 1500s, that illustrated fortifications, such as Lorini[195], mining science, such as Biringucci's *Pirotechnia*[196], or operating machines, such as Ramelli[197] and Zonca[198]. It was

---

equal, or less, intervals in time while traversing greater distance]" (*I dieci libri dell'Architettura* cit., p. 259; the italics are ours). Note how that *minor peso* [lesser weight] recalls the Leonardian *di minor momento* [of lesser moment].

189 The first Latin version was, in fact, performed on the one given by Fausto (cf. *supra*, note 149).

190 Numerous pages of the *Codice Atlantico*, the *Codice Arundel*, the *Codice di Madrid I*, the *Ms. L* of the Institute of France, contain Leonardo's notes on balances. Cf. R. Caverni, *Storia del metodo sperimentale* cit., IV, Ch. I; P. Duhem, *Les origines de la Statique* cit., and *Études sur Leonard de Vinci*, Première Série cit.; M. Clagett, *Leonardo da Vinci, D.S.B.* cit.

191 *Quesiti et inventioni diverse* cit., c. 78*v*. Tartaglia, in examining the question, distinguished two ways of considering it: that of "matematico [mathematician]", which ignores material imperfections, and that of the "fisico [physicist]", which instead takes them into account.

192 *Idiota. de sapientia. de mente. de staticis experimentis*, Edited by L. Baur, *Nicolai de Cusa omnia iussu et auctoritate Academiae Litterarum Heidelbergensis*, V, Lipsiae 1937.

193 *De re aedificatoria*..., Florentiae 1485; cf. *Alberti index, De re aedificatoria, Index verborum*..., von Hans-Karl Lücke, 2 vol., Prestel Verlag, München, II, pp. 825-826. The term *momento*, in a "technical" sense, is also absent in the *Ludi matematici* (see them in Alberti L. B., *Opere volgari*, edited by Cecil Grayson, 3 vol. Laterza, Bari 1968-1973, III, pp. 131-173).

194 Cf. *supra*, paragr. 1.

195 Lorini Buonaiuto, *Delle fortificazioni. Dove con facilissime dimostrazioni si dichiarano le Scienze delle Meccaniche*..., Rampazzetto, Venezia 1596.

196 *De la Pirotechnia libri X, dove ampiamente si tratta non solo di ogni sorte et diversità di miniere, ma anchora quanto si ricerca intorno a la prattica di quelle cose di quel che si appartiene a l'arte della fusione over gitto de' metalli*..., Venturino Ruffinello, Vinegia 1540.

197 Agostino Ramelli, *Le diverse et artificiose machine... nelle quali si contengono varii et industriosi movimenti, degni di grandissima speculatione.... Composte in lingua italiana et francese*, in the author's house, a Parigi 1588.

198 Vittorio Zonca, *Novo teatro di machine ed edificii per varie et sicure operationi*..., Published by

also missing from Giuseppe Ceredi's *Tre discorsi sopra il modo d'alzar acque da'luoghi bassi*[199], in which we find, however, some very interesting passages. Ceredi appears decidedly influenced by the *de ponderibus* tradition, of which he repeats the characteristic terminology of "heaviness according to position", showing that he understood and shared the "virtual" speeds approach[200]. In an elegant passage of the *Discorso Secondo*, Ceredi clearly distinguished, in the weight placed on a steelyard, the effect of the "gravezza naturale [natural heaviness]" (i.e., the weight) from that produced by the "anima motrice [driving spirit]". As Ceredi stated, the greater effect produced by the smaller weight, further from the fulcrum,

> non è mai col peso della *gravezza naturale* solamente, la quale sempre mira il centro del mondo, ma con *la forza dell'anima motrice*, o di qualche altra cosa *che non sia impetuosa per lo peso solo*. Et queste tali non hanno per centro altro che il centro cui sopra si muove il peso istesso, sia pure situato come si voglia; che se più da questo centro s'allontana un braccio che l'altro, minor forza od impulsione si ricerca a fare il giro[201]. [it is never with the weight of *natural heaviness* alone, which always aims at the center of the world, but with the *force of the driving spirit*, or of some other thing *that is not impetuous due to the weight alone*. And these have nothing as their center other than the center above which the weight itself moves, even if it is positioned however you want; that if one arm moves further from this center than the other, less strength or impulse is needed to make the turn.]

In Gerolamo Cardano, despite many pages dedicated to statics and the behavior of bodies on the balance, I have not found a single example of *moment* in our meaning. In the *Liber I de rerum subtilitate*, the Pavian scientist discussed, among other things, **88**
"de vacuo, motu naturali et loco [the void, natural motion and position]", constantly proclaiming the superiority of the pseudo-Aristotelian *Questions of Mechanics* over the writings of Jordanus. And from the *Questions*, he took up the method of justifying the greater effectiveness of bodies hanging further away from the fulcrum in terms of greater speed.

> Generalis igitur ratio haec sit, pondera quo plus distant a meta seu linea descensus per rectam aut obliquam, id est per angulum, eo sunt graviora[202]. [The general rule then is this, the farther the weights

---

Pietro Butelli, in Padova 1607 (2nd ed., 1621; 3rd ed., 1627).

[199] Published by Seth Viotti, in Parma 1567; on Ceredi see the recent essay by Stillman Drake, in Bert S. Hall, Delno C. West (Editors), *On pre-modern technology and science: A volume of Studies in the honor of Lynn White Jr.*, University of California, Los Angeles 1976.

[200] "Imperò che questi effetti non accadono per altra cagione che per lo moto delle statere, nelle quali il peso che si trova nel raggio maggiore, quanto sarà più lontano dal centro che lo sostiene, sempre alla proportione supererà *secondo il sito* quello che si trova nel raggio minore. Ho detto secondo il sito, perché il cerchio che fa il peso nel maggiore, quanto sarà più lontano dal centro, occuperà più gran giro; et la parte di lui in lunghezza di linea a quella che fa il raggio minore sarà meno obliqua et andrà più rettamente verso il centro del mondo... [For these effects do not occur for any other reason than the motion of the steelyard, in which the weight which is in the greater arm, the further it is from the center which supports it, will always exceed in proportion *according to the location* that which is in the lesser arm. I said according to the location, because the circle which makes the weight in the greater, the further it is from the center, will occupy a greater arc; and the part of it in length of a line to that which makes the lesser arm will be less oblique and will go more directly toward the center of the world]" (*ibid., Discorso secondo*, p. 45); the italics are ours.

[201] *Ibid.*; the italics are ours.

[202] *Hieronymi Cardani de rerum subtilitate libri XXI*..., Norimberga 1550, p. 38. For a general

are from the turning point or the line of descent, whether straight or oblique, that is, by an angle, the heavier they are.]

Cardano reiterated the same principle in the *Opus Novum de proportionibus*, this time introducing *impetus*:

Quanto magis remotum fuerit pondus ex eodem centro a recta linea, tanto maiore impetu agetur, ut ultra locum medium feratur non aequali, sed producta proportione[203]. [The more the weight is removed from the same center in a straight line, the greater the impetus urge, so that it is carried beyond the middle place not equally, but in a product proportion.]

Julius Caesar Scaliger, who contrasted Cardano's *De subtilitate* with over a thousand pages of his *Exotericae exercitationes*[204], used *momentum* only once, in the *Exercitatio de ictu*:

Equidem didiceram motum, sicut pulsum, addere ponderi. Nam et absque ictu, sola impressione plus affertur *momenti* quam quantum eius pondus efficere valet[205]. [Indeed, I had learned that movement, like a strike, adds weight. For even without a blow, more is brought about by the push of *moments* alone than what its weight alone can achieve.]

Motion produces and adds *moment* to weight. The thesis alluded to here by Scaliger is the same supported in the pseudo-Aristotelian *Questions of Mechanics*. And the distinction between *pondus* and *momentum* appears clear. We will find the term used in this meaning by Galileo: the effect of weight augments with speed[206].

Giovanni Battista Benedetti's *Speculationum liber* contains an entire section *De*
89 *mechanicis*[207]. The Venetian mathematician never used *momentum*. He expressed himself in terms of "magis aut minus grave [more or less heavy][208]" and mentioned the difference between the weight (constant) and its effect (variable) by introducing the expression *quantitas ponderis* [quantity of weight][209]. Benedetti sometimes also

orientation on Cardano's scientific interests, see the article by M. Gliozzi, in *D.S.B.*, III (1971), pp. 64-67.

[203] *Opus novum de proportionibus numerorum, motuum, ponderum, sonorum aliarumque rerum mensurandarum… Praeterea Artis Magnae, sive de regulis algebraicis liber unus... Item de Aliza regula liber...*, Ex officina Henricpetrina, Basileae 1570, *Lemma Quartum*, p. 240.

[204] *Exotericarum Exercitationes Lib. XV de Subtilitate ad Hieronym. Cardanum…*, Apud Andream Vechelum, Francofurti 1576.

[205] *Ibid., Exercitatio CCCXXXI* (*De ictu*), p. 1060; the italics are ours.

[206] Cf. *Le Mecaniche*, in G.G., II, p. 164.

[207] *Speculationum liber, in quo mira subtilitate haec tractata continentur. Theoremata Arithmetica. De rationibus operationum perspectivae. De Mechanicis. Disputatio de quibusdam placitis Arist. In Quintum Euclidis librum. Physica et Mathematica responsa per epistolas*, Apud Baretium Baretium et socios, Venetiis 1559.

[208] "Pondus magis aut minus sit grave quo magis aut minus a centro pendet aut eidem nititur; atque haec est causa proxima et per se, qua fit ut unum idemque pondus in uno eodemque medio magis aut minus grave existat [A weight is more or less heavy the more or less it hangs or rests from the center; and this is the proximate and intrinsic cause by which one and the same weight exists more or less heavy in one and the same medium]" (*ibid.*, p. 142).

[209] "Est ergo quod proposuimus verum quantitatem alicuius ponderis respectu ad eam quae est alterius debere deprehendi a perpendicularibus, quae a centro librae ad lineas inclinationis exiliunt [It is, therefore,

resorted to the terminology of the *de ponderibus* tradition, using *situs* [position]: "omne pondus positum in extremitate alicuius brachii librae maiorem aut minorem pro diversa ratione situs ipsius brachii [every weight positioned at the extremity of any arm of the scales is greater or less according to different positions of the arm itself]"[210].

Bernardino Baldi's *Exercitationes*, published posthumously, in which *momentum* is never used, must also be assigned to the tradition of the pseudo-Aristotelian *Questions of Mechanics*. Of the abbot of Guastalla, it will also be worth remembering his significant work of disseminating Hero's writings, of no small interest for the history of mechanics. However, in Baldi's versions of Hero's texts, *momentum* never occurs[211].

5. "Momentum" as a Natural Inclination in the Aristotelian Tradition

a) *The text of the fourth book of* "Physics"

Momentum, meaning the inclination to motion produced by weight or lightness, is present, or rather common, in some 15th- and 16th-century translations of the **90**
fourth book of Aristotle's *Physics*. Given the influence and diffusion of the work, it will be appropriate to focus on these places, which constitute a central point of the story of *moment*.

The text of the fourth book of *Physics* is inserted in the Aristotelian discussion on the void and the possibility of admitting it. The Stagirite attempts to demonstrate the impossibility of the void with arguments drawn from natural movements, which have precise directions and precise scope. In the void, without low or high, devoid

---

that we have proposed that the true quantity of a certain weight, in relation to that of another, must be found by the perpendiculars, which emerge from the center of the scale to the lines of inclination]" (*ibid.*, p. 143).

[210] *Ibid.*, p. 141.

[211] *Bernardini Baldi Urbinatis... in Mechanica Aristotelis problemata exercitationes...*, typis et sumptibus Viduae Joannis Albini, Moguntiae 1621. For Baldi's Heronian interests, see his vernacularized *Di Herone Alessandrino de gli Automati, overo machine semoventi*, Girolamo Porro, in Venetia 1589 (2nd ed. "con ogni diligenza ricorretto [corrected with all diligence]", Gio. Battista Bertoni, Venezia 1601) and *Heronis Belopoeca, hoc est Talifactiva...*, Typis David Franci, Augustae Vindelicorum 1616. For the influence of Hero, the versions of the *Spiritalia* carried out by the Ferrarese Giovan Battista Aleotti (*Gli artificiosi e curiosi moti spiritali di Herone...*, Vittorio Baldini, Ferrara 1589) and by the Urbinate Alessandro Giorgi (*Spiritali di Herone Alessandrino ridotti in lingua volgare...*, Bartholomeo and Simone Ragusij brothers, Urbino 1592). No occurrences of *momentum* are found in the two versions. On Baldi, on the Heronian "renaissance" and on its importance for the history of mechanics, see P. L. Rose, *The Italian renaissance of mathematics* cit., Ch. II, pp. 243-279. Baldi was also responsible for the version of the eighth book of Pappus's *Collectiones Mathematicae*, which however remained unpublished (*ibid.*, p. 247). Lastly, it is worth remembering that Hero is the author of a *Mechanica*, probably significantly influenced by the pseudo-Aristotelian, which is known to us through the Arabic version, the Greek original having been lost, except for a short fragment. Hero's *Mechanica* was not known in the Middle Ages and the Renaissance (see M. Clagett, *La scienza della meccanica*, trad. it. cit., pp. 33-38).

of any place, natural motion cannot occur. Emptiness is, therefore, impossible. It is at this point that Aristotle introduces another argument against the void. It is a famous text, of which commentators have paid particular attention for centuries. Galileo himself will dedicate numerous pages of his work to the refutation of the thesis that the Stagirite puts forward. In essence, Aristotle attempts to demonstrate that the admission of the void leads to absurd consequences, either if one supposes that the different natural speeds of bodies depend on the density of the media they pass through, or, instead, if one assigns the cause to proportion of their weights. As for the different thicknesses of the media (water and air, for example), they proportionally hinder the speed of the body. But the proportion of the void to any other medium is infinite, just as the proportion between 0 and a finite number is infinite. The proportion between the speed in a vacuum and that in any other medium will therefore also be infinite. It is concluded that, if the vacuum exists, the body must move with infinite speed[212].

If the other criterion for explaining the different speeds of bodies is admitted – that is, the difference in weight or lightness – the existence of the void would lead to an equally absurd situation: all bodies would have equal speed. The greater speed of heavier bodies depends, in fact – states Aristotle – on the superior ability they have to divide the medium. But, if the medium is the void, there is no division; that is, the cause of the speed differences disappears. The most interesting part of the Aristotelian text from our point of view is the one in which the refutation of motion in vacuum is introduced, starting from the second criterion of explanation of speed differences. Let's see it in the Greek text:

ὁρῶμεν γὰρ τὸ αὐτὸ βάρος καὶ σῶμα θᾶττον φερόμενον διὰ δύο αἰτίας, ἢ τῷ διαφέρειν τὸ δι᾽ οὗ, οἷον δι᾽
ὕδατος ἢ γῆς ἢ ἀέρος, ἢ τῷ διαφέρειν τὸ φερόμενον, ἐὰν τἆλλα ταὐτὰ ὑπάρχῃ, διὰ τὴν ὑπεροχὴν τοῦ βάρους
ἢ τῆς κουφότητος[213]*.... ὁρῶμεν γὰρ τὰ μείζω ῥοπὴν ἔχοντα ἢ βάρους ἢ κουφότητος, ἐὰν τἆλλα ὁμοίως ἔχῃ*
91 *τοῖς σχήμασι, θᾶττον φερόμενα τὸ ἴσον χωρίον, καὶ κατὰ λόγον ὃν ἔχουσι τὰ μεγέθη πρὸς ἄλληλα. ὥστε*
καὶ διὰ τοῦ κενοῦ. ἀλλ᾽ ἀδύνατον· διὰ τίνα γὰρ αἰτίαν οἰσθήσεται θᾶττον; ἐν μὲν γὰρ τοῖς πλήρεσιν ἐξ
ἀνάγκης· θᾶττον γὰρ διαιρεῖ τῇ ἰσχύϊ τὸ μεῖζον· ἢ γὰρ σχήματι διαιρεῖ, ἢ ῥοπῇ ἣν ἔχει τὸ φερόμενον ἢ τὸ
ἀφεθέν[214]. [for we see the same weight and body moving faster for two reasons: either because of the difference in the medium through which it moves, such as through water, earth, or air, or because of the difference in the moving object itself, if all other conditions remain the same, due to the superiority of the burden or the lightness…. for we see that larger bodies, whether due to heaviness or lightness, if other conditions are the same in terms of shape, have greater momentum over the same distance, and in proportion to their sizes relative to each other. This would be the case even through a void. But it is impossible; for by what cause would it move faster? In a plenum, it is by necessity; for the larger body divides more quickly by its strength. It either divides by its shape, or by the momentum that the moving body or the thing released has.]

As per usual, our attention focuses on *ῥοπή*, which appears twice in the second part of the Aristotelian passage. It is worth immediately emphasizing the close relationship that *ῥοπή* presents, in this context, with *βάρος* (burden) and with

---

212 *Aristotelis opera*, Edidit Academia Regia Borussica... ex recognitione Immanueli Bekkeri..., I-V, Apud Georgium Reimerum, Berolini 1831-1870, I, Δ, 214^b-217^b.

213 215^a, 25-29.

214 216^a, 13-20.

κουφότης (lightness). It will also be noted that, in the first part of the quote, we talk about *ὑπεροχὴ τοῦ βάρους ἢ τῆς κουφότητος*, namely, "dominance", "supremacy" of weight and lightness. *'Ροπή* is clearly equivalent to an "inclination", "propensity" to move with a given speed. That is, presenting substantially the same meaning that we observed in the *Commentary* of Eutocius.

Let's see how the translators and commentators reacted when faced with this passage. The most successful and widespread medieval translation of Aristotelian *Physics* was undoubtedly that of William of Moerbeke, who had revised the previous translation on the Greek text, recorded in the *Aristoteles Latinus* as *Physica Veteris Translationis* (12th century)[215]. The Dominican's translation, indicated in the *Aristoteles Latinus* as *Physica Novae Translationis*, was also used in the first printings of the Aristotelian text. It provided, for example, the text for the editions of famous commentaries such as those by Walter Burley, Giles of Rome, Albertus Magnus, Thomas[†,216], etc. This is how Moerbeke rendered the Greek text cited above.

Amplius autem et ex his manifestum est quod dicitur. Videmus enim idem et grave et leve corpus velocius ferri propter duas causas: aut in differendo per quod fertur, ut per aquam, aut per terram, aut per aerem, aut in differendo quod fertur, si alia sunt eadem, *propter excellentiam gravitatis aut levitatis*...[217].

Videmus enim maiorem inclinationem habentia aut gravitatis aut levitatis sic et alia similiter se habent **92**
figuris citius ferentia per equalem spacium finitum, et secundum rationem quam habent magnitudines ad invicem. Quare et per vacuum. Sed impossibile est. Propter quam enim causam fertur velocius? In plenis quidem ex necessitate. Velocius enim dividit ex fortitudine maior, aut enim figura dividit aut *inclinatione* quod fertur aut proiectum[218].

[Moreover, it is further evident from these things that which is said. For we see indeed that the same heavy and light body is carried more quickly for two reasons: either in the difference through which it is carried, as through water, or through earth, or through air, or in the difference in what is being carried, if other conditions are the same, *due to the superiority of heaviness or lightness*...

For we see indeed that those having greater inclination either of heaviness or of lightness so also other things similarly related in shape will carry more swiftly through an equal finite space, and according

---

[215] *Aristoteles Latinus*. Codices descripsit G. Lacombe in societatem operis adsumptis A. Birkenmajer, M. Dulong. Aet. Franceschini, "Corpus Philosophorum Medii Aevi", La Libreria dello Stato, Roma 1939 (*Pars Prior*), pp. 51-52. For the most part Moerbeke accepted the text of the *Vetus translatio* (probably 12th century), which is generally believed to be the work of James of Venice. We limited ourselves to examining Moerbeke's version, ignoring both the *Vetus Translatio* and the so-called *Physica Cremonensis* (translated from Arabic by Gerard of Cremona in the 12th century), given that it was the only one to have a certain circulation in print. On Moerbeke's notable work and for a judgment on the most relevant characteristics of his work as a translator, see the useful and recent entry *William of Moerbeke* of the *D.S.B.*, IX (1976), pp. 434-440, edited by L. Minio Paluello. In the entry there is a list of translations attributed to the Flemish Dominican and a rich bibliography.

[†] Tommaso, or Thomas, refers to St. Thomas Aquinas. Galluzzi frequently refers to him as "Tommaso" and has been literally translated throughout, particularly in the discussions on pp. 107-112. Care should be taken to prevent confusion with Thomas Bradwardine, Tommaseo, and Tomeo

[216] For specific information, cf. *infra*.

[217] I quote from the printed edition of Walter Burley's Commentary, which presents the text of the *Physics* in Moerbeke's *versio*: *Expositio excellentissimi Gualterii de Burley Anglici in libros octo de physico auditu Ar. Stageritae dilligentissime emendata*, Impressa arte et expensis Jeronimi de Duranti..., Papiae 1488, cc. non num., text 71; the italics are ours.

[218] *Ibid.*, text 74; the italics are ours.

to the ratio which magnitudes have to one another. Likewise, through a vacuum. But it is impossible. For what cause, indeed, would it be carried more swiftly? In a plenum, indeed, out of necessity. For it divides more swiftly out of greater strength, either by shape or by *inclination* of what is being carried or projected.]

The solutions adopted by the Dominican translator deserve comment. He rendered *ὑπεροχή* [superiority] with *excellentia* [excellence], while he proposed *inclinatio* [inclination] for *ῥοπή*. It would be interesting to know whether Moerbeke worked on the revision of *Physics* before or after the *versio* of Archimedes's works, in which, as we have seen, he preferred to transliterate *ῥοπή* with *rope* (*reptio*). The *inclinatio* adopted does not appear entirely satisfactory, not preserving the idea of the cause of the movement with the same significance as *ῥοπή*[219].

The most important humanistic translation of *Physics*, and also the most widespread in print, was undoubtedly that conducted by Argyropoulos, of which two editions are preserved, the first dedicated to Cosimo, dating back to 1460, the second to Piero di Cosimo, certainly after 1460[220]. Argyropoulos had drawn inspiration for
93 the new translation from the courses he held on Aristotle in the Florentine Studium

[219] Another medieval version translated from Arabic enjoyed considerable success in print. According to the prevailing opinion, it was carried out by Michael Scotus – but there are very strong doubts about the attribution – in the first half of the 13th-century. Scotus, in addition to the text of the *Physica*, translated Averroes's commentary. The influence of Scotus's version is due to the famous edition of Aristotle's Works with Averroes's commentary, produced by the Giunti of Venice around the middle of the 16th century. The editors, however, deemed it appropriate, alongside Scotus's Arabic version, to present a second one from Greek (Moerbeke's carefully revised version). Here is how Scotus translated our two texts: "Videmus enim idem corpus velocitari in motu propter alterum duorum aut propter hoc, quod illud, mediante quo movetur, diversatur, v.g. ut motus eius fit in aqua, aut in terra, aut in aere, aut in igne, aut quia motum diversatur. si igitur alia fuerint eadem *ex differentia gravitatis et levitatis*, illud mediante quo erit motus, erit causa.... Dicamus, igitur, quod videmus corpora quae sunt *magis ponderosa*, sive gravia, sive levia, cum fuerint in aliis dispositionibus in capitulo figurae, eodem modo moveri in aequali loco velocius secundum proportionem eorum adinvicem. ergo sequitur ut talis sit dispositio eorum in vacuo etc. Sed hoc est impossibile. Quoniam non potest aliquis dicere, qua de causa expellitur in eo velocius: quoniam hoc in pleno est necessarium, quod enim est fortius velocius dividit illud. illud enim quod movetur, aut illud quod cedit, *aut per suam figuram, aut per suum pondus* . Ex quo sequitur ut motus omnium corporum sit aequalis in velocitate. quod est impossibile [For we see that the same body speeds up in motion due to one of two reasons: either because the medium through which it moves varies, for example, its motion occurs in water, or in earth, or in air, or in fire, or because the moved object varies. Therefore, if other things are equal *due to the difference in heaviness and lightness*, the medium through which the motion occurs will be the cause… Let us say, therefore, that we see bodies which are *more hefty* (i.e., *more voluminous*), whether heavy or light, when they are in different dispositions in terms of shape, move more quickly in an equal place in proportion to their relationship to each other. Therefore, it follows that their disposition in a vacuum would be the same, etc. But this is impossible. For no one can say for what reason it is expelled more quickly in it: for in a full space, it is necessary that what is stronger divides it more quickly. For that which is moved, or that which yields, does so either *by its shape or by its weight*. From this, it follows that the motion of all bodies would be equal in speed, which is impossible]" *(Aristotelis opera... tum Averrois Cordubensis... Commentariis... Quartum Volumen. Aristotelis de Physico Auditu libri octo... Marci Antonii Zimarae Contradictionum in eosdem libros solutiones...*, Apud Iunctas, Venetiis 1562, pp. 158-164; the italics are ours). The use of *declinatio* and that *magis ponderosus* referring not to weight but to volume appear to be the relevant data in Scotus's version. For the attribution to Scotus, see *Aristoteles Latinus* cit., *Pars Prior*, pp. 104-105.

[220] The two redactions are preserved in two Laurentian Codexes (respectively *Plut.* 84, 7 and *Plut.* 81, 1; the first is the exemplary dedication to Cosimo).

and from the need to base them on a reliable text. He was also motivated by the aim of making the pages of the Stagirite more appealing and clear[221]. This is how the humanist translated our passage from the fourth of the *Physics*.

Videmus enim idem pondus corpusve celerius duas ob causas ferri. Aut quia id differt per quod movetur, ut si per terram, vel aquam, vel aerem moveatur, aut quia id quod fertur differt, si sint eadem caetera: *ob ponderis aut levitatis excessum…*[222]. [For we see that the same weight or body is carried more quickly for two reasons: either because the medium through which it moves differs, as if it moves through earth, or water, or air, or because that which is carried differs, if other things are the same: *due to the excess of weight or lightness…*]

Haec igitur eveniunt ob eorum differentiam, per quae corpora motu cientur. Cernimus enim ea corporum quae plus aut *gravitatis aut levitatis* habent *momenti*, modo eadem ipsis insit figura, celerius per aequale spatium ferri, aequa ratione quam inter sese habent. Quare et per vacuum ita ferentur. At fieri profecto nequit quam enim ob causam celerius movebuntur; in plenis enim id necessario fit; maius enim celerius vi sua medium dividit. Nam id quod fertur, aut est iactum, figura dividit, aut *momento* quod habet[223]. [These things, therefore, happen because of the difference in the things by which bodies are moved. For we see that bodies which have more of either *moments* of *heaviness or lightness*, provided they have the same shape, are carried more quickly through an equal space, in equal proportion to what they have among themselves. Therefore, they will also be carried in the same way through a vacuum. But it surely cannot happen, for what would cause them to move more quickly; for this necessarily takes

[221] On Argyropoulos and his Aristotelian versions, in particular, as well as on the humanistic versions of Aristotle in general, cf. E. Garin, *Le traduzioni umanistiche di Aristotele nel sec. XV*, in "Atti dell'Accademia Fiorentina di Scienze Morali 'La Colombaria'", 1950, especially pp. 28-32. A previous version of *Physics* had been produced by George of Trebizond, but it did not have a printed edition (see the *incipit ibid.*, p. 46).

[222] I quote directly from the Laurentian Codex (*Plut.* 84, 1), containing the version dedicated to Piero di Cosimo (c. 38*v*; the italics are ours). In fact, it was this that spread through prints. The text appeared different in the previous version dedicated to Cosimo: "Cernimus enim ipso plane periculo duas ob causas celeriores tardioresque fieri motus, aut ob mediorum densitatem, aut si per aquam, aerem et huiuscemodi fiant, aut ob differentiam corporum quae feruntur in gravitatis ac levitatis excessu… [For we clearly see that movements become faster or slower for two reasons: either due to the density of the mediums, or if they occur through water, air, and the like, or due to the difference between bodies that are carried in excess of heaviness and lightness…]" (*Plut.* 84, 7, c. 60*v*). On the point we have insisted on, Argyropoulos had initially used *gravitas* instead of *pondus*.

[223] *Plut.* 84, 1, c. 39*v*; the italics are ours. Here too, we give the text of the first draft: "Videmus enim ea corporum quae plus aut gravitatis aut levitatis habent momenti, si modo eadem cunctis adsit figura, celerius semper per idem vel aequale spacium ferri.... Quam enim ob causam gravitatis aut levitatis quicquam celerius unquam transibit per vacuum. In mediis enim plenis idipsum necessitate quadam fieri solet. Cum eorum quae per ipsa motu cientur; quod plus habent momenti, id vi sua medium celerius dividat. Nam omne quod aut suopte nutu aut externa vi pulsum movetur figura *vel suo momento roboreve* medium dividere perrumpereque videtur [For we see that those bodies which have more moment of either heaviness or lightness, if only the same shape is present for all, are always carried more quickly through the same or equal space… For what reason indeed will anything ever pass more quickly through a vacuum due to heaviness or lightness? For in a plenum, the same thing usually happens by necessity. When it comes to those things that are moved by themselves, that which has more moment divides the medium more quickly by its force. For everything that is moved either by its own impulse or by an external force seems to break through the medium by its shape *or its own moment strength*]" (*Plut.* 84, 7, cc. 62*v*, 63*r*; the italics are ours). Also from this case it can be deduced that Argyropoulos subjected the first draft to a fairly careful revision. In regard to our passage, it seems that in the first version Argyropoulos felt less sure of the perspicuity of *momentum* and felt the need to clarify its meaning with other terms (*nutus*, *robur*).

place in a plenum; for the greater divides the medium more quickly by its force. For that which is being moved, either thrown, is divided by its shape, or by the *moment* that it has.]

The translation of Argyropoulos introduced substantial terminological changes compared to that of the *Nova Translatio*, attributed to Moerbeke. Argyropoulos showed, among other things, a different sensibility towards *ῥοπή*. In fact, to render
94 the Greek term, *momentum* was introduced. Our term seemed to Argyropoulos more suitable to express the dynamic consequences implicit in the Aristotelian concept of the generic *inclinatio* [inclination] or *pondus* [weight] of the previous versions[224]. It should be recalled that, around the same time, Iacopo da Cremona had substituted *momentum* for the *rope* in the transliteration proposed by Moerbeke for the initial passage of Eutocius's *Commentary* on the *Equilibrium of Planes*[225]. This is a chronological coincidence that is worth reflecting on. In fact, it does not seem – as we have already noted – that *momentum*, in this specific meaning, was a term in use in the Middle Ages[226]. Around the middle of the 16th-century, it started to recur with a certain frequency.

Another important version, due to its circulation, is that of Agostino Nifo, published in his famous *Expositio* of Aristotelian *Physics*. In the Scotti print, for example, the text of the Stagirite was given "cum duplici translatione, antiqua videlicet et nova [with a dual translation, namely the old and the new]". The *antiqua* was the *Nova Translatio*, previously examined, while the *Nova*, based on the Greek text, was due to Nifo, but appeared strongly conditioned by the Argyropoulos version, whose model emerges clearly under the few corrections of the Aristotelian from Sessa[227]. The similarity between Nifo's version and that of Argyropoulos is very close in the fourth passage of the *Physics*. The solutions adopted are, in fact, the same. For the passages of the translation of Argyropoulos that we have given above in italics, Nifo proposed these three solutions:

a) *ob ponderis vel levitatis excessum*[228] [due to excess weight or lightness]
b) *quae in gravitate aut levitate plus momenti habent* [which have more moment in heaviness or lightness]
c) *aut sua figura dividit, aut momento*[229] [either it divides by its shape, or by moment].

Nifo's commentary on this text is of great interest, as we will see later.

The Giunti edition of Aristotle's works with the commentary by Averroes presented (in the fourth volume containing the *Physics*), in addition to the *versio* from Arabic attributed to Michael Scotus[230], a second translation, probably due

---

[224] *Pondus* constituted Michael Scotus's solution (cf. *supra*, note 219).
[225] Cf. *supra* pp. 49-50.
[226] Cf. *supra*, p. 47. The only exception is Grosseteste (Cf. *supra*, note 26).
[227] *Augustini Niphi... Expositio super octo Aristotelis Stagiritae libros de Physico Auditu, cum duplici textus translatione, Antiqua, videlicet, et Nova eius...*, Apud Hieronymum Scotum, Venetiis 1559.
[228] *Ibid.*, p. 217.
[229] *Ibid.*, p. 221.
[230] Cf. *supra*, note 219.

to the care of the editors[231]. The translation presents notable importance for the original solutions adopted in our context:

Videmus enim pondus atque corpus velocius ferri propter duas causas, aut quia id differt per quod, 95
ut per aquam, aut terram, aut aerem, aut quia id differt quod fertur, si alia sint eadem, *propter excessum gravitatis et levitatis*...[232]

Videmus enim ea quae *maiorem impetum* habent aut gravitatis aut levitatis, si quo ad alia similiter se habeant figuris, citius ferri per aequale spatium, et secundum rationem quam habent magnitudines ad invicem. quare et per vacuum. Sed impossibile est. Propter quam enim causam feruntur velocius? in plenis enim ex necessitate velocius enim dividit fortitudine maius. Aut enim figura dividit *aut impetu* quem habet id quod fertur, aut proiectum est[233].

[For we see that weight and body are carried more swiftly for two reasons: either because that through which it is carried differs, as through water, earth, or air, or because that which is carried differs, if other things are the same, *due to the excess of heaviness and lightness*...

For we see that those things which have *greater impetus*, through either of heaviness or lightness, if they are similarly configured in other respects, are carried more quickly through an equal space, and in proportion to the size they have relative to each other. Therefore, they would move in the same way through a vacuum. But this is impossible. For what reason would they move more quickly? In a plenum, indeed, by necessity, the larger strength divides more quickly. For either the shape divides *or the impetus* with which it is carried, or is projected.]

The translator of this passage clearly shows that he has kept the *Nova Translatio* in mind. Many characteristic choices are in fact common, such as, for example, the use of *fortitudo* [fortitude]. But, in reworking the previous version, he did not limit himself to a few formal adjustments. Our translator, for example, does not believe it is correct to render *ῥοπή* with the *inclinatio* of Moerbeke. And he replaces it with *impetus*. It's no small matter. It is evident that the meaning of the modification is to be identified in the effort to more explicitly express the *dynamic* value of *ῥοπή*, clearly illustrating that weight and lightness generate an inclination "to move" according to different speeds. What divides the medium more or less easily cannot be *inclination*, nor *pondus*, but rather *impetus*. If we reflect on the fact that the great Giuntina edition came out of the Paduan environment and on the awareness of the men who edited it, the significance of the *impetus* proposal cannot escape us[234]. Thus, alongside *momentum*, only in this version (but of what diffusion!)[235], *impetus* also appeared, with the clear meaning of "cause" of motion, or rather of speed.

---

[231] I don't know who the author is. In any case, rather than a version, it will be better to talk about a revision of Moerbeke's translation. Girolamo Bagolino himself could have been the author, who had a significant role in the Giuntina edition of Aristotle with Averroes's commentary (on Bagolino, cf. the entry by Cesare Vasoli in *D.B.I.*, V, 1963, p. 267).

[232] *Aristotelis opera... cum Averrois Cordubensis* cit., IV, p. 157; the italics are ours.

[233] *Ibid*., p. 164; the italics are ours.

[234] Cf. E. Garin, *Storia della filosofia italiana*, vol. I-III, Einaudi, Torino 1966, II, Ch. I.

[235] This version circulated, in fact, not only in the great Giuntina edition, but also offered the text in many commentary editions. This is the case, for example, of Domingo de Soto's Commentary on *Physics* (cf. *infra*, note 311). It will also be necessary to take into account that the declarations on the title pages cannot always be trusted. The "modern" version of the Giuntina edition appears, in fact, in a print of Aristotle's works whose title page declares that the text of *Physics* is "Io. Argyropulo Byzantio interprete" (*Operum Aristotelis*..., Excudebat G. Laemarius, s. 1., 1597, pp. 502-503 and 504).

The edition of Simplicius's commentary on *Physics*, published by Scotti in Venice in 1566, presented the Aristotelian text in the same *versio* now illustrated[236]. In regard to our passage, the agreement of the Scotti edition with that of Giunti
96 was complete, except in one significant detail: the editor deemed it appropriate to replace the two occurrences of *impetus* with *momentum*[237]. It's an interesting fact, one worth reflecting on. We do not know who presided over the edition, which among other things was not original, neither for the translation of Aristotle's text nor for that of the commentary[238]. However, the unknown editor must have noticed the awkward contrast that emerged from the "assembly" of two different translations, between the terminology of the version of the Aristotelian text, where *impetus* was used, and that of the version of the corresponding commentary by Simplicius, in which he used, as we will see, *momentum*. Deeming it untenable, he decided to suppress the inconsistency by replacing the two occurrences of *impetus* with *momentum*.

It may also be that a solution in this sense was suggested by the Aristotelian text published in the previous Giuntina edition of 1551. The Giuntina edition, which offered Simplicius's commentary "cum ipso Aristotelis textu a Gentiano Herveto Aurelio nova ac fideli interpretatione donati [with Aristotle's own context given by Gentian Hervetus Aurelius in a new and faithful interpretation]", proposed the following solutions for the three cases that interest us:

a) *propter exuperationem gravitatis vel levitatis*[239] [because of the excess of heaviness or lightness]
b) *ea quae habent maius momentum vel gravitatis vel levitatis* [those which have a greater momentum either of heaviness or of lightness]
c) *aut* (*enim dividit*) *momento ac inclinatione quam habet*[240] [or (for it divides) by the moment and inclination that it has].

Even the translation by Girolamo Scotti of the Aristotelian text published with the famous John Philoponus commentary, in 1558 in Venice, was presented as an original version, as work of Giovanni Battista Rasario[241]. In reality, the text offered by Rasario literally reproduced the previous *versio*, published under the name of Guglielmo Doroteo, in 1553[242]. On the other hand, even in the translation

[236] *Simplicii Philosophi perspicacissimi Clarissima Commentaria in octo libros Arist. de Physico Auditu...*, Apud Hieronymum Scottum, Venetiis 1566.

[237] *Ibid.*, p. 246.

[238] In fact, it took up the text offered by the first printed edition of the translation of Simplicius's commentary, the work of Lucillo Maggi, known as the "Filalteo" (*Simplicii Peripatetici acutissimi Commentaria in octo libros Aristotelis de Physico Auditu... Lucillo Philaltheo interprete...*, Parisiis 1544); cf. the Publisher's *Supplementum Praefationis*, in *Simplicii in Aristotelis Physicorum Libros...*, Edidit H. Diels, 2 vols. (*CAG*, IX), Typis et expensis G. Reimeri, Berolini 1882, I, pp. XXI-XXII.

[239] *Simplicii commentarii in Octo Aristotelis Physicae Auscultationis Libros cum ipso Aristotelis textu a Gentiano Herveto Aurelio nova ac fideli interpretatione donati*, Apud Iuntas, Venetiis 1551, c. 117*r*.

[240] *Ibid.*, p. 118*v*.

[241] *Aristotelis Physicorum Libri Quatuor, cum Ioannis Grammatici cognomento Philoponi commentariis, quas nuper ad graecorum codicum fidem summa diligentia restituit Ioannes Baptista Rasarius*, Apud Hieronymum Scotum, Venetiis 1558.

[242] Apud Octavianum Scottum, Venetiis 1554 (*Guilelmo Dorotheo Veneto Theologo interprete*).

of Doroteo, who however had the decency to mention it[243], it is not difficult to recognize the substance of the translation of Argyropoulos, which in many contexts is followed precisely.

Evidently, around the middle of the 16th century, there arose a need, certainly 97
also driven by the harsh laws of editorial competition, to present "new" versions, as the humanistic ones were considered obsolete. But, with few exceptions, rather than actual remakes, there were skillful and certainly not laborious superficial revisions: some terms changed, the sequence of words reversed, the syntactic construction modified, etc. This is the case of the text presented by Rasario in 1558. He repeated Doroteo and must not have been aware of Doroteo's reliance upon Argyropoulos in all the demanding choices. However, he did not hesitate to mix up the cards to earn the merit and honors of *nova interpretatio* with little sweat. See, in the 1558 edition, the usual passages of the Aristotelian text:

a) *ob gravitatis vel levitatis excessum*[244] [due to excess of heaviness or lightness]
b) *ea corpora quae plus aut gravitatis aut levitatis habent momenti* [those bodies which have more heaviness or lightness of moment]
c) *aut momentum quod habet id quod fertur, aut figura dividit*[245] [either the momentum that which is conveyed, or the shape divides].

In points a) and b), the proposal of this version fully coincides with Argyropoulos's solution. In point c), we instead find an eloquent example of how superficial the review was. In fact, Rasario (Doroteo) limits himself to changing phrasing, leaving the rest unchanged. Rasario redeemed himself, however, a few years later, effectively performing an original translation of the Aristotelian text and commentary[246]. In his version he never used *momentum*[247].

In addition to those on which we have focused, other translations or, better yet, revisions, of Aristotle's *Physics* circulated in the 16th and early 17th centuries.
Pace[248], Périon, later revised by Grouchy[249], and many others whose authorship 98

[243] Cf. *Ioannis Philoponi in Aristotelis Physicorum Libros... Commentaria*, Edidit Hieronymus Vitelli, 2 vol. (CAG, XVII), Typis G. Reimeri, Berolini 1888, I, pp. XIX-XXI.

[244] *Aristotelis Physicorum... restituit Ioannes Baptista Rasarius* cit., p. 217.

[245] *Ibid.*, p. 221.

[246] *Ioannis Grammatici cognomento Philoponi in Aristotelis Physicorum libros quatuor explanatio... Io. Baptista Rasario Novarensi interprete*, Apud Vincentium Valgrisium, Venetiis 1569.

[247] For the three points we have insisted on he proposed the following solutions: *a*) "Differat aut levitate aut pondere [it differs either in lightness or in weight" (*ibid.*, p. 387); *b*) "quae maiorem vel ponderis vel levitatis praestantiam habent [which have a greater value either in weight or in lightness]" (p. 393); *c*) "movetur aut iactum est, vel figura, vel pondere suo [it is moved or thrown, or by its shape, or by its own weight]" (p. 394).

[248] *Aristotelis Stagiritae... Naturalis auscultationibus libri VIII. Jul. Pacius a Beriga... latina interpretatione auxit et commentariis analyticis illustravit....*, Apud heredes Andreae Wecheli, Claudium Marnium et Ioannem Aubrium, Francofurti 1596. For our three passages, Pace proposed the following solutions: *a*) "differt ob exuperantiam ponderis vel levitatis [it is different because of the excess of weight or lightness]" (p. 132); *b*) "quae maius momentum habent aut ponderis aut levitatis [which have a greater momentum either in weight or in lightness]" (p. 135); *c*) "vel figura dividit, vel momento [either shape divides, or by moment]" (*ibid.*).

[249] Joachim Périon (*Librorum omnium naturalis philosophiae Aristotelis... Interprete Ioanne Argyropulo, addita e regione Ioachimi Perionii versione... ex castigatione Nicolai Grouchij*, Tomus I-II, Apud

is not always clear[250]. It is only worth mentioning, in addition to the vernacular version by Antonio Brucioli in which *momento* appears[251], the text of the version of Aristotle's *Physics* published by the Conimbricenses in their famous commentary, given its widespread circulation. Even in this case, of course, it was a translation-revision[252]. As for our context, no innovations are recorded:

a) *id quod fertur... dispari est pondere aut levitate*[253] [that which is carried... is unequaled in weight or lightness]
b) *ea quae ponderis aut levitatis momentum maius obtinet* [that which obtains the greater momentum of weight or lightness]
c) *aut figura dividit aut momento quod obtinet*[254] [either the shape divides, or the moment it obtains].

Only in the first case do we have an original proposal, but at the cost of deviating from the letter of the Greek text, which speaks of *ὑπεροχή*, of "superiority", not of "difference". In cases b) and c), the model followed, even for the Conimbricenses, the one introduced by Argyropoulos.

b) *The Greek Commentators*

The success of *Physics*, as well as its translations, depended in no small measure on the numerous comments, expositions and paraphrases that accompanied it. In fact, significant annotations were added to Aristotle's text, some of which had decisive weight and influence. This is the case of at least three great Greek commentaries on *Physics*, those of Themistius, Simplicius and Philoponus.

Themistius was the author of a very influential *Paraphrase*, even though it does not appear to have been translated in the Middle Ages. The first Latin version was performed around 1480, by Ermolao Barbaro, who dedicated it to Sixtus IV[255]. Barbaro, as Schenkl states, "verba Themistii adumbravit magis quam vertit [he outlined the words of Themistius rather than translated them]". Furthermore, he

---

haeredes Arnoldi Birkmanni, Coloniae 1568) proposed the following solutions: *a*) "res quae movetur... dissimilis disparisque est aut levitate aut pondere [a thing that moves... is unlike and disparate either in lightness or in weight]" (p. 195); *b*) "quae maiorem vel ponderis vel levitatis praestantiam habet [which has a greater value either in weight or in lightness]" (p. 197); *c*) "vel figura, vel pondere suo dividit [either by shape or by weight it divides]" (*ibid*).

250 For a general idea of the diffusion of the various translations through print, see E. Cranz, *A bibliography of Aristotle editions* cit.

251 *La Fisica di Aristotele, tradotta di greco in volgare italiano...*, Per Bartholomeo Imperatore, Venetia 1551. Here are Brucioli's solutions: *a*) "per la disparità del peso e della leggerezza [due to the disparity in weight and lightness]" (c. 88*r*); *b*) "que' corpi che hanno più momento di gravità o di leggerezza [those bodies which have more moment of heaviness or lightness]" (c. 90*r*); *c*) "o per la figura divide, o per la inclinazione che ha [either by the figure it divides, or by the inclination it has]" (*ibid.*).

252 *Commentarium Collegii Conimbricensis Societatis Jesu in octo libros physicorum Aristotelis...*, Apud Vincentium et R. Amadinum, Venetiis 1606.

253 *Ibid.*, p. 73.

254 *Ibid.*, p. 74.

255 *Themistii Peripatetici Lucidissimi Paraphrasis in Aristotelis... Physica... Hermolao Barbaro Patricio Veneto Interprete...*, Apud Hieronymum Scotum, Venetiis 1542.

committed quite a few errors and provided a rather verbose text[256]. The main edition of the Greek text did not appear before 1534[257], while the Barbaro version was first 99
published for the first time in 1481, without the text of Aristotle[258].

Themistius's *Paraphrase* of the *Physics* texts on which we have focused upon is of great interest. The commentator, like in other places, distances himself from the Stagirite, rejecting his arguments on the absurdity of the motion of bodies in the void. To the Aristotelian thesis of an infinite speed of bodies in the void – which would result in instantaneous motion, in other words, an absurdity – Themistius opposes a different conception: all bodies will move in vacuum in a given time, and their speeds will *all* be the same. This depends, according to Themistius, on the fact that in a vacuum – unlike what happens in a plenum – the differences in heaviness and lightness do not produce any variation in speed, given that they are not exerted against any resistance. Therefore, it is as if these differences were abolished, with the consequent neutralization of the different speeds.

Adding interest to these famous pages is the certainty that Galileo had perfect knowledge of them and had to meditate on them for a long time. We find a specific quote from this very place of Themistius's *Paraphrase* in the formative *De motu*[259]. And this is a passage where Barbaro employed *momentum*. Furthermore, the Pisan scientist, like anyone who dealt with the *de motu* questions between the end of the 1400s and the beginning of the 1600s, dedicated quite a few pages to the analysis of the void, of the possibility of movement occurring in it and at what speeds. And always, in such discussions, the reference to these passages from Themistius appears explicit. The citation of Themistius's *Paraphrase* in *De motu* takes on particular significance, given that in that writing Galileo rejected the Greek commentator's thesis regarding the uniformity of all speeds in vacuum. Not many years later, Galileo retracted and recognized that Themistius was right.

In the Greek text of *Paraphrase*, Themistius, unlike Aristotle, does not use *ῥοπή* to express the inclination to move with different speeds downwards or upwards produced by heaviness or lightness. In fact, Themistius limits himself to talking about *διαφορὰ τῶν δι' ὧν φερέται*[260] [difference of those for which it is carried], indicating the blurring differences in weight and lightness that occurs in a vacuum and the consequent disappearance of speed differences with this simple expression:

> *ἀνήρηται γοῦν τὸ βαρύτερον καὶ κουφότερον καὶ πάντα ἰσοταχῶς*[261]. [indeed it is the heavier and the lighter, and all equally fast.]

100

[256] "Multa male intellexit, multa loquacius quam accuratius expressit [He misunderstood many things, he expressed many things verbosely than accurately]" (*Themistii in Aristotelem Physica Paraphrasis*..., Edidit Henricus Schenkl, *CAG*, V, Pars II, Typis G. Reimeri, Berolini 1900, pp. X and XIV).
[257] Aldo, Venetiis 1534, edited by Trincavelli.
[258] Tarvisii 1481, Aristotle's text is also absent in subsequent editions.
[259] Cf. *infra*, Part Two, Ch. I, p. 172.
[260] *CAG*, V, Pars II, p. 132, 15-16.
[261] *Ibid.*, p. 132, 34.

Barbaro's version of this passage of *Paraphrase* was, in turn, a paraphrase. Let's see the parts corresponding to those in which the Greek terms mentioned above occur:

Hanc itaque motus inaequalibus causam de duabus alteram tollunt ii qui vacuum intuuntur asserere, sed neque illam tuentur, quae *de ponderum discrimine et momentis corporum* nascitur[262]. [Therefore, those who assert that the void is the other cause of unequal motion remove one of the two causes, but do not defend the other one, that arise from *the difference of weights and the moments of bodies.*]

Barbaro was not satisfied with the literal Greek text and considered it appropriate, in translating, to make its meaning more explicit. The differences in weights generate the *moments* of bodies, that is, the different inclinations to move more or less quickly. Barbaro probably has Argyropoulos's translation of *Physics* in mind, from which he derives *momentum*. The term also returns in the passage in which Themistius contrasts the Aristotelian thesis of the instantaneity of motion in the void with his own conception of the equality of all speeds.

In pleno autem necesse est moveri velocius quicquid est gravius, quia scindit vi et nisu maiore subiectum, sive aquam id, sive aërem statuamus. At in vacuo non ocyus ferrum quam suber corruit? *Quum vacuum utrique cedat aequaliter, sed neque cedat quidem: quum enim id nihil fit, subtilis hominis est putare vacuum cedere: ita fit ut differentia gravium et levium, id est rerum momenta tollantur, et quod sequens est, omnibus quae moventur aequalis et indiscriminata volucritas accidat*[263]. [But in a plenum it is necessary for whatever is heavier to move faster, because it splits the medium with greater force and effort, whether we consider water or air. But in a vacuum does not iron fall as quickly as cork? *The vacuum yields to both equally, and yet it does not yield at all: for since the void is nothing, an overly acute person would think that a vacuum yields: thus the difference between the heavy and the light is made, in other words the moments are removed, and what follows is, all things move with equal and undifferentiated speeds.*]

The passage we italicized corresponds precisely to Galileo's citation from Themistius in *De motu*[264], who, according to his early works, knew of Barbaro's version of *Paraphrase*. Here, *momentum* is characterized with great clarity as a term indicating the inclination of all things to move at a specific speed. Proportional to weight or lightness, *moment* indicates the dynamic effect, which becomes uniform when – as in the case of motion in a vacuum – there are no resistances to overcome. *Momentum* and "speed" therefore appear to be closely connected. Speed, indeed, is a consequence of *moment*[265].

Simplicius also takes different positions on the problem of the void and motion in the void than the Stagirite. The influence of Simplicius's commentary – truly enormous in the 16th century, especially in the Paduan environment[266] – was greatly aided by the fact that he reported the opinions of Alexander of Aphrodisias

262 *Themistii Peripatetici Lucidissimi* cit., p. 139; the italics are ours.
263 *Ibid.*
264 Cf. *G.G.*, I, pp. 294 and 412.
265 The Barbaro version sometimes renders the same concept with *nutus* (p. 129) and with *differentia gravium et levium* (p. 138).
266 In Padua there was at a certain point a notable movement of "Simplicians", cf. E. Garin, *Storia della filosofia italiana* cit., II, Ch. I, paragr. 10.

whose commentary on *Physics* has been lost. Galileo certainly knew Simplicius's **101** commentary[267].

We have already mentioned the versions of Aristotle's text that accompany Simplicius's commentary in the three 16th-century prints. The text of the commentary published in those editions corresponded to two different translations, carried out by Lucillo "Filalteo" Maggi (Paris 1544 and Venice 1558) and by Gentian Hervet (Venice 1551)[268]. In the Greek text, the use of *ῥοπή* is frequent[269], which both translators typically rendered with *momentum*.

In the Filalteo version, for example, *σχῆμα* [shape] and *ῥοπή*, which consists of the two "virtues" with which the mobile overcomes the resistance of the medium, dividing it, are rendered respectively with *figura* [shape] and *momentum*. The different speeds of motion – adds Simplicius, expounding Aristotle's argument – depend on the difference in *momentum*. On the other hand, *moments* in a vacuum cannot produce any difference in speed, since there is no division to be made in a vacuum. It is concluded, therefore, that the defenders of the void are obliged to maintain that "aeque igitur velocitate ferentur ea quae moventur, quamquam differant quo ad momentum et pondus [those things which are in motion will be carried with equal speed, although they differ in momentum and weight]"[270]. This consequence, however, is unsustainable because – as Alexander also emphasized – the *substantia* [essence] of the *ponderosius* [heavier] is to go *deorsum* [downward] faster. The supporters of the void and the *sectatores* [followers] of Democritus, Leucippus and Epicurus are therefore wrong. But the absurdity of their position can be demonstrated even more easily. If, in fact, greater heaviness does not make motion faster, "neque principio causa erit motus deorsum ipsa gravitas [nor will heaviness itself be the principal cause of downward motion]"[271]:

> Si vero *momenta* in corporibus sunt causae motus in vacuo, neque moverentur in ipso initio corpora in vacuo. Neque enim est quaedam alia causa lationis naturalibus corporibus, praeter eam quae est *proprii momenti et inclinationis*[272]. [But if *moments* in bodies are the causes of motion in a vacuum, bodies in a vacuum would not be moved in the very beginning. For there is no other cause of motion in natural bodies than that which is due to their *own moments and inclination.*]

Therefore *momentum et inclinatio* [momentum and inclination] (the Greek text simply has *ῥοπή*)[273] are the only causes, both of the motion and of the different speeds of the motion. According to Alexander – continues Simplicius – Epicurus considered heaviness the only cause of motion and its *maius momentum* [greater momentum] the cause of faster motion. But in a vacuum, all speeds are the same; therefore, a body "neque ob pondus moveretur, si vero non ob hoc, nec penitus

[267] Galileo already refers to Simplicius in *De motu*.
[268] Cf. *supra*, the note 238 and 239.
[269] Cf. the *index verborum* at the bottom of Simplicius's edition in *CAG*, IX.
[270] *Simplicii Philosophi perspicacissimi* cit. [Venezia 1566], c. 247*r*.
[271] *Ibid*.
[272] *Ibid*.
[273] *Τῆς οἰκείας ῥοπῆς* (Simplicius, in *CAG*, I, p. 679, 32).

moveretur [would not move because of the weight, if indeed it were not because of this, it would not move at all]"[274].

102 At this point, Simplicius distances himself from Alexander and also from the method of proof adopted by Aristotle to demonstrate the absurdity of motion in a vacuum. He states that if any advocate of the void listened to Alexander's arguments,

> dicet statim, existimo, quod si quidem est momentum causa motus corporum et plus momenti est causa celerioris, cur non movebuntur corpora per vacuum? Si enim habeat aliquid quod facile vel non facile dividi valeat, illud etiam per quod motus fit ad illud, ut motus sit celerior et tardior, quae sunt differentiae motus. sed hoc non est causa momenti; sed cum fuerit illa differentia, addit aliquid differentiae momenti, secundum naturam et erit bene divisibile, vel non. Ut enim aliae qualitates similes genere in maioribus sunt maiores, sic etiam consentaneum est ut habeant momentum. Et generatim momentum est magis causa divisionis quam divisio momenti, et si alicubi spatia subiecta aliquid impedierint et alicubi minimo[275]. [he will immediately say, I believe, that if moment is indeed the cause of the motion of bodies and more momentum is the cause of greater speed, why would bodies not move through a vacuum? For if it has something that can be easily or not easily divided, that also through which the motion occurs will make the motion faster or slower, which are differences of motion. But this is not the cause of momentum; rather, when that difference exists, it adds something to the difference in momentum according to nature, and it will be easily divisible or not. For just as other qualities similar in kind are greater in larger objects, so it is fitting that they have momentum. And generally, momentum is more a cause of division than division is of momentum, and if in some places the intervening spaces hinder somewhat and in other places only minimally.]

The demonstrative procedures employed by Aristotle and Alexander on this point are unacceptable. Both authors make the different *moments* the cause of motion and of greater or lesser speed; from here, they illegitimately deny that there are *moments* (i.e., the causes of motion) if there is no division (as in a vacuum). Simplicius points out the paralogism: first they affirm *momentum*, that is, the difference in weight or lightness, as the cause of the division of the medium; then they introduce the division of the medium as the cause of *momentum*. In reality, as the commentator specifies, the more or less penetrable medium modifies the natural *momentum* of bodies, but certainly does not determine their existence.

Here, *momentum*, inserted in the context of the Aristotelian conception of movement, indicates the natural tendency, the "appetite" of bodies to reach their "proper" place with speeds proportional to the greater or lesser quantity of matter. *Momentum*, like *ῥοπή*, indicates the spontaneous tendency towards a goal, the natural disposition to seek it. Even in the version of Simplicius's commentary thus emphasized the close connection between *momentum* and "speed"[276].

The text of the third Greek commentator that we intend to examine, John Philoponus, also presents a very similar situation. His contribution to the development of

---

[274] *Simplicii Philosophi Perspicacissimi* cit. [Venezia 1566], c. 247*r*.

[275] *Ibid.*

[276] In Gentian Hervet's version, the expression *momentum celeritatis* is even used, which we will see reappear frequently in the Latin pages of Galileo's *Discorsi e Dimostrazioni Matematiche* (*Simplicii Commentarii* cit., c. 118*v*). In general, Hervet, despite making extensive use of it, showed some doubts when faced with *momentum*. In fact, he never presented it alone, preferring to accompany it with other terms with a more evident meaning: "pondus ac momentum", "momentum et inclinatio" etc. (cf. above all, cc. 118*r*-*v*).

dynamics partially independent from Aristotle and the importance that his doctrines **103**
had in determining the birth of the so-called *impetus* physics are too well known to require illumination[277]. The Greek text of the Grammarian's commentary on *Physics*, of which only the first four books were known, saw the light for the first time in 1535, edited by Trincavelli[278]. During the 16th-century, Latin versions of Philoponus's commentary were published. The first two editions presented Aristotle's text and Philoponus's commentary in the translation by the Venetian patrician Guglielmo Doroteo[279]. In 1569, the third edition of the commentary was published, this time effectively due to the efforts of Rasario, whose version appears original and independent, at least to some extent, from the previous one[280].

After expounding Aristotle's theses on the possibility of the void and motion in the void, Philoponus follows, in his commentary, a long digression, in which he discusses the Stagirite's arguments acutely and severely criticizes them. Precisely in the pages of the digression, *ῥοπή* appears numerous times[281]. The translators reacted differently when faced with the Greek term. Doroteo, in fact, resorted, as we will see shortly, to *momentum*, while Rasario did not use it even in one case. For this reason we will focus our attention on Doroteo's version.

Philoponus's criticisms of Aristotle's thesis regarding motion in a vacuum, delivered in the *Digressio de inani*, are not few, nor negligible[282]. The commentator articulates his objections in three *arguments*. The first calls into question the principle stated by Aristotle that, in a vacuum, the differences – the *moments* – of heavy and light things disappear. The diversity of *moments* – claims Philoponus – is preserved in the void. In fact, it is not clear why the void should cancel out the specific *substantia* [essence] of all bodies, that is, that of being differently heavy or light. If the Aristotelians then intended to escape the trap of this simple reasoning, stating that the *inequalitas motus* [inequality of movement] depends solely on the different densities of the medium, they would find themselves having to admit destructive consequences for their own *positio* [proposition]:

> Nam si intervallum per quod motus absolvitur foret solum causam inaequalitatis motus, accideret profecto inaequalia corpora *secundum momentum movendi* per unum et idem corpus mota aequali
> velocitate moveri, cum nulla sit differentia propter ipsa *momenta et inclinationes* movendi corporum et **104**
> solum fiat motus inaequalis ob differentiam interstitiorum per quae fiunt motus. At hoc plane praeter evidentiam fit, quippe inaequalia mota per idem intervallum inaequali celeritate moveantur.[283] [For if the distance through which the motion is completed were the only cause of the inequality of motion, it would certainly happen that unequal bodies moved *according to the moment of motion* moved by one and the same body moved with equal speed, since there is no difference due to the actual *moments and inclinations* of the bodies in motion, and unequal motion only occurs because of the difference of

[277] I simply refer to the entry *John Philoponus* of the *D.S.B.*, VII (1973), pp. 134-139 (by S. Sambursky).
[278] *Ioannis Grammatici in primos quatuor Aristotelis de naturali auscultatione libros Commentaria...*, In aedibus Bartholomaei Zanetti, Venetiis 1535.
[279] Cf. *supra*, note 241 and 242.
[280] Cf. *supra*, note 246.
[281] Cf. the *index verborum* in *CAG*, XVII, p. 970.
[282] *Aristotelis Physicorum libri quatuor* cit., pp. 226 ff.
[283] *Ibid*., p. 226; the italics are ours.

the intervals through which the motion occurs. But this is plainly beyond the evidence, since unequal motions move through the same distance with unequal speed.]

In the Greek text, the expression rendered by Doroteo with *secundum momentum movendi* [according to the momentum of motion] is simply *κατὰ τὰς ῥοπὰς*[284], while *propter ipsa momenta et inclinationes movendi* [due to the actual moments and inclinations of the motion] corresponds to *παρὰ τὰς ῥοπὰς*[285]. The translator did not think that *momentum* was sufficient to indicate that it was an inclination to motion and preferred to specify it with the addition of *movendi* [to move].

In the *Secundum argumentum*, Philoponus recalls that Aristotle posed two causes for *inequalitas motus* [inequality of movement]. The cause dependent on the *moments* of bodies does not disappear in the void, "nam gravitas non est de numero quae habent respectum ad aliud, sed est qualitas ipsa per se ipsa inexistens corporibus [for heaviness is not a matter of the number of things which have a relation to another, but is a quality which by itself does not exist in bodies]", not being conferred by the medium. It must, then, be concluded that, "etsi vacuum fuerit, non tollentur inaequales motus [even if it were a vacuum, unequal motions would not be eliminated]"[286].

The third argument is more malignant. If bodies – states Philoponus – did not move in a vacuum with different speeds, it would not be understood why they should do so in plenum, for example, in air. In fact, the air is always uniform, as far as it is in itself, and the bodies, on the other hand, were unequal even in vacuum:

Nam unus ac idem cum sit aer, ipsorum causa debuit esse, constat igitur quod *fluxus et momenta corporum* quorum unum aptum est celerius deorsum ferri, alterum vero segnius, hanc differentiam inaequalitatis motuum faciunt[287]. [For since the air is one and the same, it must have the same cause for both of them; therefore, it is evident that the *flows and moments of bodies*, of which one is suited to be carried downwards more quickly and the other more slowly, produce this difference in the inequality of motions.]

Again, Doroteo did not consider it appropriate to translate *αἱ ῥοπαὶ τῶν σωμάτων* [the moments of the bodies][288] of the Greek text simply with *momenta corporum*, but preferred to accompany our term *fluxus* [flow]. The use of *fluxus* tells us something about Doroteo's sense of the semantic value of *momentum*: intrinsic, original quality. This is clearly seen from the rest of Philoponus's argument. That which has the greatest natural inclination to descend downwards divides the medium most easily: "magis etiam aerem impellit vi sui impetus [it pushes the air even more with the force of its impetus]"[289]. The speed of motion is therefore determined by the different innate *sursum* [upwards] or *deorsum* [downwards] *momentum* of

284 *CAG*, XVII, p. 678, 6.
285 *Ibid*., p. 678, 7.
286 *Aristotelis Physicorum* cit., p. 227.
287 *Ibid*.; the italics are ours.
288 *CAG*, XVII, p. 679, 6.
289 *Aristotelis Physicorum* cit., p. 227.

the individual bodies, which is variously weakened by a plenum, but retains all its 105
strength only in vacuum:

Neque enim *maius momentum maiusque pondus* habens cum magis dividat magis est deorsum mobile, sed e contra quia magis deorsum mobile est, magis dividit. Nam latio deorsum non sequitur divisionem interstitii, sed e contra divisio interstitii sequitur ipsam lationem[290]. [For it is not, *in having greater momentum and greater weight*, with greater division, the greater it is downwardly mobile, but rather, because the greater it is downwardly mobile, the greater it divides. For the downward movement does not follow the division of the interval, but rather, the division of the interval follows the movement itself.]

Again we find *momentum* "assisted" by another term – this time *pondus* – while in the Greek text it is simply found *ῥοπή*[291]. The translator's awareness of using a term that is not entirely perspicuous is evident. The argument presented here by Philoponus is the same one that we have already seen used by Simplicius to denounce Aristotle's paralogism: the greater or lesser ease of division of the medium depends on the *moment*, not the *moment* of the division, as Aristotle seems to affirm. Therefore the *momentum innatum* [innate momentum], the *momentum ac pondus innatum* [innate momentum and weight] – to use the expression of the translator of Philoponus – can only be fully exercised in the void. In a vacuum, therefore – if it exists – the speeds will not be uniform, as Themistius had maintained, challenging Aristotle that this would be absurd, but, on the contrary, they fully correspond to the different *moments*. Only in a vacuum will bodies be able to move with speeds exactly proportionate to their natural inclinations.

Nam densitate substantiae non habentia aerem insertum ipsa sursum pellentem ac trahentem feruntur per aquam deorsum, ab ipsa gravitate innata nullo impediente, sic ergo undique *innatum pondus ac momentum* fertur ad proprium locum per nihil aliud quam per se ipsum nisi impediatur. Si ergo deorsum lationem, vel sursum lationem habent ipsa *momenta et pondera insita* non per aliud respectum ac aliam positionem ad aliud quam per se ipsa, igitur cum fuerit diversa gravitas in corporibus, necesse erit ut etiam deorsum latio fiat non per aliud quippiam quam per eandem. Nullum enim momentum est de his quae ad aliquid sunt. Eadem etiam dicemus de levitate. Quod quidem igitur et per vacuum fiat motus et tamen non necesse sit omnia aequa celeritate moveri, atque argumentum non habet aliquod quod necessarium per quod videtur Aristote. probare hoc satis demonstravimus[292]. [For substances with a density that do not have embedded air which drives it upwards, drag through water downwards with its innate heaviness, so therefore from every direction the *innate weight and momentum* is carried toward its proper place by nothing other than itself unless it is impeded. Therefore the downward extension, or upwards extension, have *innate moments and weights* that are not dependent on anything else nor in position other than themselves, thus when the weights carry different heaviness, then the extension must also be made downward, not through anything else but through the same. For there is no *momentum* about these things which are of any use. We will say the same about lightness. And it is true therefore that motion takes place through a vacuum and yet it is not necessary for all things to be moved with equal speed, and the argument does not have anything necessary that Aristotle seems to propose. The proof is satisfactorily demonstrated.]

Once again, Doroteo didn't feel like leaving *momentum* alone and placed *pondus*

[290] *Ibid*.; the italics are ours.
[291] *CAG*, XVII, p. 679, 15.
[292] *Aristotelis Physicorum* cit., p. 227; the italics are ours.

next to it. The translation performed using two different expressions (*gravitas innata*, *innatum pondus ac momentum* [innate heaviness, innate weight and momentum]) for the same concept corresponds adequately to the Greek text which also presents two different terms *ἡ ἔμφυτος βαρύτης* [or innate burden] and *ἡ ἔμφυτος ῥοπή* [or innate moment][293].

106 The use of *momentum* in these pages of the *versio* of Philoponus's commentary is very fitting. And the connotation of the term is very clear. As we have already seen in Simplicius, *moments* indicate more or less intense "inclinations" to move towards its proper place. The speeds of movement are proportional to the *momenta movendi* [moments of movement]. Every body, according to its greater or lesser heaviness, has an *innatum momentum*, an innate tendency to move with a certain speed; a condition that can only occur in a vacuum. *Momentum movendi* [moments of movement], *momentum innatum* [innate momentum], *fluxus et momenta* [transience and moments], are expressions of great interest. Galileo knew these pages perfectly. It will be enough to remember that, in *De motu*, just as he rejected Themistius's opinion on the uniformity of all movements in the void, he accepted, without citing the source, Philoponus's thesis: "in vacuo etiam solo velocitatum discrimina vera et naturalia contingere, dubio procul est asserendum [only in a vacuum do speeds differ truly and occur naturally, it is asserted beyond doubt]"[294]. And it will also be appropriate to emphasize that *momentum movendi* [moments of movement] closely recalls Galileo's "momento del discendere [momento of descent]"[295], while in *De motu* he cites "Philoponus" as one of Aristotle's opponents on the problem of motion in a vacuum, in a context explicitly referring to the Aristotle's text to which we have returned many times[296].

c) *The Latin Commentators*

The results we have reached by examining the translations of the text of *Physics* and those of the Greek commentators indicate with a certain consistency that the use of *momentum* to translate the term *ῥοπή* in the two places in *Physics* became habitual no earlier than the Argyropoulos version, that is, around in the middle of the 15th-century. We have seen, in fact, that previously the translators oriented themselves differently to convey the concept of inclination to movement expressed by the Greek term. The confirmation of this periodization – which is not intended to be absolute, but relative only to the influence of the texts we have examined – also comes from the examination of the translations of the Archimedean *corpus*: the entry of *momentum* as a correspondent to *ῥοπή* with analogous meaning, at least in one case (Eutocius), is not recorded before the *versio* of Iacopo da Cremona, that is, around 1450.

---

293 *CAG*, XVII, p. 680, 17-18.

294 *G.G.*, I, p. 296.

295 In *Mechaniche*, above all, the expression is frequent; cf., e.g., *G.G.*, II, p. 182.

296 *G.G.*, I, p. 284.

In the Latin commentators on *Physics*, we observe the same phenomenon. The medievals repeat, in general, the same terms with which the author of the *Nova Translatio* had translated our passages from the fourth book.

Albertus Magnus, for example, employs in his *Commentum in libros physicos*
expressions like "quidam enim maiorem motum habent ascendendi vel descendendi **107**
[for some have a greater motion of ascending or descending]", or "quae sunt maioris gravitatis vel levitatis [which are of greater heaviness or lightness]"[297]. Burley repeats the *inclinatio aut gravitatis aut levitatis* [inclination or heaviness or lightness] of the *Nova Translatio*, or resorts to the equivalent *differentia in gravitate et levitate* [difference in heaviness and lightness][298]. The usual expressions also appear in Giles of Rome's *Interpretatio fidelissima*[299].

Thomas Aquinas's commentary on Aristotle's *Physics* had notable influence. Divided into *lectiones*, it offered a simplified reading of the difficult text of the Stagirite and allowed an orderly consultation of the most important questions. The *Physics* text used by Thomas, and reproduced in the first printings of his commentary, was – as we have already said – Moerbeke's *Nova Translatio*. And the terminology of the great Dominican master in the *Commentary* on the *De vacuo* of the fourth book of *Physics* appears strongly influenced by that used by his colleague brothers to translate it. Thomas, in expounding Aristotle's arguments against motion, speaks of "differentia mobilis, quia est gravius vel levius [mobile difference, because it is heavier or lighter]" and frequently resorts to the *inclinatio* with which Moerbeke thought he could render *ῥοπή*[300]. Thomas is not convinced of Aristotle's arguments against the void and does not fail to raise some objections, in which it is not difficult to recognize the influence of the themes of Themistius, Simplicius and Philoponus. To the Aristotelian arguments against the vacuum *ex parte medii* [on behalf of the medium], Thomas objects that the absence of resistance in the vacuum medium does not suppress the differences in *potentia* [power] in the mobiles, differences on which the various speeds depend[301]. In Thomas's arguments, the motifs of Philoponus seem to re-emerge, although he is not cited. The impression of the Grammarian's influence is strengthened by the arguments used by Aquinas to contest the Aristotelian theses against the void, derived *ex parte ipsius mobilis* [on behalf of the mobile itself]. In the *expositio* of Aristotelian arguments on this point, we see Thomas repeating Moerbeke's terms almost literally:

Si enim motus fiat per spatium plenum aliquo corpore, potest assignari causa maioris vel minoris velocitatis, secundum aliquam praedictarum causarum; hoc enim est, quia illud quod movetur maius existens, ex sua fortitudine velocius dividit medium; vel propter aptitudinem figurae, quia acutum est

---

[297] *Commentum Doctoris excellentissimi Alberti Magni O. P. in libros Physicos Aristotelis…*, Per Iohannem de Forlivio et Gregorium fratres anno 1494, die ultimo Ianuarii, c. 55*r*.

[298] *Expositio excellentissimi philosophi Gualterii de Burley* cit., c. non num. (commentary on the text 74).

[299] *Praeclarissimi summique philosophi Aeg. Rom. ...in octo physicos libros interpretatio fidelissima*, Impensis ac diligentia solertissimi viri Hieronimi Durantis, Padova 1493.

[300] *In octo libros Physicorum Aristotelis Expositio*. Cura et studio P. M. Maggiolo O.P., Marietti, Taurini-Romae 1954, *Lectio* XII, p. 256.

[301] *Ibid*., p. 257

108 penetrabilius, aut propter inclinationem maiorem quam habet, vel ex gravitate, vel ex levitate, vel etiam propter violentiam prohibentis[302]. [For if the motion is made through a plenum towards some body, it may be assigned the cause of a greater or lesser speed, according to any of the aforesaid causes; this indeed happens, because the movement being greater, divides the medium quickly due to its own strength; either because of the suitability of the shape, because it is sharp and more penetrating, or because of a greater inclination that it has, or from heaviness, or from lightness, or even because of something preventing it.]

Also in this case, the commentator is dissatisfied with the Aristotelian argument, of which he highlights the paralogism already indicated by Simplicius and Philoponus. According to Thomas, in fact, Aristotle

> videtur enim supponere quod differentia velocitatis in motibus non sit nisi propter differentiam divisionis medii; cum tamen in corporibus caelestibus sint diversae velocitates, in quibus non est aliquod plenum medium resistens, quod dividi oporteat per motum corporis caelestis[303]. [seems to suppose that the difference of speed in the movements is only due to the difference of the division of the medium; however, there are different speeds in the heavenly bodies, in which there is no resisting plenum that must be divided by the motion of the heavenly body.]

Aristotle's paralogism is evident: as Philoponus states, *moment* is the cause of division, but division is not the cause of *moment*. Thomas, however, never uses *momentum* in the sense of inclination to motion[304].

In addition to the commentators already mentioned, *momentum* is absent in Buridan[305], Albert of Saxony[306], Pierre Tartaret[307] and John of Jandun[308]. It seems therefore that in the medieval commentators, as well as in the medieval translations of Aristotle's text, *momentum* is never found in the sense of the propensity to move according to speeds proportional to weight or lightness as expressed by *ῥοπή* in

[302] *Ibid*., p. 259; the italics are ours.

[303] *Ibid*.

[304] It also appears in the *Index Thomisticus* cit. by Father Busa, *sub voce* (vol. XIII).

[305] *Acutissimi Philosophi... Ioannis Buridani Subtilissimae quaestiones super octo physicorum Aristotelis, diligenter recognitae et revisae a Magistro Johanne Dullaert de Gandave*..., Parigi 1513. The expression generally used by Buridan is "*inclinationes gravium et levium* [heaviness and lightness inclinations]" (cf., e.g., f. LXXVII*r*).

[306] *Acutissimae quaestiones supra libros de physica auscultatione ab Alberto de Saxonia editae*, Iacobum Pentium, in inclitis Venetiis 1504. Albert also uses *inclinatio* (cf., e.g., c. 49*r*).

[307] *Petri Tatareti... Ioan. Duns Scoti Doctoris Subtilis fidelissimi in Aristotelem philosophiam, naturalem, divinam et moralem, exactissima commentaria*, Apud Haeredes M. Sessae, Venetiis 1581. Here is how Tataretus commented on the second part of our text: "Secunda ratio sumitur ex parte mobilis et est haec si in vacuo fieret motus; tunc magis grave non moveretur velocius quam minus grave: nec magis leve quam minus leve. Consequens est impossibile; igitur probatur quod mobile velocius movetur eo quod citius et fortius dividit [The second reason is taken in part from the mobile and if the motion were to take place in a vacuum; the more heavy would not move faster than the less heavy, nor more light than the less light. The result is impossible; therefore it is proved that a mobile moves faster because it divides faster and more strongly]" (cc. 81*v*-82*r*).

[308] *Ioannis de Ianduno... super Octo Libros Aristotelis de physico auditu subtilissimae quaestiones*.... Apud Haeredem Hieronymi Scoti, Venetiis 1575. In *Question X*, dedicated to examining the opinions of Aristotle and his opponents on the possibility of motion in a vacuum, the preferred expression used by John of Jandun to express the tendency to move according to speeds proportional to weight (or lightness) is "naturalis et propria inclinatio [natural and proper inclination]" (cf., for example, p. 114).

the Greek text. Argyropoulos's translation and the choice he made by rendering
ῥοπή with *momentum* seem to have truly marked an important stage in the history **109**
of our term. As we have already observed in the versions of Aristotle's text, after the translation of Argyropoulos, *momentum* also begins to be registered in the commentaries.

The term appears, in fact, in Agostino Nifo's annotations to *Physics*. Nifo presented the version of the Aristotelian text as his own and original, while in reality – as we have already seen – he kept the Argyropoulos translation in mind, of which he was substantially dependent. But the lively Aristotelian displayed a peculiar ability of blowing smoke into the eyes of naive readers. To better conceal the imitation – which, in the case of the text of the fourth book, was exact – he used the commentary. Here he is indeed declaring, taking inspiration from *momentum*, how much effort and expertise this original translation had cost him:

> Quo vero ad verba attinet verbum illud momenti graece est rope, latine est inclinatio, sive propensio, ego vero momentum transtuli magis ad verbum[309]. [As for the words, the term *moment* in Greek is "rope," in Latin it is "inclinatio," or "propensio," however, I have translated it more literally as "momentum."]

In Nifo's commentary on the passages of Aristotle's *Physics*, the objections of previous commentators against the Stagirite's arguments are not found. Nifo limits himself to paraphrasing the text, using *momentum* several times in the usual meaning: the different *moments* correspond to the *potentiae motivae multitudo* [the multitude of motivating powers][310].

It does not seem, however, that the term *momentum*, used in the translations and commentaries relating to the fourth book of *Physics* starting from the middle of the 15th-century, imposed itself with full authority. The commentators and expositors closest in time to Galileo made, in fact, very limited use of it, often avoiding using the term in their observations. Thus *momentum*, in some cases, appeared in Aristotle's texts, but did not appear in the commentaries.

This is the case of the annotations to the text of Domingo de Soto's *Physics*, which never uses *momentum* in this sense[311]. The absence of our term is also noted in Pereira's *De communibus principiis*, whose work, although not a real commentary,

---

[309] *Expositio* cit., p. 326; the italics are ours.

[310] "Velocitatis esse velocem medii divisionem, id enim omne velocius fertur quod velocius... dividit. Huius vero velocioris divisionis causae duae sunt, qualitas figurae… et potentiae motivae multitudo… [the speed of the division of the medium determines speed, for everything that moves faster… divides faster. There are indeed two causes for this faster division, the quality of the shape… and the multitude of motivating power…]" (*ibid.*).

[311] *Dominici Soto Segobiensis, Philosophi ac Theologi praestantissimi, in inclyta Salmanticensi Academia Professoris... Super Octo libros Physicorum Aristotelis praeclarissima commentaria…*, Apud Franciscum Zilettum, Venetiis 1582. The text of Aristotle that accompanies the commentary repeats the "modern" one published in the grand Giuntina edition of Aristotle, with the commentary of Averroes (cf. *supra*, pp. 94-95). The only difference – highly significant – consists in the fact that the two occurrences of *impetus* in that version appear replaced with *momentum* and *inclinatio* respectively. On De Soto and his function as a "precursor" of modern dynamics, cf. P. Duhem, *Études* cit., III, pp. 271 ff.

110 often contains precise expositions of the questions of *Physics*, as stated in the title[312]. Pereira, who tries to defend the Aristotelian opinions on the void and motion in the void from the accusations of Avempace, Philoponus and Thomas, speaks of "maior (aut minor) virtus motiva [major (or minor) motive force]"[313].

Continuous references to *Physics* are also found in Achillini's *De proportione motuum*[314], in which *momentum* is absent. In some pages of this writing, Achillini addresses the question of motion in the void, explicitly referring several times to our passage in the fourth book. Against Aristotle, Achillini believes that motion in a vacuum would not be instantaneous. In the Paduan commentator, we see *declinatio* [decline] appear, that is, the same term used by Michael Scotus in the Arabic version of the Aristotelian text[315]:

> si causa rerum diversitatis in velocitate est diversitas eorum in *declinatione*, id est in gravitate et levitate, sequitur quod quanto magis fuerit grave aut leve, tanto magis erit velox et manifestum est quod hoc convertitur. idest quod quanto magis fuerit velox, tanto magis erit grave et leve[316]. [If the cause of the difference of things in speed is the difference of their *declination*, that is, in heaviness and lightness, it follows that the more heavy or light it is, the more it will be swift, and it is evident that this is reversed. That is to say, that the more it is swift, the more it will be heavy and light.]

Not even in Toledo does *momentum* appear in the meaning of inclination to motion produced by weight or lightness. Toledo's commentary, which did not give Aristotle's text, had considerable importance and enormous diffusion. In fact, in just a few years, six editions were released[317]. Furthermore, it was the official school book adopted by the Society of Jesus for the teaching of physics and was
111 certainly known to Galileo[318]. Toledo's commentary is also of considerable interest for its wealth of information relating to previous commentators and for its use of language clearly oriented towards terminological preferences. On the question of the possibility of motion in a vacuum, the Jesuit reports the opinions of numerous

312 Benedicti Pereiri S.I., de communibus omnium rerum naturalium principiis et affectionibus libri quindecim. Qui plurimum conferunt ad eos octos libros Aristotelis, qui de Physico auditu inscribuntur intelligendos..., Apud Franciscum Zanettum et Bartholomaeum Tosium socios, Romae 1576.

313 *Ibid.*, *Liber XI*, p. 177.

314 *Alexandri Achillini Bononiensis... Opera omnia in unum collecta cum annotationibus… Pamphili Monti...*, Apud Hieronymum Scotum, Venetiis 1568.

315 Cf. *supra*, note 219.

316 *Ibid.*, p. 323; the italics are ours. Achillini, to express the Aristotelian concept contained in our text of the fourth book, also uses the expression "excessum potentiae motoris ad rem motam [excess power of the mover to the thing being moved]" (*ibid.*, p. 329).

317 *D. Francisci Toleti S.I... Commentaria una cum quaestionibus in octo libros Aristotelis De Physica auscultatione. Nunc quinto in lucem edita*, Apud Iunctas, Venetiis 1550. Other editions: J. Dupuis, Parigi 1581; Marsilius Lucensis, Lugduni 1581; in officina Birkmannica, Coloniae Agrippinae 1585; J. Veyrat, Lugduni 1598; apud Juntas, Venetiis 1606.

318 In the *Epistola ad lectorem*, indicating the many Greek and Latin commentaries used, he specifies that he devoted himself with enthusiasm to this burdensome effort, convinced that it was necessary to create an instrument capable of facilitating the task of teachers and students. In regard to Galileo's knowledge of the commentary, it was proven by Carugo and Crombie, who identified passages directly from that text in the *Juvenilia* (v. A. C. Crombie, *Sources of Galileo's early natural philosophy*, within *Reason, Experiment and Mysticism in the Scientific Revolution*, M. L. Bonelli and W. R. Shea, editors, Science History Publications, New York 1975, pp. 157-175).

commentators, expressing his preference for that of Thomas, who – as should be recalled – had challenged Aristotle on the instantaneous nature of motion in a vacuum. And precisely in this context, Toledo offers a significant example of his familiarity with the expressions of the terminists, addressing the problem of the different heavinesses and lightnesses, which produce different speeds, in terms of *intensio qualitatum* [intensity of quality] which are divided into degrees. Referring to the opinions of Avicenna and Avempace on the effects of the differences in *moments* or those of the medium, he states in fact:

> Dicunt enim velocitatem non solum sumendam ex spatio, sed ex ipsa *virtutis*, puta, gravitatis, et levitatis *intentione*, ita, ut ex se santae virtuti tanta respondeat velocitas nullo medio impedie te: et hanc velocitatum vocat naturalem Avempace. reliquam vero, quae a subtilitate† medii sumitur, accidentalem. Nec solum gravitatem aut levitatem penes resistentiam mensurant, sed penes intrinsecum *gradum intentionis*[319]. [For they say that speed must not be taken only from space, but from the very *intesnity* of the *power*, such as heaviness and lightness, in such a way that to a power of itself so great there corresponds such a speed, with not medium impeding it; and this kind of velocity Avempace calls natural. The other, however, which is taken from the subtlety of the medium, he calls accidental. Nor do they measure heaviness or lightness solely according to resistance, but according to the intrinsic *degree of intensity*].

In Contarini's *De Elementis*[320] – another work which, although not a commentary, discusses some questions, especially of *Physics* and *Heavens* – *momentum* is absent. Contarini addresses the questions discussed by Aristotle in the fourth book of *Physics* in *Liber I*. He maintains that the moving principle is intrinsic to bodies; they move, that is, "propria levitate vel gravitate [according to its own lightness or heaviness]". The cause of motion is, moreover, the "inclinatio ad locum ad quem unaquaeque res movetur; inclinatio vero in elementis proprio eorum gravitas seu levitas est qua deorsum sursumve tendunt… [the inclination toward the place to which each thing moves; the inclination in the elements is their own heaviness or lightness by which they tend downward or upward…]"[321]. As for motion in a vacuum, Contarini is

---

† This passage included several errors and has now been corrected. Aside from minor typographical errors, an instance of "velocitatum" was rendered as "voluntatem", and "subtilitate" was rendered as "velocitate".

[319] *Ibid.*, c. 127*r*; the italics are ours. Toledo, confirming the thesis of Thomas (but also of Simplicius and Philoponus), added that in a vacuum, even if the resistance of the medium is suppressed, the internal resistance of the mobile remains: "Nota quod praeter illam resistentiam medii et praeter intentionis gravitatis vel levitatis differentiam quae faceret velociorem vel minus velocem motum, etiam in vacuo est resistentia intrinseca mobilis ipsius [Note that in addition to the resistance of the medium and the difference in the extension of heaviness or lightness that would make the movement faster or slower, there is also the intrinsic resistance of the mobile itself in a vacuum]" (c. 127*v*).

[320] *Gaspari Contareni Cardinalis... de Elementis et eorum mixtionibus libri quinque…*, Apud Andream Wechelum, Parisiis 1564.

[321] *Ibid.*, *Liber I* c. 12*v*. Yet another evident echo of the Aristotelian text can be perceived in another passage of *De elementis*, where the natural inclination to the motion of bodies is defined by Contarini in *forma*: "elementa gravia et levia naturali motu non moveri ab ullo quod extra sit, quodque sit efficiens proximum illorum motus; imo moventur a propria forma, propriaque levitate vel gravitate, tamquam a principio motus, quae in maiori mole maior est et efficacius movet, id est, velocius. Hoc vero est adeo perspicuum, ut sensui etiam sit exploratum, quamvis nulla ratione comprobaretur. moventur igitur elementa a principio intrinseco [the heavy and light elements are not moved in natural motion by anything

112 fully with Aristotle, who he defends against the objections of the Arab and Greek commentators. Clearly referring to the passage in the fourth book, in which the two causes of the greater or lesser speed of bodies are discussed, the commentator indicates the one dependent on the varying heaviness or lightness, with expressions such as the following: "in gravibus et levibus corporibus excellentiam gravitatis sequitur velox descensus, levitatis praestantiam velox ascensus [in heavy and light bodies, the excellence of heaviness is followed by rapid descent, the superiority of lightness by rapid ascent]"[322].

Not even in the famous *Commentarius* of the Conimbricenses is *momentum* used, despite it being present in the version of Aristotle's text that accompanies the commentary in the prints[323]. The term appears, however, in the *Commentarius* of Pace, but not with frequent use[324]. *Momentum*, lastly, is not recorded either in the *Compendium*[325] or in the *Epitome in octo libros Physicorum*[326], both very influential, by the Dominican Crisostomo Javelli.

Other occurrences of *moment* can be traced back, more or less directly, to the Aristotelian tradition and to this passage in *Physics* in the writings of authors of the second half of the sixteenth century, which are particularly significant in the role attributed to them as sources or points of reference for the young Galileo.

This is the case of Giovan Battista Benedetti, whose *Demonstratio proportionum motuum localium* had two partially different editions produced within a year[327]. The Venetian mathematician, in both editions, cited a part of the Aristotelian passage from the fourth of *Physics*, which has been discussed many times. The quotation was precise and corresponded to Argyropoulos's version. Therefore, the expression
113 *momentum gravitatis, aut levitatis* appeared there[328]. Outside of the quote, however,

that is external, that being the proximate cause of their motion; nay, they are moved by its own form, by their own lightness or heaviness, as it were from a principle of motion, which is greater in large bodies and moves more effectively, that is, faster. This, indeed, is so clear that it has been examined even by perception, although it has not been proved by any reasoning. Therefore the elements are moved from an intrinsic principle]" (c. 12*v*).

322 *Ibid.*, c. 15*r*.

323 *Commentarii Collegii Conimbricensis* cit. In the commentary on our texts, it speaks of "inaequalibus ponderis et levitatis [unequal weight and lightness]" (p. 75).

324 *Aristotelis... Julius Pacius latina interpretatione* cit. Pace, in his commentary, limited himself to paraphrasing and clarifying Aristotle's text, without taking a position.

325 *Totius rationalis, divinae et moralis Philosophiae compendium, innumeris fere locis castigatum...*, I-II, Lugduni 1567. In the first volume there is *Question XVII* dedicated to the problem of the void. Javelli, who follows the opinion of his brother Thomaeus, often uses "magis (minus) grave (gravare) [more (less) heavy (burden)]".

326 *Epitome Chrisostomi Javelli Canepitii in universam Aristotelis philosophiam*, Apud Hieronymum Scotum, Venetiis 1552.

327 *Demonstratio proportionum motuum localium contra Aristotelem et omnes philosophos. Per Ioannem Baptistam de Benedictis inventa*, Venetiis 1554. The second edition is from the same year and appears to have been reworked. I use the critical edition provided by Carlo Maccagni, *Le speculazioni giovanili 'de motu' di Giovanni Battista Benedetti...*, Domus Galilaeana, Pisa 1967.

328 "Aristotelis primo in 4. Physicorum, in capitulo de vacuo, ubi scilicet intendit demonstrare quod si detur vacuum tollatur motus etc., ibi autem dicit: cernimus enim ea corporum quae plus gravitatis aut levitatis habent momenti… [Aristotle first in the 4th Physics, in the chapter on the void, where he intends to demonstrate that if a vacuum is given, motion is impossible etc., and there he says: for we

Benedetti, who argued with the Stagirite's opinions regarding motion in a vacuum, preferred to express the same concept using "*proportiones gravitatis et levitatis* [proportions of heaviness and lightness]"[329]. There were no other occurrences of *momentum* in the writing, confirming Benedetti's reluctance for the term, already observed in the *Speculationum Liber*[330].

Also traceable to the Aristotelian tradition is a very interesting page from *De motu gravium et levium* by Girolamo Borro, a text cited by Galileo in *De motu* and certainly well known to him, since Borro was among his teachers in Pisa[331]. On this page, *momentum* appears alongside some terms that Borro presents as equivalent. The peculiar philosopher from Arezzo – paraphrasing a passage from *Heavens*[332] – states that the true progress of science is from what is known "per noi [by us]" to what is known "per natura [by nature]". "By us", nothing is clearer than the actions of the "facoltà [faculties]". It is therefore a question of first explaining these *facultatum actiones* [actions of ability], and then proceeding with their "contemplazione [contemplation]". However, this cannot be achieved immediately, given that

harum facultatum actionibus a maioribus nostris non sint imposita nomina, nisi forte quis levitatis et gravitatis actionem esse existimet *ῥοπή*, id est momentum, inclinationem, pondus, sive nutum, quo efficitur ut gravia et levia corpora susque deque concitentur; quod quanquam absurdum omnino non est, nobis tamen ad has definitiones inveniendas vel parum quidem proderit; ideo hoc momentum, inclinationem, pondus, sive nutum prudentes praeterimus et ad alia, quae hanc nostram causam magis iuvent, omissis ambagibus properamus[333]. [since names have not been assigned to the actions of these faculties by our predecessors, unless perhaps someone considers that the action of lightness and heaviness to be *ῥοπή*, that is, momentum, inclination, weight, or nod, by which heavy and light bodies are caused to be moved by them; which, although it is not at all absurd, will nevertheless be of little use to us in

---

see those bodies which have more heaviness or lightness have more moment]" (C. Maccagni ed., *Le speculazioni* cit., p. 26). The same passage returns in the second edition (ibid., p. 69).

329 Cf. for example, *ibid.*, p. 42.

330 Cf. *supra*, note 207.

331 *De motu gravium et levium ad Franciscum Medicem...*, In officina Georgii Marescotti, Florentiae 1575.

On Borro, as a precursor of Galileo, cf. E. A. Moody, *Galileo and Avempace. The dynamics of the Leaning Tower experiment*, "Journal of the History of Ideas", XII, pp. 163-193 and 375-422 (now reprinted in *Studies in Medieval Philosophy* cit., pp. 203-286). On that curious character that was Borro, cf. the entry in *D.B.I.*, XIII (1971), pp. 13-17 (edited by G. Stabile).

332 *On the Heavens*, 307^b, 28-308^a, 1. A reference to the same passage and a paraphrase very close to this one by Borro can be seen in the *Libri ad scientiam naturalem* by Francesco Piccolomini: "Arist. inquit operationibus gravitatis et levitatis non esse imposita nomina, nisi quis *ῥοπήν*, hoc est inclinationem appellat... nam est excitatio quaedam ad motum... [Arist. says that there are no names assigned to the operations of heaviness and lightness, unless one calls it *ῥοπήν*, that is, inclination... for it is a kind of excitation towards motion...]". However, Piccolomini, unlike Borro, did not use *momentum* to explain *ῥοπή* (*Librorum ad Scientiam de natura attinentium...*, I-II, Apud Haeredes Francisci de Franciscis, Venetiis 1600, II, *De Elementis, Liber II, Ch. XIX*, c. 91*v*). Borro's passage is repeated in the *Absolutissima questio de motu gravium et levium* (Ms. in B.N.F., Conv. Soppressi, B, 10480) attributed by the Catalog to Filippo Fantoni (cf. Ch. B. Schmitt, *The faculty of Arts at Pisa at the time of Galileo*, "Physis", XIV, 1972, fasc. 3, pp. 259 ff.). In the *Quaestio absolutissima*, *ῥοπή* was presented as equivalent to *momentum*, *inclinatio* and *propensio*.

333 *Ibid.*, p. 29.

finding these definitions; therefore we prudently pass over this momentum, inclination, weight, or nod, and quickly proceed to other things that better serve our purpose, omitting the ambiguities.]

114 In addition to suggesting its relationship with ῥοπή, Borro assimilated *momentum* precisely to those terms by which we have learned to see it accompanied or replaced: *inclinatio*, *pondus*, *nutus*. Furthermore, he explicitly clarified that the *momentum* of bodies is a *facultas* [ability], the intensity of which determines the direction (*sursum* [upward] or *deorsum* [downward]) and speed of the motions. In *De motu*, Borro did not use *momentum*, but showed that he was familiar with the distinction introduced by Avempace between *velocitas naturalis* [natural speed] and *accidentalis* [accidental], referred by many of the commentators to our passages in the fourth of Aristotelian *Physics*[334].

Jacopo Mazzoni, professor at the University of Pisa and linked to Galileo by close ties of friendship[335], in some pages of his *De comparatione Platonis et Aristotelis*, in which he discusses the void and the possibility of movement being given to it, promptly cites our passage from the fourth book. From the quotation, we can deduce that Mazzoni uses Argyropoulos's version[336]. In Mazzoni's exposition, alongside the passages on *Physics*, the references to Archimedes's *De insidentibus aquae* (in other words, to the *Galleggianti* [Floating Bodies]) assume a privileged role, which he certainly knew through Tartaglia's edition[337]. According to the Cesenate professor, Aristotle's statements on the impossibility of motion in a vacuum are false, as demonstrated by the *mathematicians*, who are, fundamentally, Plato and Archimedes. They rightly deny the existence of positive lightness and, therefore, of *sursum* [upward] natural motion. As for the *deorsum* [downward] natural motion, it depends on *gravitas* [heaviness]. In all of Mazzoni's work, we never find *momentum* in our meaning. Here is an example of his terminology:

Virtus autem motrix deorsum impellens corpora est gravitas, quemadmodum et illa quae rursus attollit corpora gravia est vis corporis gravioris extrudens minus grave[338]. [The impelling power that moves bodies downward is heaviness, just as that which lifts heavy bodies upward is the force of a

---

[334] *Ibid.*, p. 250 ff.

[335] On the relationships between Jacopo Mazzoni and the Pisan scientist, cf. F. Purnell Jr., *Jacopo Mazzoni and Galileo*, "Physis", XIV (1972), fasc. 3, pp. 273-294.

[336] "Velocitas ergo et tarditas motuum a duabus causis pendent, idest ab efficacia virtutis moventis et a resistentia medii, quae quidem secundum magis et minus velociorem et tardiorem producunt motum. Haud veritatem et ipsemet Aristoteles cognovit in 4. Physicorum in illis verbis: Eorum autem quae plus aut gravitatis aut levitatis habent momenti… [Therefore, the swiftness and slowness of motions depend on two causes, the effectiveness of the moving force and the resistance of the medium, which according to whether it is more or less dense, produce motion that is faster or slower. Aristotle himself recognized this truth in Book 4 of Physics, in these words: For we see that those bodies which have more or less heaviness or lightness have more or less moment…]" (*In universam Platonis et Aristotelis philosophiam Praeludia* cit., pp. 191-192).

[337] *Ibid.*, pp. 190-191. For the edition of Archimedean writings edited by Tartaglia (only the first book), cf. *supra*, paragr. 1, note 3; the *De insidentibus* was reprinted, with the addition of the second book, and always in Moerbeke's version, in 1565 (Curzio Troiano, Venice).

[338] *Ibid.*, p. 190.

heavier body pushing out the less heavy].

It should be emphasized that Mazzoni uses the expression *nullum nutum habere* [have no nod] to indicate a state of absolute indifference in which a body would find itself in perfect equilibrium with an equivalent volume of water[339].

A notable attention to the problem of motion in a vacuum is given by Francesco 115
Buonamici, another Pisan professor, with a clear Aristotelian orientation. Of the more than a thousand pages of his *De motu*[340], the entire *liber quartus* addresses the question of motion in the void and the natural place of bodies. Buonamici and his *De motu* have been considered by many scholars, especially Alexandre Koyré, to be very important sources for the study of Galileo's scientific and philosophical training[341]. Buonamici's exposition is of absolute Aristotelian conformism in regard to the problem of motion in a vacuum. In fact, he defends the Stagirite from the accusations of the Greek and Latin commentators, essentially relaunching the entire series of his arguments as valid and truthful. In these pages, the term *momentum* appears on very few occasions (four in total), and – as we will see – not always with the same meaning. The terms preferred by Buonamici to express the dynamic effects of weight and lightness are *propensio ad locum*[342] [propensity toward the (proper) place], *inclinatio*[343] [inclination], *propensio innata*[344] [innate propensity], *moveri ab innato principio* [moved by innate principle], *principium sui motus internum*[345] [internal principle of self-movement].

The first occurrence of *momentum* is found in *Caput XXIV* of the *liber quartus*[346]. Buonamici intends to discuss what would happen to a mixture that did not have a dominant element, that is, a mixture in which the *moments* of all the elements were perfectly "temperati [tempered]". According to Averroes – as reported by Buonamici – such a mixture could not exist, because in that case it would not tend towards any place. The author of *De motu* defends, however, the Aristotelian position, assigning a perfectly "temperato [tempered]" body its natural place and its specific movement. In the passage in question, the expression *elementorum momentum* [momentum of the elements] clearly indicates the strength with which each element tends towards its own place – a sort of "appetito [appetite]", as Buonamici himself will later define it – ultimately, the speed with which it is led to move by one's own nature[347].

---

339 *Ibid.*, p. 193.

340 *De motu libri X quibus generalia naturalis philosophiae principia summo studio collecta continentur*..., Apud Bartholomaeum Sermartelium, Florentiae 1591.

341 Cf. A. Koyré, *Studi galileiani*, trad. it. by Maurizio Torrini, Einaudi, Torino 1979, pp. 19-40.

342 *De motu libri X* cit., p. 360.

343 *Ibid.*, p. 363.

344 *Ibid.*, p. 366.

345 *Ibid.*, p. 374.

346 From the title "Sit ne par elementorum momentum in aliquo misto, ubi nonnulla Averrois dicta perpenduntur [there is not equal momentum of elements in some mixture, in which some of Averroes's comments are considered]" (*ibid.*, pp. 383-385).

347 "Explicatum est antea quemadmodum mixtum moveatur secundum dominans. Verum ex eo dicto quandam sensum abstrusiorem elicebat Averroes, nempe unum indicari quod in quolibet mixto dominaretur; itaque concludebat ex hoc effici non posse ut par elementorum omnium momentum in

116 *Momentum* is used, therefore, in the meaning that we have seen it undertake in the versions of the fourth book of *Physics*. The same concept is expressed by Buonamici also in terms of *elementorum robur* [strength of the elements] and *elementorum virtus* [power of the elements][348].

The second occurrence of *momentum* is recorded, again in the *liber quartus*, on a page in which Buonamici addresses the problem of the increase in speed that occurs, in natural motion, towards the end. It goes without saying how interesting this theme may have appeared to Galileo. Buonamici assigns the increase in speed that occurs towards the end of natural motion to the increase in the *degree* of *form* of motion. He presents *gravitas* as *forma motus* [form of movement], and, as such, considers it capable of *intentio* [intensity]. From the *intentio* of *forma*

> sequitur ut efficacior sit elementorum functio, atque perfectior, perfectior autem functio est motus celerior et iam instans consecutio loci, donec ad summum perfectionis gradum veniat, cumque protinus est locus et quies, et quasi imitetur illum progressum quem in qualitate notavit Galenus, inquiens quaedam *exiguum consecuta momentum maximas efficere mutationes*, efficacius multo moveatur[349]. [it follows that efficacy is the function of the elements, as well as perfection, however perfect function is swift motion and immediate approach to the proper place, until it reaches the highest degree of perfection, where there is immediate place and rest, and just like the progression noted by Galen, who says that a *small gained amount of momentum causes a great change*, effectively greater motion.]

Momentum here presents a meaning very close to that of the previous passage. Our term indicates, in fact, a tendency to move according to a certain speed. In this context, which is not very clear, it would however seem that *momentum* is also capable of *intentio* [intensity], and that the increase in the "grado di velocità [degree of speed]" depends on this. It should be noted that Buonamici emphasizes here the closeness between *momentum* and *gradus* [degree], on which Galileo will insist so much. Indeed, one has the impression that the Pisan scientist followed very similar reasoning at a certain point. If *gravitas* [heaviness] is divided into "grado [degrees]", which *intenduntur* [are intensified] as the mobile approaches its natural place, and if the *momentum* of a mobile – i.e., its "natural" speed – depends on its *gravitas* [heaviness], it is clear that at each *gradus gravitatis* [degree of heaviness] corresponds to a different *momentum*, that is, a different *gradus velocitatis* [degree of speed]. In natural motion, the heaviness of each mobile increases as it approaches its proper place. The result is a continuous increase in *momentum* and speed.

A third occurrence of *momentum* is recorded in a significant context of *Caput XXXIX*. The discussion is still related to the problem of acceleration, which is observed in the final part of natural motion. Buonamici refers to the Aristotelian explanation of the increase in speed as a consequence of the increase in heaviness as the body approaches its natural place. Some have objected that, if the Aristotelian

---

mixto unquam inesset… [It has been explained previously how the mixture is moved according to the dominant. It is true that Averroes elicited from this statement a certain more abstruse meaning, namely, that one is indicated to dominate in any mixture; and therefore he concluded from this that it is impossible for equal momentum of all elements to ever exist in a mixture.]" (*ibid.*, p. 383).

348 *Ibid.*

349 *Ibid.*, *Cap. XXXVIII*, p. 414; the italics are ours.

thesis were true, the same weight, placed on the scale at different distances from
the center of Earth, should weigh differently, while experience shows that this does 117
not occur. Buonamici believes that this objection can be answered conclusively in a few words:

> Nam pondus in trutina probatur per contrapensionem. quare tantundem momenti ponderi atque illi quod ex altera parte appenditur, ubicunque illa sint accedit[350]. [For the weight on the balance is tested by the counterweight. Therefore, a weight has the same momentum as the one which hangs from the other side, wherever they may be.]

If one has the mistaken impression that the weight on the scale does not increase as it approaches the center of the world, this happens because its increase corresponds to the exactly proportional increase of the counterweight, by which it is balanced. Their *moments*, if they were equal at a given distance from the center, will be equal *everywhere*. Here *momentum*, despite referring to the balance, does not have – we must be careful here – the technical meaning of "forza motrice [driving force]" of the same weight, variable according to the different points of application. In our context, *momentum* indicates "inclination" to move at different speeds, an inclination that depends *exclusively on weight* (Buonamici has in mind, in fact, a balance of equal arms): *momentum* grows because weight increases. That is, we are faced with a semantic value of *momentum* that is absolutely indistinguishable from the previous one. Buonamici assigns our term the delicate function mediating between weight and natural speed.

This reading is confirmed even more explicitly by the fourth occurrence of *momentum*, again in *Caput XXXIX*, where Buonamici, concluding his exposition on this point, presents the causes (efficient and material) of the natural motion of bodies:

> Nam sicut ille qui est in cursu, vel *paullo momento* praecipitatur, sic in unum opus convenientibus efficiente quod movet secundum naturam, et materia quae secundum naturam movetur...[351]. [For just as one who is in the middle of running is hurled headlong, by even the *slightest moment*, so too when there come together into one work both the efficient cause which moves according to nature, and the matter which is moved according to nature...]

It is difficult, in fact, to distinguish in this case whether *paullum momentum* [small amount of momentum] indicates the small weight that causes the modest speed or, instead, the modest speed itself.

In the few occurrences of *momentum* found in Buonamici's *De motu*, we can therefore recognize the clear prevalence of the meaning of inclination to move according to different speeds, which we have seen our term constantly undertake throughout the Aristotelian tradition (text and commentary) of the fourth book of *Physics*. Of particular interest is Buonamici's use of *momentum* in the discussion of
acceleration. He had, in fact, recognized that *momentum* is not only different from 118

[350] *Ibid*., p. 416.

[351] *Ibid*., p. 417; the italics are ours.

body to body, but in the same body it does not remain constant, varying according to the distance from the end of the motion. Buonamici had entrusted the explanation of *intentio velocitatis* [intensity of speed] to the growth of *momentum*[352].

[352] Buonamici also made the explanation of the force of impact depend on this same principle. And in fact the speed, as Galileo will say in *Mecaniche*, increases the *momento* of weight: "si vero extollens securim percutiat, illud scindit cum ceteroqui minus habeat ponderis id quod percutit, quam id quod premit; namque intervenit motus qui adauget pondus et duplici gravitate seu propensione movetur, altera nativa et propria atque altera quae percutientis est [but if he raises an axe and strikes, it cleaves in its strike with an otherwise lesser weight, than that which would otherwise press; for the intervening movement increases the weight and is moved by a doubled heaviness or inclination, one is native and proper while the other strikes]" (*Ibid.*, p. 415).

CHAPTER IV

"MOMENTUM" AS INDIVISIBLE MOTION IN ARISTOTLE'S "PHYSICS"

1. THE TEXT

In the sixth book of Physics, in the translations of the Aristotelian text and the Greek commentators, as well as in the same previously examined Latin commentaries, we also find *momentum* used in a different meaning from those of inclination to motion according to speed determined by weight (or lightness) and variable effectiveness of weight in mechanical systems. In two passages of the Aristotelian text of the sixth book, to which we will refer, the Latin translators, this time substantially in agreement since the Middle Ages, generally resorted to *momentum* to render the Greek *τὸ κίνημα* [the movement]. 119

As is well known, the central theme addressed in the sixth book of *Physics* is that of the divisibility of the continuum. After having defined "continuity", "contact" and "consecutiveness"[1], Aristotle focused his attention on the "continuum", to determine whether it was composed of indivisibles or, on the contrary, divisible into ever further divisible parts. The subjects of his analysis were magnitude, time and motion, previously defined as "continuous magnitudes". Aristotle denied that the continuum was composed of indivisibles, showing the absurdities to which the assumption of such a hypothesis would have led. In the case of magnitude (*μέγεθος*) and motion (*κίνησις*), if the former is composed of indivisibles, the latter must also be considered as such: the motion that passes through an indivisible magnitude will also be indivisible. On the other hand, there are two inviolable logical conditions: that which moves is in motion, and that the moving (*τὸ κινεῖσθαι*) and the having been moved (*τὸ κεκινῆσθαι*) are not contemporaneous. If we admit that magnitude is composed of indivisible points and motion of indivisible motions, these two conditions are violated: indeed, there is either movement without motion, or absolute simultaneity between the "moving" and the "having been moved". So, magnitude, movement and time cannot be resolved into indivisibles, but rather into 120
ever further divisible parts.

---

[1] Z 226^{a}, 18-229^{a}, 6.

P. Galluzzi, *Momento: Galilean Studies*, Archimedes 78,
https://doi.org/10.1007/978-3-032-35049-7_4

In the initial pages of the sixth book, Aristotle used a specific example to illustrate this reasoning. He stated, in fact, that if the quantity ΑΒΓ is composed of indivisible parts (Α, Β, Γ), the movement ΔΕΖ that passes through it will also consist of indivisible parts. Therefore, if an object Ω travels ΑΒΓ, it will surpass Α with a motion Δ, Β with a motion Ε, and Γ with a motion Ζ, all indivisible. In Δ, Ε and Ζ, since they are indivisible, the moving and the having been moved cannot be distinguished, which therefore appear simultaneous. This is absurd and unacceptable, and it would be like admitting – states Aristotle – that someone is walking towards Thebes and simultaneously has already completed the journey towards Thebes[2]. It is at this point that the passage is interesting for us. Let's see it, first of all, in the Greek text:

εἰ δὲ τὴν μὲν ὅλην τὴν ΑΒΓ *κινεῖταί τι, καὶ ἡ κίνησις ἣν κινεῖται τὰ* ΔΕΖ *ἐστί, τὴν δ' ἀμερῆ τὴν* Α *οὐθὲν κινεῖται ἀλλὰ κεκίνηται, εἴη ἂν ἡ κίνησις οὐκ ἐκ κινήσεων ἀλλ' ἐκ κινημάτων καὶ τὸ κεκινῆσθαί τι μὴ κινούμενον· τὴν γὰρ* Α *διελήλυθεν οὐ διεξιόν. ὥστε ἔσται τι βεβαδικέναι μηδέποτε βαδίζον· ταύτην γὰρ βεβάδικεν οὐ βαδίζον ταύτην*[3]. [but if the whole ΑΒΓ is moved, and the motion in which it is moved is ΔΕΖ, and if the part A is not moved at all but has been moved, then the motion would not be of motions but from movements, and the having been moved would be something not moving; for this assumes A has passed through but does not go through. It would be like saying that someone has walked without walking; for this has walked, not walking.]

Also in the sixth book of *Physics*, in the concluding pages, there is a passage of similar interest. The context is similar to the one now examined; the only difference is that the emphasis, in this case, is placed not on magnitude and movement, but on movement and time. The Stagirite, in fact, wants to demonstrate that the indivisible cannot move. If a partless object moves or changes from ΑΒ to ΒΓ, let Δ be the time during which the movement or change occurs. During time Δ, it is necessary for the object to be either in ΑΒ or ΒΓ, or partly in ΑΒ and partly in ΒΓ. It cannot be part in ΑΒ and part in ΒΓ, because the indivisible does not consist of parts. Nor can it be in ΒΓ, since it would have already completed the change that, by assumption, it is undergoing. It remains, therefore, that it must be in ΑΒ; but in this case, it will be at rest, since being at rest means remaining in the same place for a certain time. We can, therefore, conclude

ὥστ' οὐκ ἐνδέχεται τὸ ἀμερὲς κινεῖσθαι οὐδ' ὅλως μεταβάλλειν· μοναχῶς γὰρ ἂν οὕτως ἦν αὐτοῦ κίνησις, εἰ ὁ χρόνος ἦν ἐκ τῶν νῦν· ἀεὶ γὰρ ἐν τῷ νῦν κεκινημένον ἂν ἦν καὶ μεταβεβληκός, ὥστε κινεῖσθαι μὲν μηδέποτε, κεκινῆσθαι δ' ἀεί. τοῦτο δ' ὅτι ἀδύνατον, δέδεικται καὶ πρότερον· οὔτε γὰρ ὁ χρόνος ἐκ τῶν νῦν οὔθ' ἡ γραμμὴ ἐκ στιγμῶν οὔθ' ἡ κίνησις ἐκ κινημάτων· οὐθὲν γὰρ ἄλλο ποιεῖ ὁ τοῦτο λέγων ἢ τὴν κίνησιν ἐξ ἀμερῶν, καθάπερ ἂν εἰ τὸν χρόνον ἐκ τῶν νῦν ἢ τὸ μέγεθος ἐκ στιγμῶν[4]. [thus, it is not possible for the indivisible to be in motion or to change in any way; for only one condition could have made it possible for it to have motion, namely, time is composed of nows, in which case at any now it would have moved or changed, so that it would never be in motion, but would always have been moving. This is impossible, as has been demonstrated before: time is not composed of nows, nor is the line composed of points, nor

[2] 231^b, 18-232^a, 5.
[3] 232^a, 6-11.
[4] 240^b, 30-241^a, 6.

is motion from movements; for the one who says this does nothing other than to make motion from indivisibles, just as if time consists of nows or magnitude from points.]

In the two passages cited above from the sixth book of *Physics*, the entire 121
range of terms used by the Stagirite to indicate the different aspects of movement appears. Furthermore, the expression *οὐκ ἡ κίνησις ἐκ κινημάτων* [motion is not from movements] is recorded twice. These are the only two occurrences of *τὸ κίνημα* [the movement] in *Physics*[5]. And it is precisely this Greek term that the Latin translators generally believed they could render with *momentum.*

In examining the Latin translations of the two Aristotelian passages, we will follow the same criterion that we followed in the previous chapter for the passages of the fourth book. Therefore, we will begin with William of Moerbeke's *Nova Translatio*. The Flemish Dominican introduced *momentum* to translate *τὸ κίνημα* [the movement] in the two passages of the sixth book:

Si vero secundum totum ABC moveatur aliquid, et motus quo movetur DEZ est, secundum autem impartibile A nihil movetur, sed motus est; erit utique motus *non ex motibus, sed ex momentis*. Et *motum esse* aliquid non motum; secundum enim A transivit non transiens[6]. [Indeed, if something is moved according to the whole ABC, and the motion by which it is moved is DEZ, however, according to the indivisible A, nothing is moved, and yet it is moved; therefore, the motion will be *not of movements, but of moments*. And the *completed motion* is not moved; for it has traversed A without traversing.]

Quare non contingit impartibile moveri, neque omnino mutari. Solum enim sic esset ipsius motus si tempus esset ex ipsis nunc. Semper enim in ipso nunc motum esset et mutatum; quare moveri quidem nequaquam est, motum autem esse semper. Hoc autem quod impossibile sit ostensum est prius. Neque enim ex ipsis nunc tempus, neque linea ex ipsis punctis, *neque motus ex momentis* est[7]. [Therefore, it does not happen that the indivisible can be moved, nor can it be changed at all. For motion itself would exist only if time were composed of the instants themselves. For it would always be in the instant both moved and changed; therefore, it is by no means possible to be moved, but always to have been moved. That this is impossible has been demonstrated before. For neither is time made up of instants, nor is a line made of points, *nor is motion made of moments*.]

It will be appropriate, first of all, to recall that the Flemish Dominican had not used *momentum* to translate *ῥοπή*. Instead, he now uses our term to render *τὸ κίνημα* [the movement]. Furthermore, Moerbeke uses *motus* [motion] for *κίνησις* [motion], which in the Aristotelian text indicates the action of moving, while *τὸ κίνημα* [the movement] expresses the motion already completed. Therefore, *momentum* here means an already completed, indivisible movement. It stands to *motus* [motion] as the "point" is to the "line" and *nunc* [instant] to *tempus* [time]. The *motus*, according to what Aristotle states in the two passages of the sixth book, is not composed of *moments*, because the collection of *moments* is a *non-movement*, in other words, a succession of "rests".

[5] In truth, these are the only two occurrences of *τὸ κίνημα* in the entire Aristotelian corpus; cf. H. Bonitz, *Index aristotelicus*, Akademie Verlag, Berlin 1955, p. 394, *sub voce*.
[6] I use the text published in the *Expositio… Gualterii de Burley* cit., cc. not num.; the italics are ours.
[7] *Ibid.*; the italics are ours.

Momentum here presented the meaning of minimum movement, which we have seen recorded in some medieval lexicons[8]. It is needless to emphasize that this semantic value was not easily attributable to that connected to *momentum* =
122 inclination (*ῥοπή*). In any case, for the reader of *Physics* in Moerbeke's version, there were no problems, since *momentum* was only used with this meaning[9].

In the version conducted by Argyropoulos, no original solutions are found in relation to these passages. In fact, he follows Moerbeke in rendering *τὸ κίνημα* [the movement] with *momentum* and *κίνησις* [the motion] with *motus* [motion]. Thus, in Argyropoulos's version, our term was used as the equivalent of both *ῥοπή* and *τὸ κίνημα*. In the two passages in which the Greek text presented this last term, Argyropoulos translated:

a) Motus profecto non ex motibus, sed *ex momentis* constabit[10]. [Motion surely consists not of motions, but *of moments*.]

b) Neque enim tempus ex nunc ipsis, neque linea ex punctis, neque motus *ex suis momentis*[11]. [For neither is time made up of instants themselves, nor a line of points, nor motion *of its own moments*.]

[8] Cf. *supra*, Ch. I, pp. 14-15.

[9] It will be useful to indicate – as we did in the previous chapter (cf. note 219) – the solutions adopted by Michael Scotus, to whom the version of Aristotle's works with Averroes's great commentary is attributed. In the version attributed to Scotus, no *momentum* was used. Let's see the two passages: "Et si aliquid movetur per spatium ABC, spatium quo movetur est DHZ; sed nihil movetur omnino per aliquid quod non habet partem, et est pars A, sed iam movebatur; motus igitur erit compositus *non ex motibus* et aliquid moveretur et iam movebatur absque eo quod semper movebatur.... Impossibile est igitur impartibile moveri, aut transmutari omnino. Posset enim hoc modo moveri solum si tempus componeretur ex instantibus. quoniam semper moveretur et transmutaretur in instanti et sic nunquam moveretur et esset iam motum. Et iam declaravimus hoc esse impossibile. tempus enim non componitur ex instantibus, neque linea ex punctis, *neque motus ex finibus*. dicens enim hoc nihil magis dicit quam quod motus componitur ex partibus indivisibilibus, ac si diceret quod tempus componitur ex instantibus et longitudo ex punctis [And if something moves through the space ABC, the space through which it moves is DHZ; but nothing moves at all through something that does not have a part, and it is part A, but it was already being moved; therefore, the motion will be composed *not of motions*, and something would be moved and already was being moved without always being moved.... Therefore, it is impossible for the indivisible to be moved or to be changed at all. For it could only be moved in this way if time were composed of instants. For it would always be moved and changed in an instant, and thus it would never be moved and would already be moved. And we have already declared this to be impossible. For time is not composed of instants, nor a line of points, *nor motion of indivisibles*. For saying this is no different from saying that motion is composed of indivisible parts, as if one were to say that time is composed of instants and length of points]" (*Aristotelis Opera... cum Averrois Cordubensis... Commentariis* cit., IV, pp. 250 and 302). In the first passage, Scotus limited himself to defining what the ultimate component of motion was not (*non motus*), without further specifications. In the second, he introduced *finis*, with which he intended to indicate the indivisible constituent of motion, what remains when the division ends, as it is not possible to proceed further.

[10] I quote from the Laurentian Codex (*Plut.* 84, 1, c. 54*v*), which provided the text for the prints; the italics are ours. In the first draft of the version (dedicated to Cosimo the Elder), Argyropoulos had given a slightly different text: "non ex motibus tum motus, sed *ex momentis* profecto, quae sunt ultima motus finesque constabit [Motion is not composed of motions, but indeed *of moments*, which are the ultimate parts and termini of motion]" (*Plut.* 84, 7, c . 91*r*). In the first version, there is a need to specify the meaning of *momentum* by placing a clear expression alongside it.

[11] *Ibid*. (*Plut.* 84, 1, c. 64*v*); the italics are ours. Also in this case, the version of the first draft presented some negligible differences: "Nec enim tempus ex ipsis nunc individuas, nec ex suis linea punctis, nec ex suis denique motus constat *momentis* [For neither is time composed of individual instants, nor a line of its points, nor finally motion of its *moments*]" (*Plut.* 84, 7, c. 108*v*).

Nifo also presented Argyropoulos's solution verbatim, limiting himself to replacing *sane* [reasonably] with *profecto* [certainly] in the first passage[12], and omitting, in the second, the *suis* [its own] that precedes *momentis*[13]. The version of **123**
Moerbeke was repeated in the second translation published in the Venetian edition of Averroes's commentary on *Physics*[14]. As we have seen previously, the versions of *Physics* that accompanied both the translation of Simplicius's commentary published by Scotti[15] and the commentary by Domingo de Soto[16] depended on the text of this Giunti print. In fact, in both passages of the two commentaries, the solutions adopted by Moerbeke were exactly reproduced[17].

A slightly different translation, yet attributable to Argyropoulos's model, was found in the edition of Simplicius's commentary on *Physics* published in 1551 by the Giunti in Venice[18]. Here are the solutions adopted:

a) si autem non sunt motus, *motus non esse ex motibus*[19]. [but if there is no motion, *motion is not of motions*.]
b) Neque enim tempus ex ipsis nunc, *neque motus ex momentis*[20]. [For neither is time of instants itself, *nor motion of moments*.]

On the contrary, the Aristotelian Giulio Pace, for the two passages of the sixth book, avoided using momentum to render *τὸ κίνημα* [the movement][21]. The solution adopted by Pace in the first context consisted in simply introducing the expression *terminus motus* [terminus of motion]:

a) erit utique motus non ex motibus, sed *ex terminis motuum*; et aliquid motum erit quod non movebatur[22]. [indeed, motion will not be of motions, but *of the termini of motions*; and something will be moved which was not moving.]

Here, in a certain sense, the task was facilitated by the fact that in the Greek text, alongside *τὸ κίνημα* [the movement], there was an equivalent expression, *πέρας τῶν κινήσεων* [the limit of the movements]. Pace limited himself to translating the latter. In the second passage, *τὸ κίνημα* was the only term used to indicate the indivisible of motion. It could therefore not be ignored:

Uno enim modo sic esse potest eius motio, si tempus constet ex momentis; semper enim in momento res mota est et mutata, adeo ut numquam moveatur, sed semper mota sit. Hoc autem esse impossibile probatum antea fuit, quia nec tempus ex momentis, nec linea ex punctis, nec motus ex terminis qui

[12] *Augustini Niphi... Expositio* cit., p. 443. Alongside this version, which he defined as *nova* [new], Nifo provided an *antiqua* [old], corresponding to Moerbeke's *Nova Translatio*.
[13] *Ibid*., p. 517. In this case, Nifo only gave the *versio nova*.
[14] Cf. *supra*, Ch. III, note 231; see also 251*r* and 302*v*.
[15] Cf. *supra*, Ch. III, note 236.
[16] Cf. *supra*, Ch. III, note 235.
[17] For Simplicius, cf. ed. cit., pp. 333 and 364. For De Soto, cf. ed. cit., pp. 308 and 349.
[18] *Simplicii Commentarii... Gentiano Herveto... interpretatione donati* cit.
[19] *Ibid*., c. 161*v*; the italics are ours.
[20] *Ibid*., c. 177*v*; the italics are ours.
[21] *Aristotelis... Jul. Pacius... illustravit* cit.
[22] *Ibid*., p. 199; the italics are ours.

124 vocantur *κινήματα*. Nihil enim aliud facit qui hoc dicit, quam motum constare ex iis quae partibus carent, perinde ac si tempus constitueretur ex momentis aut magnitudinem ex punctis[23]. [For there is only one way in which the motion can be, if time consists of moments; for the thing is always moved and changed in a moment, so that it is never to be moved, but may it always be moved. However, this has been proven to be impossible previously, because neither time consists of moments, nor a line of points, nor motion of termini, which are called *κινήματα* [movements]. For he who says this does nothing other than make motion consist of things that lack parts, just as if time were constituted of moments or magnitude of points.]

Pace, who had used *momentum* in the meaning of "instant" of time[24], could not repeat it to indicate the indivisibility of motion. He therefore considered it legitimate to translate somewhat freely with *terminus motus* [terminus of motion], adding the Greek word for accuracy. Pace's version closely resembles that of Argyropoulos, to which, in essence, those of Périon and Brucioli can also be attributed: both used *momentum* (*moment*) in the sense of the indivisible aspect of motion[25].

Pace's model is followed in the version that accompanies the Conimbricenses' commentary on *Physics*[26], which chose the following solutions for the two passages from the sixth book:

a) motus profecto non ex motibus constabit, sed *ex terminis motus*[27]. [motion certainly will not consist of motions, but of the termini of motion]

b) Hoc enim tempus ex momentis est, nec linea ex punctis, nec motus ex eius *divisionibus*, quae *κινήματα* appellant[28]. [For this reason, time is not made of moments, nor a line of points, nor motion of its *divisions*, which are called *κινήματα* [movements].]

Even in the case of the Conimbricenses' *versio*, the orientation was probably influenced by the fact that *momentum* had already been presented with the value of an atom of time[29].

It can, therefore, be concluded that, in the translations of these locations in *Physics*, we find almost complete unanimity. The only exception is, in fact, Pace. It is therefore a different situation from what we observed for *ῥοπή*.

For the medieval translators of Aristotle's *Physics*, the legitimate meaning of *momentum* – the only one, one might say – therefore seems to have been that of indivisible motion; a meaning that is very close to that "quasi movimentum [almost movement]", which we have observed in some of the medieval lexicons

---

23 *Ibid.*, pp. 236-237.

24 The translation of *τὸ νῦν* [the now] with *momentum* is constant throughout the sixth book.

25 *Librorum omnium naturalis philosophiae Aristotelis... Perionis versione* cit.: "Nam nec tempus ex punctis temporis constat, nec linea ex punctis, nec motus ex momentis [For neither is time composed of points of time, nor a line of points, nor motion of moments]" (p. 328). *La Phisica... tradotta di greco in volgare* cit.: "e il moto consisterà non di moti, ma di momenti e sarà mossa alcuna cosa, che avanti non era mossa [and the motion will consist not of motions, but of moments, and something will be moved that was not moved before]" (c. 133*r*).

26 *Commentarium Collegii Coninbricensis* cit.

27 *Ibid.*, col. 235; the italics are ours.

28 *Ibid.*, col. 310-311; the italics are ours.

29 Cf., for example, *ibid.*, coll. 285-286.

noted above[30]. The fifteenth and sixteenth century translators limited themselves to confirming Moerbeke's choice and therefore legitimized the double – on the other hand, original – meaning of *momentum*: "inclination" to move produced by the **125**
weight (or lightness) and "indivisible" movement. In examining the comments on these texts, we will see, however, that the term and the discussion in which it was inserted must have appeared, even in this case, anything but perspicuous and easy to understand.

2. The Greek Commentators

In Themistius's *Paraphrase* of Aristotle's *Physics*, there are numerous interesting ideas in the pages dedicated to the divisibility of motion, magnitude and time. The author of the translation was, as we have seen, Ermolao Barbaro[31]. In the entire paraphrase of Themistius, there is only one occurrence of *τὸ κίνημα* [the movement][32]. Barbaro did not use *momentum* to translate the Greek term. The famous humanist, whom we have seen use our term to express the dynamic inclination produced by weight or lightness[33], used *momentum*, as well as the more usual *nunc* [instant], to translate the Greek *τὸ νῦν* [the now]: often employing the expression *nunc sive momentum* [instant or momentum][34]. In the *Paraphrase* of the fourth book, Themistius posed, among other things, the problem of the division of motion through time. And he wondered if what resulted from the division could be considered, with respect to movement, as the parts of time are to time:

> Cum ergo dividimus motum et in plura momentis secamus aliud quam tempus et temporis partes asserimus? Praeteritum enim et futurum ex quibus constat tempus, nihil aliud sunt quam series prioris et posterioris in motu, quae *momentis* definiuntur, et tamquam aliud atque aliud numerantur[35]. [When therefore we divide motion and cut it into several moments, do we assert something other than time and the parts of time? For the past and the future, from which time is made up, are nothing other than a series of earliers and laters by means of motion, which are defined by *moments*, and are counted as distinct from each other.]

Momentum, clearly denoted here with the meaning of instant of time, was therefore suitable for measuring motion. The reality of *momentum* seemed in this passage to present itself on the border between time and motion (in the Greek text, the term used is *τὸ νῦν* [the now]). As Themistius had the opportunity to clarify a little later, *momentum* is, "per se", *extremum temporis* [the terminus of time].

[30] Cf. *supra*, Ch. I, pp. 14-15.
[31] *Themistii… Paraphrasis* cit.
[32] This is what appears in consulting the *Index verborum* in *CAG*, V.
[33] Cf. *supra*, Ch. III, p. 100.
[34] Cf., for example, *Themistii... Paraphrasis* cit., p. 150.
[35] *Ibid*., p. 148; the italics are ours.

However,

> quia tempus nihil aliud est quam partes motus secundum prius et posterius numeratae, quotiens momentum et numeratum accipitur, tunc motus quoque ipsius extremum et terminus fuerit. Ita, per accidens, erit terminus motui... sed per se terminus tempori est, quod numero motus describitur[36]. [because time is nothing other than parts of motion counted according to earliers and laters, whenever a momentum is taken and counted, then there would also be boundaries and termini of motion. Thus, incidentally, there will be a terminus to the motion... but intrinsically, there is a terminus to time, which is described by the quantity of motion.]

126 The ultimate constituent of time, *momentum*, as a quantity of motion, was therefore a terminus *per accidens* [incidentally]. In reality, as Themistius specified, motion can only be divided into motions, just like line and time into lines and times respectively[37]. Throughout this part of the exposition, the Greek term that Barbaro rendered as *momentum* was still *τὸ νῦν* [the now][38].

In the final comment on the fourth book of the *Paraphrase*, Barbaro returned to the problem of the division of motion, recalling that it can be conceived in different ways. One of these consists in the division according to time, whose termini, the *moments*, define the termini of motion. As for the specific termini of motion – those that correspond to points for the line or *moments* for time – Barbaro declared that *appellantur movimina* [they are called *movimina*][39]. The term *movimina* is quite interesting, even if it is not recorded in lexicons and dictionaries. It is, in fact, difficult to escape the impression that Barbaro attributes to the noun *movimen* the meaning of indivisible movement, similar to the value assigned to *momentum* in the versions of Moerbeke and the other translators of the two passages of *Physics* examined above.

Another characteristic term is used by Barbaro in the opening pages of the sixth book of the *Paraphrase*, in which it is discussed whether the continuum is composed of indivisibles or continuously divisible elements:

> si magnitudo individuis constaret, ipsum quoque motum ex individuis et impertibus motulis constitutum[40]. [if magnitude were composed of indivisibles, then the motion itself would be constituted from indivisibles and inseparably small motions.]

Motulus [lesser motion] is not difficult to decipher, although it is not a common term. *Impers motulus* corresponds in fact to *ἀδιάρετος κίνησις* [inseparable motion][41]. Following the example of previous translators of *Physics*, it would have been logical to translate it with *momentum*. But Barbaro evidently believes he has already abused the term, burdening it with the responsibility of two very distant meanings ("inclination" and "instant"). And he absolutely wants to avoid

---

[36] *Ibid.*, p. 151.
[37] *Ibid.*
[38] *Ibid.*, pp. 150-159.
[39] *Ibid.*, p. 160.
[40] *Ibid.*, p. 172; the italics are ours.
[41] *CAG*, V, p. 183, 16-17.

assigning it an even different third meaning. For this reason, he falls back on *motulus* [small motion] and also subsequently avoids using *momentum* in the meaning of the indivisible terminus of movement.

This can be seen clearly in the parts of the *Paraphrase* that correspond to our passages from the Aristotelian text of the sixth book. In the first, the position of those who intend to maintain that motion is made up of indivisibles is declared absurd:

> Quis uno te excepto conniveat motum eum qui super magnitudinem illam ex individuis tribus A,
> B, C compositam exercetur, *non ex tribus motibus* constitutum, sed *ex tribus terminis motus* et (quod **127**
> cum venia dixerim) *ex tribus motum mutatumque esse*?[42] [Who, except you alone, would concede that motion over that magnitude composed of three indivisibles, A, B, and C, constituted *not of three motions* but *of three termini of motion* and (if I may say so) *from those three it is moved and completed change*?]

Barbaro's translation is, in this case, faithful to the letter of the Greek text, which presents respectively *ἐκ περάτων κινήσεως καὶ τοῦ κεκινῆσθαι* [from the termini of motion and the been moved][43]. If anything, it is worth noting the hesitation that accompanies the substantivization of the passive perfect infinitive of *moveo* [to set in motion] and *muto* [to change].

The refusal to use *momentum* in the meaning of indivisible motion in the second passage is more explicit:

> Atque una tantum ratio est qua per se moveri possent impertia, si videlicet constaret ex momentis tempus. Quum enim res momento quolibet moveretur, quod necesse utique est, si de momentis tempus existeret, constitui posset punctum et individuum momentis singularis motum fuisse atque mutatum, et (quod sequens est) toto tempore in motu et agitatum fuisse. Ita eveniret ut motus quoque ex iis quae mutatum esse appellaverim, id est ex terminis motus, componeretur, ex momentis tempus coalesceret, ex insectilibus atque punctis magnitudo consurgeret. Praeterea mutatum esse non semper priusquam mutari aut moveri contingeret[44]. [And there is only one reason by which the indivisibles can be moved by themselves, namely, if they were constituted of moments of time. For indeed, if a thing were to move in any given moment, which is certainly necessary if time were to consist of moments, a point or indivisible could be established as having been moved and changed in those singular moments, and (as a consequence) it would have been in motion and agitated throughout the whole time. Thus, it would follow that motion itself would also be composed of what I would call completed change, that is, of the termini of motion, that time would coalesce from moments, and magnitude would arise from indivisibles and points. Furthermore, the completed change does not always occur before something is being changed or moved.]

Here, Barbaro had to contend with the extreme complexity of Greek terminology. Meanwhile, the choice of *momentum* to render *τὸ νῦν* [the now] had already been made, and it was too late to retract it.

Our term, therefore, could not be used to translate *τὸ κίνημα* [the movement] which appears – this is the only occasion in the entire *Paraphrase* of Themistius – in this text[45]. Barbaro resorted to *mutatum esse* [a completed change], prudently

---

[42] *Themistii... Paraphrasis* cit., p. 173; the italics are ours.
[43] *CAG*, V, 184, 15-16.
[44] *Themistii... Paraphrasis* cit., p. 184.
[45] *CAG*, V, 202, 17-18.

proposed in the previous passage and now introduced again with some caution. Thus, *τὸ κίνημα* [the movement] and *τὸ κεκινῆσθαι* [the having been moved] was rendered with a single expression, which flattened the distinction between the two terms: the first, in fact, indicated a completed motion, but minimal, or rather indivisible; while the second was devoid of this infinitesimal connotation and generically expressed the movement already achieved.

In the version of Themistius's *Paraphrase* by Ermolao Barbaro, *momentum* appeared, then, both in the meaning of inclination to movement and in that of indivisible of time and, *per accidens* [incidentally], of motion, but it was never used in that of indivisible of motion *per se*.

In the version of Simplicius's commentary on *Physics* performed by Filalteo[46],
128 there are numerous occurrences of *momentum* in the sense that now interests us. Before examining the commentary on the passages of the sixth book, it will be appropriate to focus on some parts of the fourth book, in which the translator showed considerable quandary and confusion with certain Greek terms, which he decided to render arbitrarily with *momentum*.

This is the commentary on the passage from the fourth book in which Aristotle examines the analogies between time and movement. Having already previously excluded the idea of concurrence between time and motion, the Stagirite suggests considering time a property of motion. Time would therefore be the quantity of movement, according to before and after. Now, it is evident that the movement follows magnitude (the mobile moves, in fact, from one point to another point) and is followed by time, which quantifies it according to before and after. Since magnitude is continuous, movement and time will necessarily be continuous. The "point" is followed by "that which moves" (the mobile, the moved), which is followed by the "instant". "Point", "mobile" and "instant" are, therefore, corresponding terms. Just as the mobile can be said to be identical and changing (it is always the same, but it changes location and, therefore, is changing), so the instant – Aristotle continues – is always identical to itself, but always different as the quantity of always changing mobiles[47].

The intention of the analysis conducted by the Stagirite in these pages is clear. He aims to demonstrate that the instant, that is, the indivisible, is not time and, however, time could not exist without the instant, just as movement without the mobile is not a movement. The instant measures the before and after of the mobile. It, being indivisible, is the cause of the continuity of time, just as the mobile is the cause of the continuity of movement and the point of that of the line. In the Greek text, the instant is indicated with *τὸ νῦν* [the now] and the point with *στιγμή* [point], while the mobile is indicated with *τὸ φερόμενον* [the carried] or with *τὸ κινούμενον* [the moving], depending on whether we are talking about movement (*κίνησις*) or about displacement (*φορά*).

---

[46] *Simplicii... Commentaria* cit. (Venetiis 1566; but the version is the same as published in Paris in 1544; cf. *supra*, Ch. III, note 238).

[47] 219^{b}.

Simplicius's commentary on these passages is quite accurate and precise. He insists upon one question above all: whether the mobile can really be said, in motion, to be equivalent to the point in the line and the instant in time. Precisely in the course of this analysis, we will see Simplicius resort to *τὸ κίνημα* [the movement], which never appears in the Aristotelian text of the fourth book.

Filalteo, in translating these pages of Simplicius's commentary, committed, at least on two occasions, a serious error. It may have probably been a sort of choice, instead of an error, to make a text coherent that, in his eyes, appeared unsustainable. However, we are faced with another case of translation error in which our term is involved. The two passages, in which the incorrect use of *momentum* is recorded, contains Simplicius's paraphrase of a portion of Aristotle's argument, in which the **129**
analogy between the *τὸ νῦν* and the *τὸ κινούμενον*, that is, the "instant" and the "moving", is emphasized. Let's see them in the Filalteo version:

> Sed cum ostendisset quod *id quod fertur* subiecto idem sit et ratione diversum, similiter inquit se habere etiam ipsum nunc. siquidem sequitur et eadem patitur tempus quae motus, ut diximus prius, *et nunc ut ipsum momentum*[48]. [But when demonstrated that *which is carried* is the same in subject and different in reason, it is similarly said for the instant itself. For indeed it follows and undergoes the same time as motion, as we said before, *and the instant as momentum itself.*]

> Quare quando ipsa nunc fuerint eadem et numerus et tempus erit idem, sed quando diversa et reliqua erunt diversa. Sed cum ostenderit ex priore et posteriore nunc sequi *ipsum momentum*, ostendit etiam haec ambo nota esse per haec, motum videlicet per *id quod movetur* et tempus per ipsum nunc. Cognoscitur enim motus per *id quod movetur* et tempus per ipsum nunc[49]. [Therefore, when the instants themselves are the same, the quantity and time will be the same, but when they are different, the rest will also be different. But when demonstrated that *momentum itself* follows from the earlier and later instants, it demonstrates also that both of these are known by these means, namely motion by *that which is moved* and time through instant itself. For motion is known by that which is moved and time by the instant itself]

In the first passage, the portions of the translation in italics correspond respectively, in the Greek text, to *τὸ φερόμενον* [the carried][50], and *τῷ δὲ κινουμένῳ τὸ νῦν* [for that which is moving, the now][51]. In the second, we have *τὸ κινούμενον* [the moving][52] for all three italicized expressions. Let us observe, first of all, that the Filalteo correctly maintained the distinction between *τὸ φερόμενον* [the carried] and *τὸ κινούμενον* [the moving], rendering the two expressions with *id quod fertur* [that which is carried] and *id quod movetur* [that which is moved]. As for *τὸ κινούμενον* [the moving], in two cases, he however preferred to translate it with *momentum*. Based on the conclusions we have already reached previously regarding the meaning of *momentum* in these contexts, it is not difficult to recognize that the term is introduced inappropriately. *Τὸ κινούμενον* [the moving] is in fact the "mobile" or, if preferred, the "moved object", while *momentum* expresses the

[48] *Simplicii... Commentaria* cit. (Venetiis 1566), p. 264; the italics are ours.
[49] *Ibid.*; the italics are ours.
[50] *CAG*, IX, 723, 21.
[51] *Ibid.*, 723, 23-24.
[52] *Ibid.*, 724, 9, 11 and 12.

completed movement, the "moved". It should be noted that Filalteo did not intend *momentum* to mean "moving" object, "object in motion". A few lines later, he will indeed use it correctly in the sense of atom of movement, and he will do the same several times in the translation of the sixth book. Why, then, did he introduce our term twice in this context?

One should keep in mind that Filalteo was translating directly from the Greek text, as no previous versions were known[53]. Therefore, it seems highly unlikely that the use of *momentum* to render the Greek term twice, which in the same context is then correctly translated as *id quod movetur* [that which is moved], is attributed to carelessness. It is therefore extremely probable that it was a conscious choice. What could it have suggested? Determining this with absolute precision is difficult, and yet one gets the impression that the capable translator was not entirely convinced by the analogy between the *mobile* and the *nunc* [instant]. It is not improbable that,
130 instinctively, he felt the instant and the point were analogous not to the mobile, but to the atom of movement, that is, to *momentum*. This would explain why he decided, in translating, to amend the Greek text, probably believing that he was faced with an incorrect reading. It will be observed that *momentum* was introduced precisely in the two expressions in which it is strongly suggested that, if the *nunc* [instant] is the ultimate component of time, the *τὸ κινούμενον* [the moving] is the ultimate component of motion. And we have already seen, in examining the versions of the text of the sixth book of *Physics*, that Aristotle himself strongly emphasizes the analogy between *στιγμή* [point], *κίνημα* [movement] and *τὸ νῦν* [the now], terms which the translators generally rendered as *punctum*, *momentum* and *nunc* respectively.

Filalteo did not understand the specific nature of Aristotle's analysis in these pages of the fourth book, where the inquiry is not about the ultimate composition of time and motion, but rather on the differences between motion and time, and he read them, so to speak, with the conclusions that the Stagirite reached in the sixth book in mind.

It should be noted that Gentian Hervet's translation was correct. He always made *τὸ κινούμενον* [the moving] with *id quod movetur* [that which is moved][54].

---

[53] Cf. *supra*, Ch. III, note 238.

[54] This is how the two passages from Simplicius cited above appeared in Hervet's version: "Cum autem ostendisset quod *id quod fertur* subiecto quidem idem est, ratione autem aliud dicit, nunc quoque similiter habere siquidem motum tempus sequitur et eodem modo afficitur, sicut prius dictum est, nunc autem *id quod movetur*.... Quare quando Nunc quidem fuerint eadem et numerus et tempus erit idem: quando autem alia, reliqua quoque alia. Postquam autem ostendit huius Nunc, ad *id quod movetur* consequentiam ex priori et posteriori ostendit etiam ex eo quod ambo itidem nota sint, motus quidem *ex eo quod movetur*, tempus vero ex eo quod est nunc. Cognoscimus enim motus per *id quod movetur* et tempus per nunc [But when it was shown *that what is carried* is the same in subject, but different in reason, it is similarly said that the same holds for the instant, since time follows motion and is affected in the same way, as previously stated, the instant, however, *that which is moved...* Therefore, when the instants are the same, both the quantity and the time will be the same; but when they are different, the remaining will also be different. After demonstrating these instants, he demonstrated the consequence in relation to *what is moved* from the prior and posterior. He also showed that both are known by these means: motion *by that which is moved*, and time by the instant. For we know motion by *that which is moved* and time by

From a careful reading of the rest of Simplicius's commentary on these passages in the fourth book, Filalteo could have obtained all the elements necessary to clarify his misunderstanding. A few lines later, in fact, the Greek commentator posed the problem of the analogy between point, instant and mobile in the form of a question. And he indicated that the hesitation is natural and legitimate in admitting that the mobile is "proportional" to the instant and the point:

Existimo quaerendum esse cum tria haec sint, magnitudo, motus, tempus, et sic se habeant ut unum alterum sequatur, ut, inquit, et magnitudo a punctis terminetur et tempus ab ipso nunc, quid erit quod habeat proportionem, cum his in motu, dico scilicet cum his puncto et ipsi nunc? Nam oportet esse alium terminum motus, sed non *id quod movetur*. Quamnam enim habet hoc proportionem ut consequatur nunc et punctum?[55]. [It forms the question with those three – magnitude, motion, time – and they relate to each other so that one follows the other, as it says, and just as magnitude is bounded by points and time by the instant itself, what will have proportion with these in motion, namely with the point and the instant? For there must be another terminus for motion, but not *that which is moved*. For how does this have proportion so that it follows the instant and the point?]

In this case, Filalteo correctly translated *τὸ κινούμενον* [the moving] as *id quod movetur* [that which is moved]. Simplicius's answer to the question is quite
interesting. He denies that the analogy between the mobile, the point and the instant **131**
could appear as a, to use the translator's own terms, *violenta* [violent] and *dura similitudo* [harsh similarity]:

Dicendum quod *mutatum aut momentum motus* est finis motus, non tamen principium motus, neque causa. Sed quia velut causam punctum et nunc petit (nam aliud est lineae et aliud temporis) merito id quod sit his respondens et proportionem cum his habens est *id quod fertur*[56]. [It is said that the *change or the momentum of motion* is the end of motion, but not the beginning of motion, nor the cause. But because the point and the instant are required as if they were a cause (for one pertains to the line and the other to time), it is justly concluded that what corresponds to these and holds proportion with them is that which is moved.]

What Filalteo translated with *mutatum aut momentum motus* [the change or the momentum of motion] is the Greek term *τὸ κίνημα* [the movement][57], which in the Latin translations of the Greek text of *Physics* we have seen generally rendered with *momentum*, either accompanied or related to *mutatum esse* [the completed change], as it also happens here. The meaning of *momentum* in this context is very clear: *momenta motus* [moments of motion] are what the movement is broken down into, when the terminus of divisibility is reached. They are the indivisible termini into which movement is divided, but they are neither movement nor cause of movement. Since the point and the instant are "causes" of magnitude (the passing of the point causes the line) and of time (without the instant – which quantifies the

---

the instant]" (*Simplicii... Commentaria* cit. [Venetiis 1551], cc. 127*r*-*v*; the italics are ours).

[55] *Simplicii... Commentaria* cit. (Venetiis 1566), p. 264; the italics are ours. Hervet's version is, in relation to this passage, practically identical (*Simplici... Commentaria* [Venetiis 1551]*, c. 127*r*).

[56] *Ibid.*; the italics are ours. Hervet offers the same solutions as Filalteo; in the Giuntina edition, however, it is highlighted that the *momentum*, not *τὸ κίνημα* [the movement], is *finis motus* [end of motion] (*Simplicii... Commentaria* cit. [Venetiis 1551]*, c. 127*r*).

[57] *CAG*, IX, 725, 25.

mobile according to before and after – time would not exist), the corresponding term to these will be not *momentum*, but *id quod movetur* [that which is moved], which is the cause of the movement. In fact, there would be no movement without first having the moving object. *Momentum*, therefore, is that according to which the movement is divided, but it is neither movement nor part of the movement. As Simplicius will say later, reporting the opinion of Aspasius,

divisionem appellat terminum motus tamquam secundum ipsum dividatur, quem quidem in libris de motu vocat *mutatum vel momentum*[58]. [he calls division the terminus of motion as if it were divided according to itself, which indeed he calls *the changed or momentum* in the books on motion.]

The analogy between *divisio* (*διαίρεσις* [division]) and *momentum* (*κίνημα* [the movement]) depends on the fact that the division of motion occurs according to the *moments*, which are therefore the very condition of the division (i.e., what remains when dividing). It should be noted that in the Conimbricenses version, the term used to render *τὸ κίνημα* [the movement] was precisely *divisio* [division][59].

We still find *momentum*, in the meaning that interests us now, in the sixth book,
132 in the version of Simplicius's commentary on the two texts in which Aristotle uses *τὸ κίνημα* [the movement]. In translating Simplicius's paraphrase of the first of the two contexts, Filalteo avoided using *momentum*:

Porro hoc ponens, sequentia absurda ex tali intelligentia adducit. Aut motus erit compositus non ex motibus, sed *ex mutatis*, idest ex terminis motus et ex ipsius *moveri*[60]. [Moreover, assuming this, absurd consequences are brought forth from such an understanding. Either motion will be composed not of motions, *but of changes*, that is, from the termini of motion and from the state of *being moved* itself.]

Again, in this passage, the translator shows confusion with the Aristotelian terminology and creates a mess. The error is not in rendering *mutatum* [the changed] with the *τὸ κίνημα* [the movement] of the Greek text, but rather in translating *ἐκ τοῦ κεκινῆσθαι* [from the having been moved] with *ex ipsius moveri* [of being moved itself]. Actually, *moveri* [being moved], in the following passages, is correctly used to translate *τὸ κινεῖσθαι* [the being moved]. *Τὸ κεκινῆσθαι* [the having been moved] instead corresponded to *motum esse* [the completed motion], as Filalteo himself recognized immediately afterwards[61], and indicated not "moving", but rather the

---

[58] *Simplicii... Commentaria* cit. [Venetiis 1566]*, p. 265; the italics are ours. Hervet, translating, had, once again, felt the need to clarify the term present in the Greek text: "Divisionem autem vocat finem motus, ut qui per ipsum dividatur. quod quidem in libris de motu vocat τὸ κίνημα, id est momentum [He calls division the end of motion, as something that is divided through it. This term he indeed calls *τὸ κίνημα* [the movement] in the books on motion, that is, momentum.]" (*Simplicii... Commentaria* cit. [Venetiis 1551]*, c. 127*v*).

[59] Cf. supra, p. 124.

[60] *Simplicii... Commentaria* cit. [Venetiis 1566]*, p. 334; the italics are ours. Again, Hervet offered a corrected version instead: "Cum hoc ergo posuisset, subiunget absurda quae hanc coniecturam. Nam et motus erit compositus non ex motibus sed ex momentis, hoc est ex finibus motus et ex eo quod est *motum esse* [Since he has thus posited this, he will add absurdities to this conjecture. For motion will be composed not of motions but of moments, that is, of the termini of motion and of what is *the completed motion*]" (*Simplicii... Commentaria* cit. [Venetiis 1551]*, c. 162r).

[61] The difference between *τὸ κινεῖσθαι* [the moving] and *τὸ κεκινῆσθαι* [the having been moved] will be

motion already completed. Filalteo evidently did not understand Aristotle's analysis on this point and failed to grasp the meaning of the subtle terminological distinctions employed therein. Having used *mutatum esse* [the completed change] to translate *τὸ κίνημα* [the movement], he thought he could avoid the awkward repetition by rendering *τὸ κεκινῆσθαι* [the having been moved] with *moveri* [being moved]. Truly Lucillo Maggi was not a translator up to the task!

In the version of Simplicius's commentary on the second Aristotelian passage of the sixth book, Filalteo, however, rendered *τὸ κίνημα* [the movement] with *momentum*:

quod si ex ipsis nunc componitur motus[62], sequitur etiam ut tempus constet *ex mutatis et momentis*, quae sunt termini motus impertibiles. Si enim in unoquoque nunc motum est impertibile, erit tempus ex ipsis nunc et motus ex totidem *mutatis aut momentis* et terminis motus compositus[63]. [but if motion is composed from instants themselves, it follows that time also consists *of changes and moments*, which are the indivisible termini of motion. For if in each instant the motion is indivisible, then time will be composed of these instants themselves and motion from the *same changes or moments* and termini of motion.]

The *mutatum esse* [completed change] is added by the translator, as nothing else than *κίνημα* [movement] found in Simplicius's text[64]. Probably, for Filalteo, *momentum* was a far from clear term in this sense, and he preferred to follow it with an expression with a much clearer meaning.

We have already seen that *momentum* was present, with the meaning of **133**
inclination produced by weight, also in both versions of John Philoponus's commentary on *Physics*[65]. Of Philoponus's commentary, only the portion relating to the first four books of the Aristotelian text is preserved. It is therefore not possible to compare the passages of the sixth book that we have examined for the other Greek commentators. However, some pages of the commentary on the fourth book allow interesting observations, where the same Aristotelian texts and the same problems examined in Simplicius's commentary are analyzed. Neither Doroteo nor Rasario – the two translators of Philoponus's commentary – employed *momentum* in the meaning of indivisible movement, despite the fact that the Greek text of Philoponus's commentary presents, in the fourth book, two occurrences of *τὸ κίνημα* [the movement][66].

---

appropriately clarified by Filalteo a few lines later: "Eodem quoque modo ipsius moveri partes erunt non ipsius moveri [*τὸ κινεῖσθαι*]*, sed *moti esse* [*τὸ κεκινῆσθαι*]* et dividetur moveri in partes quae non sunt ipsius *moveri*, sed ipsius iam *moti esse* [In the same way, the parts of being moved [τὸ κινεῖσθαι] will not *be moving* [*τὸ κινεῖσθαι*], but *having been moved* [*τὸ κεκινῆσθαι*], and being moved will be divided into parts that are not being moved, but are parts of having been moved]" (*Simplicii... Commentaria* cit. [Venetiis 1566]*, p. 334).

62 In the print we read *tempus* [period of time], but this is evidently an error (the Greek text in fact has *ὁ χρόνος* [time]).

63 Ibid., p. 365.

64 *CAG*, IX^2, 1027, 21-22.

65 I am referring to those of Doroteo (Cf. *supra*, Ch. III, note 242) and of Rasario (Cf. *supra*, Ch. III, note 246).

66 *CAG*, XVII^2, *Index verborum*, 949.

In the two passages where the Greek term appears, the instant of time (*τὸ νῦν*) and its nature as a quantity of movement according to before and after are discussed. Let's see how Doroteo and Rasario reacted when faced with these texts:

DOROTEO	RASARIO
verum haec aliquid aliud sunt et non motus, sed velut aliqua passio motus; si autem prius et posterius non in partibus ipsius motus acceperimus sed *in ipsis terminis et mutatis*, ratio apertior erit, quoniam ambo sunt in subiecto motu, quamquam ab eodem sunt diversa, quemadmodum puncta lineae et superficies corporis[67] [but these are something else and not motion, but as if some endured motion; however, if we do not take the earlier and the later in the parts of the motion itself, but rather *in the termini and changes themselves*, the reasoning will be clearer, since both are in the subject motion, although they are different from the same, just as points are of the line and surfaces of the body]	verum haec aliud quid sunt, et non motus, sed velut quaedam motus passio. Si vero prius et posterius non in partibus motus, sed *pro terminis et incitamentis* capiamus, apertior erit ratio: ut ambo sint in motu, ut in subiecto, diversaque ab ipso motu sint: sicut puncta a linea, ut superficies a corpore diversa est[68]. [but these are something else, and not motion, but as if a certain motion endured. But if we take the earlier and the later not in the parts of the motion, but taken *as termini and excitations*, the reasoning will be clearer: so that both are in motion, as in the subject, and are different from the motion itself: just as points are different from the line, as the surface is different from the body.]
nam cum fuerit in ipso prius et posterius (ut diximus) cum hoc definierimus et dixerimus aliud quidem esse nunc in quo coepit motus et aliud in quo cessat; intermedium sane horum diversum ab extremis hoc dicimus esse tempus, quemadmodum intermedium *mutatum* prioris et posterioris dicimus esse motum. Nam ut punctis definitur linea est et *mutatis* motus, ita et momentis et nunc definitum est tempus[69]. [for when there has been in it the earlier and the later (as we have said), when we have defined this and said that one is the instant in which motion begins and another in which it stops; certainly the intermediate of these, different from the extremes, we say is time, just as we say the intermediate *change* between the earlier and later is motion. For just as a line is defined by points and motion by *changes*, so too time is defined by moments and instants.]	cum enim in ipso quoque prius posteriusque sit, ut dictum est, cum hoc definiamus ac dicamus, aliud esse punctum temporis, in quo motus incepit, et aliud in quo cessavit, quod interiectum est, quod ab extremis diversum est, id esse tempus dicimus: sicut etiam quod inter *incitamenta* motus 134 et prius et posterius interiectum est, id motum esse asserimus, nam ut quod punctis definitum est, linea est, quodque *incitamentis*, est motus, ita et quod puncto temporis definitum est, tempus est[70]. [for when there is also in it the earlier and the later, as has been said, when we define this and say that one is the point in time in which motion began, and another in which it ceased, what lies between, which is different from the extremes, we call time: just as we assert that what lies between the motion *excitations*, both earlier and later, is motion. For just as that which is defined by points is a line, and that which is defined by *excitations* is motion, so too that which is defined by a point in time is time.]

The words in italics correspond to the occurrences of *τὸ κίνημα* [the movement]

[67] *Physicorum... cum Ioannis Grammatici... commentariis Dorotheo... interprete* cit., p. 241; the italics are ours.
[68] *Ioannis Grammatici... Explanatio... Rasario Novarensi interprete* cit., col. 429; the italics are ours.
[69] *Physicorum... cum Ioannis Grammatici Commentariis* cit., p. 241; the italics are ours.
[70] *Ioannis Grammatici... Explanatio* cit., col. 430; the italics are ours.

in Philoponus's Greek text. In the first case, the italicized expression corresponds to *ἐπὶ τῶν περάτων καὶ τῶν κινημάτων* [concerning the limits and the motions][71]. The solution adopted by Doroteo – essentially correct – does not require any particular comment.

Rasario's choices are much more curious and worthy of reflection, in whose version it is easy to recognize Doroteo's model. However, Rasario proposes a surprising solution for *τὸ κίνημα* [the movement]: *incitamentum* [the excitation]. It is the first time that we see *incitamentum* advanced as an adequate term to render *τὸ κίνημα*. And it is not easy to understand the translator's intentions, who in any case must have seriously confused readers. Rasario probably, without worrying too much about the context, rendered the Greek term according to one of its actual meanings. Consulting a good Greek dictionary reveals that *το κίνημα* has, in addition to that of movement, also the meaning of "excitation" and "sensible impression"[72]. Rasario must have used a dictionary, where he found the meaning of *incitamentum* recorded, and did not hesitate, perhaps for originality, to coin the expression *incitamenta motus* [excitations of motion]. This, obviously, doesn't make much sense in this context, nor does it certainly contribute to making it understandable.

Even in the case of the versions of Philoponus's commentary on *Physics*, in which *momentum* does not appear in the meaning of indivisible movement, we observe, therefore, a certain confusion on the part of the translators when faced with *τὸ κίνημα*. In a way, the same curious situation we observed when examining the Latin translations of *ῥοπή* seems to repeat itself[73].

---

[71] *CAG*, XVII^2, 721, 11.

[72] Cf. for example, Liddell-Scott, cited several times. In the Greek-Latin lexicons of the 15th and 16th centuries cited in Ch. I (cf. *supra*, Ch. I, note 18), *τὸ κίνημα* [the movement] is never found, which is anything but a common term. There we encounter, however, *κινητικός* [kinetics, pertaining to movement], which is equated with *excitativus* [pertaining to excitation]. This parallelism is proposed in the Conv. Suppr. 580, in Gadd. 15, and in the Conv. Suppr. 181 (compiled by Filelfo), all from the Biblioteca Medicea Laurenziana of Florence.

[73] For the non-Latin commentaries, we should add that Averroes, whose version from Arabic into Latin, was probably edited by Michael Scotus in the 13th-century. The translation of Averroes's commentary on the Aristotelian texts of the sixth book presents notable terminological interest. Scotus never used *momentum* to express the concept of the indivisible element of motion, which was central in those pages. He preferred, in fact, to resort to other expressions. In the commentary on the first text, for example, Scotus translated the passage in which Averroes confirmed the absurd consequences reached by the proponents of the possibility of movement along indivisible magnitudes, using the expression *finis motus* and contrasting it with *motus*. *Finis motus* indicated the indivisible aspect of movement (*Aristotelis opera ... cum Averrois Cordubensis* cit., p. 250*v*). A little further on, the contradictions of those who support motion *per indivisibilia* [through indivisibles] were denounced even more explicitly: "ex quo sequitur ut motus qui est per magnitudinem compositam ex indivisibilibus, sit compositus *ex ultimis motuum*, non ex *motibus*, *sed ex quietibus*, et secundum hoc erit aliquid quod iam movebatur absque eo quod prius movetur [From which it follows that a motion, which is through a magnitude composed of indivisibles, is composed of the *ultimate parts of motions*, not of *motions*, *but of rest*, and according to this, there will be something that has already been moved without it first being in motion]" (*ibid.*) *Ultimum motus* is a very beautiful expression, and also effective in indicating what remains when the division is taken to the extreme. The use of the contrast between "rest" and "motion" also appears successful in clarifying the difference between *momenta vel mutatum esse* [moments or changes] and *motus* [motion]. *Momentum* is a movement already completed, in other words, a *rest*.

3. The Latin Commentators

135 We have already observed that the use of *momentum* in the meaning of inclination produced by weight or lightness is never recorded by the medieval commentators on *Physics*. Our term instead appears in many of these texts in the sense of an indivisible movement.

In Walter Burley's commentary, who must have used Moerbeke's version, *momentum* is never used in our meaning. Illustrating the first of the two passages of the sixth book, Burley, referring to the supporters of the indivisibles, declares that they are forced to admit that "motus componitur ex finibus motuum [motion is composed of the ends of motion]"[74].

Albertus Magnus provided more of a paraphrase than a true commentary on *Physics*. In the sixth book, while analyzing the two Aristotelian texts, he uses *momentum*, essentially repeating Moerbeke's version[75]. However, Albertus Magnus also uses *momentum* in the meaning of instant of time, alternating it with *nunc* [instant][76].

Giles of Rome's commentary on the Aristotelian passages of the sixth book
136 is quite extensive[77]. Giles uses *momentum* in the meaning of atom of motion, accompanying it with *mutatum esse* [completed change] to better clarify its meaning. In the commentary on the first of the two Aristotelian passages of the sixth book, he reaffirms with the Stagirite that motion is composed of motions. If anyone maintains that it is composed of indivisibles,

> sequitur hoc inconviniens, quod erit utique motus non ex motibus sed *ex momentis, vel ex mutatis esse*. Notandum autem quod sicut partes magnitudinis componunt magnitudinem, sic motus facti in partibus huiusmodi componunt ipsum motum; quare si in qualibet illarum partium non est moveri, sed mutatum esse, sequitur *motum* esse *ex mutatis esse*, quod est inconviniens[78]. [it follows from this inconsistency that there would indeed be motion, not of motions, but *of moments, or of completed changes*. However, it should be noted that just as the portions of a magnitude compose the magnitude, so the motions occurring in such parts compose the motion itself; therefore, if in any of those portions there is not motion, but the completed change, it follows that *motion* is composed of *completed changes*, which is inconsistent.]

In Egidio, we see the correct deployment of all the necessary terminology

---

[74] *Expositio excellentissimi... Gualterii* cit., c. non num.

[75] "Quorum unum est, quod motus componitur ex non motibus sed ex momentis indivisibilibus... Neque enim ex ipsis nunc, neque linea ex punctis, neque motus ex momentis… [One of which is that motion is composed not from motions but from indivisible moments... For neither from instants themselves, nor a line from points, nor motion from moments…]" (*Commentum... Alberti Magni* cit., cc. 79*v* and 89*v*).

[76] At the beginning of the fourth book, for example, he states that the terminus of time (*nunc*) can also be considered, *per accidens*, as terminus of motion. In this context, alongside *nunc*, whose use is prevalent, Albertus (Magnus) occasionally introduces *momentum* in the sense of instant of time: "Et ideo primum etiam falsum est quod temporis non sit nisi momenta; sicut continuus est motus inter quaelibet duo puncta ante et post in spacio accepta [And therefore it is also false that time consists only of moments; just as motion is continuous between any two points taken before and after in the given space]" (*ibid.*, c. 65*v*).

[77] *Praeclarissimi... Aegidii Romani* cit.

[78] *Ibid.*, c. non num. (commentary on the first Aristotelian text); the italics are ours.

to adequately render the Aristotelian arguments on the divisibility of movement: *moveri* [being moved] (in the sense of movement in progress), *motum esse vel momentum* [completed motion or momentum] (in the meaning of infinitesimal movement already completed)[79]. The same terminology is used by Giles of Rome in the commentary on the parts of the fourth book in which Aristotle analyzes the analogies between time and movement. In those pages, however, *momentum* is never recorded, to which Egidio prefers *mutatum esse* [completed change].

There are numerous occurrences of *momentum* in the meaning of indivisible movement in Thomas's commentary on *Physics*[80]. They are almost all concentrated in *Lectio II* of the sixth book, corresponding to the first of the Aristotelian texts examined. A further clarification of the meaning of *momentum* and its relationship with *mutatum esse*, with which we have often seen it confused, is provided by Thomas in a passage in which – following the Aristotelian example – he explains what would happen if a body moved along a magnitude ABC composed of three parts A, B, C, indivisible, according to three motions D, E, Z, also indivisible.

> Ergo relinquitur quod D non sit motus, sed sit *momentum, a quo denominatur motum esse*, sicut a motu denominatur moveri; et quod ita se habet ad motum, sicut punctum indivisibile ad lineam. Et eadem **137**
> ratio est de aliis partibus motus et magnitudinis. Ex necessitate ergo sequitur, si magnitudo componitur ex indivisibilibus, quod *motus ex indivisibilibus componatur, idest ex momentis*[81]. [Therefore, it follows that D is not motion, but a *momentum, from which motion is named*, just as being moved is named from motion; and that it stands in relation to motion as an indivisible point stands to a line. And the same reasoning applies to other parts of motion and magnitude. Therefore, it necessarily follows that if magnitude is composed of indivisibles, *motion must be composed of indivisibles*, that is, *of moments*.]

This time, *momentum* is explicitly defined as a term that indicates an infinitesimal of completed motion, and which is contrasted with *motus* [motion], which indicates

[79] Cf., for example, *Lectio XVIII* (sixth book), in which Egidio reports Aristotle's argument aimed at demonstrating that there is no motion in the indivisible: "Dicendum quod ut patet ex questione premissa si mobile esset quid impartibile spatium per quod fieret motus esset compositum ex impartibilibus. Cum ergo in impartibili non possit esse nisi mutatum esse si aliquod spatium per quod fieret motus esset compositum ex impartibilibus, oporteret motum illum qui fieret in illo spatio componi *ex momentis vel ex mutatis esse*. Rursus quia mutatum esse mensuratur instanti ut supra ostendebatur si aliquis motus componeretur ex mutatis esse oporteret tempus mensurans motum illum componi ex instantibus; patet ergo quod qui ponit impartibile moveri ponit quod spatium componatur ex impartibilibus et quod motus componitur ex mutatis esse et quod tempus componatur ex instantibus; est ergo hoc ratio philosophi, quod cum omnia haec sint impossibilia impossibile est impartibile moveri [It must be said that, as is evident from the preceding question, if a body were to move, then the space through which the motion occurs would be composed of indivisibles. Since in an indivisible there can only be a completed change, if any space through which motion occurs were composed of indivisibles, it would be necessary for that motion to be composed of moments or of a completed change. Furthermore, because a completed change is measured in instants, as was shown above, if any motion were composed of a completed change, then the time measuring that motion would have to be composed of instants. It is clear, therefore, that one who posits the movement of the indivisible also posits that space is composed of indivisibles, that motion is composed of a completed change, and that time is composed of instants. Hence, this is the philosopher's reasoning: since all these things are impossible, it is impossible for the indivisible to move]" (*Ibid.*; the italics are ours).

[80] *In octo libros Physicorum... Expositio* cit.

[81] *Ibid.*, p. 378; the italics are ours.

movement in progress. *Momentum* is to *motus* as the point is to the line.

The identity relationship between *momentum* and *mutatum esse* [completed change] is reaffirmed by Thomas in *Lectio XII*, in which the commentator focuses on the second of our examined Aristotelian passages:

> Ostensum est enim supra quod neque tempus componitur ex ipsis nunc, neque linea ex ipsis punctis, neque motus componitur ex momentis (ut per momentum intelligamus hoc quod est mutatum esse)[82]. [For it has been shown above that neither is time composed of the instants themselves, nor is a line composed of the points themselves, nor is motion composed of moments (so that by momentum we understand that which is a completed change).]

Thomas's insistence on explaining the meaning of *momentum* in both the *lectiones* in which he used it cannot be justified otherwise than by Aquinas's awareness of using a term that was not completely perspicuous.

We do not find *momentum* in the meaning that now interests us in either the *Quaestiones* of Buridan[83] and Albert of Saxony[84], nor in the commentary of Tartaret, who uses *mutatum esse* [completed change] in the exposition of the two passages of the sixth book[85]. *Momentum* is instead frequent in the *Quaestio V* of the sixth book of John of Jandun[86]. In the *Quaestio*, John states that the "pars motus correspondens parti magnitudinis indivisibilis aut est motus, aut momentum [part of motion corresponding to the part of indivisible magnitude is either motion or momentum]"[87]. If it is *motus*, it follows – as Aristotle stated – the absurd consequence that the mobile moves by an indivisible. If instead,

> illa pars motus sit momentum sive motum esse, eadem ratione pars motus correspondens alteri parti magnitudinis erit momentum et non motus, et sic *motus erit compositus ex momentis, quae sunt indivisibilia motus*, sicut punctus est indivisibile lineae[88]. [that part of motion is momentum or the completed motion, by the same reasoning, the part of motion corresponding to the other part of magnitude will be momentum and not motion, and *thus motion will be composed of moments, which are indivisibles of motion*, just as a point is an indivisible of a line.]

138 It must, therefore, conclude that, for an indivisible quantity, there is no *moveri* [being moved], but only *momentum, seu mutatum esse* [or that which has completed change]. John explicitly defines *momentum* as *indivisible motus* [indivisible

---

[82] *Ibid.*, p. 441.

[83] Buridan habitually resorts to *mutatum esse* [completed change]: "Sicut se habet punctum ad lineam item mutatum esse ad mutationem… [As a point relates to a line, so a completed change relates to change…]" (*Subtilissimae quaestiones* cit., f. *Cr*).

[84] *Mutatum esse* [completed change] is still the expression used by Albert (of Saxony): "Tale potest imaginari in linea sicut punctum, vel in tempore sicut instans, vel in motu sicut mutatum [Such a thing can be imagined in a line as a point, or in time as an instant, or in motion as a change]" (*Acutissimae quaestiones* cit., c. 643*r*).

[85] *Petri Tatareti.... commentaria* cit.

[86] *Ioannis de Ianduno... subtilissimae quaestiones* cit.

[87] *Ibid.*, p. 157.

[88] *Ibid.*, the italics are ours.

motion][89]. In his commentary, *momentum* is never used in the sense of inclination produced by weight[90].

Some pages of Agostino Nifo's commentary appear particularly interesting for our investigation[91], particularly those of the fourth book, in which the analogies and differences between time and motion are illustrated, emphasizing the analogies and differences of their ultimate components. In Nifo's commentary on these passages by Aristotle, we see the reemergence of the hesitations and uncertainties that we highlighted in the examination of Filalteo's translation of Simplicius's commentary. Let us begin with the section of Nifo's commentary that analyzes the causes that the *unitas numero*, the numerical identity, of the different *nunc temporis* [instants of time] depend upon. Nifo first reports the opinion of Themistius and Averroes, according to whom the unity of time, in terms of quantity, is guaranteed by the same unity of the *nunc* [instant]. Furthermore, in every time, the *nunc* remains the same "in quantity" and "in subject", while being, however, always different *secundum esse* [according to its being]. The unity of *nunc* [instants] does not – contrary to what Aristotle claims – necessarily depend on the unity of the mobile. Time can, in fact, be divided "division mobilis et division spatii [by the division of the mobile and by the division of space]". Let's assume that it is divided according to the space along which the motion occurs. Its unity, in this case, cannot be conferred by the mobile, but rather directly by the *nunc*[92]. At this point Nifo presents a different opinion:

Alii vero dicunt quod ut tempus se habet ad motum, ita nunc temporis ad mobile, tempus autem se habet ad motum, quod illa res quae est motus est res quae est tempus, definitione autem est quid diversum, ita similiter nunc temporis et ipsum mobile subiecto sunt idem, definitione differunt, ut enim ipsum fluens accipitur ut subiectum motui, dicitur *mobile et momentum*. Mobile et subiectum, *momentum ut motus continuativum*, ut vero terminat, inchoat aut continuat prius et posterius in motu, dicitur nunc igitur mobile, momentum et nunc sunt subiecto idem, definitione differunt. Mobile enim dicitur subiectum motus, ut fluxum suscipit, *dicitur momentum ut motus inchoat et continuat et finit*, dicitur nunc ut inchoat, continuat et finit temporis partes[93]. [Others, however, say that just as time is related to motion, so the instant of time is related to the mobile; however, time is related to motion in that the thing which is motion is the same thing which is time; by definition, however, they are something different; similarly, the instants of time and the mobile are the same in subject, but differ in definition, for as that which is in flux is understood as the subject of motion, it is called *mobile and momentum*. The mobile and the subject, *momentum as the continuation of motion*, but insofar as it terminates, initiates, or continues before and after in motion, it is called the instant; therefore, the mobile, momentum, and

[89] John, in *Quaestio XV*, also from the sixth book, specified that *mutatum esse* [a completed change] (*momentum*) can properly be called the end of motion only *afterwards*; in fact, to anyone who maintains that the *motus* is composed only of *mutatum esse* one could object "verum est a parte post, sed a parte ante bene terminatur per aliquod, quod non est mutatum esse proprie respectu illius motus, sed est indivisibile ante nihil erat actu de illo motu [It is true from the part of what comes afterwards, but from the part of what comes before, it is indeed terminated by something, which is not which has completed change properly in respect to that motion, but is an indivisible point before which nothing existed in actuality of that motion]" (*ibid.*, p. 167).

[90] Cf. *supra*, Ch. III, note 308.

[91] *Augustini Niphi... Expositio* cit.

[92] *Ibid.*, pp. 345-346.

[93] *Ibid.*, p. 346; the italics are ours.

the instant are the same in subject, but differ in definition. The mobile is called the subject of motion, as it undertakes flux, *it is called momentum as it initiates, continues, and ends the motion*, and it is called the instant as it initiates, continues, and ends the parts of time.]

139 The thesis presented here by Nifo is that of Aristotle. Time follows motion. And the instant – which is the indivisible terminus of time, and also the cause of its continuity – follows the mobile, of which it is a quantity according to the before and the after. Without magnitude, there would be no motion (the space along which motion occurs); without a mobile, there would be no motion; without motion, there would be no time, nor instant of time. There is, therefore, a correspondence between motion and time and between termini of motion and termini of time. As for the subject, the termini of motion and time (*momenta* and *nunc*) are identical, as both appear to be attributes of the mobile, but, as for the definition, they are different. The mobile is, in fact, defined as the subject, i.e., the cause, of the motion; *momentum* defined as *motum continuativum* [the continuation of motion], that is, an indivisible movement, from which the continuity of motion depends; finally, *nunc* as the quantity of before and after in motion. The mobile – as Nifo states – "suscipit fluxum [undertakes flux]", *momentum* "motum inchoat et continuat et finit [initiates, continues, and ends the motion]" and *nunc* quantifies the *moments*, thus constituting time.

Unlike what we saw done by Simplicius[94], Nifo does not explicitly focus on the problem of the ultimate termini of magnitude, motion and time. However, it is already clear in this passage that he is inclined to believe that the *motion-time* relationship is expressed, at the level of ultimate components, by the *momentum-nunc* relationship. The terminus of magnitude, proportional to those of motion and time, is then *punctum* [point]. The *mobile* (*τὸ φερόμενον*), which Aristotle seemed to present in the fourth book as the terminus of motion, is looked at by Nifo exclusively as the subject.

Nor does Nifo, unlike Simplicius[95], question whether the analogy between *punctum*, *momentum* and *nunc* is legitimate, given that *punctum* and *nunc* are "termini" and at the same time "causes" of magnitude and time, while *momentum* is only a "terminus" of motion, the *mobile* being the "cause". Simplicius – if we recall – had concluded, following Aristotle, that the terminus of motion analogous and proportional to the point for the magnitude and to the instant for the time is not *τὸ κίνημα* [the movement] but *τὸ κινούμενον* [the moving], that is, *mobile*, not *momentum*[96]. And it will also be remembered that the translator of Simplicius, showing that he was not convinced, had amended the text, in some cases replacing *momentum* with *id quod movetur* [that which is moved], with an operation not very different from that performed by Nifo in these pages[97].

The only hint of awareness of the Aristotelian interpretation of *mobile*, instead of *momentum*, as the terminus of motion occurs in Nifo's admission that *per accidens*

[94] Cf. *supra*, pp. 128 ff.
[95] Cf. *supra*, p. 128.
[96] Cf. *supra*, p. 129.
[97] Cf. *supra*, pp. 129-130.

[incidentally] the mobile can be considered *terminus temporis* [terminus of time].

Concedo quod mobile per accidens est terminus temporis, quia ipsum nunc pro subiecto est ipsum **140**
mobile, modo mobile est terminus per accidens temporis[98]. [I concede that the mobile is incidentally the terminus of time, because the instant as a subject is the mobile itself; therefore, the mobile is incidentally the terminus of time.]

With the same reasoning – which is that of Aristotle, reiterated by Simplicius – Nifo could have *conceded* that the mobile is the terminus of motion, at least *per accidens*.

That this was Nifo's reading of the Aristotelian text seems to be confirmed by the commentary on the subsequent text of *Physics*. This is the Aristotelian passage in which it is stated that, just as motion follows magnitude and time follows motion, so *id quod fertur* [that which is carried] (the *mobile*) follows the "point" and is followed by the *nunc* (which is the quantity of the *prius et posterius* [before and after] of the mobile itself). Aristotle clearly maintains in this passage that the ultimate termini into which magnitude, motion and time are divided are respectively *punctum* (*στιγμή*), *id quod fertur* (*τὸ φερόμενον*) and *nunc* (*τὸ νῦν*)[99].

In the commentary on this passage, Nifo begins by stating that magnitude, motion and time can be evaluated both as effects and from the point of view of their causes:

Ut effectus sunt *mobile sive magnitudo*, *motus* et *tempus*, quod sic se habent, quod mobile comitatur motus, hunc tempus sequitur. Ex parte causae tria alia *punctum* quod est ipsum mobile indivisibile secundum speciem et numerum, sive ipsa unitas mobilis individua specie et numero; est *momentum*, quod est unitas motus; est et *nunc*, quod est unitas temporis. Illa tria effectus esse videntur, haec autem causae cum punctum, sive unitas fluens ad imaginationem mathematici fecit mobile sive magnitudinem; momentum fluens motum; nunc autem ipsum fluens tempus. Tunc arguo, ut se habet punctum ad mobile sive magnitudinem ita momentum ad motum, et ipsum nunc ad tempus. Sed punctum unum subiecto fluens secundum diversa esse, et ipsum nunc unum subiecto manens fluendo facit tempus secundum diversa esse[100]. [As effects, there are *the mobile or magnitude*, *motion*, and *time*, which are related in such a way that the mobile accompanies motion, and time follows this. On the part of the cause, there are three other things: *the point*, which is the mobile itself, indivisible according to species and quantity, or the very unity of the mobile, individual in species and quantity; there is *momentum*, which is the unity of motion; and there is the *instant*, which is the unity of time. The former three appear to be effects, but the latter are causes, since the point, or the fluxing unity in the imagination of the mathematician, makes the mobile or magnitude; the fluxing momentum makes motion; and the instant itself makes the fluxing time. Then I argue, as the point relates to the mobile or magnitude, so the momentum relates to motion, and the instant itself relates to time. But the point is one in subject, fluxing according to different being, and the instant itself, one in subject, remaining while fluxing, makes time according to different being.]

As can be seen, Nifo's exposition is quite unfaithful to the Aristotelian text. Nor is the reader warned of the modification introduced. The distortion of the Aristotelian text is based on the identification of the mobile with "magnitude". The Aristotelian text, however, clearly presents magnitude, not as the spatial extension of the mobile, but as the space *along which the motion is carried out*; and, on the

[98] *Augustini Niphi... Expositio* cit., p. 346.
[99] 219^{b}, 16-17.
[100] *Augustini Niphi* cit., p. 347; the italics are ours.

other hand, the mobile simply indicates the moved object. From Nifo's distorted reading derives the possibility of conceiving *momentum*, not only as a "terminus", but also as a "cause" of motion, an opinion decisively rejected by Aristotle in the sixth book. For Aristotle, the cause of motion cannot be anything other than the
141 "mobile", which is therefore proportional, as a cause, to the "point" and to the *nunc* [instant]. Only as an "indivisible terminus of motion" is *momentum* analogous to the point for magnitude and the *nunc* for time.

In reality, it cannot be excluded that Nifo made this distortion due to an incorrect reading of the text. This is suggested by a clarification that our commentator introduces immediately after the text cited above. Repeating the words of the Aristotelian text, he declares how they should be interpreted, showing that he is not animated by absolute certainty:

> *Inquit* [*Arist.*]* *nam quemadmodum diximus motus magnitudinem sequitur hunc* (*ut diximus*) *tempus, punctum etiam sequitur id quod fertur, quo et motu et ipsum prius et posterius in ipso motu cognoscimus.* Per omnia haec (ut mihi videtur) habere vult maiorem, videlicet ut punctum ad magnitudinem, ita momentum ad motum et nunc se habet ad tempus[101]. [*He says* [*Arist.*] *for, as we have said, motion follows magnitude which* (*as we have said*) *follows time, the point also follows what is carried, by which both the motion and the before and after in the motion are known.* Through all these things (as it seems to me), he primarily wants to consider, namely, that just as the point relates to magnitude, so momentum relates to motion, and the instant relates to time.]

The part in italics corresponds to the precise citation of the Aristotelian text, in the version of which Nifo declares the author. Interpreting the *id quod fertur* as *magnitudo* [magnitude] (i.e., as the spatial extension of the mobile), he considered it necessary to add at least *ut mihi videtur* [as it seems to me].

There's more. In the *Recognitiones* that follow these pages, Nifo declares that, after drafting the commentary, he carefully examined Aristotle's Greek text to see if the meaning was not different. He thus developed the conviction that it should be read in another way[102]. He therefore represents Aristotle's text according to this new interpretation, which is based on the conception of *magnitudo* [magnitude] not as the spatial extension of the mobile, but rather as the space along which motion occurs:

> Et inquit [Arist.]*: sequitur enim, ut dictum est, magnitudinem quidem motus, hoc est spatium supra quod fit; sumitur enim prius et posterius in motu a priori et posteriori in magnitudine, ut dictum est; hunc autem motum sequitur tempus ut diximus, nam prius et posterius in tempore sumuntur a priori et posteriori in motu, *similiter autem et punctum sequitur ad id, hoc est ad spatium, supra quo fertur aliquid, quo puncto motum et prius et posterius in ipso cognoscimus....* Ex his vult habere ipsum nunc sequi ad tempus, ut punctum ad spatium, supra quo fit motus[103]. [And he says [Arist.]: for, as it has been said, magnitude follows motion, that is, the space over which it occurs; for the before and after are

[101] *Ibid.*; the italics are ours.
[102] "At nunc diligentius verba graeca considerans dico ipsum nunc esse unitatem temporis... sumitur autem haec unitas *ab unitate magnitudinis supra qua fit motus*, quae est punctum... [But now, considering the Greek words more carefully, I say that the instant is the unit of time... This unit is taken *from the unit of magnitude over which motion occurs*, which is a point...]" (*ibid.*, p. 347; the italics are ours).
[103] *Ibid.*, p. 349; the italics are ours.

taken in motion from the before and after in magnitude, as has been said; but time follows this motion as we have said, for the before and after in time are taken from before and after in motion, *similarly, a point follows this, that is, the space over which something is carried, by which point we know motion and what is before and after in it…* From these things, he wants to have the instant follow time, just as a point follows the space over which motion occurs.]

Without a doubt, this reading of Nifo corresponded more closely to the Aristotelian text, at least for the interpretation of *magnitudo* [magnitude]. It justified the *subiecto* [subject] identity of the *nunc* [instant], as a consequence 142
of the *subiecto* identity of the *puncta* [point]. However, it is evident that a distortion was still being carried out. By comparing magnitude (point) and time (instant), the intermediate function of motion (mobile), clearly indicated by the Aristotelian text, remained omitted. This emerges from the corrections introduced in the version of the Aristotelian passage, and suggested – if we are to believe Nifo – by the careful reading of the Greek text. The *punctum sequitur id quod fertur* [the point follows that which is carried] of the Aristotelian text, used in the commentary (where *id quod fertur*, the *mobile*, is "subject" and *punctum* is "object") is replaced with *punctum sequitur ad id, hoc est ad spatium, supra quod fertur* [the point follows along this, that is, to the space over which it is carried] (where *punctum* becomes "subject" and the *mobile* disappears).

Nifo explicitly admits that the difference between the two "readings" depended on this, as he concludes the *Recognitiones* for this part of the commentary:

Haec exposito, quae occurrit nunc, differt autem ab ea, quae in commentariis, quia haec per magnitudinem spatium intelligit, et non mobile, quod movetur, sed magnitudinem supra qua fit motus[104]. [This explanation, which occurs now, differs from the one in the commentaries, because this one understands space through magnitude, and not the mobile that is moved, but the magnitude over which the motion occurs.]

Thus, due to Nifo's inability to understand the exact meaning of this Aristotelian passage, undoubtedly favored by his reticence to consider the *mobile* homogeneous in motion to the point in the line and to the instant in time, *momentum* was once again presented as fully proportional to the *punctum* [point] and the *nunc* [instant], that is, "cause" and "terminus" of the motion. Here is another disconcerting episode of the particular fate of our term, which we are now accustomed to finding at the center of errors and misunderstandings[105].

In the commentary on the Aristotelian passages of the sixth book, which we have analyzed many times, Nifo will show that he shares Aristotle's opinion on the impossibility that motion could be composed of *moments*, that is, of indivisibles of

104 *Ibid.*

105 It should be emphasized that these observations seem to confirm Paola Zambelli's judgment regarding Nifo's approximate knowledge of Greek ("il Nifo resterà sempre un grecista principiante e velleitario… [Nifo will always remain a novice and an overly ambitious Greek scholar…]"); cf. *I problemi metodologici del necromante Agostino Nifo*, "Medioevo", 1975, pp. 129-171, especially pp. 134-135, where Zambelli effectively argues against the thesis of Nifo's notable philological competence, supported by Mahoney, Risse, Cranz and Ch. B. Schmitt).

motion. Also in the sixth book, there is an interesting attempt by Nifo to reconcile the contrasting opinions of Aristotle, on the one hand, and of Plato and Epicurus, on the other, regarding the composition of the continuum. The attempt consists in distinguishing the *compositio integrativa* [integrative composition] from the *compositio definitiva* [definitive composition].

> Cum enim Arist. asserit nullum continuum componi ex indivisibilibus, loquitur de indivisibilibus quantitatis et de compositione integrativa; cum Plato et Epicurus asserunt continuum componi ex
> 143 indivisibilibus, loquuntur de indivisibilibus quantitatis sed de compositione definitiva. Cum enim nos ignoremus rerum differentias quodlibet continuum describitur per propria indivisibilia, ut corpus per superficies..., tempora per ipsa nunc et motus per momenta[106]. [For when Aristotle asserts that no continuum is composed of indivisibles, he is speaking of indivisibles of quantity and of integrative composition; when Plato and Epicurus assert that a continuum is composed of indivisibles, they are speaking of indivisibles of quantity but of definitive composition. Since we are ignorant of the differences among things, any continuum is described by its own indivisibles, such as a body by surfaces..., times by the instants, and motions by moments.]

Moments, according to this interpretation, cannot be conceived as constituting motion through multiplication (as Aristotle intends), and yet, human ignorance *de rerum differentiis* [of the difference among things] forces us to define motion as a continuum articulated in indivisible *moments* (as intended by Plato and Epicurus).

In Domingo de Soto's commentary on *Physics*, *momentum* is never used in the meaning of atom of motion[107], despite that the term is present in the version of the Aristotelian text that accompanies the commentary[108]. Instead, De Soto uses our term in the temporal sense[109]. The terminus of motion is indicated by the Spanish commentator with *mutatum esse* [completed change], which is explicitly declared indivisible. *Mutatum esse*, in motion, corresponds to the *punctum* on the line; both exist potentially:

> quemadmodum in linea, aut revera sunt, aut cogitatione considerando puncta, ita sunt et in motu mutata esse indivisibilia. Unde quemadmodum linea, puncta continuativa non sunt actu, sed potentia... ita et mutata esse continuantia motum sunt in potentia, quando vero mutatum esse actu terminat motum, tunc est in actu[110]. [just as in a line, either they are, or by thought, considering the points are either actual or potential, so too in motion the completed changes are indivisible. Hence, just as in a line, the points are not actually continuous but potential, so too the completed changes that continue the motion are in potentiality; when, however, the completed changes actually terminate the motion, then it is in actuality.]

Neither in Achillini's *De proportione motuum*[111], nor in Pereira's work[112], nor

[106] *Augustini Niphi* cit., p. 446.
[107] *Dominici Soto... commentaria* cit.
[108] Cf. *supra*, p. 123.
[109] Cf., for example, p. 343: "nam si inferatur mobile in aliquo tempore non moveri, assumendum est quod in nulla parte talis temporis movetur; momenta vero non sunt partes ex quibus tempus constat... [For if it is inferred that the mobile does not move at some time, it must be assumed that it does not move in any part of such a time; but moments are not parts of which time is composed...]". Generally, however, De Soto prefers to indicate the instant of time with *nunc*.
[110] *Ibid.*, p. 331.
[111] *Alexandri Achillini... opera* cit.
[112] *De communibus* cit.

in Contarini's *De elementis*[113] are there any *momentum* in the meaning that interests us now. Our term is not used there even in the temporal meaning. Pace, in whose commentary there are no examples of *momentum* as an atom of motion, uses it to indicate the instant of time[114]. *Momentum* is, on the contrary, used with the value **144**
of indivisible motion in the *Epitome* by Crisostomo Javelli[115].

The absence of *momentum* with this value is also noted in Toledo's commentary, which expresses the concept of indivisible motion with the expression "mutatio instantanea vel mutatum esse [instantaneous change or completed change]"[116]. *Mutatum esse* is also used by the Conimbricenses, presenting it as the terminus of motion:

> Hoc refellit [Arist.]* quia sequeretur motum non constitui ex motibus sed ex terminis motus, quae dicuntur mutata esse[117]. [This refutes Aristotle because it would follow that motion is not composed of motions but of the termini of motion, which are said to be completed changes.]

The Conimbricenses employ *momentum* elsewhere, alternating it with *instans*, meaning minimum of time[118]. The opinion of the Coimbra commentators regarding the relationship between *mutatum esse* [completed change] and the instants, on the one hand, and the parts of motion and time, on the other, is that "partes motus copulantur mutatis esse; partes autem temporis instantibus [parts of motion are connected with the concept of the completed changes; however, parts of time are connected with instants]"[119].

An echo of discussions relating to the Aristotelian texts of the fourth and sixth books on the analogies of motion and time and on the composition of the continuum is also found in works that cannot strictly be defined as commentaries on *Physics*. In some pages of Giorgio Valla's *De expetendis*[120], for example, explicit reference is made to the themes we have examined above. Valla, in fact, discussed the problem

---

[113] *De elementis* cit.

[114] *Aristotelis ... Jul. Pacius... illustravit* cit. Throughout the sixth book Pace uses *momentum* to indicate the instant of time. As for the ultimate components of motion, he repeatedly states that the *motus* is composed "non ex motibus, sed ex terminis motuum [not from motions, but from the termini of the motions]" (*ibid*., p. 755).

[115] "Si magnitudo continua componitur ex punctis et motus qui est super eam componitur ex momentis, quae sunt indivisibilia, sicut punctum in linea et instans in tempore, sed hoc est impossibile [If a continuous magnitude is composed of points and motion, which is upon it, is composed of moments, which are indivisible, just as a point in a line and an instant in time, then this is impossible]" (*Epitome Chrisostomi Javelli* cit., c. 56*r*). Javelli consistently uses *momentum* with this meaning, alternating it with *mutatum esse*: "mutatum esse, vel momentum, quod est idem [the completed change, or momentum, which is the same]" (*ibid*., p. *60v*).

[116] *Francisci Toleti... Commentaria* ed. cit., c. 172*r*; cf. also cc. 167*v* and 191*r*.

[117] *Commentarii Collegii Conimbricensis* cit., col. 239.

[118] "Quare cum tempus momentis dividatur, ut linea punctis, quotiens tempus in quo motus fit secari poterit, toties mobile dicetur motum [Therefore, since time is divided into moments, just as a line is divided into points, whenever the time in which motion occurs can be divided, then the mobile will be said to be in motion as many times as it can be divided]" (*ibid*., col. 286).

[119] *Ibid*., col. 151.

[120] *De expetendis* cit., *Liber* XXII.

of the indivisible termini of motion and time. In Valla's argument, it is easy to recognize a paraphrase of the Aristotelian text:

Ideo etiam tempus in quo qualemcunque mutationem subivit cum illo in quod mutavit quidquid non est maius illo in quod solum mutavit, quandoquidem non in tempore mutavit, sed in limite temporis in quo mutavit, quod cum limes quantitatis expers sit, neque accumulatus augetur, neque ablatus minuitur; sicut igitur linea limes et principium puncta sunt magnitudinis expertia, et sub sectione non cadentia, eodem modo etiam motus principium omne et limes omnis insecabilia quaedam et indivisibilia sunt,
145 vocentur sane, differentiae causa, hae motiones, ac quemadmodum linea secatur in punctis, nec punctum de linea quidquid aufert, ita etiam motus dividitur in motiones et motus nihil aufert motio[121]. [Therefore, also the time in which any change whatsoever has occurred, together with that into which it has changed, is not greater than that into which it has merely changed, since it did not change in time, but at the terminus of time in which it changed, and since the terminus is devoid of quantity, neither does it increase when added, nor does it decrease when taken away; just as, therefore, the beginning and end of a line are points devoid of magnitude, and do not fall under division, in the same way also the beginning and end of every motion are certain indivisible and uncuttable things, which, for the sake of distinction, are these motions; and just as a line is divided at points, and the point does not take anything from the line, so too motion is divided into motions, and motion takes nothing from the motion.]

Valla here repeated the Aristotelian conception of the impossibility of conceiving the continuum composed of indivisibles and of the necessity, nevertheless, of dividing it according to indivisibles. The author of *De expetendis* used *motio* [the motion] to express the concept that we have often seen indicated with *momentum*, which, in this meaning, is not present in *De expetendis*.

Momentum is recorded in a very interesting page of Francesco Patrizi's *Naturalis Magia*[122], in which the composition of mathematical space is discussed. Patrizi supports the thesis that the *vetus opinio* [old opinion], according to which "omnem quantitatem seu quantum dividi posse in infinitum [every quantity or quantum can be divided infinitely]"[123], is false. For the renowned philosopher, the opposite is true: every division necessarily stops at a minimum limit. Beyond this limit, we would arrive at something other than the thing divided, as would happen if, by dividing space, we arrived at a certain point where "non-space" is isolated. This "non-space" would be exactly what the ancient geometers called a "point", which has no dimension and, therefore, has no parts. If, as many of the ancients admitted, the line was composed of the motion of a point, it would be necessary to admit that the motion is a natural affection of the point, from which the production of the line arises. But this proves to be false, and therefore the line must be defined differently[124].

During this discussion, Francesco Patrizi used *momentum* to indicate the indivisibles of time and motion and referred to the Aristotelian thesis of motion, which leads to magnitude, and of time, which is subsequent to magnitude and

---

[121] *Ibid., Ch. VI* (*De motu et quiete*).
[122] *Naturalis magia, sive innumerabilium naturae arcanorum aurea explicatio*..., I-II, Ex typographia Benedicti Mammarelli, Ferrariae 1640 [ediz. contraffatta]*.
[123] *Ibid., De spacio mathematico*, I, c. 66*r*.
[124] *Ibid.*, c. 66*r*-*v*.

motion:

> Dixere veteres momentum temporis, momentum motus esse indivisibilia. Sunto, sed puncto sunt posteriora. Tempus enim post motum, vel cum motu est. Motus post corpora et cum corporibus[125]. [The ancients said that momentum of time and momentum of motion are indivisibles. They are, but are subsequent to the point. For time is behind the motion, or it is with motion. Motion is behind the bodies and with the bodies.]

In the *Discussiones Peripateticae*[126], Patrizi strongly reiterated the absurdity of those who claim to apply the methods of investigation of mathematicians to natural things, which consist in considering quantities infinitely divisible. Curiously, the **146**
specific accusation of excessive mathematical formalism was mainly directed by Patrizi against Aristotle and his thesis of the infinite divisibility of the continuum. Neither time nor motion can be considered infinitely divisible. There will be a minimum of division and this minimum will be "time" and "motion", contrary to what the Stagirite claimed[127]. According to Aristotle – continues Patrizi – there is only the *nunc* [instant] of time, which is not time. He stated the same thing about movement:

> Neque etiam motus qui succedit aliquid est.... Et post docet nihil posse accipi in tempore praeterquam *τὸ νῦν, instans*, sicut in motu *τὸ κίνημα, motio*. Haec autem esse ait indivisibilia, ut punctum in magnitudine ex quibus nec magnitudo, nec motus, nec tempus conficitur. Sic enim habet sexto physico, capite 10, *οὔτε γὰρ ὁ χρόνος ἐκ τῶν νῦν, οὔτε ἡ γραμμὴ ἐκ στιγμῶν, οὔτε ἡ κίνησις ἐκ κινημάτων Neque enim tempus ex ipsis nunc, neque linea ex punctis, nec motus ex motionibus*[128]. [Neither is the motion that succeeds something.... And later he teaches that nothing can be perceived in time except *τὸ νῦν, the instant*, just as in motion, *τὸ κίνημα, the motion*. He states that these are indivisible, like the point in magnitude, from which neither magnitude, nor motion, nor time is made. For it is so in the sixth book of Physics, chapter 10: *οὔτε γὰρ ὁ χρόνος ἐκ τῶν νῦν, οὔτε ἡ γραμμὴ ἐκ στιγμῶν, οὔτε ἡ κίνησις ἐκ κινημάτων* [For just as time is not composed of nows, nor the line from points, nor motion from movements] *For neither is time from the instants themselves, nor the line from points, nor motion from movements.*]

Here, then, appears *τὸ κίνημα* [the movement], the term that we have often seen translated as *momentum*[129]. Patrizi indicates the Latin correspondent as *motio* [the motion].

If these Aristotelian principles are accepted, it is not possible – continues Patrizi – to escape absurd conclusions: time is non-divisible, non-continuous, and

---

[125] *Ibid.*

[126] *Discussionum Peripateticarum Tomi IV. Quibus Aristotelicae philosophiae universa historia atque dogmata cum veterum placitis collata eleganter et erudita declarantur...*, Ad Perneam Lecythum, Basileae 1581.

[127] *Ibid.*, *Tomus IV*, *Liber IV* (*De mundi temporisque aeternitate*), p. 404. Interesting, for the application of this point of view, are the *Della nuova geometria di Francesco Patrizi libri XV...*, Vittorio Baldini, in Ferrara 1587.

[128] *Discussionum* cit., p. 408.

[129] It is unnecessary to emphasize that Patrizi was citing one of the Aristotelian texts on which we have based our analysis.

non-quantity. The same can be said for movement:

> Motus namque iisdem rationibus non est quantum continuum, quia partes nullas habet existentes. Et si aliquas habeat, nullas tamen quae positionem habeant habet. Quia pars praeterita non amplius est. Et ea quae secutura est nondum est. Et illud *κίνημα*, *motio*, et illa *διαίρεσις*, *divisio*, quae respondet *τὸ νῦν*, *nunc temporis* et *puncto*, *στιγμή*, *lineae* est indivisibilis, atque ideo non est pars ipsius motus....[130] [For motion, by the same reasoning, is not a continuous quantity because it has no existing parts. And if it has any parts, it has none that have position. For the past part no longer exists. And that which is to come does not yet exist. And that *κίνημα*, *the motion*, and that *διαίρεσις*, *division*, which corresponds to *τὸ νῦν*, the *instant of time* and the *point*, *στιγμή*, *of the line*, is indivisible and therefore is not a part of motion itself…]

In truth, the consequences would appear even more paradoxical. In fact, one would be forced to even conclude that time does not exist. The same naturally applies to motion, when one insists on maintaining – as Aristotle does – that its indivisible terminus is not part of motion.

147 Eadem ergo ratione eadem falsitas in quanto vel divisibili vel discreto consequetur motum... Motio vero et *τὸ κίνημα*, ipsi nunc correspondens indivisibilis sit, et motus pars non sit, nec motus. Consequitur necessario ut motus omnino nullus sit, nihil est[131]. [Therefore, by the same reasoning, the same falsity will follow in motion, whether considered as a quantity, divisible, or discrete... But motion and *τὸ κίνημα*, corresponding to the instant, are indivisible and are not a part of motion, nor are they motion itself. It necessarily follows that all motion is non-existent; it is nothing.]

Also in the *Libri ad scientiam de natura attinentes* by Francesco Piccolomini[132], the Aristotelian conception of the composition of the continuum is also illustrated. Piccolomini explicitly refers to the sixth book of *Physics*, in which he believes that the Stagirite has conclusively demonstrated the continuity of motion. The Paduan professor maintains, in fact, that motion and *ultimus eius terminus* [its ultimate terminus] are distinct:

> motus enim omnis praeditus est suis terminis, dum est finitus, qui a motu sunt distincti; termini motus sunt mutatum esse, sive *momenta*, quae individua sunt, quare motus necessario est continuus; omne enim individuo termino terminatum continuum esse debet, quoniam individuum et non continuum, non est praeditum termino.... Itaque *nullum momentum, sive mutatum esse* debere iudicari motum; et in mutationibus id quod est indivisibile et non continuum dicitur mutatum esse; igitur, quod est indivisibile et totum simul non est motus, et per conversionem cum contrapositione omnis motus est divisibilis et continuus...[133]. [for every motion is endowed with its own termini, as long as it is finite, which are distinct from the motion; the termini of motion are the completed change, or *moments*, which are indivisible, therefore motion is necessarily continuous; for everything terminated by an indivisible terminus must be continuous, since that which is indivisible and not continuous does not have a terminus... Therefore, *no moment, or completed change*, should be considered motion; and in changes, that which is indivisible and not continuous is called the completed change; therefore, that which is indivisible and whole at once

---

[130] Ibid.

[131] *Ibid.*, p. 410.

[132] *Francisci Piccolomini Senensis... Librorum ad Scientiam de natura attinentium*, I-II…, Apud Haeredes Francisci de Francescis, Venetiis 1600; I, *De motu libri duo.*

[133] *Ibid.*, I, p. 200*v*; the italics are ours.

is not motion, and by conversion with contraposition, all motion is divisible and continuous…]

In this passage by Piccolomini, we find the Aristotelian argument correctly restated. *Momentum* is the indivisible of motion, which can also be indicated with *mutatum esse*. As such, it is not motion. In the same pages, *momentum* was used by Piccolomini to mean an instant of time[134].

Neither in Borro's *De motu*[135], nor in the concordist work of Jacopo Mazzoni[136], nor in Buonamici's *De motu*[137], was *momentum* used in the meaning of an atom of motion. Buonamici dedicated numerous pages to the problem of the analogy between motion and time and to that of the continuity of motion[138]. He showed, among other things, that he shared the opinion expressed by Aristotle, in the fourth book, regarding the correspondence or proportionality between "mobile", "point" and "instant", so tenaciously ignored by Nifo[139]. However, he never used *momentum* **148**
to indicate the indivisible of motion, which he preferred to define as *terminus motus* [the terminus of motion][140], *mutatum esse* [the completed change][141], or *mutatio* [the change][142]. Instead, *momentum* was sometimes used to express the instant of time[143].

In the analysis of these texts, we have seen *momentum* frequently used in a meaning not attributable to those examined in the previous chapter. The term proves to be central to the discussion on the infinite division of the continuum and continually takes on infinitesimal connotations. Furthermore, it does not always appear clearly characterized as a terminus of motion, but is often introduced with bivalence of meaning: halfway between motion and time. This is basically a normal phenomenon, considering that, according to Aristotle, time is a measure of motion: the quantity and the measured often get confused. Moreover, we have seen that this corresponds to the very way in which Isidore, for example, conceived the term[144].

We will soon see that, for a good number of years, Galileo used *momentum*-

---

[134] Cf., for example, *ibid.*, I, p. 218 ("nunc sive momenta [instants or moments]").
[135] *De motu* cit. In his work, Borro does not focus on the problem of the ultimate composition of motion.
[136] *In universam* cit.
[137] *De motu libri X* cit.
[138] Especially in *Liber II* and in *Liber IX*.
[139] "Iam enim conspicuum factum est haec sese consequi indissolubiter magnitudinem, motum et tempus. Motus autem ipse oritur ex processu mobilis super suo spatio, veluti linea ex processu puncti super plano...; ex quo sequitur proportio inter motum et tempus, inter mobile et instans [For now it has become evident that magnitude, motion and time are inseparably connected. However, motion itself arises from the progression of a moving object over its space, just as a line arises from the progression of a point over a plane...; from which it follows the proportion between motion and time, and between the moving object and the instant]" (*ibid., Liber IX*, Cap. *XXXVII*, pp. 874-875).
[140] Cf., for example, *ibid., Liber IX, Cap. XLI*, p. 882.
[141] Cf., for example, *ibid., Liber II, Cap. XXX*, p. 196.
[142] Cf., for example, *ibid., Liber II, Cap. XXXII*, p. 199.
[143] "Instans, quod nos momentum sive punctum dicimus, in tempore, mutatum esse motu… [Instant, which we call momentum or point, in time, completed change of motion…]" (*ibid., Liber II, Cap. XXXI*, p. 197).
[144] Cf. *supra*, Ch. I, p. 13.

momento exclusively in the sense of natural inclination to motion produced by weight. However, at a certain point, the Pisan scientist began to supplement the use of *momentum-momento* of this value by also employing the term to denote the indivisible aspect, not of motion, but of speed.

The transformation of *momentum* from the meaning of the ultimate component of motion to that of the ultimate component of speed, or even "instantaneous" speed, had already been prepared in a certain way by the *calculators*, often cited as important precursors of Galileo. In the writings of the *calculators*, however, *momentum* is never used. They undoubtedly extended to speed the infinitesimal analysis proposed by Aristotle for motion, as seen from the discussions on the *intentio et remissio formarum* [intensity and remission of forms] and on the specific *intentio velocitatis* [intensity of speed]. Take, for example, Bradwardine's *Liber de continuo*[145]. In the *Definitiones*, the conditions were established for the analysis of the *intentio velocitatis*, according to the method of resolution in terms developed by
149 Aristotle for movement. With this operation, it becomes possible to present speed as articulated into indivisible *momenta*, to use Galileo's terms. Here are some of the definitions of Bradwardine:

> 10) Instans est certus athomus temporis [An instant is a certain atom of time]
>
> 11) Motus est continuum successivum tempore mensuratum [Motion is a successive continuum measured by time]
>
> 12) "Motum esse" est indivisibile finis motus [The "completed motion" is the indivisible limit of motion]
>
> 13) Materia motus est quod per motum acquiritur [The matter of motion is that which is acquired through motion]
>
> 14) Gradus motus est illud materiae motus suscipientis magis et minus, quod acquiritur per aliquod *motum esse*[146]. [The degree of motion is that aspect of the matter of motion, which receives more and less, that is acquired through a certain *completed motion*.]

We have now learned to consider *motum esse* [the completed motion] as fully equivalent to *momentum*, with which it has been confused in many texts. *Gradus motus* [degree of motion] is an expression equivalent to *gradus velocitatis* [degree of speed], very common among terminists, which indicates, in uniform motions, the magnitude of the speed, while, in the case of non-uniform motions, it signals the instantaneous speed. *Gradus velocitatis* [degree of speed] is commonly used by Galileo as a full equivalent of *momentum velocitatis* [momentum of speed]. Here, in Bradwardine's text, it is explicitly stated that the *gradus motus* [degree of motion] is acquired through a certain *motum esse* [the completed motion] (i.e., *momentum*), which is the indivisible terminus of motion.

As we will see in the following chapters, Galileo initially used *momento*, in mechanical speculations, with the "technical" meaning of inclination produced by weight and other factors. Subsequently, he strove to build, hinging upon the concept of *momento*, a rigorous (geometric-mathematical) science of movement. It was after

[145] *Liber Th. Bradwardini de continuo* cit., by M. Clagett, *La scienza della meccanica*, trad. it. cit., pp. 263-265.

[146] *Ibid*., p. 264.

the dramatic failure encountered in this attempt that we see, first, the emergence, and then the predominance (for reasons that will be illustrated in Part Two) of Galileo's use of *momentum* with a meaning not far from that which the term presented in the intense discussions on the composition of the continuum originating from the Aristotelian texts of *Physics*: that is, with a precise infinitesimal connotation and with a very close analogy with time.

A significant portion of the texts on which we have focused in this chapter was presumably known to Galileo, who undoubtedly knew both values of *momentum*, and was able to derive considerable profit from the skillful use of both. Many of the difficulties historians face in accounting for the meaning and role of *momento*, which was truly decisive in the science of the Galilean *De motu*, derive from having overlooked both the duplicity of meaning that the term presents in the scientific-philosophical tradition and the ambiguous reunification that Galileo attempted.

We have seen that the history of *momento* is largely a history of uncertainties, misunderstandings and errors. We are therefore already, in a certain sense, prepared to consider the unrestrained use that Galileo was able to make of the term.

PART TWO

GALILEO

CHAPTER I

BEFORE "MOMENTO"

The analysis conducted in the previous chapters highlighted the presence of *momentum-moment* in a set of texts whose importance for the formation of Western culture is fundamental. Galileo certainly knew of these texts, many of which he had studied in depth on several occasions. The "godlike" Archimedes, Aristotle's *Physics*, the *de ponderibus* texts, the pseudo-Aristotelian *Mechanica*, the works of Guidobaldo and Commandino, not to mention Buonamici, Benedetti or Mazzoni, indeed constituted – as has been emphasized several times – a valuable and constant point of reference for the scientific and philosophical reflections of the young Galileo. 153

In these texts we have seen *momentum-moment* used fundamentally with three distinct meanings: *a*) a minimum quantity in general, an infinitesimal of weight (*Isaiah* and *Wisdom*) or of motion (Aristotelian *Physics*, sixth book); *b*) the natural inclination to motion, produced by weight or lightness and proportional to them (*Physics*, fourth book, Eutocius's *Commentary*, *Questions of Mechanics*, Buonamici etc.); *c*) "mechanical" inclination, produced not only by weight but also by other factors (Archimedes, *Equilibrium of Planes*, Maurolico, Vitruvius etc.). This last value, of which we have recorded very few occurrences, must be clearly distinguished from *b*) for the reasons previously illustrated, with which it has quite often been confused[1]. While *b*) refers, indeed, to a sphere of *dynamic* interests, *c*) rather pertains to the field of *static-mechanical* investigations.

As demonstrated by the occurrences of our term (cf. *Appendix*), throughout Galileo's works, we find examples of the three *momentum-momento* values indicated above, in addition, naturally, to the temporal and phraseological ones, which our entire investigation has overlooked. However, as will become evident later, the different meanings of *momento* are distributed in Galileo according to proportions that are absolutely not corresponding to those observed in the First Part. Furthermore, 154
although we ended up excluding the possibility that a direct relationship could be established between certain semantic values of *momento* (ῥοπή and κίνημα, for example)[2] in the group of texts examined previously, in Galileo's case, it is difficult

[1] Cf. *supra*, Part One, Ch. III, pp. 61 ff.
[2] Cf. *supra*, Part One, Ch. IV, p. 121.

P. Galluzzi, *Momento: Galilean Studies*, Archimedes 78,
https://doi.org/10.1007/978-3-032-35049-7_5

to maintain a similar certainty.

Highlighting now the main aspects of our research, it can be said that in Galileo's writings, especially in the pages dedicated to the analysis of statics problems and in those in which the laws of naturally accelerated motion are discussed, two types of use of *momento* are encountered, which, at least seemingly, cannot be reconnected. On the one hand, in statics and mechanical questions, we observe, in fact, the frequent use of *momento* with the meaning of variable effectiveness of weight: this is the use already experimented by Maurolico, Vitruvius and other authors analyzed in previous pages. This value, which we will conventionally define as "mechanical" *momento*, has predominantly been the focus of attention for science historians. It is generally to this concept and this lexical value that one refers, in fact, when discussing the Galilean notion of *momento*. In other words, the term indicates the non-uniform ability of a given weight to balance another, greater or lesser, depending on the distance from the fulcrum of the mechanical system. It also expresses the variable speed with which a body tends to descend along differently inclined planes. In this latter sense, *momento* appears, as we will see, capable of functioning as a "link" between statics and dynamics.

On the other hand, in the pages dedicated by Galileo to the study of the laws governing the free fall of bodies and naturally accelerated motion, *momento* appears with a lexical value that does not seem to be attributable to the "mechanical" one. The use of this second *momento* is very dense, especially in *De motu locali* and appears, in general, fixed in the expressions *momentum velocitatis*, *momentum seu gradus velocitatis*, which indicate the minimum (indivisible) increment of speed that occurs in naturally accelerated motion, parallel to the flow of indivisible minimums of time. It is unnecessary to emphasize – as abundantly evident – how close the relationship is between this second *momento* and the value with which many translators of Aristotelian Physics used the term to render *τὸ κίνημα* [the movement] in the two texts of the sixth book, on which we previously focused[3].

Much of our attention, in the following pages, will be dedicated to clarifying the relationships that exist in Galileo between the "mechanical" *momento* and the "minimum speed increment" *momento*. In the previous chapters, the examination of the Archimedean tradition, on the one hand, and the Aristotelian one, on the other, did not allow us to establish any precise relationship between the "natural" *momentum* and the "mechanical" *momentum* (in Greek, *ῥοπή*) and the indivisible *momentum* of motion (*τὸ κίνημα*). It will be necessary to verify whether, even in the
155 case of Galileo, we should speak of two perfectly isolated and independent values, or if, on the contrary, some relationship can be recognized.

Momento, as a central term of mechanics and the Galilean science of movement, has repeatedly attracted the attention of numerous scholars, many of whom have credited the Pisan scientist with being the first to introduce and definitively establish it in scientific language[4]. Despite the interest shown in our term and in the very

---

[3] Cf. *supra*, Part One, Ch. IV.

[4] It would be impossible to provide in a footnote the list of works that reference our term and its use in Galileo's work. Practically all the endless literature on Galilean mechanics and the science of movement

broad and differentiated use made of it by Galileo, there still lacks a study that focuses on the different values of *momento* in the work of the Pisan scientist. In the best case scenario, we have so far been limited to recording the use of the two different *momenti* specified above, clearly favoring the static-mechanical one as characteristic of Galileo's science and the fruit of his most original contribution. Thus, the rather intriguing problem of the relationships between the "mechanical" *momento* of heaviness and the *momentum seu gradus velocitatis* has often been completely overlooked.

This attitude can already be found in the classic work of Olschki, whose observations refer exclusively to the "mechanical" *momento*[5]. Altieri Biagi, on **156**
the other hand, in the pages of her highly esteemed *Galileo e la terminologia tecnico-scientifica* dedicated to *momento*, seems to note only the mechanical use of *momento*, completely ignoring the different use of the term in *De motu locali*[6].

---

contains, in fact, some considerations dedicated to *momento*. Many of the essays in which somewhat significant judgments are expressed will be reviewed in the course of subsequent chapters. We will therefore limit ourselves now to a few essential points. In addition to the works of the authors on whom we will explicitly focus in these very pages, numerous contributions deserve to be highlighted, due to the influence they have exerted and continue to exert: R. Caverni, *Storia del metodo sperimentale in Italia* cit., IV (*Del metodo sperimentale applicato alla scienza del moto dei gravi*), pp. 86-87, 180 and 184 ff.; P. Duhem, *De l'accélération produite par une force constante, Notes pour servir à l'histoire de la dynamique*, *in* "Congrès International de Philosophie", IIme Session..., Genève... 1904, *Rapports et Comptes Rendus*, Genève 1905, pp. 896-897; E. Wohlwill, *Galilei und sein Kampf für die kopernikanische Lehre*, Voss, Hamburg und Leipzig 1909, I, p. 115; E. A. Moody, *Galileo and Avempace* cit., pp. 256 and 285; Galileo Galilei, *Discorsi e dimostrazioni matematiche intorno a due nuove scienze*, edited by A. Carugo and L. Geymonat, Boringhieri, Torino 1958, pp. 724-726; A. C. Crombie, *Da S. Agostino a Galileo*, trad. it. di V. Di Giuro, Feltrinelli, Milano 1970, pp. 270-271 and 342-345; R. S. Westfall, *The problem of force in Galileo's Physics*, in *Galileo Reappraised*, edited by L. Golino, University of California Press, Berkeley-Los Angeles 1966, pp. 71-72 and 86-88; by the same author see also *Force in Newton's Physics. The science of dynamics in the 17th century*, London-New York 1971, pp. 11, 25-27 and 38; *Les Mécaniques de Galilée... traduites de l'italien par le P. Marin Mersenne*, Édition critique par B. Rochot, P.U.F., Paris 1966, p. 29; S. Moscovici, *Torricelli's* Lezioni Accademiche *and Galileo's theory of percussion*, in *Galileo Man of Science*, Edited by E. McMullin, Basic Books, New York-London 1967, p. 448, note 6; M. Clagett, *La scienza della meccanica*, trad. it. cit., pp. 729-730; Max Jammer, *Storia del concetto di forza. Studio sulle fondazioni della dinamica*, trad. it. di Enrico Bellone, Feltrinelli, Milano 1971, pp. 110-113; W. R. Shea, *La rivoluzione intellettuale di Galileo*, trad. it. di P. Galluzzi, Sansoni, Firenze 1974, pp. 33-35.

5 L. Olschki, Galileo und seine Zeit, Halle 1927, pp. 243-254. According to Olschki, Galileo definitively established the term *momento* in the mechanical meaning.

6 M. L. Altieri Biagi, *La terminologia tecnico scientifica* cit., pp. 44-55. For *momento*, Altieri Biagi depends on both Olschki and on the previously cited edition of the *Discorsi* with extensive commentary by L. Geymonat and A. Carugo, whose observations regarding the history of the term (Cf. above all pp. 724-727) present quite a few inaccuracies. They state, for example, that "*momentum* fu introdotto nel linguaggio scientifico dal Commandino, nel 1566, per tradurre il termine ῥοπή, nel commentario di Eutocio all'*Equilibrio dei piani* di Archimede [*momentum* was introduced into scientific language by Commandino, in 1566, to translate the term ῥοπή, in Eutocius's commentary on Archimedes's *Equilibrium of Planes*]" (p. 724). In addition to the date error (the translation of the Commandino was published, in fact, in 1558, in Venice, by Paolo Manuzio), it should be noted that the *Equilibrium of the Planes* was not among the Archimedean writings included in the collection (Cf. *supra*, Part One, Ch. III, note 79), nor was it ever translated by the Urbinate. In regard to the opinion of Altieri Biagi, who identifies *momento* as a term typical of the emerging language of technicians and practical operators (*La Terminologia* cit., p. 48), I have already discussed it in my work *A proposito di un errore dei traduttori*

Alexandre Koyré, however, repeatedly focuses on both *momenti*, which he clearly distinguishes as two irreducible values. In the expression *momentum velocitatis*, our term would be “point”, “small quantity”[7]. As Koyré explicitly emphasizes, this would be the same meaning with which *momento* was used by the translators of the pages of Aristotelian Physics dedicated to the discussion of the composition of continuous magnitudes[8]. Koyré also adds that the term, in this sense, was also used by Descartes[9]. As for Stillman Drake, he shows that he believes in only one *momento*, the “mechanical” one, and justifies the use of the term in the expression *momentum velocitatis* by arguing that, in this case, it means “efficacy of”[10]. In other words, it would not be attributable in any way to the mechanical value.

The opinion of other scholars appears different. Clavelin, for example, who
157 deserves the credit of having first understood the enormous importance of *momento* not only in Galileo’s statics but also in his conception of accelerated motion, strove to establish a relationship between the two *momenti*. He argued that the *momentum velocitatis* “derives” from the mechanical one[11]. We will return to Clavelin’s interpretation later, as well as emphasize some of its limitations and inaccuracies. It seems fair, however, to acknowledge that the French scholar was the first to note the extreme importance of the problem of the relationship between the two *momenti*[12].

In truth, two years before the publication of Clavelin’s book, Thomas Settle, in a doctoral dissertation, which unfortunately remained unpublished, had dedicated two entire chapters to the enigma of *momento*[13]. Through a very detailed analysis, Settle concluded that the “mechanical” *momento* and that of *De motu locali* (defined

---

di Vitruvio nel ‘500 cit., pp. 87-88. Regarding *momento*, among language historians, Migliorini had already drawn attention to it in his *Galileo e la lingua italiana*, in *Lingua e Cultura*, Rome 1948, p. 146. Still on the subject of the interest of linguistic historians in the Galilean *momento*, see also the recent and important work of G. Mombello, especially useful for the relationships between *momento* and *talento* (*Les avatars de* ‘talentum’. *Recherches sur l’origine et les variations des acceptions romanes et non romanes de ce terme*, S.E.I., Torino 1976, especially pp. 279-280).

[7] “Senza dubbio, si potrebbe dire che anche in Galileo il movimento si concentra nell’istante; e che è anzi uno dei suoi grandi meriti l’aver saputo formulare la nozione di *momento*, velocità istantanea, elemento (o differenziale) del moto [Without a doubt, it could be said that, even in Galileo, movement is concentrated in the instant; and that it is indeed one of his great merits to have been able to formulate the notion of *momento*, instantaneous speed, as an element (or differential) of motion]” (*Studi galileiani*, trad. it. cit., pp. 337-338).

[8] *Ibid*., p. 143, note 23 and 24.

[9] “*L’intensità della velocità* o *il grado della velocità* è la velocità istantanea del mobile. Descartes la chiamerà “momento” o “punto di velocità” [*The intensity of the speed* or *the degree of the speed* is the instantaneous speed of the mobile. Descartes will call it “moment” or “point of speed”]” (*ibid*., p. 138, note 5).

[10] Galileo Galilei, *Two new sciences. Including centers of gravity and force of percussion*. Translated with Introduction and Notes by Stillman Drake, The University of Wisconsin Press, Madison, Wisconsin 1974: *Glossary*, *sub voce*, pp. XXXV-XXXVI.

[11] M. Clavelin, *La philosophie naturelle de Galilée*. *Essai sur les origines et la formation de la mécanique classique*, Librairie Armand Colin, Parigi 1968, pp. 362 ff. and 372-373.

[12] Cf. *infra*, p. 364, note 2.

[13] Th. Settle, *Galilean Science*. *Essays in the mechanics and dynamics of the* Discorsi. *A thesis presented to the Faculty of the Graduate School of Cornell University for the degree of Doctor of Philosophy* (1966). The chapters dedicated to *momento* are IV and V. I was able to consult Settle’s Dissertation at the Library of the Institute and Museum of the History of Science in Florence, which has a copy.

by Settle as "mathematical")[14] served a single purpose: the application of static principles in the development of the science *of naturally accelerated motion*[15]. Settle also made an effort to explain why Galileo decided to maintain *momento* in the treatment of naturally accelerated motion in the *Discorsi*, apparently detached from any reference to statics[16]. Lastly, he also attempted to establish the different functions performed by *momentum* and *gradus*, often competing within the *De motu locali*[17].

Wisan's excellent work, recently published, conducted primarily on Galilean fragments relating to the *De motu locali*, essentially confirms the conclusions reached by Settle on the relationships between the two *momenti*[18]. However, Wisan raises some doubts about some of Settle's observations relating to the function of *momentum* in relation to *gradus*[19].

We will have numerous opportunities to revisit the opinions of the aforementioned scholars, elaborating on them further. It will now suffice to emphasize that **158**
we are faced with conflicting interpretations and that, despite these being valuable contributions, the question appears to be far from conclusively resolved, with many ambiguities and contradictions remaining.

For this reason, in the following chapters, we will try to follow with the utmost attention and precision the occurrences of *momento* in Galileo's work, not only in their presence but also in the no less significant absences and omissions. Naturally, we will not limit ourselves to the printed works, but will also search carefully in those that have remained unpublished, the drafts, and in the bulk of the *Frammenti*. Assisted by the awareness that comes from reconstructing at least the main variations of the semantic field of *momento*, in the technical meanings, up to the end of the 16th century, we will endeavor to provide satisfactory answers to a series of fundamental questions that have remained unanswered so far. When did Galileo start using "mechanical" *momento*? With what function and in what contexts? When and for what reason did he introduce *momentum velocitatis*? Is there a relationship between the two *momenti*? If so, why did Galileo always carefully avoid highlighting the connection between the "mechanical" *momento* and that of speed? Was he aware that he was using a term that had found fruitful use both in statics and in discussions on the ultimate composition of time and motion?

Being able to give a reasoned answer to these questions is tantamount – if we are not mistaken – to shedding new light on the troubled path undertaken by the Pisan scientist to journey toward the discovery of the law of falling bodies.

---

[14] *Ibid.*, p. 159.
[15] *Ibid.*, p. 246.
[16] *Ibid.*, pp. 245 ff.
[17] *Ibid.*, pp. 164-165.
[18] W. Wisan, *The new science of motion. A study of Galileo's* De motu locali, in "Archive for History of Exact Sciences", Ed. by C. Truesdell, XIII (nov. 1974), nn. 2-3. The reference to Settle's work and its merits for the study of *momento* is found on p. 226, note 21. For a critical evaluation of W. Wisan's work, cf. the extensive review by R. Naylor in "Annali dell'Istituto e Museo di Storia della Scienza di Firenze", I (1976), fasc. 2, pp. 89-97. See also the reasoned reply from Wisan, *ibid.*, II (1977), fasc. 2, pp. 149-160 (*Mathematics and Experiment in Galileo's science of motion*).
[19] *The new science of motion* cit., pp. 292-293.

The problem of the "path" followed by Galileo in acquiring the law of fall is currently the subject of very lively discussions. While some of the steps taken by Galileo are known, it is not yet incontrovertibly clear which element of Galileo's investigation had the decisive role in this undertaking: rational deduction or observation, geometric-mathematical analogy or experiment[20]. In an effort to
159 achieve a reconstruction of the Galilean "path", scholars have examined and continue to examine the many undated notes that Galileo dedicated to analyzing the problem of naturally accelerated motion, which were published by Favaro as *Frammenti attenenti ai Discorsi*. Fundamental in this investigation is the effort to restore the chronological order of this vast collection of fragments. The study of watermarks, the analysis of handwriting and the establishment of connections between concepts stated in fragmentary notes and dated documents (letters or writings of which the year of composition is known) contribute to this[21].

Even the careful observation of lexical variations can make a significant contribution to the logical and chronological reordering of the *Frammenti*. *Momento* offers us a typical example. Galileo's lexicon is initially devoid of *momento*, which he replaces with a series of different terms. *Momento* is introduced at a certain point, for non-random reasons, that should be highlighted. Subsequently, it seems to split into two values, which in the *Discorsi* now appear as irreducible. The reasons for the duality and the final, very clear distinction of the two meanings must be

[20] In recent years, thanks primarily to North American scholars, numerous investigations have been conducted on Galilean *de motu* science to demonstrate that the image of Galileo proposed first by Cassirer, and then by Koyré, in the very successful *Studi galileiani* cited above, is strongly distorted. To the "Platonic" philosopher, whose conquests depended exclusively on geometric-mathematical reflections and deductions, portrayed by Koyré as somewhat remiss in observation and experiment, recent investigations have contrasted, often with considerable evidence, a Galileo who was entirely anything other than detached from the practice of the experiment, which, in some cases, would have played a decisive role (for example, in the discovery of the correct law of natural fall). The debate is ongoing. If a review of Koyré's important work is appropriate, which has had so much influence, especially with an "anti-positivist" function, it nevertheless seems that the "re-evaluation" of the experimentation in Galileo leads to, or risks leading to, excesses that are no less dangerous, albeit of opposite value. Settle was the first to speak out against Koyré's interpretation in 1961 (*An experiment in the History of Science*, "Science", 133 (1961), pp. 19-23); he later returned to the topic for a more organic discussion of the role of the experiment in Galilean science (*Galileo's use of experiment as a tool of investigation*, in *Galileo Man of Science* cit., pp. 315-337). More recently, Stillman Drake, an indefatigable and meritorious scholar of Galileo, focused his attention on a page of Galileo's manuscript notes (B.N.F., Ms. Galileiano 72, c. 114) in which he believed he had found proof that Galileo reached the discovery of the law of fall on the basis of an experiment (*Galileo's New Science of Motion*, in *Reason, Experiment and Mysticism in the Scientific Revolution* cit., pp. 131-156 and 302-303). According to Drake, Galileo refrained from presenting experimental evidence in the *Discorsi* so as not to excessively bore his readers. A new contribution to the re-evaluation of Galileo's experimental ability has recently come from Ronald Naylor's research, who was primarily concerned with explaining why Galileo chose not to present the experimental evidence, in the *Discorsi*, he had at his disposal to confirm the laws of motion (*Galileo's simple pendulum*, in "Physis", XVI, 1974, fasc. I, pp. 23-46 and *Galileo: Real Experiment and Didactic Demonstration*, in "Isis", 67, Sept. 1976, n. 238, pp. 398-419).

[21] Exemplary, for the combination of all the tools useful for dating of individual notes, is the aforementioned work by Winifred Wisan, who also provides in the Appendix a table in which the themes addressed by Galileo are related, by chronological periods, to the methods and procedures used and the characteristic terms (*The new science of motion* cit., pp. 296-297). Although not everything in Wisan's table is convincing, it is a laudable attempt and should be developed further.

sought mainly in the *Frammenti*. Bringing them out from these documents also means proposing a precise order of succession between what constitutes the starting point of Galileo's research on movement, the subsequent developments and the final maturation. In addition to clarifying some aspects of Galileo's conception of movement, an accurate investigation of *momento* can therefore also contribute excellently in reconstructing the actual chain of reflections and observations (a chain, mind you, that is anything but straight and logical, but rather extremely intricate and entangled) that led Galileo from the shaky dynamics of *De motu* to the laws of *motu naturaliter accelerato* presented in the *Discorsi*.

1. The "Theoremata"

The *Theoremata circa centrum gravitatis solidorum*, which can be assigned **160**
between 1585 and 1587[22], are, in all probability, Galileo's first original work[23]. With very few, non-substantial modifications[24], the *Theoremata*, over fifty years later, were published as an *Appendix* to the *Discorsi*, in the 1638 edition[25]. The early

---

[22] Antonio Favaro's opinion is that we must stick to Galileo's explicit declaration, who, in a letter to Elia Diodati dated 6 December 1636, promised to send him "alcune dimostrazioni di certe conclusioni *de centro gravitatis solidorum*, trovate da me essendo d'età di 21 anni e di 2 anni di studio di geometria [some demonstrations of certain conclusions of the *de centro gravitatis solidorum*, found by me when I was 21 years old and had studied geometry for 2 years]" (cf. *G.G.*, I, *Avvertimento* ai *Theoremata*, p. 182). For Galileo's Editor, despite the differing testimony of Vincenzo Viviani, according to whom the *Theoremata* were written by Galileo at the age of 24, that is, in 1588 (cf. *ibid*., p. 182), the Latin writing should be assigned to 1585. In truth, Favaro, when later publishing the aforementioned letter from Galileo to Diodati, of which only four copies by Viviani and his amanuensis are preserved, preferred the reading of Viviani's handwritten copy, which in the passage cited above gave "trovate da me essendo d'età di 22 anni e di 2 anni di studio di geometria [found by me being 22 years old and having studied geometry for 2 years]" (*G.G.*, XVI, pp. 523-524). For this reason, Favaro proposed moving the composition date to 1586. The fact that the drafting of the *Theoremata* should be assigned to the two-year period of 1586-1587 is also suggested by the exchange of opinions that Galileo had, at the beginning of 1588, with Guidobaldo and Christopher Clavius regarding some propositions of his writing (*G.G.*, X, pp. 22 ff.). This is, moreover, the prevailing opinion today among Galileo scholars; cf. *Discorsi e Dimostrazioni matematiche*, edited by Carugo and Geymonat cit., pp. 840-841 and Galileo Galilei, *Two new sciences...* by Stillman Drake cit., p. 261.

[23] Favaro considers it the second in chronological order, preceded only by the so-called *Juvenilia* (assigned to 1584), which, however, show no originality, being merely a compilation. Regarding the composition dates of the various treatises that make up the *Juvenilia*, and for a different evaluation of the importance of the writing, see the specifications of A. C. Crombie, *Sources of Galileo's early natural philosophy* cit., pp. 157-175 and of W. A. Wallace, *Galileo and the Thomists*, in *St. Thomas Aquinas 1274-1974 commemorative studies*, Pontifical Institute of Medieval Studies, Toronto 1974, 2 vols., I, pp. 293-330.

[24] The handwritten manuscript of the *Theoremata* has not reached us, nor are manuscript copies preserved, with the exception of an example belonging to Giovanni Vincenzo Pinelli, which is missing the first part (*G.G.*, I, *Avvertimento* ai *Theoremata*, p. 182). The text is thus guaranteed by the printed edition. The variations of the Pinelli manuscript compared to the print appear modest and concern exclusively the expressive form; however, Favaro records them precisely (*ibid*., pp. 204-208).

[25] The full title given to the *Theoremata* in the Elzeviri edition (Leiden 1638) was *Appendix in qua continentur theoremata eorumque demonstrationes, quae ab eodem Authore circa centrum gravitatis*

origin of Galileo's reflections on the center of heaviness of solids was correctly declared by Salviati, whose contribution also contains other valuable indications:

> Queste sono alcune Proposizioni attenenti al centro di gravità de i solidi, le quali in sua gioventù andò ritrovando il nostro Accademico, parendogli che quello che in tal materia aveva scritto Federigo
> 161 Commandino non mancasse di qualche imperfezione. Credette dunque con queste proposizioni, che qui vedete scritte, poter supplire a quello che si desiderava nel libro del Commandino; ed applicossi a questa contemplazione ad instanza dell'Illustrissimo Sig. Marchese Guid'Ubaldo Dal Monte, grandissimo matematico de' suoi tempi... ma incontratosi, dopo alcuno tempo, nel libro del Sig. Luca Valerio[26], massimo geometra, e veduto come egli risolve tutta questa materia, senza niente lasciar in dietro, non seguitò più avanti, ben che le aggressioni sue siano per strade molto diverse da quelle del Sig. Valerio[27]. [These are some propositions concerning the center of heaviness of solids, which our Academician discovered in his youth, as it seemed to him that what Federigo Commandino had written on this subject was not without some imperfection. He therefore believed that with these propositions, which you see written here, he could supplement what was lacking in Commandino's book; and applied himself to this contemplation at the request of the most illustrious Signor Marquis Guidobaldo Dal Monte, the greatest mathematician of his time... but encountered, after some time, in the book of Signor Luca Valerio, a great geometer, and seeing how he resolved this entire matter, leaving nothing behind, he did not continue further, although his approaches were very different paths from those of Signor Valerio.]

We already know the writing by Federico Commandino to which Salviati refers. In fact, it is the *Liber de centro gravitatis solidorum* of 1565, which we previously discussed[28]. It should be recalled that the Urbinate proposed two different definitions of the center of heaviness of solids: the first followed the model of the definition of Pappus of Alexandria in the *Collectiones Mathematicae*[29]; the second, in which *momentum* appeared, can be reduced to the enucleation of a concept implicit in Archimedes's *Equilibrium of Planes*, which had not proposed a definition of the center of heaviness[30].

That Galileo, as Salviati states in the *Discorsi*, dedicated himself to the revision and improvement of Commandino's work "ad instanza [at the request]" of Guidobaldo seems to be denied by a letter from the Marquis dal Monte to Galileo, dated 16 January 1588. Although there are some reservations about Commandino's work on the centers of heaviness[31], the letter demonstrates that only at that time did Guidobaldo become aware of the young Galileo's studies on this subject[32]. In

---

solidorum olim conscripta fuerunt.

26 Galileo refers to Luca Valerio's *De centro gravitatis solidorum libri tres* (Typis Bartholomaei Bonfadini, Romae 1604).

27 *G.G.*, VIII, p. 313.

28 Cf. *supra*, Part One, Ch. III, pp. 59 ff.

29 *Ibid.*

30 "Centrum gravitatis uniuscuiusque solidae figurae est punctum illud intra positum, circa quod undique partes aequalium momentorum consistunt [The center of heaviness of any solid figure is that point within which all parts are in equilibrium, with equal moments on all sides]" (see *supra*, p. 60).

31 "Io le scrissi che l'ultima del Commandino De centro gravitatis solidorum, non era buona per non esser universale [I wrote to him that the last (definition) of Commandino's De centro gravitatis solidorum was not good because it was not universal]" (*G.G.*, X, p. 26).

32 "E perché nella sua mi dice di haver altre cose sopra i centri della gravezza, a me farà sempre favor grande a farmi partecipe delle sue cose, che per questo saggio che mi ha mandato, non possono se non essere di esquisita dottrina [And because in his (work) he tells me that he has other things on the centers

any case, as we have previously observed, Guidobaldo showed appreciation for the definition of *de centro gravitatis* [the center of heaviness] in terms of "momenti eguali [equal momenti]" proposed by Commandino, including it in the *Paraphrasis* on the *Equilibrium of the Planes* that he published in 1588[33].

Galileo, just over twenty years old, knew these texts perfectly. This is an important piece of data for us. They contained, in fact, one of the few pre-Galilean examples of *momento*, with "mechanical" value, even if the use – as we have already emphasized – was limited to the definition of the center of heaviness. Neither **162**
Commandino nor Guidobaldo ever used *momento* in the "mechanical" sense outside of this context[34].

It is not within the scope of this work to judge the value of the *Theoremata*, which today are considered significant not so much for the originality of their conclusions, but rather because they demonstrate how the interests of the young Pisan were decidedly oriented towards Archimedes and towards the techniques of geometric demonstration of the mechanical problems, typical of the Syracuse scientist[35]. Setting aside the discussions Galileo had with Guidobaldo and Clavius regarding some propositions[36], we will limit ourselves to observing that *momentum* is never used in the *Theoremata*. It should be noted that Galileo does not provide any demonstration of the center of heaviness. This omission can be explained by the fact that, intending to review and complete Commandino's study, he presupposed the Urbinate's definitions. Furthermore, Galileo adopts strictly geometric techniques in his demonstrations. Although these are theorems on the centers of heaviness of solids (magnitudes far more similar to real bodies than Archimedean planes), Galileo speaks of *magnitudines* [magnitudes], without ever resorting to *pondus* [weight] and limiting the use of *gravitas* [heaviness] to the expression *centrum gravitatis* [center of heaviness]. The demonstration of the center of heaviness of a complex mechanical system (a balance to which different *grandezze* [magnitudes] are applied at regular intervals, as in the case of the initial *Lemma*) is conducted by breaking down the complex system into a series of simple subsystems in which identical *magnitudines* [magnitudes] are applied at equal distances from the fulcrum-center of heaviness[37]. With this procedure, Galileo was able to avoid dealing with the problem of the varying effectiveness of the weight (or greatness) at different distances from the center of the mechanical system. There was therefore no need in any way for a term that would indicate this change in effectiveness.

The method used by Galileo in the *Theoremata* therefore appears to be purely Archimedean. We will see that the Pisan scientist will make use of it again in the

---

of heaviness, it would be a great favor to me to participate in his (work), which, for this essay he sent me, can only be an exquisite doctrine]" (*ibid.*, p. 25).

[33] Cf. *supra*, Part One, Ch. III, p. 60.

[34] *Ibid.*, p. 61.

[35] Cf., for example, Th. Settle, *Galilean Science* cit., pp. 190 ff.

[36] *G.G.*, X, pp. 22 ff. Clavius's discussion focused on the legitimacy of the demonstrative procedure adopted by Galileo in the initial Lemma.

[37] Cf., for example, the Galilean demonstration of the initial *Lemma* of the *Theoremata* (*G.G.*, I, pp. 187-188).

Mecaniche, although in his later writing he will feel the need to support it with a second demonstration, based on the weight and variation of its effectiveness[38]. In the strictly geometric universe of the *Theoremata*, there is therefore no room for *momento*, which we have now learned to recognize as inevitably connected to the concept of "weight". In any case, for our scientist, so carefully engaged in the study
163 of the center of heaviness of solids, the use of *momentum* in the definition of the center of heaviness proposed by Commandino and taken up by Guidobaldo was certainly not overlooked. We will soon see that he will be able to take advantage of the suggestion.

2. The "Bilancetta"

The feat of Archimedes, in which he managed to expose the greedy goldsmith who had impudently dared to sell a crown made of gold and silver alloy as pure gold to the tyrant Hiero, was already a widely known *topos* in Galileo's time. It effectively expressed both the possibility of applying scientific principles to the solution of practical questions and the absolute superiority of the scientist, equipped with geometric-mathematical analysis tools, compared to the skilled but crude craftsman. As the initial reference "a chi di leggere gli antichi scrittori cura si prende [to those who take care to read the ancient writers]"[39] seems to suggest, Galileo may have drawn it directly from Vitruvius's *De architectura*, in which it is first mentioned[40].

The starting point from which Galileo begins in the *Bilancetta*, certainly written between 1585 and 1586[41], is the well-founded suspicion that the Archimedean process, not preserving its direct exposition by the author, was arbitrarily reconstructed and distorted by those who handed it down. Galileo was led to doubt by the lack of rigour in the techniques that, according to sources, were used by Archimedes to expose the fraud. A procedure completely devoid of the mathematical exactness to which the reader of the works of the "godlike" Archimedes is accustomed, and

[38] Cf. *infra*, p. 208.

[39] *De architectura*, IX, 3. The Vitruvian exposition of the episode of the crown and the invention of Archimedes was certainly known to Galileo, who mentions it in the dialogue part of *De motu* (written a few years later), referring to the *Bilancetta*: "Veram igitur rationem qua dicitur Archimedes aurificis furtum in corona aurea depraehendisse, te quoque excogitasse asseris? At nonne eadem scripta a plurimis praesertim a Vitruvio? [Therefore, do you also claim to have devised the true method by which Archimedes is said to have caught the goldsmith's theft in the golden crown? But were not the same writings by many, especially by Vitruvius?]" (*G.G.*, I, p. 379).

[40] *Ibid.*, *La Bilancetta*, p. 215.

[41] *La Bilancetta*, which was not published during Galileo's lifetime, is assigned by Viviani to 1586 (*Racconto istorico della vita del Sig. Galileo Galilei*, in *Fasti Consolari dell'Accademia Fiorentina* di Salvino Salvini, nella Stamperia di S.A.R. per Gio. Gaetano Tartini e Santi Franchi, Firenze 1717, p. 403; see now the *Racconto* in *G.G.*, XIX, pp. 597-632). Favaro adheres to Viviani's opinion (*G.G.*, I, *Avvertimento*, p. 211) and most scholars. A. C. Crombie has recently proposed moving the dating of the *Bilancetta* forward, promising to provide the reasons in a work to be soon published (*Sources of Galileo's early natural philosophy* cit.).

consisting of "mettere tal corona dentro a l'aqqua, avendovi prima posto altrettanto di oro purissimo e di argento separati, e che dalle differenze del far più o meno ricrescere o traboccare l'aqqua venisse in cognizione della mistione, di che tal corona era composta [putting the crown into the water, having first placed in it an equal amount of pure gold and silver separately, and which from the differences in making the water rise more or less or overflow, he came to know the alloy of which the crown was composed]"[42].

After a careful re-reading of the books "Delle cose che stanno nell'aqqua" and **164**
those "Delle cose che pesano egualmente", Galileo finally believes he has managed to develop a method to solve

esquisitissimamente il nostro quesito: il qual modo crederò io esser l'istesso che usasse Archimede, atteso che, oltre all'esser esattissimo, depende ancora da dimostrazioni ritrovate dal medesimo Archimede[43]. [our question most exquisitely: which method I will believe to be the same that Archimedes used, given that, in addition to being very exact, it also depends on demonstrations found by Archimedes himself.]

The method consists in using the principle of the different specific weight that the various solid bodies exhibit in water. In fact, every solid body weighs in water exactly as much as the difference between its absolute weight (i.e., in air) and the weight of an equivalent volume of water. Given that a volume of gold weighs twenty times an equivalent volume of water, it follows that the metal will weigh twenty times less in water (than in air). If, however, a volume of silver weighs 12 times more than water, its weight in water will only decrease by one twelfth. That is, it will lose proportionally less weight than gold. Take a balance (in Figure 1, it appears with equal arms, but Galileo does not explicitly require this) and hang from its ends, on one side, a certain volume of pure metal, and on the other, a counterweight that achieves perfect equilibrium.

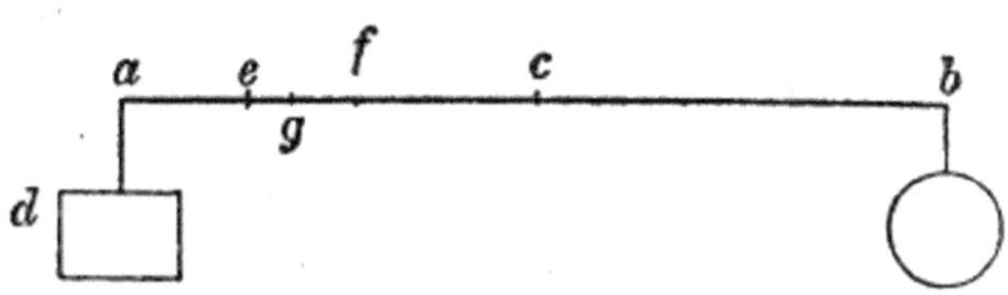

Figure 1

Mettendo il peso *b* [il metallo]* nell'aqqua, il peso *d* in *a* peserebbe di più: però, acciò che pesasse ugualmente, bisognarebbe ritirarlo verso il perpendicolo *c*, come v.g. in *e*; e quante volte la distanza *ca* supererà la *ae*, tante volte il metallo peserà più che l'aqqua. Poniamo dunque che il peso in *b* sia oro e che, pesato nell'aqqua torni il contrapeso *d* in *e*; e poi, facendo il medesimo dell'argento finissimo, che il suo contrapeso, quando si peserà poi nell'aqqua, torni in *f*: il qual punto sarà più vicino al punto *c*, sì come l'esperienza ne mostra, per esser l'argento meno grave dell'oro; e la differenza che è dalla distanza

[42] *G.G.*, I, p. 216.
[43] *Ibid.*

af alla distanza *ae* sarà la medesima che la differenza tra la gravità dell’oro e quella de l’argento. Ma se noi aremo un misto di oro e di argento, è chiaro che, per partecipare di argento, peserà più che il puro argento: e però, pesato in aria, e volendo che il medesimo contrapeso lo contrapesi, quando tal misto sarà tuffato nell’aqqua, sarà di mestiero ritirar detto contrapeso più verso il perpendicolo *c* che non è il punto *e*, il quale è il termine dell’oro, e medesimamente più lontano dal *c* che non è l’*f*, il quale è il termine
165 dell’argento puro; però cascherà tra i termini *e*, *f*, e dalla proporzione nella quale verrà divisa la distanza *ef* si avrà esquisitamente la proporzione dei due metalli che tal misto compongono[44]. [By placing the weight *b* (the metal) in the water, the weight *d* at *a* would weigh more; however, in order for it to weigh equally, it would need to be pulled back toward the perpendicular *c*, for example, to *e*. And as many times as the distance *ca* exceeds *ae*, the metal will weigh that many times more than the water. Let us assume then that the weight in *b* is gold and that, when weighed in the water, the counterweight *d* returns to *e*; and then, doing the same with fine silver, its counterweight, when weighed in the water, returns to *f*, which point will be closer to point *c*, as experience shows us, because silver is less heavy than gold. And the difference from distance *af* to distance *ae* will be the same as the difference between the weight of gold and that of silver. But if we have a mixture of gold and silver, it is clear that, because it contains silver, it will weigh more than pure silver. And so, weighed in air, and wanting the same counterweight to balance it, when this mixture is submerged in water, it will be necessary to pull the counterweight closer toward the perpendicular *c* than the point *e*, which is the endpoint of gold, and similarly farther from *c* than *f*, which is the endpoint of pure silver; thus, it will fall between the endpoints *e* and *f*, and from the proportion by which the distance *ef* is divided, one can determine precisely the proportion of the two metals that make up the mixture.]

The long passage from the *Bilancetta* that we have quoted is extremely significant. In fact, it demonstrates the consummate mastery with which Galileo now handles the lever theorems and knows how to apply them to the resolution of practical problems. Unlike what happens in the *Theoremata circa centrum gravitatis solidorum*, where Archimedes was not mentioned, here Galileo pays ongoing, deferential homage to the supremacy of the great scientist. Indeed, he informs us that, eager to reconstruct the true method followed by the Syracusan to demonstrate the goldsmith’s deception, he carefully “reread” his works. The date of composition and these explicit references bring the *Theoremata* closer to the *Bilancetta*, which appears to derive from the same type of readings and interests. In the *Theoremata*, Galileo limited himself to a strictly geometric approach, ignoring weights, in other words, real objects, and limited himself to developing the indications contained in Archimedes’s *Equilibrium of Planes*; in the *Bilancetta*, however, the young scientist uses a more sophisticated technique. First of all, the objects on which he focuses his attention are, this time, real objects (not generic *magnitudines* [magnitudes]), endowed with weights, which vary not only with volume and material but also with the medium in which they are found. Secondly, to solve the problem, he skillfully combines the investigative methods used in the two aforementioned Archimedean writings. The method of immersing one of the weights and the use of the theoretical foundation of the difference in the weights of the various metals in the water undoubtedly refer to the *On Floating Bodies*, while from the *Equilibrium of Planes* depends the principle of the inverse proportionality of the weights to the distances of the points of application from the “perpendicolo [perpendicular]”, to use the Galilean term from the *Bilancetta* (i.e., the fulcrum of the balance).

[44] *Ibid.*, pp. 217-218.

Galileo is now master of Archimedean science, and by fruitfully combining its different aspects, he manages to broaden its sphere of application. However, he still uses a terminology that is essentially scholastic, which begins to prove incapable of following the rigorous progression of reasoning, which is increasingly inspired by the essentiality and apodictic nature typical of geometric demonstrations. Consider, in the aforementioned passage, the ambiguity of expressions such as "il peso *d* in *a* peserebbe di più [the weight of *d* in *a* would weigh more]". The need for a profound terminological reform was beginning to emerge. In the *Bilancetta*, the awareness of a clear distinction between the concept of "weight" and that of the "mechanical" effect of weight did not yet correspond to a similar terminological clarity.

Thus, not even in the *Bilancetta* is there a trace of *momento*. Yet, unlike what **166**
we found for the *Theoremata*, more than one occasion for the use of our term, in the mechanical sense, was offered by the development of Galileo's arguments. In fact, the Pisan scientist uses, with great ease, those laws that he himself a few years later will define as the laws of *momenti* on the balance.

It was probably due to an anachronistic overlap that the Archpriest of Palma, the Sicilian Giovan Battista Odierna, the first editor of the *Bilancetta* in 1644[45], gave the writing such a peculiar title. While respecting the text, in which *momento* was never used, Odierna felt authorized, in light of the subsequent developments of Galilean terminology, to rename the *Bilancetta* (with much greater solemnity!) the *Archimede redivivo con la Stadera del Momento*.

3. The "De motu"

Of much different interest and complexity is the third Galilean writing, known as *De motu*, or, to respect the title originally recorded on the cover of the codex which preserves the handwritten, *De motu antiquiora*[46]. The complications arise, in this case, not so much from the dating, which can be traced back to 1589-1592 without excessive difficulty[47], but from the very arrangement of the various *capita* **167**

[45] *Archimede redivivo con la stadera del momento* cit. The *Bilancetta* was republished in the collection of Galilean works edited in two volumes in Bologna (by the Heredi del Dozza) in 1655-1656, under the care of Viviani and Rinaldini. In the Bolognese edition, the *Bilancetta* appeared with some *Annotazioni* by Domenico Mantovani. On the peculiar person that is Odierna, cf. *Della vita e delle opere di Giovan Battista Odierna, Astronomo, fisico e naturalista del sec. XVII*. Memoria del Prof. F. Napoli (Tip. E. Ferrigno e F. Andò, Palermo 1881).

[46] This can be gleaned from the description provided to us by Vincenzo Viviani: "Un manoscritto del Galileo, in più quinternetti, in ottavo, intitolato fuori sulla coperta De motu antiquiora… [A manuscript by Galileo, in several quinternions, in octavo, titled outside on the cover De motu antiquiora]" (*Quinto libro degli Elementi d'Euclide, ovvero scienza universale delle proporzioni, spiegata colla dottrina del Galileo…*, Alla Condotta, Firenze 1674, p. 104). Today the Codex, preserved in the Biblioteca Nazionale di Firenze (Ms. Galileiano 71), no longer has this title. It is probable that the paper on which it was originally recorded was torn when the Galilean Collection was sorted around the middle of the 19th century, by Vincenzo Antinori on behalf of Leopold II (cf. *G.G.*, I, *Avvertimento* al *De motu*, pp. 246-247).

[47] In truth, certain documents are lacking, given that Galileo never explicitly mentioned these early

[chapters] in which the manuscript is articulated[48]. The arrangement adopted by Favaro in the Edizione Nazionale[49] had long appeared unsatisfactory, and Drabkin had already provided suggestions for a reorganization[50]. Lastly, in recent years, Raymond Fredette, with a truly excellent work, has reconstructed the structure of the Latin writing based on highly convincing internal and external elements, concluding that we are faced not with an incomplete treatise[51], but with a "concluded" work, for parts of which different versions are preserved[52].

In essence, the Canadian scholar asserts that the dialogue part represents the oldest version[53]. Subsequently, Galileo dedicated himself to the drafting of a treatise divided into two *Libri*[54], in the first of which (corresponding to the first thirteen Chapters published by Favaro) he attempted to develop a general concept
168 on which to base the proportionality, in free fall, of the speeds of mobiles to their weights. In the second (Chs. 14-23 of the Ed. Naz.), three types of problems can be isolated: *a*) the attempt to apply the dynamics to specific fields elaborated in *Liber*

---

works of his. However, they certainly all belong to the Pisan period and are subsequent to the *Bilancetta*, to which explicit reference is made (*ibid*., pp. 379-380). For a comprehensive explanation of the reasons that lead to assigning the *De motu* to the years 1589-1592, cf. R. Fredette, *Galileo's* De motu antiquiora, "Physis", XIV (1972), fasc. 4, 331-333. Indications on the chronology can also be found in Galileo Galilei, *On Motion and on Mechanics...*, Translated with introduction and notes by I. E. Drabkin [*De Motu*]... Stillman Drake [*Le Meccaniche*], The University of Wisconsin Press, Madison, Wisconsin 1960; and in S. Drake, *The Evolution of* De motu (*Galileo Gleanings XXIV*), "Isis", LXVII (June 1976), n. 237, p. 240, who states that the *De motu* was entirely drafted between 1587 and 1590.

[48] The Ms. Galileiano 71 contains several writings. It opens (a) with a Latin dialogue on motion (cc. 4*r*-35*v*), which is followed by a fragment *De motu accelerato* (cc. 38*r*-42*v*), which has nothing to do with *De motu* (cf. *infra*, pp. 334 ff.), (b) a *de motu* essay in 10 chapters (cc. 43*r*-60*v*), (c) another essay in 23 chapters (cc. 61*v*-124*v*), traditionally referred to as the true *De motu*, and, lastly (d), attempts at a new draft of the first two chapters (cc. 133*r*-134*v*).

[49] Favaro, "dopo mature considerazioni [after careful considerations]", believed he had identified in 23 chapters (c) the original draft, in two chapters (d) a later revision and in ten chapters (b) a third treatment. And he followed this order in the Edizione Nazionale, placing at the end the Latin dialogue between Alexander and Dominicus, without, however, clarifying whether he was led to do so by considering it as having been written last (cf. *G.G.*, I, pp. 251-417 and *Avvertimento*, pp. 247-248).

[50] Drabkin proposed, in fact, a series of very effective reasons to demonstrate that the dialogue part represents the oldest surviving testimony of Galilean studies on movement (I. E. Drabkin, *A Note on Galileo's De motu*, "Isis", 1960, 51, pp. 271-277). However, he limited himself to translating the 23 chapters (c), the dialogue part (a) and only two of the ten chapters (b), thus avoiding conclusively addressing the crux of the relationships between (b) and (c); cf. Drake and Drabkin, *On Motion* cit.

[51] According to Favaro, these writings presented a "fragmentary" appearance (cf. *Avvertimento* cit., p. 247). The thesis of the "incompleteness" of *De motu* is strongly affirmed by Koyré, *Studi galileiani*, trad. it. cit., p. 55.

[52] R. Fredette, *Galileo's* De motu cit. The remarkable essay summarizes the results that Fredette reached through a close analysis of the Latin writing in a doctoral dissertation: *Les De motu "plus anciennes" de Galileo Galilei: prolégomènes*. Thèse présentée en vue de l'obtention du grade de Philosophiae Doctor, Montréal, Département d'Études Médiévales de l'Université de Montréal, 1969. I was able to consult Fredette's dissertation at the Institute and Museum of the History of Science in Florence, a copy of it is kept in the library.

[53] *Galileo's* De motu cit., pp. 331-332.

[54] That the 23 chapters of the original version were divided into two books is deduced by Fredette based on the internal references used by Galileo. In chapter 19, he, in fact, states: "cum certo sciamus, ex demonstratis *in primo libro*... [since we know for certain, from what has been demonstrated *in the first book*...]" (*G.G.*, I, p. 318); cf. *Galileo's* De motu cit., pp. 334-335.

I (inclined plane, projectiles); b) the development of critique of some concepts of Aristotelian dynamics, from which Galileo was distancing himself; c) the effort to justify the mismatch between the theoretical part of his conception of movement and the "observations" (i.e., the acceleration problem)[55]. Galileo, dissatisfied with the achieved results[56], subsequently returned to *Liber I* to attempt a refoundation, which is attested by ten chapters (published by Favaro following the 24 mentioned above)[57]. The reorganization proposed by Fredette appears quite convincing, despite Stillman Drake's recent attempt to demonstrate that the writings in question were written according to a different chronological sequence[58]. The analysis of *De motu* that we will carry out in the next pages, following the disposition of the *capita* indicated by the Canadian scholar, brings – if we are not mistaken – quite a few confirmations to the validity of his thesis.

For anyone intending to conduct an investigation into the formation of Galileo's
scientific and technical vocabulary, *De motu* constitutes a valuable document[59]. In **169**
the original draft, as well as to an even greater extent in the revision attempts, it is

---

[55] *Ibid.*, p. 343.

[56] *Ibid.*, pp. 344-345. Fredette maintains that the dissatisfaction was primarily related to the explanation of acceleration as an accidental phenomenon. In fact, experience showed that natural motions of falling were all accelerated.

[57] According to Fredette, the modifications introduced in the new draft would have led to the development of foundations incompatible with the explanation of acceleration provided by Galileo in *Liber II*. Hence the abandonment of *De motu* (*ibid.*, pp. 347-348, especially note 48).

[58] *The Evolution of* De motu cit. According to Drake, the dialogue part (a) would have been written first. Galileo would then have written the 10 chapters, while the composition of the 23 chapters should be assigned to an even later date (c). The correct order of drafting would correspond, in substance, to the arrangement of the writings in the Galilean codex. I confess that I am not at all convinced by the arguments with which Stillman Drake supports his thesis. It seems to me, to tell the truth, very difficult – if not impossible – that Galileo's *de motu* reflections could have followed the development proposed by Drake. Regardless of any "logical" evaluation, there are, in fact, also material proofs that are difficult to ignore. Some marginal notes to the text of (c), relating to the need to correct or add something (cf., for example, *G.G.*, I, 264), in fact appear to be inserted and developed in the ten chapters (b). Rightly, the American scholar did not pretend to ignore the weight of these elements, but the justification he provided (*ibid.*, p. 249) is far from convincing. The hydrostatic model-balance analogy – present only in (c) – is, for Drake (pp. 245 and 248), one of the most evident proofs of the previous draft of (b). I will address this problem later, trying to demonstrate that the analogy is part of the theoretical framework of the initial draft, while Galileo was subsequently forced to abandon it for very serious reasons.

[59] On *De motu*, in addition to the numerous works already mentioned, other important contributions deserve mention. Fundamental, for example, is E. A. Moody, *Galileo and Avempace* cit., to whom we owe the identification of an important source of Galileo's dynamics at the time of *De motu*: Avempace's theory, which our scientist most likely drew from *De motu gravium et levium* cit. by Girolamo Borro; it is worth remembering that Moody's analysis, which used the Albèri edition (XV vols., Soc. Editrice Fiorentina, Florence 1846-1857) of the works of the Pisan scientist, was based exclusively on the dialogue part, as the other chapters had not been included in the Edizione Granducale. See also Th. Settle, *Galileo's Use of Experiment* cit., important for the attempt to determine whether Galileo could have carried out some experiments (the tower and the inclined plane) to confirm his dynamic conception; M. Clagett, *La scienza della meccanica*, trad. it. cit., especially pp. 728-729; M. Clavelin, *La philosophie naturelle de Galilée* cit., pp. 127-147. The only Italian work on *De motu* to which one can refer is that of R. Giacomelli, *Galileo Galilei giovane e il suo* De motu, Domus Galilaeana, Pisa 1949, which is now decidedly outdated. There is a lack of extensive works on Galileo's Latin lexicon; however, one can refer to G. B. Pighi, *The Latin of Galileo Galilei*, in *Saggi su G. G. ...* Edited by C. Maccagni, Barbèra, Firenze 1972, pp. 541-550.

not difficult to recognize the ongoing confusion in which the young scientist found himself due to the use of terminology that was absolutely inadequate to express the new concepts with which he was approaching the analysis of the movement[60]. If, even before preparing to write *De motu*, he had already developed the awareness that the Aristotelian conception of movement had to be contrasted with a radically different vision, it was during the troubled drafting of the Latin writing that he must have convinced himself of the inevitable failure of any effort in that direction, unless a profound reform of the terminology was undertaken first. Given the nature of the present work, in the analysis of Galileo's first reflections on movement that have come down to us, our attention will focus primarily on the pages where the damage caused by the use of a lexicon incapable of univocally capturing fundamental conceptual distinctions emerges most clearly.

With no less attention, we will evaluate all the elements that show how Galileo, during the course of writing, became increasingly more lucidly aware of the need to reform his own lexicon. This reform, in addition to causing the introduction of a series of new terms (including our *momento*), will rigorously establish its criteria of validity: all terms are equally valid in theory, provided that they are distinctly "declared" in advance and that, in practice, such "declarations" are adhered to.

Far from being mere "matters of words", research of this type often allows us to reconstruct the scope and "intellectual" meaning of subsequent modifications that
170 would otherwise not be easily appreciable[61]. In our case, they will also highlight the importance Galileo attributed to the development of univocal and unambiguous expository tools.

In reality, the concern with "declaring" terms appears evident from the first lines of the Latin treatise. After having stated what he intends to demonstrate ("lationem naturalem, sive deorsum, sive sursum illa sit, a proprio mobilis gravitate vel levitate fieri [the act of natural motion, whether downward or upward, is caused by the intrinsic heaviness or lightness of the moving object]"), Galileo entrusts the task of clarifying the first of the substantial differences between his own conception of movement and that of the Aristotelian one to a "declaration of terms": "quomodo quid alio levius vel gravius dicendum sit [how something is to be called lighter or heavier than something else]". Sometimes, in fact – Galileo continues – we consider

---

[60] Fredette emphasized, very appropriately, how the terminological concerns are continuous and very evident in *De motu*.

[61] See Vailati's fine considerations on the importance of a careful inspection of the terms used to express scientific theories, for the purposes of a perfect understanding of their meaning: "non è raro il caso di importanti e decisivi progressi scientifici che ci appariscono come determinati [...]* dall'esplicarsi di controversie che a noi sembrano riferirsi al significato che si doveva o no dare a una determinata frase o parola, di controversie cioè [...]* che non avrebbero potuto aver luogo [...]* se quelli che vi presero parte si fossero dati fin dal principio la pena di ben definire i termini di cui si servivano… [it is not uncommon for important and decisive scientific progress to occur which appear to us to be determined... by the unfolding of controversies which to us seem to refer to the meaning that should or should not be given to a specific phrase or word, of controversies that is... which could not have taken place... if those who took part in it had taken the trouble from the beginning to clearly define the terms they used…]" (*Alcune osservazioni sulle Questioni di Parole nella Storia della Scienza e della Cultura*, Prolusione al Corso Libero di Storia della Meccanica, 1898, ora in *Scritti* cit., pp. 203-228).

a large piece of wood *gravius* [heavier] than a small fragment of lead "cum tamen plumbum ligno simpliciter gravius existat [although lead is simply heavier than wood]". To avoid such ambiguities, he therefore proposes to define "inter se aeque gravia, quae, cum fuerint aequalia in mole, erunt etiam aequalia in gravitate [equally heavy with respect to each other those things which, when equal in bulk, will also be equal in weight]"[62]. Thus, the first Archimedean correction to the Aristotelian dynamic conception is introduced: in natural motion, the decisive element is not constituted by the "absolute" weights, but by the "specific" ones[†]. The second correction is not as clearly stated for the time being, perhaps because it is considered an obvious consequence of the first. It could, however, be read between the lines in the definition of *levius* [lightness]:

> Converso demum modo de levioribus est statuendum: illud namque levius est censendum, cuius pars accepta, alterius parti in mole aequalis, *in gravitate* minor esse invenietur: ut si partes duas, alteram quidem ligni, alteram vero plumbi, accipiamus quae in mole aequales sint minus autem *gravet* lignea quam plumbea, tunc merito lignum plumbo levius esse, est statuendum[63]. [Lastly, the converse case of light things is established: objects are considered lighter, whose part, when taken, compared to an equal part in bulk of another, is found to be less *in heaviness*: so if two parts, one of wood, the other of lead, taken equal in bulk, the wood part is less *heavy* than the lead part, then justly the wood is lighter than the lead, as established.]

The meaning of Galileo's statements, if one pays attention to the italicized words, appears evident: positive lightness does not exist; everything is "heavy", everything has "heaviness". "Light" is only a relative term, in fact, it simply indicates the lesser heaviness compared to the comparison term. Therefore, Aristotle is wrong to present lightness as a positive reality.

Having clarified that "heavy" and "light" are intended to be legitimately used to indicate relations between equal volumes, and that "light" actually indicates "lesser in heaviness", Galileo attempts to propose, in the following chapter, an explanation different from that of Aristotle, and less arbitrary, of why "gravia in inferiori loco, levia vero in sublimi a natura constituta esse [heavy things are naturally situated in lower places, while light things are situated in higher places]"[64]. The solution is **171**
offered by the "antiquiores philosophi [ancient philosophers]"[65], who believed that "una omnium corporum sit materia et illa quidem graviora sint quae in angustiori spatio plures illius materiae particulas includent [there is one matter common to all bodies, and indeed those bodies are heavier which contain more particles of that matter in a smaller space]". It follows that *graviora* [heavier] bodies present

---

[62] *G.G.*, I, p. 251.

[†] *I.e.*, the specific gravity or density.

[63] *Ibid.*, p. 252; the italics are ours.

[64] *Ibid.*

[65] These are the atomists, in particular Democritus (cf. 68A 135 Diels-Kranz), whose theses Galileo certainly draws from Aristotle, who attempts to refute them in the fourth book of *On the Heavens* (308^{b}, 29-36-309^{a}, 1-18). Galileo, referring to the opinions of these philosophers, explicitly mentions the place in *On the Heavens* in which the Stagirite tried to prove them untenable: "iidem philosophi, immerito fortasse ab Aristotele 4 Caeli refutati [the same philosophers, perhaps unjustly refuted by Aristotle in the 4th of the On the Heavens]" (*G.G.*, I, pp. 252-253).

a greater concentration: that is, they occupy less space than *leviora* [the lighter], whose matter is "espansa [expanded]" into larger spaces[66]. It therefore seems logical that, "nella sfera del mondo [in the sphere of the world]"[67], the narrowest places, that is, those near the center, are *per natura* [by nature] occupied by the heaviest bodies, while the "light" are located in larger spaces, such as those that open up as one moves further away from the center. Here, then, the starting point for an analysis of natural motions is clearly established: if, in fact, *per natura* the heavy necessarily remain *under* the light, and vice versa, nothing will be able to prevent – again naturally – that the heavy "supra leviora posita, sub leviora ferantur [when placed above the lighter, are carried below the lighter]", or that "levia sursum ferentur, cum fuerint sub gravioribus [the lighter are carried upward, when they are beneath the heavier]". If, in fact, "a levitate habent ut super graviora maneant, ab eadem levitate habebunt ut sub gravioribus non maneant [from lightness, they have to remain above the heavier; from the same lightness, they will not remain below the heavier]"[68].

To complete the cosmological foundation necessary to move on to the analysis of natural motions, however, another question must be resolved. In natural motion, in fact, one cannot limit oneself to taking into account exclusively "de mobilis levitate vel gravitate, sed de gravitate etiam et levitate medii per quod fit motus [on the lightness and heaviness of the mobile, but also on the heaviness and lightness of the medium through which the motion occurs]"[69]. In fact, the descent of a mobile in water only occurs if it is *gravius* [heavier] than the liquid. At this point, Galileo reiterates that *gravius* (*aeque grave or levius*) [heavier (equally heavy or light)] must be understood not in an absolute sense, but by comparing – as previously stated – equal volumes of the mobile and the medium[70]. Finally, he proposes to choose water, among the different media in which natural motion occurs, to analytically examine the behavior of bodies. In fact, he declares himself convinced that, to establish such behavior, "eadem ratio sit habenda [the same reasoning should be applied]" in all mediums; the results obtained in the case of motions in water could, therefore, be legitimately transferred to motions in air, fire, etc[71].

172 The supposition put forward by Galileo – a supposition that not everyone would have peacefully conceded to him[72] – had a decisive function. In fact, the entire attempt made by Galileo in *De motu* was based on it: to attribute all types of

[66] *Ibid.*
[67] "In sphaera [within the sphere]" (*Ibid.*).
[68] *Ibid.*
[69] *Ibid.*, p. 254.
[70] "In memoriam revocandum est quod Cap. [...]* adnotavimus: nempe lapidem quidem aqua maris graviorem esse, si tantam aquae molem accipiamus quanta est moles lapidis [It must be recalled that in Chapter... we noted: namely, that a stone is indeed heavier than seawater, if we take the same bulk of water as the bulk of the stone]" (*ibid.*).
[71] "Cum vero media, per quae motus contingunt, sint plura, ut ignis, aër, aqua, etc., et in omnibus eadem ratio sit habenda, supponemus medium in quo fieri debet motus esse aquam… [Since there are various mediums through which motion occurs, such as fire, air, water, etc., and the same reasoning should be applied in all, we will assume that the medium in which the motion should occur is water…]" (*Ibid.*)
[72] It presupposed, among other things, the attribution of perfect uniformity to all mediums.

movement to the Archimedean hydrostatic model. The assumption of this hypothesis entailed a very important consequence, which Galileo was careful not to highlight: if the motions in water constitute a model that can be transferred to motions in any other type of medium, this means that there is no vacuum[73].

Making use of Archimedean hydrostatic laws, Galileo, in Chapters IV and V[74], **173**
demonstrates that

primo... ea corpora quae aeque gravia sunt ac ipsa aqua, in aquam demissa, demergi quidem tota, non tamen adhuc magis deorsum quam sursum ferri; 2) quae leviora sunt aqua, in aquam nedum descendere, verum etiam nec tota demergi posse; 3) ...quae sunt aqua graviora, deorsum necessario ferri[75]. [first... those bodies which are equally heavy as the water itself, when placed in the water, indeed sink entirely,

---

[73] Indeed, Galileo seems convinced that, at least in the context in which the natural motions occur and upon which he focuses his attention, there is a "plenum". For this reason, he assumes the hydrostatic model to interpret such motions and ignores the void: "in pleno, *ut apud nos*, nulla ponderantur secundum eorum propriam naturalemque gravitatem; sed semper eo erunt leviora quo in medio graviori extiterint [in a plenum, *as it is for us*, nothing is weighed according to its own proper and natural weight; but they will always be lighter to the extent that they exist in a heavier medium]" (*ibid.*, I, p. 281; the italics are ours). However, Galileo does not shy away from the task of indicating what would determine the motion of the mobiles, "si vacuum esset [if there were a vacuum]". In fact, he dedicates two entire chapters of *Liber I* (X and XI) to this problem. In a vacuum, essentially, the speeds of the motions would be directly proportional to the "absolute" weights, not to the "specific" ones, as in any resistant medium. Galileo is also concerned with refuting the Aristotelian thesis of the instantaneous motion in a vacuum, demonstrating that the Stagirite erroneously conceived the relationship between the weight of the mobile and the resistance of the medium as *geometrice* [geometric], not *arithmetice* [arithmetic] (p. 280). As we have already seen (cf. *supra*, Part One, pp. 89 ff.), Aristotle had made the speed of natural motions depend on the proportion between the heaviness (or lightness) of the mobile and the resistance of the medium. By setting the resistance offered by the motion equal to 0, he deduced that any moving object in a vacuum has an "infinite" superiority over resistance: it would thus move with infinite speed, therefore instantaneously. Galileo's invitation to consider the proportion between the heaviness of the mobile and the resistance of the *arithmetice* [arithmetic] medium, not *geometrice* [geometric], is clear. In the "arithmetic" proportion (where adding or subtracting are involved) the proportion between any quantity and zero is not infinite, but rather finite and determined, unlike what happens in geometric proportion (where dividing and multiplying are involved, and, therefore, zero cannot be related to a determined quantity). Galileo – like Benedetti (cf. *Le speculazioni giovanili* De motu cit., pp. 40-41) – clarifies that the effect of the medium must be "subtracted" from that of natural heaviness. In the case of motion in a vacuum, since there is nothing to subtract, "differentias gravitatum et motuum exacte discerni posse [the differences of heaviness and motions can be exactly discerned]" (*ibid.*, p. 294, 15). It is worth noting – previously mentioned (cf. *supra*, Part One, p. 99) – that in these chapters Galileo specifically cited a passage from Themistius's *Paraphrase* of *Physics* (from Barbaro's version) in which *momentum* appeared in the meaning of inclination to motion produced by heaviness or lightness of the mobile. Galileo tries to refute Themistius's thesis, who – if we recall – had maintained that in a vacuum all bodies would proceed with the same speed, given that the "differentiae gravium et levium id est rerum momenta [the differences between the heavy and the light, that is, the momenta]" are nullified (*ibid.*, 20-21), as there is no resistance to overcome. Curiously, many years later Galileo will have to recant and recognize that Themistius was right! It is worth emphasizing that Galileo seems to know quite well, not only the Aristotelian thesis of instantaneous motion in a vacuum, but also the different opinions of the commentators. As we observed in Chapter III, these were discussions in which the term *momento* often recurred. However, when critically analyzing Themistius's opinion, Galileo does not use our term.

[74] The Edizione Nazionale does not present a numbering of the different *capitoli* [chapters] of De motu. We introduce it, for convenience, by numbering from 1 to 23 the chapters of the original version (c) published by Favaro first. Drabkin, in the English translation of *De motu* cit., had already adopted this criterion.

[75] *G.G.*, I, p. 254.

but do not yet move more downward than upward; 2) those that are lighter than water, not only do not descend into the water, but also cannot be wholly submerged; 3) ...those that are heavier than water, necessarily move downward.]

The Galilean demonstrations do not present any originality. In fact, they follow the method of proof indicated by Archimedes in *On Floating Bodies*. The terminology, however, appears more interesting. Galileo indicates bodies with *magnitudo* [magnitude (body, in the abstract sense)] (in a few cases replaced by *solidum* [solid (body)]), as previously used in the *Theoremata*. However, this *magnitudo* [magnitude (body)] is now explicitly endowed with *gravitas* [heaviness], "qua resistit et deorsum premit [by which it resists and presses downward]". In the case that a *moles* [bulk (i.e., a volume of material)] of water equivalent to the *moles* [bulk] of the *magnitudo* [magnitude (body)][76] results in the same weight, there will be resistance and pressure of equal intensity, and therefore equilibrium, in other words, absence of motion both *sursum* [upward] and *deorsum* [downward]. With the same procedure, it is established that there will be *sursum* [upward] motion in the case of a *magnitudo* [magnitude (body)] with a specific weight less than the medium, and *deorsum* [downward] when the relationship is reversed.

Gravitas [heaviness] is used by Galileo, in these pages, to indicate not only the pressure of the *magnitudines* [magnitudes (bodies)] (whether they are more, less or equally "heavy" compared to water) but also the resistance of the water in the three possible cases. He thus uses *gravitas* [heaviness] also to indicate a force incapable of generating *deorsum* [downward] motion (as in the case of a *magnitudo* [magnitude (body)] of a specific weight lower than water). Immediately afterwards, however, Galileo, quickly drawing conclusions, returns to proposing *levitas* [lightness]:

"Patet igitur cur et quomodo motus sursum proveniat *ex levitate*"[77]. [It is therefore evident why and how upward motion results *from lightness*.]

In reality, even if they are not explicitly stated, Galileo's intentions appear clear. The preference for *gravitas* [heaviness] and the invitation to reflect on how *levitas* [lightness] is to be considered the cause of *sursum* [upward] motion indicate the concern to distance himself from the Aristotelian conception of positive lightness. In fact, Galileo uses *levitas* only for convenience and – as he has already warned – exclusively in a relative sense.

That Galileo experiences this kind of concern and feels discomforted by the use
174 of terms which, despite having been preliminarily defined, almost automatically evoke the same theories that he intends to combat, is confirmed in the following chapter, in which the hydrostatic principles are explained "minus mathematice et magis physice [less mathematically and more physically]"[78], appealing to the scale

[76] *Ibid.*, p. 255.
[77] *Ibid.*, p. 257; the italics are ours.
[78] "Omnia quae in superioribus his duobus capitibus tradita sunt, minus adhuc mathematice et magis physice declarari possunt, reducendo ea ad lancis considerationem [...]* ad uberiorem eorum quae

analogy. The meaning of "magis physice [more physically]" appears immediately evident in the abandonment of *magnitudo* [magnitude (body)], replaced by *pondus* [weight][79]. Galileo shows that, in a balance of equal arms, three conditions can be given, exactly as in natural motion: equilibrium (rest in natural motion), the descent of one of the two weights (the *deorsum* [downward] motion) or its ascent (*sursum* [upward] motion)[80]. In the balance, the first condition occurs in case of equality of the weights, the second in the case that the weight in question is greater than the counterweight, the third by reversing the ratio. Attention is placed on the third case, in which, as Galileo emphasizes, the ascent occurs not because the *pondus* [weight] "non habeat gravitatem [does not have heaviness]", but because the *gravitas* [heaviness][81] of the counterweight is greater.

"Ex quo patet quod in lance tam motus sursum quam motus deorsum proveniunt *ex gravitate*, sed diverso modo: nam motus sursum continget ipsi *e* propter gravitatem ipsius *o*, motus vero deorsum propter gravitatem suam"[82]. [From this it is evident that in the balance both upward motion and downward motion result *from heaviness*, but in different ways: for upward motion happens to *e* because of the heaviness of *o*, while downward motion happens because of its own heaviness.]

Returning to the consideration of natural motions, a universal principle can therefore be formulated: "gravius non posse attolli a minus gravi [the heavier cannot be lifted by the less heavy]"[83]. This principle, in the case of the balance, will be that "minus pondus non posse maius attollere [the lesser weight cannot lift the greater]"[84]. All natural motions and those of the balance can therefore be attributed to a single cause:

"in mobilibus etiam naturalibus, sicut et in ponderibus lancis, potest motuum omnium, tam sursum quam deorsum, causa reduci ad solam gravitatem"[85]. [for mobiles as well as in weights on a balance, the cause of all motions, both upward and downward, can be reduced to heaviness alone.]

The *sursum* [upward] motion is, in fact, produced "a gravitate medii et levitate mobilis [by the heaviness of the medium and the lightness of the mobile]", while the *deorsum* [downward] motion, vice versa, depends "a gravitate mobilis et levitate

---

tradentur cognitionem et ad exactiorem legentium cognitionem [All that has been explained in the above two chapters can be declared, less mathematically and more physically, by reducing it to the consideration of the scales... for a fuller understanding of what is taught and for a more accurate knowledge of the readers]" (*ibid.*, p. 257).

[79] "Sint autem duo *pondera* suspensa ex punctis *a*, *b*... [Let there be two *weights* suspended from points *a*, *b*...]" (*ibid.*, p. 257). It will be remembered that in the *Theoremata*, although using the balance, Galileo had resorted to *magnitudo* [magnitude], strictly avoiding the use of *pondus* (cf. *supra*, par. 1).

[80] "In pondere itaque *e* tria possunt contingere: aut quiescere, aut moveri sursum, aut moveri deorsum [In the weight *e*, therefore, three things can occur: either it remains at rest, or it moves upward, or it moves downward]" (*ibid.*, p. 257, 32-33).

[81] *Ibid.*, p. 258.

[82] *Ibid.*, the italics are ours.

[83] *Ibid.*

[84] *Ibid.*

[85] *Ibid.*, p. 259.

175 medii [on the heaviness of the mobile and the lightness of the medium]"[86]. The analogy of natural motions and those of the balance can therefore be proclaimed without uncertainty:

> "Patet igitur quomodo naturalium mobilium motus ad ponderum in lance motum congrue reducatur: ita ut, nempe, mobile naturale unius ponderis vicem gerat; tanta autem moles medii, quanta est mobilis moles, alterum in lance pondus repraesentet"[87]. [It is therefore evident how the natural motion of mobiles is appropriately reduced to the motion of weights on a scale: so that, namely, a natural mobile takes the place of one weight; and the bulk of the medium, as much as the bulk of the mobile, represents the other weight on the scale.]

The balance therefore offers an extremely precise mechanical model, capable of effectively representing the different conditions that occur in natural motion. If the directions of natural motions depend on the relationship between the specific weights of the heavy and those of the medium, what better tool than a balance to experimentally reproduce the conditions of natural fall in any type of medium[88]?

Galileo was therefore extremely clear that movement, whatever its direction, is *always* determined by "heaviness". However, as we have seen, he continues, for convenience, to use *levis* [light], *levius* [lighter], *levitas* [lightness], always with the relative meaning of "less heavy", "lesser heaviness" etc., as declared from the opening lines. Furthermore, it seems that he is not yet able to satisfactorily justify the mechanism through which the greater heaviness of the medium causes the ascent of the mobile. For this reason, he probably takes an evasive stance, consisting in presenting this ascent not as produced directly by the heaviness of the medium, but rather as an indirect consequence of its "relative lightness"[89].

176

---

[86] *Ibid.*

[87] *Ibid.*

[88] To tell the truth, Galileo does not make the slightest mention of the possibility of using the balance to experimentally establish the directions and, subsequently, the speeds of natural motions, limiting himself to emphasizing how the balance model is easier to understand and favors "exactiorem legentium cognitionem [a more accurate understanding by the readers]" (*ibid.*, p. 257). However, it is unlikely that he had not taken this use into consideration. As we will see, the mechanical model of the balance could not have been able in any case to "reproduce" that decisive characteristic of natural motion (even if Galileo considers it accidental) which is acceleration.

[89] "Si tanta moles medii, quanta est mobilis moles, gravior erit mobili, mobile autem levius tunc mobile, tamquam levius pondus, sursum feretur [If the bulk of the medium, equal to the bulk of the mobile body, is heavier than the mobile body, and the mobile body is lighter, then the mobile body, as a lighter weight, will be carried upward]" (*G.G.*, I, p. 259). It is worth observing that Galileo does not seem to want, in this first draft, to explicitly accept the thesis of the "extrusion of the medium" as the cause of the *sursum* [upward] motion of the lighter mobiles in the medium. Explicit references to the extrusion mechanism are missing in the 23 chapters, while – as we will see – he will resort to this mechanism of explanation in the attempt to reform *Liber I* completed later. To tell the truth, Galileo explicitly mentions extrusion only once even in the original version we are analyzing. The reference is included in the chapter dedicated to the analogy of the balance, immediately after the statement that the *sursum* [upward] motion depends "a gravitate medii et levitate mobilis [on the heaviness of the medium and the lightness of the mobile]", while "a gravitate mobilis et levitate medii [the heaviness of the mobile and the lightness of the medium]" determined the *deorsum* [downward] motion. "Et ex his, contra Aristotelem p. Caeli 89, facile quispiam colligere poterit, quomodo quae moventur, moveantur, quodammodo, *vi et per extrusionem medii*: nam aqua trabem, vi demersum, extrudit violenter, cum, descendendo, ad propriam redit regionem, nec pati

In the following chapters[90], Galileo extends his reflections to the analysis of fall speeds. In summary, since these are very well-known pages, the Pisan scientist argues, against Aristotle, that the speeds in the natural motions *sursum* [upward] and *deorsum* [downward] depend on the excess or deficiency of the specific weight of the mobile compared to that of the medium. The speed of mobiles of the same kind, in the same medium, will always be the same, whatever its size, and will correspond to the difference between the specific weight of the mobile and the specific weight of the medium (or, vice versa, in the case of *sursum* [upward] motion). However, when it comes to mobiles of different kinds (but of the same volume) the proportion of speeds in the same medium will correspond to the proportion of the differences between the specific weights of the mobiles and that of the medium. Finally, in the case of mobiles of different types (but of the same volume) in multiple media, the proportion of the various speeds will result from the proportion of the differences between the specific weights of the individual mobiles and those of the different media in which they move.

At this point, Galileo again proposes, also for the determination of speeds, the analogy of the balance[91], already used to build a mechanical model capable of accounting for the "directions" of natural motions. In the scale, in fact, the descent of one of the two weights is caused not by all its *gravitas* [heaviness], but only by that amount of *gravitas* by which it exceeds the other weight[92]. This means that the

---

vult ut, quod se levius est, sub se maneat; et, pari pacto, lapis extruditur ac deorsum impellitur quia gravior est medio. Patet igitur, talem motum posse dici violentum; quamvis communiter lignum in aqua sursum, et lapidem deorsum, naturaliter ferri, dicantur. Nec valet Aristotelis argumentum, dum dicit, si esset violentus in fine remitteretur, non autem, ut facit, augeretur: nam motus violentus tunc remittitur quando mobile extra manum moventis fuerit, non autem dum motori est coniunctum [And from these things, against Aristotle in On the Heavens 89, anyone can easily infer how things that move are moved, in a way, *by the power and through the extrusion of the medium*: for water, by violently extruding a submerged beam, as it descends, returns to its proper region, and does not want to allow what is lighter than itself to remain beneath it; and, in the same manner, a stone is extruded and pushed downward because it is heavier than the medium. It is therefore clear that such motion can be said to be violent; although commonly wood in water is said to be moved upward naturally, and a stone downward. And Aristotle's argument is not valid, since he says that if it were violent it would be remitted at the end, and not, as it is, increased: for violent motion is then remitted when the mobile is outside the hand of the mover, not when it is connected to the mover]" (*ibid.*, p. 259; the italics are ours). The passage, included by Favaro in the text, is, in reality, documented by a marginal addition that is certainly later (cf. *ibid.*, note 1). Furthermore, we will observe how cautiously Galileo proposes the analogy of extrusion, avoiding stating that it constitutes the true cause of *sursum* [upward] motion ("quodammodo, vi et per extrusionem medii [in a way, by power and through the extrusion of the medium]"). Probably the reasons for Galileo's initial distrust for the extrusion of the medium are to be found in our scientist's awareness that this theory excludes the possibility of considering *sursum* [upward] motion "natural". As we will see, examining *Liber II*, Galileo at this time instead considers it natural! It is no coincidence that our scientist, attempting to reform *Liber I* and excluding *sursum* [upward] motion from natural motions, invokes the extrusion of the medium to justify it.

[90] Chs. VII and VIII.

[91] Ch. IX, "in quo ea omnia, quae supra demonstrata sunt, naturali discursu considerantur, et ad lancis pondera naturalia reducuntur [In which all those things, which have been demonstrated above, through natural discourse are considered, and are reduced to the natural weights of the scales]" (*ibid.*, p. 275).

[92] "Si enim pondera aeque ponderantia fuerint, et alteri eorum aliquod grave imponatur, tunc id deorsum feretur; sed non secundum totam suam gravitatem, sed tantum ea gravitate qua alterum pondus excedit:

balance once again seems to adequately represent the mechanism of natural motion.
177 If the differences between (specific) "weights" of the mobile and the medium are together the "cause" of the direction of the movement and its speed, the balance – used according to the instructions provided (comparison of equal volumes) – in which the proportions between the two weights can be reproduced, will be able to experimentally give us the table of speed proportions[93].

Let us pause for a moment on the operation carried out by Galileo in these chapters, in which the "theoretical" part of *De motu* can be identified. The desire to oppose Aristotle appeared very clear. Equally evident is the intention to use cosmological conceptions of the atomists, against which the Stagirite polemicized, to construct a philosophical framework capable of adequately supporting the new theory of movement. However, the true "deus ex machina" of the entire Galilean operation is undoubtedly Archimedes.

The entire framework for justifying natural motion implemented by Galileo is in fact directly derived from the writings of the Syracusan scientist, especially from *On Floating Bodies*. The comparison between equal volumes of mobiles and mediums, the choice of the hydrostatic model as a foundation of demonstration, whose conclusions are transferable to other media, and finally, the use of the balance analogy, demonstrate this irrefutably. Galileo – as we have seen in the previous paragraphs – has been studying Archimedean works with the utmost attention during these years and is making an effort to extrapolate methods and procedures applicable
178 to other fields of investigation[94]. In the *Bilancetta*, our scientist explicitly declared

quod idem est ac si dicamus, pondus hoc deorsum ferri tanta vi quanto aliud pondus est eo levius. Et pari ratione alterum pondus feretur sursum tanta vi quanto alterum est eo gravius [For if the weights are equally weighted, and something heavy is placed on one of them, then it will be carried downward; but not according to its entire heaviness, but only by that portion of heaviness by which it exceeds the other weight: which is the same as saying that this weight is carried downward with such power as the other weight is lighter than it. And in the same way, the other weight will be carried upward with such force as the other is heavier than it]" (*ibid.*, p. 275).

[93] An example of the use of the balance to establish the specific weights (and, therefore, the natural speeds) of the different metals and stones in water and in the air is demonstrated by the *Tavola delle proporzioni delle gravità in specie de i metalli e delle gioie pesate in aria e in acqua* [*Table of the proportions of the specific heaviness of the metals and jewels weighed in air and in water*], to be assigned to the years of the *Bilancetta* (*G.G.*, I, pp. 223-228). Having established the specific weights of the mobiles and the medium, Galileo, by applying the proportions, could indicate the different speeds (cf. ch. VIII, *ibid.*, pp. 262-273). Needless to say, these were "uniform" speeds.

[94] The problem of the sources of *De motu* is extremely delicate. The Archimedean influence is out of the question, emphasized unanimously by all interpreters. A knot still to be unraveled – perhaps the decisive one – remains the influence that the Pisan scientist would have undergone from Benedetti, whose dynamic conception he seems to follow in many respects precisely in *De motu*. The traditional opinion, inaugurated by Wohlwill (*Galilei und sein Kampf* cit., I, p. 105) and represented, above all, by Koyré (*Studi galileiani*, trad. it. cit., pp. 55-66), is decidedly in favor of the thesis of Galileo's direct knowledge of Benedetti and a timely use of his theories. More recently, following a series of "corrections" made to the reading of *De motu* proposed by Koyré, it has been questioned whether Galileo knew and used Benedetti's texts (Galileo could have drawn Benedetti's dynamic conceptions both from the *Diversarum Speculationum liber* cit., of 1585, and from one of the two prints, partially different, of the *Demonstratio proportionum motuum localium*, both of 1554; cf. *Le Speculazioni giovanili* "de motu" *di G. B. Benedetti* cit., where both prints are reproduced). The problem, essentially, is to establish whether Galileo independently decided to apply Archimedes's hydrostatic method to the analysis of

that he has taken up *On Floating Bodies* and the *Equilibrium of Planes* to resolve

---

natural motions, or whether, instead, he limited himself to following Benedetti's example. First of all, it should be observed that the analogies between the dynamic conception of *De motu* and that supported by Benedetti are very close. Furthermore, his writings were quite widespread and well-known, even in the Pisan environment. This is demonstrated by the fact that Jacopo Mazzoni, whose close ties with Galileo are known, not only had information on Benedetti's dynamics, but gave a rather detailed exposition of it in *De comparatione Platonis et Aristotelis* cit. (pp. 190 ff.), correctly citing the source. However, such close analogies appeared even closer to Koyré as a result of a peculiar misunderstanding. He in fact saw in the *De motu* "una critica della dinamica aristotelica dal punto di vista della dinamica della forza impressa [a criticism of Aristotelian dynamics from the point of view of the dynamics of impressed force]" (*Studi galileiani*, trad. it. cit., p. 55, note 2). By identifying the dynamics of "impressed force" and that of *impetus* (*ibid.*, p. 55), Galileo's attempt could appear to Koyré to be fully consistent with Benedetti's. Now, while it is true that Galileo, in the *De motu*, does not make any distinction between *virtus impressa* [impressed force] and *impetus*, it is no less true that Benedetti's concept of *impetus* in natural motions has very little to do with that developed by Galileo in *De motu*. For the Venetian scientist, in fact, every body, whether moving by nature or by violence, receives a certain *impressio* [impression] and *impetus*, by virtue of which, detached from the mover, it can continue to move for some time. In the case of a natural motion, "suam velocitatem semper augebit, cum in eo impetus et impressio semper augeantur, quia coniuncta habet perpetuo virtutem moventem [it will always increase its speed, since in it the impetus and impression always increase, because it continuously possesses the moving force that remains perpetually connected to it]" (*Diversarum speculationum liber* cit., p. 287); and elsewhere "quae impressio et impetuositas, in motibus rectis naturalibus continuo crescit, cum perpetuo in se causam moventem, idest propensionem eundi ad locum ei a natura assignatum habeat… [which impression and impetuosity in natural rectilinear motions continuously increase, since it perpetually possesses within itself the moving cause, that is, the inclination to go to the place assigned to it by nature…]" (*ibid.*, p. 184). From the passages cited, it is clear that the theory developed by Benedetti is that of the Buridanian derivation of "proto-inertial" *impetus*, while the *impetus* or *virtus impressa*, with which Galileo justifies, in natural motions, the accidental phenomenon of acceleration is, on the contrary, "self-depleting". Koyré's error — he did not clearly distinguish, as was necessary, the two positions — certainly depends upon – as Fredette appropriately pointed out (*Galileo's* De motu cit., pp. 340-341) – the fact that his works were conceived between 1935 and 1940, that is, before the fundamental studies of Maier and Clagett clarified the gap that separates the theories of *virtus impressa* (self-depleting) and *impetus* (proto-inertial). When Galileo, in the *De motu*, uses *impetus* (or, interchangeably, *virtus impressa*) it is always to indicate – unlike Benedetti – a self-exhausting energy. In addition to recovering this substantial difference between the *De motu* and Benedetti's dynamics, it should be emphasized that Moody – who was the first to point out that the dynamics of the *De motu* had nothing to do with Buridan's *impetus* (*Galileo and Avempace* cit.) – demonstrated that the source of the dynamic conception presented by Galileo in the Latin writing is not Benedetti, but rather Avempace, whose theories our scientist was able to draw on through the exposition given by Borro (*De motu gravium et levium* cit., pp. 124 ff.), whose work is cited twice explicitly (*G.G.*, I, pp. 333 and 367) in *De motu* (*Galileo and Avempace* cit., pp. 274 ff.). The thesis of a complete dependence of *De motu* on Benedetti was also questioned by I. E. Drabkin, *G. B. Benedetti and Galileo's* "De Motu", in "Proceedings of the Xth International Congress of the History of Science" (Ithaca 1962), 2 vols., Hermann, Paris 1962, I, pp. 627-630 and by E. Grant, *Aristotle, Philoponus, Avempace and Galileo's Pisan Dynamics*, "Centaurus", 1965, II, pp. 79-95. To these considerations, which make the hypothesis of Benedetti's direct influence on Galileo much more problematic, it should be added that the Venetian mathematician is not cited by Galileo: a rather significant fact, considering that *De motu* is a private writing (Galileo never mentions Benedetti even in subsequent works, nor in other documents). Lastly, even in the pages where the analogy is most evident, the lexicon of the two authors and the order of the arguments present notable differences. Therefore, Raymond Fredette has reasonably declared that he is convinced – until someone proves otherwise – that Galileo never read Benedetti's works (*Galileo's* De motu cit., p. 338; but see also in his unpublished Dissertazione cit., vol. II, note 110, pp. 350-354, where he extensively reviews the hypothesis of the Benedetti-Galileo relationship). In truth, it appears extremely probable that *De motu* could be the result of the application of the analysis methods provided by Archimedes's statics to Avempace's dynamic concept (integrated by the theory of *virtus impressa*).

179 the problem of the deception of the crown “ad mentem Archimedis [according to the mind of Archimedes]”[95]. In the *De motu*, he works by referring to the same texts, even though the problems under discussion are profoundly different. The analogy repeatedly emphasized between natural motion (hydrostatic model) and balance motion indicates Galileo’s desire to use, in a combined manner and showing their complete equivalence, the procedures developed by Archimedes in the two aforementioned writings. If the hydrostatic model provides the criterion for explaining natural motions, the balance allows for the construction of an adequate mechanical model which greatly simplifies the explanation and allows for much easier experiments and measurements.

Such was Galileo’s desire to “reunify” and “assimilate” the different Archimedean approaches that he did not hesitate to suppress some inconsistencies. In fact, he avoided specifying that, while the hydrostatic model necessarily presupposes the comparison between mobiles and mediums of equal volume (that is, at the level of specific weights), the laws of the balance are established for absolute weights. He certainly managed to pass over this inconsistency in silence by relying on the use of ambiguous terminology and the lack of recorded different conceptual variations. Since with *gravis* [heavy], *gravitas* [heaviness] indicated both “heavy” and “weight”, both absolute and specific, he gambled on the ambiguity and avoided indicating that, in the case of the balance, the comparison was between absolute weights. The analogy was so convincing that it would hardly have aroused suspicion in the reader.

Again, to strongly emphasize the analogy between hydrostatics and the balance model, Galileo did not hesitate to commit a second abuse. It should be recalled that, according to the hydrostatic principle, “gravius non posse attolli a minus gravi [the heavier cannot be lifted by the lighter]”, a few lines later, Galileo had assimilated the principle of the balance: “minus pondus non posse maius attolli [a smaller weight cannot lift a larger]”[96]. In addition to establishing full equivalence between valid principles, one for specific weights and the other for absolute ones, here Galileo avoided highlighting a second inconsistency. If the principle in question is valid as a “norm” in hydrostatics, it can easily be denied in statics. Galileo knew very well (as demonstrated in the *Bilancetta*) that, on the balance, “minus pondus [the lesser weight]” *can* “maius attollere [raise the greater]”. In fact, it is enough to hang the smaller weight at a greater distance from the fulcrum. In the balance, that is, unlike what happens in the hydrostatic model, the relationship between the “weights” is not the only decisive element. Even in this case, Galileo could propose the analogy as very close in virtue of an equivocal terminology. If, in fact, instead of resorting
180 to *pondus* [weight], he had used a term (as in, later, *momento*) capable of indicating the combined effect of weight with distance, the very modest limits of the analogy would have appeared immediately evident.

The insistence with which – without paying too much attention to the subtle – Galileo emphasizes the analogy between the two Archimedean approaches finds

---

[95] Cf. *supra*, p. 164.

[96] *G.G.*, I, p. 258.

further justification in the chapters of *Liber II* in which he attempts a specific analysis of the different types of movement. If our scientist uses the hydrostatic model to explain natural motions and calculate their speeds, he will be able to make use of the laws of the balance to establish the proportions of the speeds with which the same mobile descends along planes of different inclination[97]. The complete analogy between the two approaches therefore allowed all types of movement to be subjected to a unified treatment, even descents along inclined planes, for the interpretation of which the hydrostatic model – at least as developed by Galileo in *De motu* – appeared unusable. However, we will see later that, in the analysis of the inclined plane, Galileo will resort to the scheme of the balance of unequal arms, a model in no way comparable to the Archimedean hydrostatic one[98].

In what we have called *Liber II* of *De motu*, Galileo endeavors to apply the principles developed in the first part to the analysis of the different types of movement and to the justification of quite a few evident inconsistencies between his theory and the observation data. Setting aside the chapter dedicated to motions along planes of different inclinations, to which we will return at the end of this paragraph, our attention will now focus primarily on Galileo's discussion of violent motion and on the explanation that our scientist provides of acceleration.

It may appear strange that the order of exposition followed by Galileo favors inclined planes and projectile motions, rather than natural fall and the phenomenon of acceleration. As we will soon see, the anticipated exposition of violent motion arises from the purpose of developing an explanatory mechanism and preparing a fundamental analogy for justifying acceleration in natural motion.

After having explained and refuted the Aristotelian conception of projectiles[99], Galileo puts forward his own opinion. He starts by proposing a definition of *virtus motiva* [motive force] that is "a proiiente in proiectum imprimitur [impressed by the thrower on the projectile]":

> Dicimus ergo, illam esse *privationem gravitatis*, cum mobile sursum impellitur; cum vero deorsum, esse *privationem levitatis*[100]. [Therefore we say that this is a *privation of heaviness* when the mobile is driven upwards; but when downwards, it is a *privation of lightness*.]

The "virtus motiva nempe levitas [motive force, namely lightness]"[101] impressed **181**
by the mover in violent motions thus acts on the heaviness of the projectile. It is capable of increasing or decreasing it, thereby altering the relationship of the projectile with the medium. In fact, in the event that a projectile, heavier than the medium in which it is immersed, is thrown *sursum* [upward], the *virtus motiva* [motive force] will equate to a *levitas* [lightness] which cancels the *naturalis gravitas* [natural heaviness] of the projectile and, by making it lighter than the medium, forces it to rise. Conversely, an additional *gravitas* [heaviness] will be conferred by

---

[97] *Ibid.*, pp. 296-302 (Ch. XIV).
[98] Cf. *infra*, pp. 194 ff.
[99] *G.G.*, I, pp. 307-309.
[100] *Ibid.*, pp. 309-310; the italics are ours.
[101] *Ibid.*, p. 310.

the mover on the projectile, in the case of a violent *deorsum* [downward] throw. The *virtus motiva* [motive force] (or *impressa* [impressed]), adds Galileo, "successive remittitur in proiecto a proiciente absente [gradually diminishes in the projectile in absence of the thrower]"[102]. The explanatory analogy is, in this case, deduced from alterative motions – the heating of an initially cold piece of iron, when placed on the fire – in which the natural quality (cold) is overcome by the contrary quality[103]. Just as heated iron, removed from the fire, *gradatim* [gradually] loses its heat, returning to its natural frigidity, so too the mobile progressively consumes the lightness or heaviness given to it by the mover, finally recovering its natural lightness or heaviness[104]. Galileo takes care to respond to any objections and clarifies that it should not be surprising if a stone heavier than the medium, when thrown upwards, becomes "light". In fact,

> "inter lapidem in actu illo motus existentem et quodcunque alium leve, nulla differentia assignari poterit.[105]" [between a stone existing in that state of motion and any other light thing, no difference can be assigned.]

Galileo is willing, at most, to concede a distinction between natural *sursum* [upward] motion, produced by a "natural" *levitas* [lightness], and violent *sursum* motion, produced by an "accidental" *levitas*:

> Ego autem dicam leve id naturaliter esse quod sursum naturaliter fertur; leve autem id praeternaturaliter aut per accidens aut vi esse, quod sursum praeter naturam per accidens et vi fertur[106]. [But I will say that something is naturally light if it is naturally carried upwards; however, something is light either supernaturally, by accident, or by force if it is carried upwards contrary to nature, by accident and by force.]

182 Throwing a stone violently upwards is no different from placing it in a medium of greater "specific" heaviness than the stone itself[107]. Even in this case, in fact,

---

102 *Ibid.*

103 "Movetur, igitur, sursum mobile a proiciente, dum in manu illius est et gravitate privatur; movetur similiter, motu alterativo ferrum ad calorem, dum ferrum est in igne et ab eo privatur frigiditate [Therefore, the object is moved upwards by the thrower, while it is in his hand and is deprived of heaviness; similarly, iron is moved with an alterative motion towards heat, while the iron is in the fire and is deprived of coldness]" (*ibid.*, p. 310).

104 "Calor remittitur in ferro, igne absente: lapis tandem reducitur ad quietem ferrum similiter ad naturalem frigiditatem redit [Heat is remitted in the iron, with the fire absent: the stone finally returns to rest, and the iron similarly returns to its natural frigidity]" (*ibid.*, p. 310).

105 *Ibid.*, p. 311.

106 *Ibid.*, pp. 311-312. Galileo specifies that the stone, despite becoming "light", returns, after the motion, to its own natural heaviness, exactly as it happens in alterative motions: "nanque neque ego dixerim lapidem, post motum, levem factum esse, sed suam primam ac naturalem gravitatem retinere: sicut etiam ferrum candens frigiditate caret; post autem calorem, suam eandemque frigiditatem resumi [For neither would I say that the stone, after motion, has become light, but rather that it retains its original and natural heaviness: just as the glowing iron lacks coldness; after the heat, it resumes its same and original coldness]" (*ibid.*).

107 "Talis autem est lapis a virtute impulsus: et in lapide sua nativa et intrinseca gravitas eo deperditur modo, quo deperditur etiam dum in mediis se gravioribus ponitur [Such is the stone moved by the force of the impulse: and in the stone, its native and intrinsic heaviness is lost in the same way as it is lost

"adeo gravitatem amittit et *levitatem induit*, ut multae etiam extrinsecae gravitati advenienti (ut si quis deorsum premat) strenue resistat [it so loses its heaviness and *takes on lightness*, that it vigorously resists even much external heaviness coming upon it (as if someone presses it downwards)]"[108].

Galileo later specifies[109] that the *virtus motiva* [motive force] is exhausted through a continuous process. For example, in the case of a violent *sursum* [upward] motion, the *virtus impressa* (*levitas*) [impressed force (lightness)] superior to the natural *gravitas* [heaviness], as the projectile moves away from the hand that threw it, decreases uniformly and continuously, so that "non in motu violento posse duo puncta assignari, in quibus eadem fuerit virtus motiva [in violent motion, two points cannot be assigned where the motive force was the same]". When the *virtus motiva* [motive force] has weakened to the point of being equivalent to the natural heaviness of the projectile, equilibrium (stillness) will be achieved and, immediately afterwards (without interruption), the projectile will begin to descend again with natural motion, prevailing in its heaviness[110].

The Pisan scientist might well have said he was satisfied. The violent motions were, in fact, interpreted and justified using the same principles used for the analysis of natural motions: comparison between the weight (lightness) of the mobile and the weight (lightness) of the medium. This comparison, in the case of violent motions, would have had to be carried out taking into account the disturbing factor constituted by the *virtus impressa* [motive force]. The violent motions and the self-exhausting *virtus impressa* developed to justify them were, in turn, about to render Galileo a great service. In fact, they will constitute the model to which Galileo will entrust the attempt to justify, to the young scholar, what must have appeared, not without reason, as the most serious threat to the stability of his *de motu* concept: acceleration.

It goes without saying — though we have not explicitly stated it so far — that Galileo's theory in *De motu*, the "Pisan dynamics"[111], as it has been defined, is a dynamics of uniform motions. In natural motion, in fact, if the two decisive elements remain constant (specific gravity of the mobile and of the medium), the
speed is also kept constant. However, as had already been emphasized by an endless **183**
number of authors since Aristotle[112] – some of whom were well known to the Pisan scientist – the most elementary of observations demonstrate that bodies, in their natural descent, present a speed that is not uniform, but rather accelerated[113]. Galileo

---

when it is placed in mediums that are heavier]" (*ibid.*, p. 312).

[108] *Ibid.*, p. 312; the italics are ours.

[109] *Ibid.*, pp. 314-315 (Ch. XVIII: "virtutem motivam successive in mobili debilitari ostenditur [It is shown that the motive force is gradually weakened in the mobile]").

[110] *Ibid.*, p. 315. We will return later to the very important problem of the "point of inflection" of violent motion in natural motion.

[111] The expression is widely used by English-speaking scholars (cf., for example, E. A. Moody, *Galileo and Avempace* cit., p. 205). It goes without saying that "Pisan" signifies developments in the years in which Galileo taught in Pisa (1589-1591).

[112] Aristotle's opinion is carefully reviewed by Galileo (*G.G.*, I, pp. 315-318). For the history of the principal opinions on acceleration, cf. the indications provided *infra*, note 123.

[113] After having indicated according to which proportions the speeds of mobiles in the different mediums

could not ignore the topical question of acceleration, nor escape the obligation to justify the inconsistency between his theory of uniform natural motions and the experimental data on acceleration[114].

Given that – Galileo emphasizes – in natural motion there is neither variation in the heaviness of the mobile nor that of the medium, the acceleration must depend on extraneous factors acting on the heaviness of the mobile[115]. The solution is offered to him[116] – as we anticipated – precisely by the *virtus* (*motiva, nempe levitas* etc.) used to justify violent movements. This is why the analysis of acceleration is, in *De motu*, deferred to that of violent motions[117].

The "virtus impressa a proiciente [the impressed force by the thrower]" is, in fact, not only capable of decreasing the heaviness of the mobile, but can make it "adeo leve ut magna velocitate sursum advolet [so light that it flies upward with

---

will be determined, Galileo warns "quod magna hic oritur difficultas: quod proportiones istae, ab eo qui periculum fecerit, non observari comperientur. Si enim duo diversa mobilia accipiet, quae tales habeant conditiones ut alterum altero duplo citius feratur, et ex turri deinde demittat, non certe velocius, duplo citius, terram pertinget; quin etiam, si observetur, id quod levius est, in principio motus praeibit gravius et velocius erit… [Here arises a major difficulty: the proportions will not be observed by the one who performs the experiment. For if he takes two different mobiles that have conditions such that one moves twice as fast as the other, and then drops them from a tower, one will not certainly reach the ground twice as quickly; indeed, if observed, the lighter one will initially precede the heavier one and will be faster…]" (*ibid.*, p. 273).

[114] Galileo states confidently, at the opening of the chapter dedicated to the discussion of acceleration (XIX) "Magis certe inventu quam explicatu difficilis est causa cur naturalis motus velocitas in fine augeatur [The reason why the speed of natural motion increases at the end is more certainly difficult to discover than to explain]" (*ibid.*, p. 315).

[115] "Quia igitur grave mobile... descendens, tardius movetur in principio, ergo necessarium est, illud minus esse grave in principio sui motus quam in medio vel in fine; cum certo sciamus, ex demonstratis in primo libro, velocitatem et tarditatem, gravitatem et levitatem sequi... Verum naturalis et intrinseca mobilis gravitas certe non est diminuta, quia nec diminuta est moles nec densitas illius: restat ergo, imminutionem illam gravitatis esse praeternaturalem et accidentariam… [Since, therefore, a heavy mobile… descending, moves more slowly at the beginning, it is necessary that it be less heavy at the beginning of its motion than in the middle or at the end; since we know for certain, from what has been demonstrated in the first book, that speed and slowness, heaviness and lightness follow each other... But the natural and intrinsic heaviness of the mobile is certainly not diminished, because neither its mass nor its density is diminished: it remains, therefore, that this diminution of heaviness is preternatural and accidental…]" (*ibid.*, I, p. 318).

[116] Galileo confesses that he has long harbored many doubts about the possibility of adequately justifying the phenomenon of acceleration. Subsequently, having intuited the right path, before reaching the goal, he more than once despaired of finding the solution: "Huius tamen non dicam admirabilis, sed necessarii effectus, causam cum interdum disquirerem (nunquam, enim, quae ab Aristotele traditur, arrisit), diu anxius fui, ne quod plene satisfaceret inveniebam. Verum, causa (meo tamen iudicio) verissima reperta, prima quidem fronte laetatus sum verum cum eam diligentius examinarem, non nulla difficultate carere suspicatus sum: sed tandem, omne difficultate temporis progressu exempta, exactam eam quidem et certissimam nunc in medium afferam [However, I will not say that this effect is admirable, but rather necessary. When I occasionally inquired into its cause (for I was never satisfied with what was handed down by Aristotle), I was long anxious, fearing that I would not find a fully satisfying answer. However, having found, in my judgment, the most true cause, I was initially pleased, but when I examined it more carefully, I suspected that it was not without some difficulty. Yet, ultimately, after all difficulties were removed with the passage of time, I now present it as thoroughly and certainly accurate]" (*ibid.*, p. 316).

[117] We will see that, in the *Discorsi*, Galileo reverses the order of succession.

great speed]":

Videamus ergo et diligenter perscrutemur, an forte virtus ista sit causa diminuendae gravitatis mobilis 184
in principio sui motus[118]. [Let us therefore see and carefully examine whether perhaps this force is the cause of the reduction of the heaviness of the mobile at the beginning of its motion.]

Galileo will affirm the legitimacy of this hypothesis in the following pages. It is, in fact, a question of analyzing the phenomenon of a violent *sursum* [upward] motion, observing what happens when the *virtus motiva* [motive force] that pushed the mobile weakens. In this process of weakening (a process that Galileo defines as uniform and continuous), the *virtus motiva* will reach the point where it "mobilis gravitatem non amplius vincet et tunc non ulterius impellet mobile [no longer overcomes the heaviness of the mobile and then will no longer push the mobile further]"[119]. In this situation, the violent motion ends and the natural motion begins simultaneously, determined by the natural heaviness of the mobile starting to prevail.

It is at this "critical point", that is, the beginning of natural motion, *necessarily* coinciding with the end of a preceding violent motion, that Galileo invites us to focus our attention, to discover the true cause of the acceleration:

Sed tamen non ideo in fine motus violenti erit adnihilata virtus illa impressa, sed tantum adeo erit diminuta, ut non amplius gravitatem mobilis superet, sed illi aequabitur; et, ut uno verbo dicam, in mobili non amplius dominabitur virtus sursum impellens, quae est levitas, sed ad paritatem cum mobilis gravitate redacta erit: et tunc, in ultimo puncto motus violenti, nec grave, nec leve erit mobile. Sed amplius, suo more decrescente virtute impressa, incipit praedominari gravitas mobilis; quare et mobile descendere aggreditur. Sed quia in principio talis descensus superest adhuc multa virtus sursum impellens (licet non amplius maior sit gravitate mobilis), quae est levitas, hinc fit ut gravitas propria mobilis ab hac levitate minuatur et, consequenter, motus tardior sit in principio. Et, rursus, quia extrinseca illa virtus amplius debilitatur, minorem gravitas resistentiam habens augetur, et mobile velocius adhuc movetur[120]. [But yet, at the end of violent motion, the impressed force will not be entirely annihilated; rather, it will only be diminished to the point where it no longer overcomes the heaviness of the mobile but is equal to it. And, to put it in one word, in the mobile the force that pushes upwards, which is lightness, will no longer dominate, but will be reduced to equality with the heaviness of the mobile. Then, at the last point of violent motion, the mobile will be neither heavy nor light. But as the impressed force further decreases in its usual manner, the heaviness of the mobile begins to predominate; therefore, the mobile begins to descend. Yet, because at the beginning of such a descent there remains much of the force that pushes upwards (though it is no longer greater than the heaviness of the mobile), which is lightness, the result is that the proper heaviness of the mobile is diminished by this lightness, and, consequently, the motion is slower at the beginning. And, again, because this external force weakens further, heaviness faces less resistance, increases, and the mobile moves ever faster.]

The acceleration, therefore, depends on the persistence of part of the *virtus impressa* [impressed force] of the opposite sign (a *levitas* [lightness], precisely), at the moment in which the violent motion, exhausted, gives rise to the natural one, and on its progressive consumption, which allows the mobile to "free" the

[118] *Ibid.*, p. 318.
[119] *Ibid.*, p. 319.
[120] *Ibid.*

185 force of its own heaviness and, therefore, to move faster[121]. Starting from a certain point, that is, when the impressed lightness is completely exhausted, the motion will continue with uniform speed[122].

Galileo, after specifying that he had arrived at this conclusion before discovering that Hipparchus had proposed a very similar one[123], had an easy time demonstrating that, even in motions "quibus non praecesserit motus violentus [in which no violent motion has preceded]", the cause of the acceleration remains the one indicated above[124]. Paradoxically, in fact, the situation described in the case of violent motions is precisely the "natural" one: every mobile (for example, a stone dropped from a tower) when it begins its natural descent presents a perfect equilibrium (on which the previous *quietis status* [state of stillness] depended) between the impressed lightness (from the hand) and natural heaviness. Having been released from the hand that supported it, the stone begins to lose "lightness" and, therefore, its heaviness progressively takes over[125]. There is, therefore, not opposition, but

---

[121] Koyré effectively emphasized how, rather than an accelerated speed, it would be appropriate to speak of "una velocità progressivamente deretardata [a progressively deretarded speed]" (*Studi galileiani*, trad. it. cit., p. 64). It is worth emphasizing that this same concept of acceleration will be proposed again by Sagredo in the *Discorsi*, but with significantly different terminology (*G.G.*, VIII, p. 201; for the analysis of the passage in the *Discorsi*, cf. *infra*, pp. 320-322).

[122] "Cum enim ideo mobile acceleretur quia virtus contraria continue remittitur, gravitas vero naturalis acquiritur, rationi quidem consentaneum erit ut tandem tota virtus contraria deperdatur, naturalisque resumatur gravitas, atque idcirco, remota causa, acceleratio desinat [For since a mobile accelerates because the opposing force is continually diminished and natural heaviness is acquired, it is indeed consistent with reason that eventually the entire opposing force will be lost, natural heaviness will be resumed, and thus, with the cause removed, acceleration will cease]" (*G.G.*, I, pp. 328-329).

[123] "Et hanc veram existimo causam accelerationis motus: quam quidem cum excogitassem, et, post duos menses, forte quae ab Alexandro de hac re scribuntur legerem, ex eo intellexi hanc quoque fuisse sententiam doctissimi illius philosophi [And I consider this to be the true cause of acceleration of motion: which, when I had thought of it, and after two months, perhaps, read what Alexander writes about this matter, I understood that this was also the opinion of that most learned philosopher]" (*ibid.*, p. 319). Galileo later specifies that, in any case, Hipparchus's theory has so far been considered capable of explaining only natural motions "cui violentus praecesserit [to which violent motions preceded]". Instead, he was the first to demonstrate that the theory is valid for all natural motions: "Nos vero et ea quae ab Hipparco explicata non sunt addemus, ostendendo quomodo etiam in motu, cui non praecesserit violentus, causa eadem locum habeat [We will also add those things which were not explained by Hipparchus, showing how the same cause applies even in motion to which no violent motion has preceded]" (*ibid.*, p. 320). To tell the truth, at least according to Simplicius's testimony (*In Aristotelis de caelo libros commentaria*, J. L. Heiberg edidit, *CAG*, VII (1894), pp. 264, 22-267, 6, Commentary on Lib. I, Ch. VIII), which informs us of Hipparchus's theory and Alexander's different opinion (Simplicius is certainly the source from which Galileo drew this information), one gets the very clear impression that Hipparchus was referring, in general, to natural motions, which he considered necessarily preceded by a violent motion. For a history of opinions about acceleration, from antiquity to the 17th century, it is always useful to consult P. Duhem, *De l'accélération produite par une force constante* cit.; cf. also M. Clagett, *La scienza della meccanica*, trad. it. cit., Ch. IX (*L'accelerazione di caduta*), pp. 577-593.

[124] *G.G.*, I, p. 320.

[125] "Quando enim etiam de manu nulla vi sursum impellente impressa decidit lapis, cum tanta virtute impressa discedit, quanta est sua gravitas. Nam quando lapis in manu alicuius quiescit, tunc non est dicendum lapidem habentem nullam vim in lapide imprimere: quis enim lapis sua gravitate deorsum premit, necesse est ut altera tanta vi a manu sursum impellatur, nec maiori, nec minori [For when a stone, even without any force pushing it upwards from the hand, falls with the same force impressed as its own heaviness, it departs with as much force as its heaviness. For when the stone is at rest in someone's

rather profound congruence between violent motion and natural motion, which seem inextricably linked[126]. Both depend on the same principles. To tell the truth, one can say about consecutiveness: "motus enim naturalis violentum sequitur et finis **186**
violenti cum principio naturalis est coniunctus [for natural motion follows violent motion, and the end of the violent is joined with the beginning of the natural]"[127].

In the following chapters, Galileo makes an additional effort to account for some phenomena of accidentally accelerated natural motion, provided by observation, which seemed unjustifiable in the light of his own theory[128]. We will not focus on these pages, which are very well known, nor will we focus on the arguments with **187**

---

hand, it should not be said that the stone has no force impressed upon it. For since the stone presses downward with its own heaviness, it is necessary that an equal force from the hand pushes it upwards, neither greater nor less]" (*ibid.*, p. 320).

[126] "Ex hoc autem discursu facile quispiam intelligere poterit quomodo isti 2 motus non vere sint contrarii, sed potius motus quidam ex violento et naturali compositus… [From this discourse, however, anyone will be able to easily understand how these 2 motions are not truly opposite, but rather a certain motion composed of both violent and natural…]" (*ibid.*, p. 322).

[127] *Ibid.*

[128] It is primarily a question of explaining why "minus gravia in principio sui motus naturalis velocius moveantur quam graviora [less heavy objects move faster at the beginning of their natural motion than heavier ones]" (*ibid.*, p. 333), as experience demonstrates. Galileo ingeniously claims that the heavier mobile, when released from the hand and beginning its natural motion, has stored a greater quantity of "lightness", compared to the less heavy mobile (the "lightness" impressed is, in fact, equal to natural heaviness). Initially, therefore, the heavier mobile is "braked" by a greater opposing force, which explains why it is overtaken by the lighter one. However, once the "lightness" is consumed, the heavier one will move faster. For his explanation to appear reliable, Galileo must warn against considering the progressive weakening of the imprinted *levitas* as "dependent" on the contrast exerted by the natural *gravitas*. If this were the case, in fact, "licet graviora plus de qualitate contraria quam leviora consumere habeant, attamen habent etiam maiorem gravitatem qua illam destruere possint [although heavier objects have to consume more of the opposing quality than lighter ones, they also have greater heaviness with which they can overcome it]" (*ibid.*, p. 334). Again resorting to the analogy with alterative motions, Galileo explains that the weakening of *levitas* is a phenomenon independent of the *gravitas* of the mobile: "in mobili qualitatem illam contrariam non ideo remitti, quia a gravitate eiusdem mobilis oppugnatur; gravitas enim cum in mobili contraria qualitate conferto nulla penitus sit, idem praestare non potest; sed qualitas illa per se debilitatur et mobile relinquit: sicut etiam dum candens ferrum frigefit, non ideo in eo calor remittitur, quia a contraria frigiditate reluctetur… [in the mobile, that opposing quality is not diminished because it is resisted by the heaviness of the same mobile; for when the heaviness is in the mobile, no opposing quality is entirely present, it cannot perform the same function; but that quality is weakened by itself and leaves the mobile: just as when hot iron cools, the heat is not diminished because it is resisted by the opposing coldness…]" (*ibid.*, p. 335). It is worth emphasizing that the discussion between supporters of early Galilean experimentalism and those who, instead, believe that his experiments are primarily "mental" has focused precisely on this chapter. Specifically in these pages, in fact, the Pisan scientist states that he has experimented several times, observing the fall of two spheres from a high tower, one of wood and the other of lead, that the wooden one initially precedes the leaden one, from which it is subsequently reached and surpassed: "et si ex alta turri demittantur per magnum spatium [la plumbea]* praecedat: *et de hoc saepe periculum feci* [and if dropped from a high tower over a great distance, (the leaden one) will precede: *and I have often tested this*]" (*ibid.*, I, p. 334; the italics are ours). It should be added that, in the *Racconto istorico*, Vincenzo Viviani states that Galileo carried out "replicate esperienze... fatte dall'altezza del Campanile di Pisa con l'intervento dell'altri lettori e filosofi e di tutta la scolaresca [replicated experiments... carried out from the height of the Bell Tower of Pisa with the involvement of other readers and philosophers and the entire school]", to demonstrate that mobiles of the same kind "si muovono tutti con pari velocità [all move with the same speed]", whatever their weight (*G.G.*, XIX, p. 603). Decidedly opposed to the possibility of admitting the experiments from the Tower of Pisa as actually conducted and recognizing Viviani's testimony as valid,

which Galileo tries to demonstrate "non dari quietem [stillness is not given]" at the inflection point of violent motion in nature[129].

At this point, Galileo's analysis could be considered complete. He had proposed a unified interpretation of all types of motion, natural and violent, *sursum* [upward], *deorsum* [downward] and along different inclinations. However, more than one reason leads us to think that Galileo was far from satisfied with the results obtained.

In truth, the inconsistencies were neither few nor insignificant. For example, in the chapter dedicated to motion along inclined planes[130], Galileo, using the model of the balance of unequal arms, had determined according to which proportions the "heaviness" of the same mobile varies as the inclination of descent varies. The "heaviness" of the mobile appeared maximal to Galileo along the perpendicular and equivalent to zero on the horizontal plane[131]. Admitting that the diversity of motions (of speed) along planes of different inclinations depended on the variation of heaviness meant supporting a thesis capable of leading to absurd consequences when compared with the hydrostatic model, to which Galileo had entrusted the explanation of all the types of movement. If, in fact, the *gravitas* [heaviness] goes from a maximum to zero, it will become, at a certain point, less than the *gravitas* of an equivalent volume of the medium. It follows that the same mobile, immersed, for example, in water, will descend until the path imposed on it has a certain inclination, but will, conversely, have to ascend when the inclination of the descent plane is less. This is obviously contradicted by the simplest of experiences.

In truth, in this chapter, Galileo proposes an analysis that holds value only if it ignores the medium entirely. The motions along the inclined planes analyzed by Galileo are, in fact, motions "in a vacuum", that is, in conditions completely different from those in which – according to Galileo's *De motu* – natural motions occur. Again an example of how the balance analogy was difficult to use fruitfully.

Furthermore, the operation carried out to justify violent motion and acceleration in natural motion must also have appeared extremely problematic and contradictory.

E. Wohlwill declared himself, with reasoned and convincing reflections, both in *Galilei und sein Kampf* cit. and, previously, in two specific essays (*Ueber einen Grundfehler aller neueren Galilei-Biographien*, in "Münchener medizinische Wochenschrift", 1903, pp. 1850-1851; and *Galilei-Studien: Die Pisaner Fallversuche*; Der Abschied von Pisa, Leopold Voss, Hamburg 1907). Wohlwill was echoed by Koyré, according to whom Galileo's "saepe periculum feci [I have often tested]" is anything but reliable (*Studi galileiani*, trad. it. cit., p. 63, note 26). Of a different opinion is Moody, who in his classic work (*Galileo and Avempace* cit.) considered it plausible that Galileo actually carried out the experiment. It is no coincidence that the subtitle of *Galileo and Avempace* is precisely *The dynamics of the Leaning Tower experiment*. With much greater conviction, Settle argued that in all probability Galileo carried out the experiments from the tower which he mentions in *De motu* (*Galileo's Use of Experiment* cit., pp. 324-326). For Settle, traces of several "real" experiments remain in *De motu*. In agreement with Settle, R. Fredette also asserts "l'expérience de la Tour Penchée de Pise, après tout, ne serait pas aussi legendaire que la critique actuelle est portée de la croire [the experience of the Leaning Tower of Pisa, after all, would not be as legendary as current critics are led to believe]" (Unpublished dissertation cit., II, note 86, p. 338).

[129] *G.G.*, I, pp. 328-333 (Ch. XXI). We will address this issue later.

[130] *Ibid.*, pp. 296-302 (Ch. XIV).

[131] We will return to the problem of Galileo's elaboration of a "proto-inertial" principle in *De motu* in the next chapter.

In essence, by evading the rule solemnly declared at the outset, Galileo had ended up making *levitas* [lightness] an autonomous reality, proposing it as the cause of both the violent ascent of heavy bodies and the progressive increase in speed in natural motion. The *virtus motiva* (*impressa*) – *levitas* of Galilean theory is not, in **188**
fact, conceivable in a relative sense. It is the "opposite" of *gravitas* [heaviness] and pushes "by nature" the projectiles *sursum* [upward]. It is not difficult to understand how this might have happened. Having to explain the violent ascent of projectiles "heavier" than the medium, how could he attribute the cause to heaviness?[132] If what is impressed on the projectile is "heaviness", how can it ascend? And how can the initial delay in natural motion (i.e., acceleration) be explained in terms of "lesser heaviness"?

Galileo pays dearly for the weakness of his conception of "force". He attempts to reduce every type of force to *gravitas* [heaviness] (also to be able to measure it) and, at the same time, attributes a natural *deorsum* [downward] tendency to heaviness. It was inevitable that, with this conception of force, justifying natural *sursum* [upward] motion would prove extremely problematic.

An evident example of such difficulties is provided by the chapter dedicated to acceleration. Galileo always and exclusively refers to the natural *deorsum* [downward] motion, carefully avoiding specifying that the natural *sursum* [upward] motion must be interpreted no differently (that is, it begins when the violent *deorsum* of a mobile less heavy than the medium ends). The reasons for the omission are not difficult to imagine. Galileo would have, in fact, been forced to attribute acceleration also to natural *sursum* motion, once again finding himself proven wrong by experience. In addition to the confusion of concepts, the negative effects produced by an inadequate vocabulary are evident in this case as well: it was necessary either to distinguish force from *gravitas* [heaviness], or to deprive *gravitas* of the natural tendency towards the center.

The new ideas could not be satisfactorily explained using traditional terminology. The previously analyzed pages offer us numerous and convincing confirmations: the impressed force, the positive lightness, the predominance of opposites, the alterative motions and the natural places. Having started with the clearest Archimedean intentions, Galileo, without realizing it, ended up back in Aristotle. Despite the many cautions and repeated warnings, the use of scholastic terms, which ended up inadvertently imposing their own values, therefore had a destructive effect.

It was certainly after having perceived the weight of these inconsistencies that Galileo returned to *Liber I* to attempt a refoundation, which, by establishing clearer and more certain principles, would protect against the errors committed. The attempts at reform preserve our scientist's awareness that it was only partly a question of concepts. In fact, he must have found it fundamental to subject the terms used to careful scrutiny, rigorously regulating their use. He had discovered that the terms, far from being "neutral", concealed very serious pitfalls.

---

[132] In reality, the only solution compatible with the general framework of *De motu* would have been to assert that *gravitas* [heaviness] is conferred on the medium, which, becoming heavier, would have caused the violent ascent of the mobile.

189 It is evident from the changes introduced in the new draft that it was primarily a matter of remedying this type of fault.

Without providing any preliminary justification, Galileo begins, in fact, by rewriting the first chapter, replacing *levis*, *levius* with *minus grave*, *minor gravitas*. To grasp the meaning of this operation, it will suffice to compare the two drafts of the opening passage of the initial chapter:

First Draft

Lationem omnem naturalem, sive deorsum sive sursum illa sit, a propria mobilis gravitate vel levitate fieri. ... Quomodo quid alio levius vel gravius, vel aeque grave dicendum sit....[133] [Every natural movement, whether downward or upward, occurs by the proper heaviness or lightness of the mobile. ... How something should be said to be lighter, heavier, or equally heavy compared to another....]

Second Draft

Lationem omnem naturalem ab excessu vel defectu gravitatis fieri.... Quid magis, quid minus, quidve quaeque grave dicendum sit...[134]. [Every natural movement occurs from the excess or deficiency of heaviness... What should be said to be more, less, or equally heavy…]

He thus introduces the second chapter with a significant "preface", from which some light can be gleaned regarding his intentions:

> Huc usque gravia et minus gravia, non autem gravia et levia, diximus; sicut centro propinquius et centro remotius, non autem deorsum: inferius enim explicaturi sumus, non esse leve, hoc est carens gravitate; nec esse locum, qui tantum sursum, non autem etiam deorsum sit. Attamen, si interdum, ut communiter loquar (parum enim interest ad propositum nostrum de nominibus contentio), grave et leve et sursum et deorsum dixero, hoc pro minus et magis grave, et pro centro propinquius et remotius intelligatur...[135]. [Up to this point, we have spoken of heavier and less heavy, not heavy and light, just as we have spoken of closer to the center and farther from the center, not downward: for we will explain below that there is no such thing as light, that is, something lacking heaviness; nor is there a place that is only upward, but not also downward. However, if sometimes, to speak more commonly (for it matters little to our purpose to argue over names), I have said heavy and light and upward and downward, let these be understood as less and more heavy, and as closer to and farther from the center.]

Still dissatisfied, he returns to the initial chapter[136], highlighting in the new draft the need for "definitions of terms"[137] and proposing a general axiom to precede the discussion: "id quod gravius est a minus gravi, *si cetera sint paria*, non posse attolli [that which is heavier than the lighter one, *if the other conditions are equal*, cannot be lifted]"[138]. This is the same principle enunciated by Galileo in the chapter in which he emphasized the analogy between the hydrostatic model and that of the

---

[133] *Ibid.*, p. 251.
[134] *Ibid.*, p. 341.
[135] *Ibid.*, p. 342.
[136] *Ibid.*, pp. 347-350 (Ch. V of the new draft).
[137] "Haec sunt quae de terminorum definitionibus dicenda erant [These are the things that needed to be said about the definitions of terms]" (*ibid.*, p. 348).
[138] *Ibid.*, I, p. 348; the italics are ours.

balance[139]. The addition now of "si cetera sint paria [if the other conditions are equal]" seems to testify to his awareness of the non-automatic applicability of the analogy.

The integration, useless in regard to the hydrostatic model used by Galileo (comparison between equal volumes of the mobile and the medium), was instead, **190**
in fact, very useful for "neutralizing" in the balance the perturbing effect of the distance from the point of application.

After adding a *Lemma*, in which he demonstrates that the heavinesses of mobiles of equal volume are proportional to the volumes[140], Galileo continues by reproposing the "Archimedean" chapters IV and V of the original draft, in which he limits himself to amending, with the technique depicted above, *levis*, *levius*, and *levitas*. At this point, we find the first striking surprise: ch. VI, in which, "reasoning" *magis physice* [more physically], Galileo had highlighted the analogy between the hydrostatic model and the balance, is dropped in its entirety[141]. We have already emphasized the arbitrariness and weakness of the analogy. Realizing this, Galileo therefore decided to abandon it. This automatically meant abandoning the project of attributing the treatment of all natural motions to unitary principles (those along planes of different inclination had, in fact, been interpreted using the laws of the balance)[142]. One of the

[139] Cf. *supra* p. 179.

[140] "Gravitates inaequalium molium corporum aeque gravium eam inter se habent proportionem quam ipsae moles [The heavinesses of unequal bulks of equally heavy bodies have the same proportion between them as the bulks themselves]" (*ibid.*, p. 348). The *Lemma* had already been stated in the original draft (p. 264)[†], but Galileo had noted in the margin that it required a demonstration ("partes enim homogeneorum sunt inter se in gravitate sicut in mole; quod demonstrari oporteret [for the parts of homogeneous bodies are in relation to each other in heaviness as they are in bulk; which should be demonstrated]"). And the demonstration is promptly given in the new draft (pp. 348-350). This is one of the elements that most clearly demonstrates how the chapters we are commenting on were written after the first 23 (cf. *supra*, note 58, for Drake's different opinions).

[†] To paraphrase Galileo's provided example: a ratio of a large volume of wood to a small volume of wood is in the same ratio of their respective heaviness of the volumes of wood; likewise, a ratio of a large volume of water (of equal volume to the large volume of wood) to a small volume of water (of equal volume to the small volume of wood) is also in the same ratio of their respective heaviness of the volumes of water; since the small and large volumes of wood and water are respectively equal, then the ratios of volumes, heaviness of wood, and heaviness of water are all the same.

[141] Cf. *supra*, pp. 173 ff. The dropping of the chapter is indisputable. One might, to tell the truth, think that Galileo tacitly intended it to be inserted after the two "Archimedean" chapters, in the same form in which it had been inserted in the original draft of *Liber I*. This hypothesis, however, must be excluded, given that the second Archimedean chapter ends with an explicit reference to the topic that is discussed in the following chapter (unlike what happened in the first draft, where the analogy with the motions of the balance was "anticipated" in the second of the two "hydrostatic" chapters).

[142] This, therefore, explains why, in the second draft, the balance analogy does not appear, without needing to reverse the order of succession of the two versions, as proposed by Drake (cf. *supra*, note 58). That this was the reason for abandoning the analogy appears even clearer from the fact that no reference to it is found in the first draft of the *Discorso intorno alle cose che stanno in su l'acqua* (cf. *infra*, pp. 229-230), in which Galileo returns to propose the same hydrostatic conceptions of the *De motu*. In the final draft, which he published, Galileo introduced an important "correction" to those principles, showing how the behavior of bodies in water cannot be adequately interpreted through a simple comparison of specific weights. He introduced the notion of *momento* (absolute weight × speed) to account for this behavior. Not by chance, Galileo immediately returned to propose the hydrostatic-balance analogy, now enabled by having developed an interpretation of both phenomena in terms of *momento* (cf. *infra*, pp.

main objectives pursued by Galileo in *De motu* appeared, at this point, unattainable.

Despite this, Galileo presses on with great determination. He dedicates an
191 entirely new chapter to the demonstration that "motum sursum nullum naturalem esse [there is no natural upward motion]"[143]. It is followed by a revised version of ch. XII[144]. The modifications here concern the title, in which the now usual removal of *leve* [light][145] is observed (however, the arguments against the Aristotelian thesis of positive lightness do not change)[146] and, above all, the insertion of a peculiar, extremely vehement outburst in which Galileo outlines the reasons that led him to delete *leve* from his lexicon:

"quia [Aristoteles et ceteri philosophi]* voluerunt (non contenti se pro levi id quod minus grave est intelligere) dari etiam quoddam corpus, quod tale simpliciter esset et omni careret gravitate, id cane peius et angue abhorrentes, omnimode et funditus usque ipsum leve evertere conati sumus[147]. [because (Aristotle and the other philosophers) wanted (not content to understand light as that which is less heavy) to even allow the existence of a certain body that would simply be such and completely devoid of heaviness, we, detesting this worse than a dog and a snake, have decided to completely and utterly overthrow this notion of lightness]"

The extraordinary impetuosity of this outburst appears less peculiar if we reflect that it served to exorcise a dangerous temptation, from which we have seen that Galileo himself was not immune.

By now, the picture of Galileo's dissatisfaction with the original draft begins to appear complete. In the following chapter, entirely new[148], we find, however, other important indications. Galileo explicitly states that *sursum* [upward] motion cannot depend on the "defect" of heaviness (which he would have originally called "lightness").

Neque dicas sicut excessus gravitatis mobilis super gravitatem medii est per se et intrinseca motus deorsum causa, ita defectum gravitatis mobilis, in respectu ad gravitatem medii esse per se et intrinseca motus sursum causam[149]. [Nor should you say that just as the excess of the heaviness of a mobile body over the heaviness of the medium is the intrinsic and inherent cause of downward motion, so the defect

---

233 ff.).

[143] *Ibid.*, pp. 352-355 (Ch. VIII, of the second draft). Galileo states that motion, to be "natural", must respect two conditions: first of all, it must be "finitus et terminatus [finite and terminated]"; furthermore, it is necessary that it is "non ab extrinseca, sed intrinseca moveatur causa [not by an external cause, but by an intrinsic cause]" (p. 353). If *deorsum* [downward] motion meets these conditions, they do not occur in *sursum* [upward] motion. In fact, it does not have a "terminus" where it stops naturally (like the center for *deorsum*), nor is it produced by an "intrinsic" cause (Galileo has in fact rigorously established the abolition of "lightness").

[144] It should be recalled that the title of Ch. XII, in the original draft, was "non esse ponendum simpliciter leve [not to be posited simply as light]" (*ibid.*, pp. 289-294).

[145] "Gravitatis corpus nullum expers esse [no body is without heaviness]" (*ibid.*, pp. 355-361).

[146] The Aristotelian "contra antiquos [against the ancients]" arguments, promptly refuted by Galileo, who declared himself "pro antiquis [for the ancients]", remain the same five in both drafts of the chapter.

[147] *Ibid.*, p. 355.

[148] "Motum sursum ex parte mobilis naturalem esse non posse [The motion upward cannot be natural from the part of the mobile]" (*ibid.*, pp. 361-363; it is Ch. X of the second draft).

[149] *Ibid.*, p. 361.

of the heaviness of a mobile body, with respect to the heaviness of the medium, is the intrinsic and inherent cause of upward motion.]

Naturale [natural] motion is, therefore, oriented in a single direction. Bodies, all of the same matter and all "heavy", are endowed with a single *naturalis inclinatio* [natural inclination]: they always tend towards the center[150]. Galileo therefore attempts to identify the reason that led many philosophers to propose lightness as the cause of natural *sursum* [upward] motion. **192**

Crediderim autem, errorem eorum qui motum a medio secundum naturam esse existimarunt, ex eo ortum duxisse, quia non potuerunt, externam causam, a qua mobilia moverentur invenire; et ideo coactos fuisse, intrinsecam ponere, eamque levitatem appellare[151]. [I would think, however, that the error of those who considered motion from the medium to be natural arose because they were unable to find an external cause by which the mobiles were moved; and therefore they were compelled to posit an intrinsic cause, which they called lightness]

Among those who shared this erroneous opinion was – as we have seen – Galileo himself. Rather, one could consider the above passage as an attempt to justify his own mistake. Not even the Pisan scientist had been able to clearly indicate an "external cause" that explained the ascent of the lighter bodies in the medium[152], finding himself forced to resort to the ironclad rule with which nature placed the heavy at the bottom and the light above. Now, however, Galileo has discovered the "external cause", that is, he has identified the mechanism on which the rise of the lesser heavy of the medium depends:

Quapropter, ut eiusmodi deleamus erratum, properemus iam ad explicandum quomodo quae sursum feruntur ab extrinseca moveantur causa, nempe ab ipso medio *per extrusionem*[153]. [Therefore, in order to eliminate such an error, let us now hasten to explain how those things which are carried upward are moved by an external cause, namely, by the medium itself *through extrusion*]

With a chapter dedicated to the explanation of the decisive mechanism of extrusion, Galileo's attempt to reform *Liber I* ends[154].

It does not appear that Galileo ever attempted to re-draft what we have defined as *Liber II*, taking into account the clarifications and new principles formulated.

---

[150] "Omnium corporum una est materia, eaque in omnibus gravis: sed eiusdem gravitatis non possunt esse contrariae inclinationes naturales: ergo, si una est naturalis inclinatio, ut contraria sit praeter naturam opus est: naturalis autem gravitatis inclinatio est ad centrum: ergo necesse est, quae a centro praeter naturam esse [All bodies have a single matter, and this is heavy in all of them: but the same heaviness cannot have contrary natural inclinations; therefore, if there is one natural inclination, for it to be contrary to nature it is necessary: now the natural inclination of heaviness is towards the center; therefore, it is necessary that what is contrary to the center]" (*ibid.*, p. 362).

[151] *Ibid.*, pp. 362-363.

[152] Cf. what we observed *supra*, note 89.

[153] *Ibid.*, p. 363; the italics are ours.

[154] "Quae sursum naturaliter moveri hucusque dicta sunt, non ab interna causa, sed ab externa, nempe ab ipso medio, per extrusionem moventur [Those which move upwards naturally, as has been said so far, are not moved by an internal cause, but by an external one, namely by the medium itself, through extrusion]" (*ibid.*, pp. 363-366).

We have already mentioned the impossibility of supporting the hydrostatic-balance analogy, upon which the possibility of exploiting the analysis of motions along inclined planes depended. However, Galileo must have been compelled by even more serious reasons to give up any attempt at reform. The new theory brought into focus could not support *Liber II*. Nor could the Pisan scientist limit himself to rewriting it while suppressing any reference to "lightness". The terminological issues were, in fact, in this case, substantive issues. We have already highlighted the
193 difficulties Galileo faced in being forced to indicate with *levitas* [lightness] the force violently impressed on a heavy object thrown upwards. His entire interpretation of violent motions was based on the hypothesis of two natural motions: *sursum* and *deorsum*. How could he support it now, after denying that there is a *levitas* [lightness] capable of pushing the heavy object *sursum* [upwards]? The least of the consequences was the disruption of the very close analogy established between violent and natural motions: it became necessary to go back to interpreting the violent motion as "contrary" to the natural. However, by accepting this point of view, Galileo would have been forced to renounce the explanation of acceleration as an incidental fact produced by the persistence in the mobile, by virtue of the previous violent motion, of a *levitas* [lightness] equivalent to its *gravitas* [heaviness].

Compared with the new conception of natural motions developed by Galileo in the chapters of the second draft, Galileo's explanation of acceleration showed another irremediable contradiction. Having established as an ironclad principle that non-natural motions depend on "external" causes, it followed that acceleration appeared not to be an accidental phenomenon, but rather a "natural" characteristic of the falling motion. The surviving *virtus impressa* (*levitas*) [impressed force (lightness)], which opposes *naturalis gravitas* [natural heaviness] during the course of motion, acts, in fact, as an absolutely "internal" cause[155]. Its effects will therefore be natural.

It is pointless to dwell on all the destructive consequences that the reform of *Liber I* entailed for the analysis of violent and natural motions along inclined planes and of the phenomenon of acceleration[156]. Galileo had now realized that all that remained of the ambitious project was merely a heap of ruins. Having started with the intention of developing unitary principles capable of accounting for the different types of motion, he had to conclude bitterly that he had failed. He dedicated himself to this project starting from Archimedes and proposing the full equivalence of the hydrostatic model and the balance model. Even on this level, he had to acknowledge a failure, demonstrated, in the revision, by the collapse of the analogy. The failure of *De motu* had clearly shown that it was necessary to pursue another path and temporarily relinquish his excessive ambitions. First and foremost, it was necessary to pay great attention to the terms, propose unambiguous definitions and rigorously

[155] Cf. R. Fredette, *Galileo's* De motu cit., pp. 347-348, especially note 48.

[156] Settle, Fredette and Stillman Drake, in their respective works dedicated to *De motu* cited several times, agree in attributing the failure of *De motu* and Galileo's renunciation of any attempt at complete revision to the difficulties he encountered in justifying acceleration and the complete lack of experimental evidence supporting "uniform" natural motions.

ensure that the use always conformed to the definitions. It was, therefore, essential to commit to fighting that "war of words" with the utmost determination, whose importance he had perhaps underestimated in *De motu*[157].

Furthermore, it was necessary to revisit some interesting ideas from *De motu* **194**
in order to deepen and develop them. Above all, the chapter on motions along planes of different inclinations[158] deserved careful consideration, as it promised rather encouraging results. Galileo had attempted an analysis of the motions of falling bodies along inclined planes, applying a "static" principle to the solution of an exquisitely dynamic problem[159]. In fact, he intended to deduce the proportion of speeds with which the same heavy object would move along planes of different inclination from the laws that govern the equilibrium of weights on the balance.

The preliminary supposition, explicitly put forward by Galileo, is quite familiar to those who know Galileo's mature works: "grave deorsum ferri tanto vi, quanta esset necessaria ad illum sursum trahendum; hoc est, fertur deorsum tanto vi quanta resistit ne ascendat [a heavy object is carried downward with as much force as would be necessary to pull it upward; that is, it is carried downward with as much force as resists its ascent]"[160]. The force with which it opposes the ascent is naturally the *gravitas* [heaviness] of the mobile, which decreases as the descent shifts from the perpendicular to that along planes of ever smaller inclination with respect to the horizon. Galileo concludes: "Tunc sciemus quantum maior vis requiratur ad sursum trahendum mobile per *bd* [perpendicular]* quam per *be* [inclined]*, quando cognoverimus quanto eiusdem mobilis maior erit *gravitas* in plano secundum lineam *bd*, quam in plano secundum lineam *be* [Then we will know how much greater force is required to pull an object upward along the perpendicular *bd* than along the inclined *be*, when we understand how much greater the *heaviness* of the same object is on the plane along the line *bd* than on the plane along the line *be*]"[161].

To discover the variation of the same "heaviness" along the different inclinations, Galileo proposes to use the balance. Let's look at Galileo's demonstration schematically.

---

[157] "Parum enim interest ad propositum nostrum de nominibus contentio [Indeed, it matters little to our purpose the contention over names]" (*G.G.*, I, p. 342).
[158] *Ibid.*, pp. 296-304 (Ch. XVI of the original draft: "in quo agitur de proportionibus motuum eiusdem mobilis super diversa plana inclinata [in which the proportions of the motions of the same object on different inclined planes are discussed]").
[159] The application of the laws of balance to the analysis of motion along inclined planes was not new. Jordanus had already attempted it in *De ratione ponderis*, however examining the motion along the arcs of a circle. Galileo considered the *gravitas* [heaviness] (later the *momento*) at the different points of the circumference equivalent to that which would occur along the tangents drawn through those points (cf. M. Clagett, *La scienza della meccanica*, trad. it. cit., pp. 180-182).
[160] *G.G.*, I, p. 297. Galileo will propose this principle again in *Mecaniche* and in the *Discorsi*.
[161] *Ibid.*, p. 297.

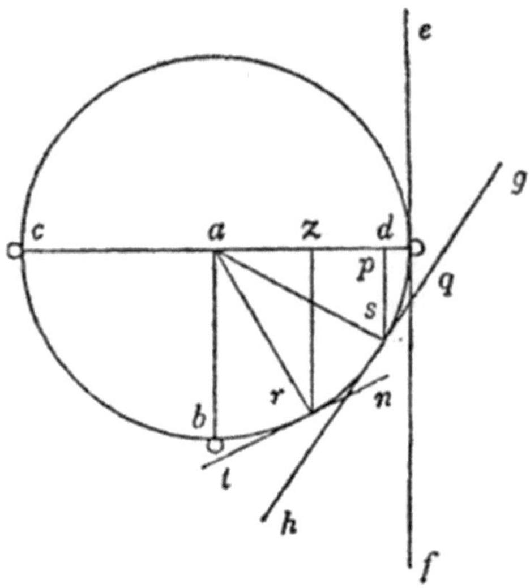

Figure 2

Let the balance of equal arms *cad* (fig. 2), inscribed in the circle *cad*, with equal weights hanging from the extremes *c*, *d*. Let *ef* be the tangent to the circle
195 through the point *d*. The descent of the mobile along the tangent *af* would occur "secundum gravitatem mobilis in *d* [according to the heaviness of the object at *d*]". Thus, if the mobile were at point *s*, its descent would occur along the tangent to the circle through *s*, that is, *gh*. Therefore, "mobilis per lineam gh motus erit secundum gravitatem quam habet mobile in puncto *s* [the object moving along the line *gh* will do so according to the heaviness that the object has at point *s*]". Similarly, at *r*, the motion will be along *tn*, and according to the heaviness of the mobile at *r*. But the heaviness at *r* is < than at *s*, and at *s* < than at *d*. The weight at *d*, in fact, balances an equal counterweight at *c*. The weight at *s* is as if it were applied at *p*. Since the suspension point is closer to the fulcrum, it will therefore be smaller. The same, of course, applies to *r*: it would, in fact, be as if the weight were applied at *z*. At this point, we can easily calculate the proportion between heavinesses at *d*, *s*, *r*, etc. They are, indeed, proportional to the distances from the fulcrum marked by their perpendicular projections on the lever arm, that is, in proportion to *ad*, *ap*, *az*[162].

Having established the proportions of the "heaviness" of the mobile along differently inclined planes, Galileo believed he could directly extend them to the speeds (*celeritates*)[163]. He observes, in fact, that in the triangle *abd*, right-angled at *b* (see fig. 3), "eiusdem mobilis in diversis inclinationibus celeritates esse inter se permutatim sicut obliquorum descensuum, aequales rectos descensus

[162] *Ibid.*, pp. 297-298.

[163] "Tanto maiori vi descendere idem grave per rectum descensum quam per inclinatum, quanto maior est inclinatus descensus quam rectus [The same weight descends with a greater force through a vertical descent than through an inclined one, by as much as the inclined descent is greater than the vertical]" (*ibid.*, p. 298)

comprehendentium, longitudine [the speeds of the same mobile on different inclinations are inversely proportional to the lengths of the oblique descents, which encompass equal vertical descents]"[164].

The pages of which we have provided a schematic explanation[165] are of great importance. And not only because the demonstrations contained therein will constitute the starting point for Galileo's subsequent investigations. The importance also depends on their documentation of a truly critical and unsustainable "terminological" situation. These same demonstrations will return in the *Mecaniche*, but will be presented again in a profoundly changed terminology. One of the modifications – the most significant – will concern the introduction of *momento*. It was probably while reflecting on these pages that Galileo felt the indelible need for a "reform" of his lexicon.

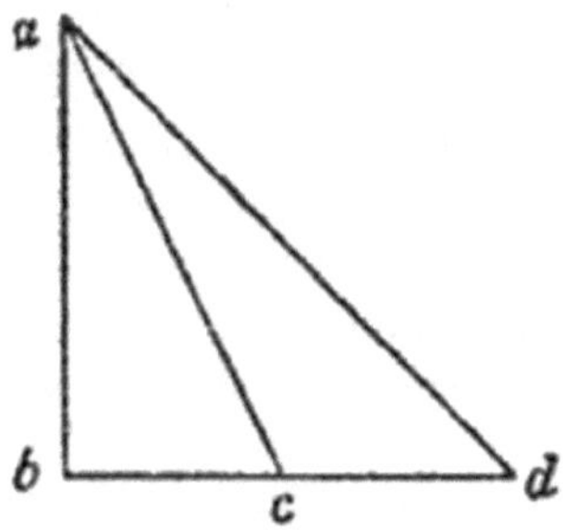

Figure 3

In the demonstration referred to above, Galileo establishes a rigorous proportionality between the *vis* [force] necessary to lift a mobile (or with which the same mobile descends) along different inclinations and the *gravitas* [heaviness] that the mobile presents on those same inclinations. The *celeritas* [speed] is still directly
proportional to this pair. As the *gravitas* of the mobile varies, proportional variations **196**
in *vis* and *celeritas* follow: "motus erit secundum gravitatem quam habet mobile in puncto *d* [the motion will be according to the heaviness that the mobile has at point *d*]"[166] – as Galileo asserts. Lastly, *pondus* is used to mean determined ("absolute") weight. Now, it is evident that, in this context, *gravitas* [heaviness] indicates both the "propensity" to descend with a certain *pondus* [weight], with an "intensity" (that is, with a "uniform" speed) dependent on the inclination of the plane of descent relative to the horizon, and the ability to counterbalance more or less effectively, according to the distance of the suspension point of the *pondus* from the fulcrum of the balance.

The use of *gravitas* [heaviness] to express this dual concept was legitimized, as we have seen previously, by an authoritative tradition. It cannot, however, be

164 *Ibid.*, p. 301.

165 We will return to many interesting aspects later.

166 *Ibid.*, p. 297.

denied that this use of the term, in the absence of precise specifications, could create misunderstandings. In fact, in *De motu* itself, on the one hand, *gravitas* [heaviness] serves Galileo to indicate the "natural" tendency of the mobile to descend, a tendency proportional to its *pondus* [weight]. In this meaning, *gravitas* ends up being confused with "weight", of which it simply indicates a function. Galileo himself, in fact, uses *gravitas* several times in the Pisan writing with the value of pure and simple "weight"[167]. On the other hand, the term expresses an "accidental" capacity (as in the theory of the natural fall presented by Galileo in *De motu*)[168] or "mechanical" capacity (as in our context), no longer proportional to *pondus*, nor precisely referable to it. Furthermore, we must not forget that Galileo had always relied on *gravitas*, without any further specification, to indicate both the concept of "specific" weight and the general cause of the *deorsum* [downward] motion of weighty bodies. Five functions, seemingly similar but essentially irreducible to one another, were decidedly too many for a single term. And, as we have documented in the previous pages, ambiguity and confusion were bound to arise.

If, as far as sensibility towards a term like *gravitas* is concerned, Galileo's *De motu* can essentially be aligned with other authors on mechanical and dynamic questions of the 16th century, however, we must not overlook an important difference. Galileo's lexicon, around 1590, appears, in fact, peculiarly poor and somewhat outdated even compared to that of many of his contemporaries or precursors. It seems, in fact, that our scientist was not familiar with two expressions, of widespread use, that would have significantly helped him to organize and promote conceptual distinctions in the pages of *De motu*: "gravità in (secondo la) specie [heaviness in (accordance to) species]"[169] and "gravità secondo il sito [heaviness

[167] See, for example, a famous passage from the same chapter, in which the inclined plane theorem is found: "Sicut dum idem Archimedes supponit, superficiem habere gravitatem, et alteram altera graviorem esse, cum tamen revera omni sint expertes gravitate [Just as when Archimedes assumes that a surface has heaviness and that another is heavier, while in reality both are devoid of heaviness]" (*ibid.*, p. 300).

[168] Cf. *supra*, pp. 182 ff.

[169] The term was used in the pseudo-Archimedean *De insidentibus humido*, which Galileo could have consulted in the 1565 print, edited by Tartaglia. The same scientist from Brescia had used it several times: "Un corpo si dice esser più grave di un altro *secondo la specie*, quando che la sostantia material di quello è più ponderosa della sostantia material dell'altro [A body is said to be heavier than another *according to the species* when the material substance of the former is more weighty than the material substance of the latter]" (*Quesiti et inventioni diverse*, riprod. anastat. cit., p. 84). The expression was also used by Buonamici in his *De motu* cit. ("conferemus aequas magnitudines, quae si fuerint ponderis aequalis, aeque idem graves esse iudicabimus; quae vincet, eam specie graviore esse censebimus [we will compare equal magnitudes; if they are of equal weight, we will judge them to be equally heavy; whichever one prevails, we will consider it to be of a specific heaviness]"), in a chapter in which, significantly, he reported Archimedes's discovery of the crown fraud, also illustrating the hydrostatic theories of the Syracusan scientist (p. 492). Galileo will later make explicit reference to this same chapter of Buonamici's *De motu*, showing that he has read it very carefully (*Discorso* cit., *G.G.*, IV, pp. 80 ff.). It is worth emphasizing, however, that *gravitas in specie* is never used by Benedetti, who usually speaks of "corpora diversarum specierum [bodies of different species]" (*Demonstratio* cit., p. 72), nor by Mazzoni, in the pages of *De comparatione Platonis et Aristotelis* cit. (191 ff.), which moreover depend on Benedetti (cf. *supra*, note 94).

according to position]" (location)[170]. Later, he will use "gravità in specie [specific 197
heaviness]"[171], while using *momento* to indicate the dynamic effect of weight.

The consequences endured, as well as the extreme poverty and approximation of his own lexicon, were heavy. It was precisely from the awareness of the incurable contradictions he had incurred by using a language incapable of adequately reflecting the articulation and distinction of concepts, largely already significantly innovative, that would inspire reformation of the terminology, to which Galileo would dedicate himself after the failure of the *De motu.* With this, we are already at the *Mecaniche* and the famous "definitions of terms" that Galileo introduced there, among which stands out in importance the one related to the meaning of *momento*, a term significantly still absent in *De motu.*

---

[170] In addition to the *de ponderibus* tradition, the expression was common in Tartaglia (cf. *supra*, p. 72) and, practically, in all the authors of mechanics of the 16th century (cf. M. Clagett, *La scienza della meccanica*, trad. it. cit.), even in Leonardo (cf. *supra*, Part One, p. 72), if not a "man without letters", certainly not a refined philologist.

[171] In *Discorso*, cf. *infra*, p. 231.

Chapter II

THE "MECHANICAL" MOMENTO

1. Le Meccaniche

Let's focus on the dates, which are quite important. According to prevailing 199
opinion, *De motu* is assigned, as we have seen, to the years 1588-1590, meaning that it is entirely a Pisan work. Some doubt might arise regarding the date to which the revision attempts of the theoretical part belong, which introduce conceptual and terminological innovations. Nor can it be ruled out, in principle, that the drafting of these chapters was carried out in a slightly later period, say between 1591 and the beginning of 1592. In the autumn of 1592, when Galileo moved to Padua bringing with him, among other things, the manuscript of the *De motu*, the Latin treatise must have by now reached the level of elaboration in which it has reached us today.

If we are to consider Vincenzo Viviani's testimony valid, the first draft of the *Mecaniche* would belong to 1593[1]. The treatise would follow, namely, the *De motu* practically without interruption and would be considered the first work to which Galileo dedicated himself after his transfer to Padua. It is certain, on the other hand – as clearly demonstrated by several chapters of the *De motu* – that already in Pisa our scientist had shown considerable attention to mechanical problems, especially in relation to questions of movement.

We will now see, with the help of other documents, that the date suggested by Viviani is to be considered largely reliable, at least in regard to the first draft of the writing.

Favaro, unfortunately after having published the volume of the Edizione Nazionale containing the text of the *Mecaniche*[2], had discovered, in 1898, the
existence of a version of the writing significantly different from all the known copies, 200

[1] "Nel 1593 scrisse le Mecaniche e altre cose [In 1593, he wrote the Mecaniche and other things]" (*Racconto istorico* cit., *G.G.*, XIX, p. 606).

[2] *Le Mecaniche* is published in vol. II of the Ed. Naz. (pp. 155-191). Since the handwritings have not survived, Favaro used the incomplete text offered by Ms. Galil 72 of the B.N.F. (cc. 9*r*-26*r*) for the edition, which he supplemented using the codex of the Palatine Library of Parma (HH, IX, 53, 184); cf. *G.G.*, II, pp. 150-152 (*Avvertimento* to *Le Mecaniche*).

P. Galluzzi, *Momento: Galilean Studies*, Archimedes 78,
https://doi.org/10.1007/978-3-032-35049-7_6

preserved in Regensburg, in a private collection[3]. In publishing the manuscript, which we will call the Regensburg copy, Favaro advanced the hypothesis that it was a first draft of the treatise, which Galileo would have used as a "guide" for both public and private lessons held during the Padua period[4]. From certain documents, also published by Favaro, we know, in fact, that Galileo devoted at least one university course to the examination of the pseudo-Aristotelian *Questions of Mechanics*[5] and gave private lessons in mechanics to his "pensionati [pensioners]"[6].

The Regensburg copy bears the following inscription: "Delle meccaniche lette in Padova dal Signor Galileo Galilei l'anno 1594 [On Mechanics read in Padua by Signor Galileo Galilei in the year 1594]". A final note indicates that the manuscript was in the possession of Monsignor Giovanni Ciampoli on 22 February 1627[7]. The date of composition recorded in the manuscript confirms Viviani's indication, albeit slightly modifying it, which, before the discovery of the Regensburg copy, had appeared unfounded to Favaro. The distinguished Galilean scholar had, in fact, suggested the years 1597-1598 as the date of composition of the *Mecaniche*, when Galileo held the course on *Questions of Mechanics* at the University of Padua[8]. Compared to the version published in the Edizione Nazionale, the Regensburg copy presents a less complete exposition, appears deliberately abbreviated in some parts and is, in general, much less accurate.

A recent, fortunate discovery demonstrates, however, that Viviani's testimony regarding the date of composition of the *Mecaniche* should be considered entirely accurate. In the summer of 1955, in fact, the North American scholar Stillman Drake, a meritorious editor and scholar of Galileo, discovered a third version of the *Mecaniche*, different from the two examined above, in the Bauer Collection of the
201 Institute of Technology in Pasadena[9]. We will call this version the Pasadena copy. The coat of arms of Count Rocco on the cover of the volume containing this version sheds light on its provenance[10]. Unfortunately, the Pasadena copy is devoid of indications regarding the date of writing. However, it presents notable similarities

---

[3] *Delle Mecaniche lette in Padova l'anno 1594 da Galileo Galilei, per la prima volta pubblicate ed illustrate da Antonio Favaro* in "Memorie del R. Istituto Veneto di Scienze, Lettere ed Arti", XXVI, 5, Venezia 1899. Cf. also, again by Favaro, *Intorno ad un inedito e sconosciuto trattato di Meccaniche di Galileo Galilei nell'Archivio di S.A. il Principe di Thurn-Taxis in Ratisbona…*, B. G. Teubner, Leipzig 1899.

[4] *Delle Mecaniche lette in Padova* cit.

[5] "Anno 1598: Exc. D. Galileus Galilaeus Florentinus Legit Euclidis Elementa et Mechanicas Aristotelis…" (Antonio Favaro, *Galileo Galilei e lo Studio di Padova*, 2 voll., Antenore [ristampa anastatica]*, Padova 1966, II, p. 115; 1st ed., Le Monnier, Firenze 1883).

[6] Cf. *Galileo Galilei e lo Studio di Padova* cit., I, Ch. VI (*L'insegnamento privato di Galileo a Padova*), pp. 139-164.

[7] *Delle Mecaniche lette in Padova* cit., p. 26.

[8] *G.G.*, II, p. 150 (*Avvertimento*).

[9] *Galileo Gleanings. V. The earliest version of Galileo's Mechanics*, in "Osiris", XIII (1958), pp. 262-290. Drake publishes the text of the Pasadena copy, giving variants of the Regensburg one. In the *Appendix*, Drake also provides the variants of another copy preserved in the Staats- und Universitätsbibliothek in Hamburg. This last copy, already indicated by E. Wohlwill (*Galilei und sein Kampf* cit., I, p. 141), does not differ much from the Regensburg one.

[10] *The earliest version* cit., p. 264.

with the Regensburg copy, though it lacks the first six chapters and presents a certain number of variants. In the introductory note to the critical edition of the text of the two Regensburg and Pasadena copies, Drake argued convincingly that the Pasadena copy represents the oldest draft of the *Mecaniche* that has come down to us[11]. The North American scholar also believes that it constitutes an "outline", prepared by Galileo so as not to lose the thread of the discussion during the oral exposition[12].

We can therefore easily assign the drafting of the Pasadena copy to 1593, as Viviani's testimony suggests. The Regensburg copy is dated 1594, while the "definitive" draft, according to the prevailing opinion today, should be assigned to the years 1598-1600[13].

We have focused carefully here on the exact determination of the dates that mark the subsequent drafts of the *Mecaniche* for a very specific reason. Precisely in this period of years, and for the reasons previously illustrated, Galileo consciously dedicated himself to reforming his lexicon, including, among other things, *momento*. Even though perhaps overly ambitious, we would like to assign a relatively precise date to this entry, the importance of which we will soon be able to appreciate.

Neither the Pasadena copy nor the Regensburg copy presents the initial "definitions" of the terms with which the final version opens. The exposition then, in both versions, is extremely cautious. Galileo had already abandoned the system, adopted in the *De motu*, of assigning different "heavinesses" to the same body applied at different points of the balance. He simply speaks of the equilibrium of the balance of equal arms and weights, and of the need, if one wants to obtain equilibrium of the instrument of unequal arms, of placing weights at different distances from the fulcrum, inversely proportional to those distances:

se nell'estremità A, B si sospendessino pesi uguali non più sarebbe la bilancia in equilibrio ma... se
vorremmo ridurre la libra all'equilibrio sarà di mestiero nella minor distanza B sospendere peso maggiore **202**
e, nella maggiore distanza A, peso minore, si come l'esperienza ci dimostra[14]. [if equal weights were
suspended at ends A and B, the balance would no longer be in equilibrium, but... if we wanted to restore
equilibrium to the scales, it would be necessary to suspend a greater weight at the shorter distance B,
and a smaller weight at the greater distance A, as experience shows us.]

Similar caution is also apparent in the treatment of the inclined plane. In fact, Galileo completely abolishes the discussion on "heaviness" according to which the same body would descend along planes of different inclination, expressing himself simply in terms of greater or lesser "force" necessary to lift a given weight along variously inclined planes:

il peso alla forza ha la medesima proporzione che ha la lunghezza del piano elevato all'altezza perpendicolare[15]. [weight has the same proportion to force as the length of the elevated plane has

---

[11] *Ibid.*, p. 265.

[12] *Ibid.*

[13] In addition to *The earliest version* cit., see, again by S. Drake (in collaboration with Drabkin), *Galileo Galilei on Motion and on Mechanics* cit., p. 137.

[14] *The earliest version* cit., p. 271.

[15] *Ibid.*, p. 277.

to the perpendicular height.]

Galileo demonstrates that he is aware of the ambiguities committed in the *De motu*; but the terminological “reform” has not yet produced satisfactory results. For now, he limits himself to avoiding the most delicate parts of the discussion, those which concern the problem of varying the “pulling” effectiveness of the same weight along different inclinations.

In both the Pasadena and Regensburg copies – and this is a very interesting detail for us – *momento* appears only on one occasion. Its presence is ambiguous, which suggests careful examination of the passage in question, in which the “lightness” in equilibrium is discussed. The two manuscripts present different versions. Let’s see them:

Pasadena Copy	*Regensburg Copy*
ma perché ogni minimo peso che s’aggiunga a quello che pareggia e contrapesa può far rimuovere; però, non tenendo conto di quel *momento* insensibile, diremo farsi il moto...[16]. [but because every minimal weight that is added to that which matches and counterbalances can cause it to dislodge; however, not taking into account that imperceptible *momento*, we will say that motion occurs…]	ma perchè ogni minimo *momento* che s’aggiunga a quello [peso]* che pareggia e contrapesa può far rimuovere; però, non tenendo conto di quel *momento* insensibile, diremo farsi il moto...[17]. [but because every minimal *momento* that is added to that (weight) which matches and counterbalances can cause it to dislodge; however, not taking into account that imperceptible *momento*, we will say that motion occurs…]

It should not escape us that the meaning assigned by Galileo to *momento* in these contexts is quite different from that with which it will be solemnly introduced and declared in the definitive version. Here, *momento* indicates, in fact, a small amount of weight. This is the most common semantic value of our term: small quantity in general (of time, weight, motion). In the two passages in question, *momento* presents, to be exact, the same value with which we have seen it used in the biblical expression *momentum staterae*: a minimum weight, sufficient, however, to tip the balance.

203 As shown by the Pasadena copy and from the structure of Galileo’s arguments, in both copies, our scientist still expresses himself in terms of “weight” and “distance from the fulcrum”, not yet having a technical term to indicate this relationship. In the two above-mentioned contexts, *momento* indicates the addition of a minimum amount of weight, so small that it can be disregarded. Even in the definitive version, while illustrating the principle of “virtual speeds”, Galileo will return to using the term with this semantic value[18].

In 1594, Galileo had therefore not yet untangled the problem. He had already made further simplifications and developed greater awareness, but the words – so

[16] *Ibid*., p. 272; the italics are ours.
[17] *Delle Mecaniche lette in Padova* cit., p. 13; the italics are ours.
[18] *G.G.*, II, pp. 163-164.

to speak – still did not help him adequately.

We can thus limit the phase of intense research, successful definition and massive introduction of the term *momento* into Galileo's lexicon to the years 1595-97, in the meaning that we find distinctly prevalent in the definitive version of the *Mecaniche*. This allows us to consider *momento* a Paduan achievement. A greater alignment is not possible, due to the lack of documents that can support it. However, it is legitimate to put forward a hypothesis. In the academic year of 1597-98, Galileo taught a course dedicated to the pseudo-Aristotelian *Questions of Mechanics* at the University of Padua. While preparing himself for this task and stimulated by extensive readings of mechanical texts (he would certainly have revisited, among others, Archimedes, Commandino and Guidobaldo's *Mechanicorum liber*), he perhaps understood precisely then all the advantages that could derive from adopting *momento*.

The definitive version of the *Mecaniche*[19] opens, as we have already anticipated, with "le diffinizioni dei termini proprii di questa facoltà [the definitions of the terms specific to this faculty]" and of the "prime supposizioni, dalle quali, come da fecondissimi semi, pullulano e scaturiscono consequentemente le cause e le vere dimostrazioni delle proprietà di tutti gl'instrumenti mecanici [initial suppositions, **204**
from which, like very fertile seeds, the causes and the true demonstrations of the properties of all mechanical instruments consequently pullulate and arise]"[20]. The suitability for such anticipation – as Galileo clarifies – is suggested by the characteristic method of the demonstrative sciences, in other words, geometric-mathematical, which he intends to take as a model for the discussion[21].

The first term defined is "gravità [heaviness]":

> Adimandiamo adunque gravità quella propensione di muoversi naturalmente al basso, la quale, nei corpi solidi, si ritrova cagionata dalla maggiore o minore copia di materia, dalla quale vengono constituiti[22]. [Let us therefore call heaviness that propensity to move naturally downwards, which, in solid bodies, is found to be caused by the greater or lesser quantity of matter from which they are

---

[19] The first printed edition of the *Mecaniche* was due to the care of Cavalier Luca Danesi (*Della scienza mecanica e delle utilità che si traggono da gl'Istrumenti di quella. Opera cavata da manoscritti dell'Eccellentissimo Matematico Galileo Galilei*..., Appresso gli Stamp. Camerali, in Ravenna 1649). Fifteen years earlier, while Galileo was alive, Marino Mersenne had published a translation or, better yet, a free adaptation of the Pisan scientist's treatise: *Les mechaniques de Galilée... avec plusieurs additions... traduites de l'Italien par L.P.M.M.*, Chez Henry Guenon, à Paris 1634 (cf. *Les mechaniques de Galilée... Edition critique* cit.). There are numerous important studies on *Le Mecaniche*. In addition to the aforementioned Galileo Galilei, *On Motion and on Mechanics* (which also offers a full English translation of the text, due to the care of Stillman Drake), it is worth mentioning the usual M. Clagett, *La scienza della meccanica*, trad. it. cit., especially pp. 172 ff.; Th. Settle, *Galilean Science* cit. (Ch. V); the works of Westfall (*The problem of force* cit., and *Force in Newton's Physics* cit.); M. Clavelin, *La philosophie naturelle de Galilée* cit., pp. 148-177; W. Wisan, *The new science of motion* cit., pp. 156 ff.; A. Koyré, *Studi galileiani*, trad. it. cit., pp. 99 ff.

[20] *G.G.*, II, p. 159.

[21] *Ibid.*

[22] *Ibid.*

constituted.]

Although there has been no shortage of "problematic" readings of this definition, it seems to me that Galileo's intentions are absolutely clear. In fact, there is no plausible reason to suspect that we are dealing with the definition of heaviness per unit of volume, i.e., specific weight[23]. The definition is divided into two parts. In the first, "gravità [heaviness]" is defined, in general, as the "cause" of the tendency of heavy bodies toward the center, while the second specifies that heavy bodies participate to a greater or lesser extent in this "propensione [propensity]" depending on the quantity of matter of which they are constituted, in other words, according to their weight (not "specific", but "absolute"). Galileo has set aside the hydrostatic model which he had used in the *De motu*. Furthermore, as we demonstrated in the previous chapter, the removal of the hydrostatic model–balance analogy, in the revision of the "theoretical" part of the *De motu*, indicates that Galileo is perfectly aware that the specific weight, upon which Archimedean hydrostatics is based, is replaced, in the balance, by the comparison between absolute weights. It would, therefore, have been truly peculiar for Galileo to have prefaced his discussion with a declaration of "gravità [heaviness]" such as "gravità in specie [heaviness in species (i.e., specific heaviness)]", which was completely unusable for the purposes of the argument he was about to undertake.

However, it is worth observing that Galileo still uses the same term to indicate both indeterminate heaviness (cause of the "propensity" towards the bottom) and determinate heaviness, i.e., weight. The reform has begun, but it does not yet appear entirely satisfactory. This lack of distinction also depends on the fact that Galileo in the *Mecaniche* uses "peso [weight]" in relation to "contrappeso [counterweight]", in other words, as "something that weighs", rather than a "quality" of bodies. He has, however, finally isolated heaviness from the variation of its effectiveness, learning to indicate it with a new term. We are presented with a passage whose extreme
205 importance must not escape us. For the first time, in fact, Galileo uses *momento* in the "technical" sense, and, not by chance, he feels the need to carefully clarify its meaning:

> Momento è la propensione di andare al basso, cagionata non tanto dalla gravità del mobile, quanto dalla disposizione che abbino tra di loro i diversi corpi gravi; mediante il qual momento si vedrà molte volte un corpo men grave contrapesare un altro di maggior gravità: come nella stadera si vede un picciolo contrapeso alzare un altro peso grandissimo, non per eccesso di gravità, ma sì bene per la lontananza dal punto donde viene sostenuta la stadera; la quale, congiunta con la gravità del minor peso, gli accresce momento ed impeto di andare al basso, col quale può eccedere il momento dell'altro maggior grave. È dunque il momento quell'impeto di andare al basso, composto di gravità, posizione e di altro, dal che possa essere tal propensione cagionata[24]. [Momento is the propensity to go downwards, caused not so much by the heaviness of the mobile, but by the arrangement that the different heavy bodies have among themselves; through this momento, one will often see a less heavy body counterbalance another of greater

[23] This hypothesis was put forward, albeit with great caution, by W. Wisan (*The new science of motion* cit., p. 157). Conversely, Stillman Drake does not dispute that Galileo defines "absolute" heaviness here (*On motion and on mechanics* cit., p. 151, note 4).
[24] *G.G.*, II, p. 159.

heaviness: just as in the steelyard one sees a small counterweight lifting another very large weight, not due to excess heaviness, but rather due to the distance from the point from which the steelyard is supported; which, combined with the heaviness of the lesser weight, increases its momento and impeto to go downwards, with which it can exceed the momento of the other greater weight. Therefore, momento is that impeto to go downwards, composed of heaviness, position and other things, from which such propensity may be caused.]

It will be necessary to point out at once that, although the explicit juxtaposition of the two terms is missing in the above-mentioned context, Galileo intends to define here the "momento della gravità [momento of heaviness]", or rather the "momento della gravità determinata [momento of determinate heaviness]" (of "weight", as we would say today), that is, the concept of the varying effectiveness of an identical "propensione di andare al basso [propensity to move downwards]" (of an identical weight), according to the conditions in which it is acting. In reality, at least in the *Mecaniche*, where the use of *momento* is limited to the field of machines that can be attributed either to the balance model or to that of the inclined plane, these are reduced to two conditions: the distance of the point of suspension of the weight from the fulcrum on the balance and the greater or lesser inclination of the planes on which the weight rests. In the definition, however, explicit reference is made only to the condition of variability determined by the distance from the fulcrum, since that produced by the different inclination of the planes can be considered – as we will see – a simple consequence.

The greater distance from the point of application adds to the weight "momento" and "impeto di andare al basso [impeto to go downwards]". *Momento*, then, is that "impeto di andare al basso, composto di gravità, posizione e di altro, dal che possa essere tal propensione cagionata [impeto to go downwards, composed of heaviness, position and other things, from which such propensity may be caused]". Here we see the emergence of the relationships between terms that we will frequently encounter in the course of our analysis: *momento* and *impeto*. In the *Mecaniche*, the phrase "impeto di andare al basso [impeto to go downwards]" occurs quite often. In the aforementioned definition, it is not difficult to recognize that the distinction is between the "cause" and the "effect". The greater distance adds "momento" to the weight and, *consequently*, "impeto di andare al basso [impeto to go downwards]". In other words, the *impeto* seems to represent the consequences more "physically", the concrete effects of the variation in the effectiveness of the weight, registered abstractly from *momento*. One could say that *impeto* represents the dynamic effect of *momento*, which remains a "propensione [propensity]", an abstract condition[25].
Through this pair of terms, Galileo probably intended to keep open the possibility **206**
of a "dynamic" use of *momento* outside of the strictly static context in which the term is used in the *Mecaniche*. It should not be forgotten that, in the *De motu*, by stating that the speeds of descent along different inclinations are proportional to the greater or lesser "gravità [heaviness]" that bodies present along those inclinations, Galileo had attempted to base dynamic considerations on static principles. With

---

[25] On the relationships between *momento* and *impeto* in Galileo, cf. M. L. Altieri Biagi, *La terminologia scientifica* cit., pp. 55 ff., and A. Koyré, *Studi galileiani*, trad. it. cit., pp. 98-100.

the *momento-impeto* parallelism, in a text where he deliberately avoids addressing the problem of "velocità [speed]", Galileo demonstrates that he has not definitively given up on the idea that the "velocità" of descent of different bodies are proportional to their "gravità [heaviness]", or rather, to the *momenti* of heaviness.

The third of the "diffinizioni [definitions]" has the task of specifying what is meant by "centro di gravità [center of heaviness]":

> Centro della gravità si diffinisce essere in ogni corpo grave quel punto, intorno al quale consistono parti di eguali momenti: si che, imaginandoci tale grave essere dal detto punto sospeso e sostenuto, le parti destre equilibreranno le sinistre, le anteriori le posteriori e quelle di sopra quelle di sotto; sì che il detto grave, cosi sostenuto, non inclinerà da parte alcuna, ma, collocato in qual si voglia sito e disposizione, purché sospeso dal detto centro, rimarrà saldo. E questo è quel punto, il quale andrebbe ad unirsi col centro universale delle cose gravi, cioè con quello della terra, quando in qualche mezzo libero potesse descendervi[26]. [The center of heaviness is defined to be that point in every heavy body, around which there exist parts of equal momenti: so that, imagining the heavy body suspended and supported by said point, the right parts will equilibrate the left, the anterior ones the posterior ones, and those above those below; so that the said heavy body, thus supported, will not incline in any direction, but, placed in any location and arrangement, as long as it is suspended from the said center, it will remain steady. And this is that point which would unite with the universal center of heavy things, that is, with that of the Earth, if it could descend through some free medium.]

The model of Commandino's definition is immediately recognized in the Galilean declaration. As we have already seen[27], the Urbinate had preceded this definition with another one, proposed by Pappus of Alexandria in the *Collectiones Mathematicae*, in which *momentum* did not appear. Guidobaldo, in the *Paraphrasis*, had also given both definitions[28]. Galileo, however, avoids referencing Pappus. The adoption of this definition, precisely in the work in which *momento* is introduced, seems to strengthen the hypothesis of an assimilation of the term from the "Archimedean" tradition, particularly from familiarity with the texts of Commandino and Guidobaldo.

The definitions are followed by three "supposizioni [assumptions]". The first is that every heavy body moves downwards in such a way "che il centro della sua gravità non esca mai fuori di quella linea retta, che da esso centro, posto nel primo termine del moto, si produce insino al centro universale delle cose gravi [that the center of its heaviness never departs from that straight line, which from this center, placed at the first terminus of the motion, extends to the universal center of heavy things]". The second consists in declaring that, in heavy things, every effect of weight can be considered concentrated in the center of heaviness:
207 "Ciascheduno corpo grave gravitare massimamente sopra il centro della sua gravità, ed in esso, come in proprio seggio, raccòrsi ogni impeto, ogni gravezza ed in somma ogni momento [Each heavy body exerts its weight principally on the center of its heaviness, and in it, as in its own seat, every impeto, every heaviness, and in short every momento is gathered]". The third, finally, assumes that "il centro della

[26] *G.G.*, II, pp. 159-160.
[27] Cf. *supra*, p. 59.
[28] Cf. *supra*, pp. 60-61.

gravezza di due corpi egualmente gravi essere nel mezzo di quella linea retta, la quale li detti due centri congiunge [the center of heaviness of two equally heavy bodies is in the middle of the straight line, which joins the two centers]"[29].

The definitions and assumptions allow us to address and explain "un comunissimo e principalissimo principio [a very common and primary principle]" of most mechanical instruments: "pesi diseguali pendenti da distanze diseguali peseranno egualmente, ogni volta che dette distanze abbino contraria proporzione di quella che hanno i pesi [unequal weights hanging from unequal distances will weigh equally, whenever said distances have an inverse proportion to that of the weights]"[30]. Here, then, Galileo proposes the principle again, which he had shown to have possessed since the *Bilancetta.* As we have previously observed, our scientist had evaded this principle in the *De motu*, introducing as a universal rule "gravius non posse attolli a minus gravi [a heavier body cannot be lifted by a lighter one]"[31]. The distinction between *gravità* [heaviness] and *momento della gravità* [momento of heaviness] now safeguards against similar errors.

Regarding the law of *momenti* in the balance, Galileo provides a twofold demonstration. Both appear particularly important because they present two different ways of understanding the *momento* of heaviness. The first demonstration, conducted with exquisitely geometric techniques, can be attributed to the Archimedean model of demonstration in the first book of the *Equilibrium of Planes*. Even in the *Theoremata*, Galileo had shown interest in this type of demonstrative procedure[32].

The demonstration is conducted by analyzing a "solido grave [heavy solid]" (a cylinder) hanging from the ends of a balance of equal arms with a fulcrum at G (see fig. 4).

Point G will be the center of heaviness of the system. However, if the cylinder is cut into two unequal segments, CISE and IDFS, the segments will not maintain their initial position if they are connected to the balance beam by the cords AC, BD. **208**
Therefore, it will be necessary to add a new rope at H, which supports the free ends of the two segments. Once the two pieces have been secured at H, we can say that, once again, point G has become the center of heaviness of the system. On the other hand, the state of equilibrium of the two segments CIES and IDSF does not change if, instead of being supported respectively by AC, HI, BD, they are connected to the balance beam by the cords MK and NL, respectively intermediate between AC, HI and between HI, BD. Even in this case, in fact, the center of heaviness of the entire system will be in G. To demonstrate the assumption, it will therefore be sufficient to prove that the two quantities CIES, IDSF are inversely proportional to the distances from the fulcrum G of their respective points of application M, N. Having demonstrated this relationship without difficulty[33], Galileo can conclude "esser verissimo che pesi diseguali pesino egualmente sospesi contrariamente da

[29] *G.G.*, II, p. 160.
[30] *Ibid.*, p. 161.
[31] Cf. *supra*, p. 174.
[32] Cf. *supra*, p. 162.
[33] MH is indeed half of HA, while NH is half of HB. MN, therefore, is half of AB, of which BG is also half. MN and BG are, therefore, equal. Subtracting the common part GN, the remaining MG will be

distanze diseguali, che abbino l'istessa proporzione dei pesi [it is very true that unequal weights weigh equally when suspended oppositely from unequal distances, which have the same proportion as the weights]"[34].

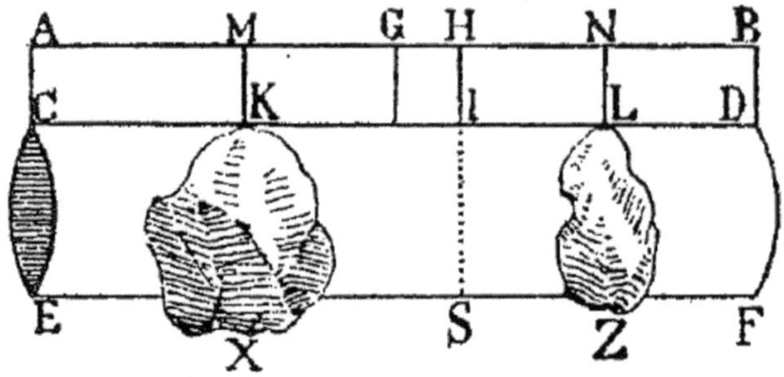

Figure 4

To this demonstration, Galileo adds some "avvertimenti [remarks]" which, essentially, present another method of proof, from "quale ci può ragionevolmente essere confermata la medesima verità [which the same truth can reasonably be confirmed]"[35]. Galileo, however, does not use the term "dimostrazione [demonstration]" this time, preferring to speak of "congruenza e probabilità [congruence and probability]". The difference between the Archimedean demonstration examined above and this new procedure lies in the type of approach: from the strictly static-geometric method of the Archimedean demonstration, we now move to a dynamic consideration. This second proof, in fact, directly examines a balance in equilibrium with unequal arms, at the ends of which two weights are applied that are inversely proportional to the distances from the fulcrum. To break the equilibrium, it will suffice to add "un minimo momento di gravità [a minimum momento of heaviness]" to one of the two weights. The addition of this minimum amount of weight will force the balance to rotate around the fulcrum, while the weights will describe the two circular arcs AD, BE (see fig. 5).

equal to the remaining NB, to which, in turn, the NH is equal. MG and NH are also equal, and, once the common part GH is subtracted, we will have MH = GN. Since – as already demonstrated – MG = HN, we can say MH: HN = KI: IL. But also = CI: ID, in other words, as the solid CS to the solid SD.

[34] *G.G.*, II, p. 163.

[35] *Ibid.*

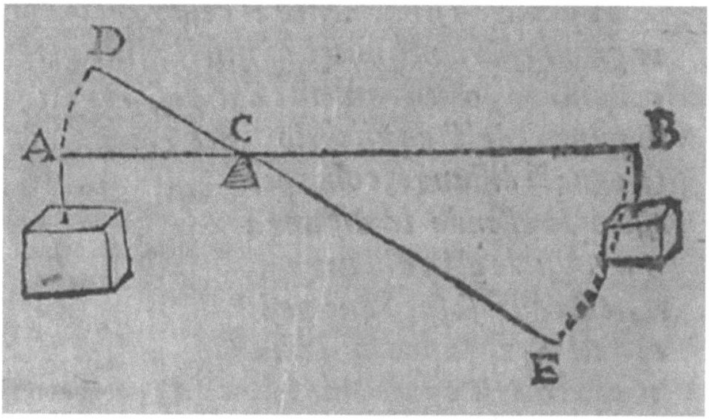

Figure 5

Before examining this very important demonstration, it will be worth emphasizing that in the expression "un minimo momento di gravità [a minimum momento of heaviness]", *momento*, as previously in the Regensburg and Pasadena copies, simply means small quantity[36]. In other words, it has a very different value from the "technical" one previously declared by Galileo. The difference in meaning of this *momento* (which we have conventionally defined as "biblical") from the "static" one, heavily used in the *Mecaniche*, is clarified by Galileo himself in the following passage, in which he declares that the added weight, despite being sufficient to tip the balance (just like the *momentum staterae*), is so small that it can be disregarded: **209**

E perché, per fare descendere il peso B, ogni minima gravità accresciutagli è bastante, però, non tenendo noi conto di questo insensibile, non faremo differenza dal potere un peso sostenere un altro al poterlo movere[37]. [And because, in order to make the weight B descend, (see fig. 5) any minimal heaviness added is sufficient for it, however, by not taking this imperceptible amount into account, we will not differentiate between the ability to support one weight and the ability to move it.]

Galileo continues by emphasizing how, in the case of this type of movement, the circular arcs described by the weights preserve the proportions of their fulcrum distances. The distances are, in fact, the radii of the circles to which these arcs belong. Since the two motions are carried out at the same time, their speeds will be proportional to the distances and, therefore, inversely proportional to the weights, as was to be demonstrated:

Viene dunque ad essere la velocità del moto del grave B discendente, tanto superiore alla velocità

[36] Cf. *supra*, p. 202. The demonstration technique is the same. To further confirm this reading, it should be noted that Mersenne, in translating this passage, did not use *momento*: "il est certain que l'un contrepesera l'autre, et qu'ils seront en équilibre; et que *si l'on adjouste* quelque chose à l'un… [it is certain that one will counterbalance the other, and that they will be in equilibrium; and that *if something is added* to one…]" (*Les Mechaniques de Galilée*, ed. crit. cit., p. 33; the italics are ours). Conversely, Stillman Drake, in his English version of the *Mecaniche*, uses *moment* to render this value of *momento*: "if to one of these [weights]* there was given the least moment of heaviness…" (*On motion and on mechanics* cit., pp. 155-156).
[37] *G.G.*, II, p. 164.

dell'altro mobile A, ascendente, quanto la gravità di questo eccede la gravità di quello; né potendo essere alzato il peso A in D, benché lentamente, se l'altro grave B non si muove in E velocemente, non sarà maraviglia, né alieno dalla costituzione naturale, che la velocità del moto del grave B compensi la maggiore resistenza del peso A, mentre egli in D pigramente si muove e l'altro in E velocemente descende[38]. [Therefore, the speed of the motion of the descending heavy body B (see fig. 5) becomes as much greater than the speed of the other mobile A, ascending, as the heaviness of the latter exceeds the heaviness of the former; and since weight A cannot be lifted to D, albeit slowly, unless the other heavy body B moves quickly to E, it will not be surprising, nor alien to natural constitution, that the speed of the motion of heavy body B compensates for the greater resistance of weight A, while one moves lazily to D and the other quickly descends to E.]

Galileo therefore believes that from this "congruenza e probabilità [congruence and probability]", a general principle can be derived, which he immediately formulates. This is an extremely important passage, on which scholars have often focused.

210 E da questo discorso possiamo venire in cognizione, come la velocità del moto sia potente ad accrescere momento nel mobile, secondo quella medesima proporzione con la quale essa velocità di moto viene augumentata[39]. [And from this discourse, we can come to the knowledge that the speed of motion is capable of increasing momento in the mobile, according to the same proportion with which the speed of motion is increased.]

Neither the demonstration nor the principle on which it is based was – as is known – original. Galileo was merely revisiting the demonstration of the laws of *momenti* of the balance recorded in the pseudo-Aristotelian *Questions of Mechanics*. The dynamic approach that was used, known as the method of "virtual speeds"[40], consisted in considering not the distances from the fulcrum of the points of application of the weights, but rather the spaces traversed, were the balance to tip with the addition of a minimal and therefore negligible weight. Guidobaldo and numerous other authors[41] had recorded this "classic" proof in their mechanical works. We also know that Galileo was remarkably familiar with the *Questions of Mechanics* and certainly knew the *Paraphrasis* of the pseudo-Aristotelian text published by Alessandro Piccolomini.

The problem that deserves a more careful and in-depth analysis is, however, another. It is indeed necessary to evaluate with the utmost care, in order to avoid truly dangerous misunderstandings and equivocation, certain interpretative hypotheses that have been formulated, in no small part, precisely on the basis of this demonstration and, above all, of the conclusion drawn from it by Galileo, namely that "la velocità del moto sia potente ad accrescere momento nel mobile [the speed of motion is capable of increasing momento in the mobile]".

---

[38] *Ibid.*
[39] *Ibid.*
[40] For the method of "virtual speeds", proposed in the pseudo-Aristotelian *Questions of Mechanics*, cf. *supra*, pp. 74 ff. See also M. Clagett, *La scienza della meccanica*, trad. it. cit., pp. 25-27 and the elegant essay by Vailati, *Il principio dei lavori virtuali da Aristotele ad Erone* cit.
[41] For example, Cardano and Scaliger, cf. *supra*, pp. 87-88.

In fact, quite a few have thought they saw in this statement clear proof of the young Galileo's adherence to the Parisian physics of *impetus*. Winifred Wisan, especially, suggested that in the increase in *momento* produced by speed, one cannot fail to recognize Buridan's doctrine[42], according to which a body begins to descend by virtue of its own heaviness, progressively acquiring an ever greater *impetus*, which translates into a constant increase in speed. This revisits the general interpretative question of the influence that the theory of *impetus* would have had on Galileo, a question which we have already examined while analyzing the *De motu*[43], to emphasize Koyré's curious misunderstanding. Now, without claiming to completely exclude interpretations or possible direct or indirect influences of the physics of *impetus*, it seems evident to me that the context cited above does not authorize similar extrapolations.

It will be useful, first of all, to reflect on the significant fact that Galileo **211**
attempts, in the passage quoted above, to formulate a principle capable of accounting for "mechanical" motions, not "natural" ones. The meaning of his statements is extremely precise: a smaller weight can "weigh" as much as a larger one (that is, equalize it in *momento*) when it moves with a speed that is inversely proportional to the speed of the other. Furthermore, in this model, the *momento* conferred by speed does not increase continuously – like Buridan's *impetus* – but remains constant. The speed that Galileo talks about in these pages of the *Mecaniche* is, in fact, absolutely "uniform", and it is not possible to find direct or indirect references to the problem of acceleration. For Galileo, acceleration is probably still an "accident [i.e., an incidental phenomenon]". When our scientist states that speed increases *momento* "secondo quella medesima proporzione con la quale essa velocità di moto viene augumentata [according to the same proportion with which the speed of motion is increased]", he is referring not to the progressive increase in speed of the same motion (acceleration) – the case that the theory of *impetus* tries to justify – but rather to different types of movements, according to different speeds which are each uniform (it should be added that, in the balance schema, where the movements of the two weights occur in equal times, these speeds are, in fact, represented by the respective distances traveled). If the space covered by one of the weights is greater, and therefore traveled with a greater "uniform" speed, this directly results from the fact that it is hanging at a greater distance from the fulcrum. On the other hand, the exposition that Galileo gave of it a few years later in the *Discorso* irrefutably demonstrates that the Galilean principle must be understood in this way[44].

The meaning of Galileo's statements, which in no way can be linked to the Parisian dynamics of *impetus*, is thus extremely clear when careful reference is made to the problematic framework in which they are inserted. In the initial chapter of the Paduan writing, Galileo illustrates the "utilità che si traggono dalla scienza mecanica e dai suoi instrumenti [utilities that can be drawn from mechanical science

---

[42] *The new science of motion* cit., p. 159.

[43] Cf. *supra*, Part Two, Ch. I, note 94.

[44] See *G.G.*, IV, pp. 68-69; the passage containing the exposition of the "principle" is cited *infra*, pp. 232-233.

and its instruments]". In these introductory pages, he strongly recommends against paying attention to the "artefici [artisans (i.e., craftsmen)]", who promise "di potere con poca forza muovere ed alzare grandissimi pesi, ingannando, in certo modo, con le loro machine la natura; instinto della quale, anzi fermissima constitutione, è che niuna resistenza possa essere superata da forza che di quella non sia più potente [to be able, with little force, to move and lift very great weights, deceiving, in a certain way, nature with their machines; whose instinct, or rather most firm constitution, is that no resistance can be overcome by a force that is not more powerful than it]"[45]. In fact – continues Galileo – if we observe the behavior of even the most effective machines, which move or lift large weights with a relatively small force, it will appear evident that the operation is carried out in such a way that the small force "più volte avrà reiterato quel moto e spazio che una sol volta sarà stato da tutto il peso misurato [will have repeated that motion and space several times which
212 will have been measured only once by all the weight]". That is to say that, during the time of the movement or lifting, the force, that is, the weight that performs the traction, travels a distance as many times greater than the weight moved or lifted, as many times the force is less than the weight. "Dal che appare la velocità della forza essere stata tante volte superiore alla resistenza del peso, quanto esso peso è superiore alla forza [From this it appears – concludes Galileo – the speed of the force has been as many times greater than the resistance of the weight, as the weight is greater than the force]"[46]. It cannot, therefore, be said that the constitution of nature has been violated in this way, since, in order to infringe, it would be necessary for "la minor forza trasferisse la maggior resistenza con pari velocità di moto, secondo il quale essa camina [the lesser force to transfer the greater resistance with the same speed of motion, according to which it moves]"[47].

By asserting that speed adds *momento* to weight, Galileo merely demonstrated that the "constitution of nature" is respected even in the case of the balance of unequal arms. Nature is not, in fact, deceived, since the lesser force (the smaller weight) lifts a greater weight while traveling, at the same time, a greater distance (i.e., moving more quickly).

After the declaration of the terms and the clarification of the assumptions, the discussion addresses the examination of the most significant mechanical instruments and the principles that regulate their functioning and effectiveness. The first group of instruments is attributable to the balance model and its properties (steelyard, lever, winch, pulley) and does not present, from the standpoint of our investigation, any particular interest.

However, fundamental for the enrichment that it allows of the meaning of *momento*, is the discussion on the "vite [screw]", to which Galileo prefaces with a brief remark. The screw, in fact, unlike the other machines discussed so far, does not depend, at least directly, on the balance model and its laws, but is instead based on the principle of the inclined plane. It is necessary, then, to determine to what

---

[45] *G.G.*, II, p. 155.
[46] *Ibid.*, p. 156.
[47] *Ibid.*

extent the effectiveness of heaviness, that is, its *momento*, varies along planes of different inclinations. Heaviness, in fact, has a "propensione di muoversi verso il centro [propensity to move towards the center]" not only along the perpendicular, "ma ancora, quando altrimenti far non possa, per ogni altra linea, la quale, avendo qualche inclinazione verso il centro, vadi a poco a poco abbassandosi [but also, when it cannot do otherwise, along any other line, which, having some inclination towards the center, descends gradually, little by little]"[48].

It should be recalled that Galileo had already addressed, in the *De motu*, the problem of the varying effectiveness of the heaviness of the same weight along planes of different inclinations[49]. In the *Mecaniche*, he revisits, with numerous enhancements, the substance of those considerations. In the Paduan writing, however – as is logical to expect – Galileo focuses his attention on the "static" problem of the force necessary to lift the same weight along different inclinations, **213**
neglecting the "dynamic" problem of the "natural" descent of the weight along the perpendicular and inclined planes.

Before moving on to the analysis of the inclined plane, Galileo puts forward a series of remarks (in fact, suppositions, even if he avoids defining them as such) from which, as from indisputable principles, the true consequences can be deduced:

> possiamo prendere per assioma indubitato questa conclusione: che i corpi gravi, rimossi tutti l'impedimenti esterni ed adventizii, possono esser mossi nel piano dell'orizzonte da qualunque minima forza. Ma quando il medesimo grave dovrà essere spinto sopra un piano ascendente, già cominciando egli a contrastare a tale salita (avendo inclinazione al moto contrario), si ricercherà maggiore violenza, e maggiore ancora quanto più il detto piano averà di elevazione[50]. [we can take this conclusion as an unquestionable axiom: that heavy bodies, having all external and adventitious impediments removed, can be moved on the horizontal plane by any minimal force. But when the same heavy body has to be pushed up an ascending plane, it will begin to counteract this rise (having a tendency towards contrary motion), greater violence will be sought, and even greater the more the said plane has elevation.]

The fundamental principle posed in this passage is that of indifference to the motion or rest of a sphere lying on a plane parallel to the horizon, for which "ogni minima forza sia bastante a muoverla, siccome all'incontro ogni pochissima resistenza, e quale è quella sola dell'aria che la circonda, potente a tenerla ferma [any minimal force is sufficient to move it, just as conversely any slight resistance, which alone is the air that surrounds it, is effective in holding it still]"[51]. That is to say, namely, the Galilean principle of inertia – "proto-inertial" or "circular inertia", as it has been otherwise defined[52] – according to which, on a plane parallel to the

---

[48] *Ibid.*, p. 179.
[49] Cf. *supra*, pp. 194 ff.
[50] *G.G.*, II, p. 180.
[51] *Ibid.*
[52] The expressions are both used by Alexandre Koyré in Chapter III of the *Studi galileiani* cit. (*Galileo e la sua legge d'inerzia*), to which we refer for the previous bibliographical indications and, above all, for the "cosmological" implications of the principle of inertia, of which Koyré provides a brief history. On Galilean "inertia", cf. also M. Clavelin, *La philosophie naturelle de Galilée* cit., pp. 373 ff.; S. Drake, *Galileo studies. Personality, tradition and revolution*, The University of Michigan Press, Ann Arbor 1970, pp. 240 ff.; ID., *A further reappraisal of impetus theory*, in "Studies in the History and Philosophy

horizon, no perceptible force is required either to move or to keep a heavy body at rest. Galileo had already formulated it in the *De motu*, where he had remarked that the *licentia* [presumptive license] of the plane parallel to the horizon (which, in reality, is an arc of a circle) should be considered in the same way as that with which Archimedes supposed "superficiem habere gravitatem, et alteram altera graviorem esse, cum tamen revera omni sint expertes gravitate [that surfaces have heaviness, and one is heavier than the other, even though, in reality, they are all devoid of heaviness]"[53]. In the *De motu*, to tell the truth, Galileo had attempted to provide a demonstration of the indifference of weight on a plane parallel to the horizon[54],
214 showing, again in the schema of the balance, that a minimal force is sufficient to lift a weight along the quarter of the circle: therefore, an infinitesimal force would have been enough to move the same weight on the horizontal plane. In the *Mecaniche*, however, the principle is placed as an axiom.

On the meaning that this type of "inertial" conception presents in the framework of Galilean scientific thought and on the "limits" that prevented the Pisan scientist from developing the modern concept of inertia (i.e., rectilinear and infinite), much authoritative commentary has been provided by Koyré and other scholars[55]. It will be worthwhile, however, to specify that, both in the *De motu* and in the *Mecaniche*, the Galilean inertial conception, which undoubtedly involves a general view of motion situated within an "infinite" cosmos, seems destined for precise purposes, strictly inherent to the discussion, while Galileo does not seem interested in emphasizing its "cosmological" significance[56].

It will be observed, in fact, that the statement of the principle is "incomplete" compared to the formulations that Galileo gave in the *Dialogo*, lacking the part in which the infinite persistence of motion and speed is declared[57]. By advancing this axiom, before introducing the discussion on the inclined plane, Galileo simply intends to prepare the terrain for the demonstrations that he is about to propose. As in the *De motu*, Galileo considers force in terms of weight; that is, he calculates the force necessary to lift a given weight along different inclinations, according to the "gravità [heaviness]" of that weight along these inclined planes. In the *Mecaniche*, after the reform of the terminology, this force will be measured by the *momento* of the weight along the different inclinations. This method of analysis is practicable, and leads to the discovery of the proportions between "forces" and *momenti*, only if a "maximum" and a "minimum" of known and calculable "forces" and *momenti* are given. The maximum of force and *momento* – as we have already seen in the

of Science", VII (1976), 4, pp. 319-336; A. Coffa, *Galileo's concept of inertia*, in "Physis", X (1968), 4, pp. 3 ff.

[53] *G.G.*, I, pp. 299-300.

[54] He had tried to show that a minimal force is sufficient to lift a weight along a quadrant of a circle; therefore, an infinitesimal force would have been enough to move the same weight on the horizontal plane (*ibid.*, p. 299).

[55] Cf. *supra*, note 52.

[56] Koyré does not seem to perceive any difference between the juvenile Galilean formulations of the principle of inertia, in the *De motu* and in the *Mecaniche*, and the more mature one in the *Dialogo* (cf. *Studi galileiani*, trad. it. cit., pp. 208-209).

[57] Cf. *Dialogo sopra i due massimi sistemi* cit., *G.G.*, VII, pp. 171 ff.

De motu – occurs in the perpendicular descent, in which the *momento* of a body is "pieno [full]". In this case, the force needed to lift it vertically corresponds to its own weight. The minimum – and this is what the axiom guarantees – occurs on the horizontal plane, where the *momento* (i.e., the tendency to move) is "nullo [null]": the force (necessary to move it) is, therefore, negligible, since it does not have to overcome any contrary inclination.

There is a second possibility that the axiom allows, and we will also find this exploited in the *Mecaniche*. The indifference to motion or to rest on the horizon plane implies, in fact, that, in the falling motion, the effect of *momento* should be calculated only in terms of "vertical" descent, neglecting the "lateral" displacement, **215**
which does not require the application of appreciable force[58].

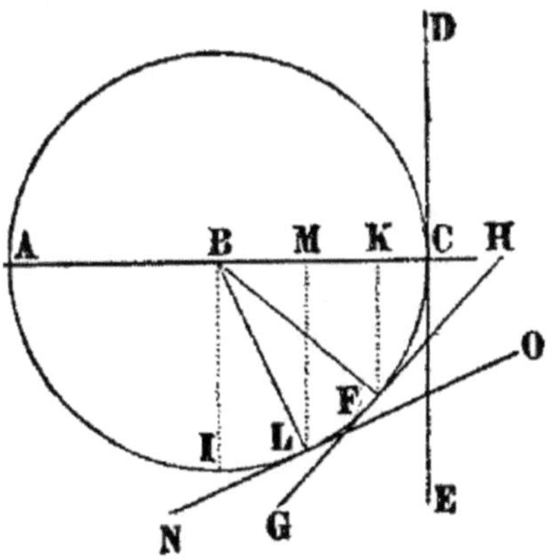

Figure 6

Having established this principle, Galileo can move on to examining the variation of the forces necessary to lift a given weight along differently inclined planes. Exactly as in the *De motu*, Galileo measures force in terms of weight (the *momento* of which, however, is variable and equivalent to the absolute weight, in the case of perpendicular lifting). The demonstration of the inclined plane theorem in the *Mecaniche* coincides with that of the Latin treatise previously illustrated. The only substantial modification is noticeable in the terms used. The *gravitas* of the *De motu*, used indiscriminately with the value of "weight" and "effectiveness of weight", corresponds in the *Mecaniche* to two different terms: "weight" and "momento of weight". Using the balance model and declaring that lifting a weight means balancing it and overcoming its tendency to descend (*momento*), Galileo demonstrates that "nell'inclinare a basso per la circonferenza CFLI il peso posto nell'estremità della linea BC viene a scemarsi il suo momento ed impeto d'andare a basso di mano in mano più, per esser sostenuto più e più dalle linee BF, BL [by tilting downwards on the circumference CFLI (see fig. 6) the weight placed at

[58] Cf. *infra*, pp. 220 ff.

the extremity of the line BC, its momento and impeto to go downward gradually diminish, as it is supported more and more by the lines BF, BL]”[59].

The demonstration in the *Mecaniche* presents, at this point, a section absent in the *De motu*. This is an addition of considerable importance, which demands to be examined with the utmost attention, given that it offers a glimpse of the possibility of applying this type of analysis to the study of purely dynamic questions: the natural motion of falling and acceleration, which constitutes its most relevant characteristic. Galileo states, in fact, that the progressive decrease in *momento* can be observed even without imagining the heavy bodies suspended at the end of the balance, but simply by imagining that the heavy body moves with natural motion along the circular arc CFLI:

> Ma il considerare questo grave discendente e sostenuto dalli semidiametri BF, BL ora meno e ora più, e constretto a caminare per la circonferenza CFL, non è diverso da quello che sarà imaginarsi la medesima circonferenza CFLI esser una superficie così piegata, e sottoposta al medesimo mobile, sì che, appoggiandovisi egli sopra, fosse constretto a descendere in essa; perché se nell'uno e nell'altro modo disegna il mobile il medesimo viaggio, niente importerà s'egli sia sospeso dal centro B e sostenuto dal semidiametro del cerchio, o pure, se levato tale sostegno, s'appoggi e camini su la circonferenza CFLI[60]. [But considering this descending heavy body and supported by the semidiameters BF and BL, one less and the other more, and constrained to travel along the circumference CFL, is no different from imagining the same circumference CFLI to be a surface thus curved and placed beneath the same mobile, so that, by leaning upon it, the mobile would be constrained to descend along it; because if in both cases the mobile travels the same path, it will not matter whether it is suspended from the center B and supported by the semidiameter of the circle, or whether, if this support is removed, it rests and travels on the circumference CFLI.]

216 If we adopt this point of view, abandoning the balance model,

> potremo affermare che, venendo al basso il grave dal punto C per la circonferenza CFLI, nel primo punto C il suo momento di discendere sia totale e integro; perché non viene in parte alcuna sostenuto dalla circonferenza, e non è, in esso primo punto C, in disposizione a moto diverso da quello che libero farebbe nella perpendicolare e contingente DCE. Ma se il mobile sarà constituito nel punto F, allora dalla circolare via, che gli è sottoposta, viene in parte la gravità sua sostenuta ed il suo momento di andare al basso diminuito con quella proporzione con la quale la linea BK è superata dalla BC... E nel medesimo modo troveremo, nel punto L, diminuirsi il momento dell'istesso mobile, come la linea BM si diminuisce dalla BC[61]. [we can affirm that, as the heavy body descends from point C (see fig. 6) along the circumference CFLI, at the initial point C its momento of descent is total and intact; because it is not at all supported by the circumference, and at this initial point C, it is not in a position for any motion different from that which it would freely make on the perpendicular and tangent DCE. But if the mobile is constituted at point F, then from the circular path to which it is subjected, part of its heaviness is supported, and its momento of descent is diminished in proportion to how much the line BK is exceeded by BC... And in the same way, we will find, at point L, that the momento of the same mobile decreases as the line BM diminishes from BC.]

It is practically undeniable that, in analyzing this phenomenon, Galileo did not realize that the falling motion of a heavy body along a quadrant of a circle cannot be

---

[59] *G.G.*, II, p. 182.
[60] *Ibid.*
[61] *Ibid.*

considered uniform. And one immediately wonders how the Pisan scientist justified the fact that the progressive increase in speed proceeded alongside the decrease in *momento*. Did he already possess the conviction that speed is conserved? Only in this case, then, could a progressive decrease in *momento* have produced an increase in speed.

These are natural questions, which we are led to answer in light of the subsequent developments of Galilean thought. We know, in fact, that, in his 1602 letter to Guidobaldo, Galileo declared that he had discovered that the motions along all the chords of a quadrant of a circle occur at equal times[62]. And it is known that the observation of pendulums was at the origin of Galilean speculations on movement. How can one escape the impression that the motion along a quarter of the circle appears to Galileo to be a pendulum motion?

Having emphasized these congruencies, which indicate that we are faced with a particularly intriguing and likely central text for the further developments of Galileo's dynamic conceptions, it will also be necessary to proceed with the utmost caution, lest we risk suggesting an interpretation "distorted" by understandings that are ours but did not yet belong to Galileo when he wrote these pages.

What I mean is (as I have already had occasion to emphasize in relation to the *impetus-momentum* relationship and the principle of inertia) that too often, in the laudable attempt to reconstruct the "origins" of subsequently established conceptions, we end up suppressing the "differences", downplaying the uncertainties and doubts initially expressed by the author, which instead appear evident when one strictly adheres to the context in which they are inserted, resisting the temptation to **217**
look to the future[63].

In our case, for example, it should not escape us how Galileo carefully avoids analyzing the motion along the quarter of the circle as a "free fall". The examination, in fact, is still conducted by abstractly "positing" the various positions of the heavy body and determining the *momento* it would present *if it began* to move from those positions. In fact, Galileo makes no reference to a fall that begins at C and continues without interruption to F, L, etc. His analysis, thus, still remains essentially "static".

Starting from this observation, it is necessary to explain why Galileo did not ask what the speed would be, that is, the *momento* of the heavy body along the quarter of the circle. I believe that the simplest and most natural explanation is also the most logical. Galileo does not question it because he knows he cannot provide an answer. Our scientist has so far strived to attribute every type of movement to constant forces that produce constant speeds. This is certainly the foundation from which he also moves in the *Mecaniche*, a foundation that, however, after the failure of the *De motu*, shows some cracks. In the analysis of motions along inclined planes, it seems, however, that this foundation "holds".

---

[62] The letter to Guidobaldo is dated 29 November 1602 (*G.G.*, X, pp. 97-100).

[63] It is, again, to W. Wisan's essay that I intend to refer. Wisan, in fact, states that this part of the *Mecaniche* demonstrates that Galileo had now reached the conviction that the motion along the quarter of the circle (and, therefore, also along the inclined planes) is accelerated.

In fact, it will be enough to suppose – as Galileo does – that, in this type of motion, the speeds are proportional to the *momenti* of the weight: the force (i.e., the determined *momento*) will therefore act constantly, imparting to the body a precise speed, uniform at all points of the plane.

Galileo is aware that, in the quarter of the circle, the *momento* instead decreases constantly (going from a maximum to zero): should, therefore, a parallel decrease in speed be deduced from this? On the other hand, the most basic of observations presents the downward motion of the bodies along the quarter of the circle as an accelerated motion. Unable to adequately justify these not insignificant difficulties, Galileo tried to hide them. This is probably why, when examining the motion of a body along the quarter of the circle, he analyzed it as if it were a "static" problem, avoiding commenting on the effects that the decrease in *momento* would have had on the speed of descent. This demonstrates that our scientist is still quite puzzled by the problem of the acceleration of natural motions: the contradictions that had led to the abandonment of the *De motu* are still largely at work in the *Mecaniche*, even if – as demonstrated by the peculiar analysis of the motions along the quarter of the circle – it appeared increasingly difficult to confine acceleration to the convenient field of "accidents [i.e., incidentals]" to be justified.

218 Let us resume the examination of Galileo's demonstration relating to the inclined plane. The Pisan scientist concludes that the heavy body presents an "integro ed assoluto [intact and absolute]" *momento* in the perpendicular descent. Such *momento* is to the *momento* along the inclined plane as "la lunghezza del piano inclinato alla perpendicolare che da esso cascherà sopra l'orizzonte [the length of the inclined plane to the perpendicular which falls from it above the horizon]"[64]. It follows that

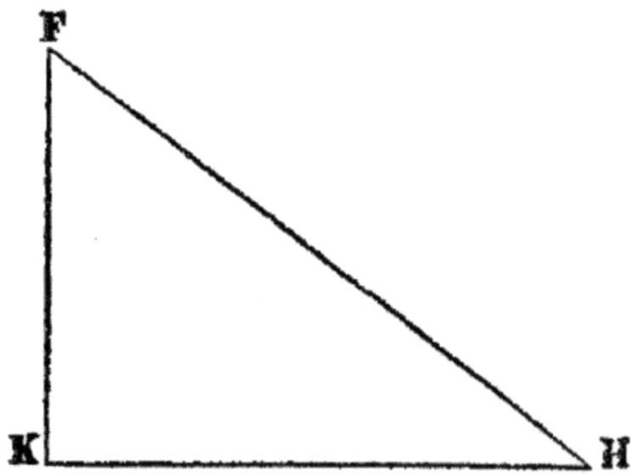

Figure 7

resta manifesto che, sì come la forza che sostiene il peso nella perpendicolare FK deve essere ad esso eguale, così per sostenerlo nel piano inclinato basterà che sia tanto minore quanto essa perpendicolare FK manca dalla linea FH, perché, come altre volte s'è avvertito, la forza per muovere il peso basta che insensibilmente superi quella che la sostiene, però concluderemo questa universale proposizione: sopra il

[64] *G.G.*, II, p. 183.

piano elevato la forza al peso avere la medesima proporzione che la perpendicolare dal termine del piano tirata all'orizonte, alla lunghezza d'esso piano[65]. [it remains clear that, just as the force that supports the weight on the perpendicular FK (see fig. 7) must be equal to it, so to support it on the inclined plane, the force will be sufficient for it to be as much smaller as the perpendicular FK is missing from the line FH, because, as has been noted on other occasions, the force needed to move the weight need only imperceptibly exceed that which supports it, however, we will conclude this universal proposition: on the elevated plane, the force to the weight has the same proportion as the perpendicular from the end of the plane drawn to the horizon, to the length of the plane.]

The law of *momenti* of a heavy body along different inclinations (already known by Galileo in the *De motu*, where he expressed it with different terminology)[66] therefore provides the basis for establishing the "forze [forces]" necessary to make a given weight rise along the same inclinations.

However, once again, compared to the Latin treatise, an important difference can be observed. It should be recalled, in fact, that, in the *De motu*, Galileo, after determining the proportions of the *vis* [power] with which the same weight tends to descend along different inclinations, had clearly stated that the speeds of descent are directly proportional to the *vires* [powers], concluding that also the speeds of the same heavy body along the perpendicular and the incline are as the incline to the perpendicular[67]. In the *Mecaniche*, the *momento di discendere* [momento of descent] of a heavy body along variously inclined planes provides the measure of the force necessary to make it ascend along those same inclinations[68]. **219**
Galileo's discourse is limited to establishing this relationship between *momento* and *forza* [force], avoiding any mention of speed. Our scientist, once again, shows that he absolutely does not want to venture beyond the terms of a strictly "static" consideration.

The reason for this extreme caution must undoubtedly be found in the awareness that the "dynamic" use of the inclined plane theorem involved considerable difficulties. It is extremely probable that, in the years of the *Mecaniche*, Galileo had, in fact, already changed his attitude towards acceleration. In any case, as demonstrated by the outcome of the *De motu*, our scientist had begun to seriously doubt the possibility of reducing it to a mere accident. On the other hand, by admitting acceleration as a characteristic of natural motion (thus, also that along inclined planes), it was necessary to explain its cause. And Galileo is unable to do so. For this reason, he deliberately abolished any reference to the *momento-speed* relationship, limiting himself to comparing *forza* [force] and *momento*. Furthermore, in posing the problem of the force necessary to make a heavy body ascend along a plane of given inclination and in making it correspond to the *momento di discendere* [momento of descent] of that same heavy body, he again avoided – certainly intentionally – clarifying that the force necessary for the purpose depends not only on the weight of the heavy body and the inclination of the plane but also on

---

65 *Ibid.*
66 Cf. *supra*, pp. 194 ff.
67 Cf. *supra*, p. 195.
68 *G.G.*, II, p. 183.

the speed with which one wants this ascent to occur.

The abandonment of the “dynamic” extrapolation of the inclined plane theorem in the *Mecaniche* should, therefore, be attributed to the awareness of its unsustainability, in the case that natural descents along inclined planes were not uniform, but accelerated. Galileo is perfectly aware that his mechanics, as well as the general conception and the laws of *momenti* on which it is based, is sustainable only if the motions that these *momenti* generate are uniform, not accelerated. For this reason, although he was now quite unconvinced that the motions were uniform, at least in the inclined plane, he suggested – without stating it explicitly – that he considered them as such, carefully avoiding highlighting the dynamic implications of the inclined plane theorem.

Therefore, in the *Mecaniche*, our scientist remains strictly within the limits of a “static” treatment, rejecting any temptation of dynamic extrapolation. The omission of any reference to speeds does not mean, however, that he is able to ignore the dynamic problem entirely. Drawing from the *De motu* the theorem that established the *momenti* of the same heavy body along different inclinations, he, in fact, used a demonstrative schema conceived essentially to determine precisely the proportions of the (uniform) speeds of descent. He omitted – it is true – the conclusion, where such proportionality was declared to correspond fully to that of the *momenti*, but the demonstrative procedure still remains all too suggestive.

220

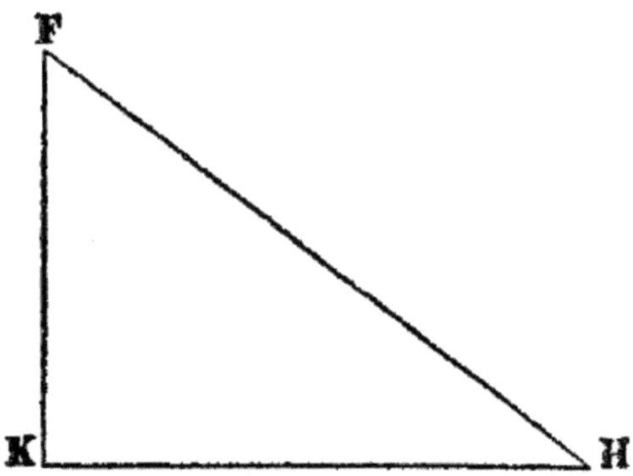

Figure 8

This becomes very clear from the law of *momenti*, with which, in the *Mecaniche*, the considerations on the inclined plane are concluded. It establishes that, in the right triangle FKH (see fig. 8), the *momento* of a heavy body along the inclined FH is to the *momento* of the same heavy body along the perpendicular FK, as FK is to FH[†,69]. Translated into dynamic terms, this means that, in a given time, the heavy

† The text originally recited “FK is to FK”, but “FK is to *FH*” appears to be intended.

69 “Sopra il piano elevato la forza al peso avere la medesima proporzione, che la perpendicolare dal termine del piano tirata all’orizonte, alla lunghezza d’esso piano [On the elevated plane, the force to the weight has the same proportion, that of the perpendicular drawn from the end of the plane to the horizon, to the length of the plane itself]” (*ibid.*).

body travels along the incline FH a distance that is to the distance it would cover if it had descended along the perpendicular FK, as FK is to FH. The *momenti* of a heavy body thus end up being conceived, or rather "depicted" geometrically, as the different distances traveled in a given time. Since the downward motions are – as we have seen – still considered uniform by Galileo, it can be concluded that the speeds are also proportional to the spaces traveled. The "total" *momento* therefore corresponds to the greatest distance traversed by the heavy body in a given time, or, if preferred, to the greatest speed with which the heavy body descends. At the opposite extreme, on the horizontal plane, there is an absence of *momento*, and therefore, for the heavy body constituted at rest, an absence of speed and space traversed. The idea of motions completed in equal times was strongly suggested by the balance model on which the entire demonstration was based.

The meaning of these implications, which depend on the principle of the indifference of the heavy body to motion and rest on the horizontal plane, appears even more evident when considering the other consequence that the affirmation of this principle entails and to which attention was already previously called: the indifference of bodies to motion or resting on the horizontal plane implies that the *momento* of a heavy body and its effects are measurable only by the "vertical" displacement, as the "lateral" motion does not require the application of appreciable force. It is precisely on this principle that Galileo's law of *momenti* of a heavy **221**
body along different inclinations, and, implicitly, that of its corresponding (uniform) speeds of descent, is founded.

The importance of the principle, which for convenience we will define as "vertical displacements", appears evident in the demonstration that Galileo proposes at the end of the brief chapter dedicated to the "screw". With this demonstration, Galileo – who had previously clarified how, in this particular mechanical instrument, the water continually rises as it descends – intends to show that the "screw" and the inclined plane on which it is based do not allow the craftsman to deceive nature. Also in this case, in fact, an increase in force corresponds to a proportional loss of time and speed[70]. Take two weights E, F (cf. fig. 9) such that the *momento* of F along the perpendicular is equal to the *momento* of E along the inclined line. Let E, at the foot of the inclined line, be connected by a string to weight F, at the vertex of the perpendicular.

With the addition of a negligible weight, the weight F, according to the usual procedure, will be able, while descending, to lift the other weight along the incline. Now, it is evident that, while E goes up the entire AC, F descends by an analogous space along the perpendicular. It would appear, then, that the inclined plane guarantees an absolute advantage, since, with a smaller weight, it is possible to lift

[70]"Finalmente non è da passare sotto silenzio quella considerazione, la quale da principio si disse esser necessaria d'avere in tutti gl'instrumenti mecanici: cioè che quanto si guadagna di forza per mezo loro, altrettanto si scapita nel tempo e nella velocità [Finally, we must not pass over in silence that consideration, which from the beginning was said to be necessary to have in all mechanical instruments: namely, that whatever is gained in force by means of them, the same is lost in time and speed]" (*ibid.*, p. 185).

a greater one, without the first having to travel a longer distance (i.e., move with greater speed). Galileo, however, unmasks the "deception".

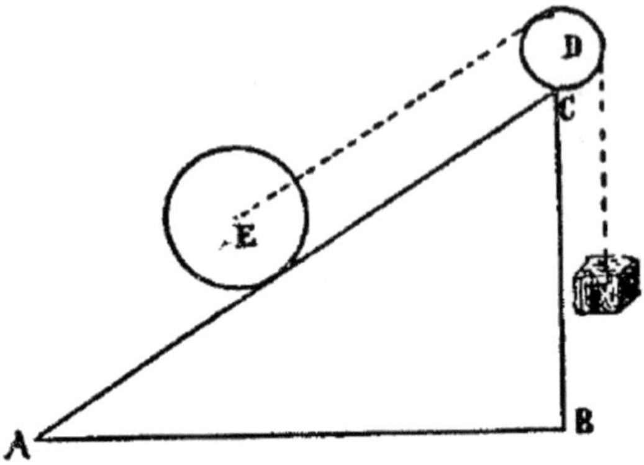

Figure 9

Ma qui però si deve avvertire che, se ben il mobile E averà passata la linea AC nel tempo medesimo che l'altro grave F si sarà per eguale intervallo abbassato, niente di meno il grave E non si sarà discostato dal centro comune delle cose gravi più di quello che sia la perpendicolare CB; ma però il grave F,
222 discendendo a perpendicolo, si sarà abbassato per spazio eguale a tutta la linea AC. E perché i corpi gravi non fanno resistenza a i moti transversali, se non in quanto essi vengono a discostarsi dal centro della terra, però, non s'essendo il mobile E in tutto il moto AC alzato più che sia la linea CB, ma l'altro F abbassato a perpendicolo quanto è tutta la lunghezza AC, però potremo meritamente dire il viaggio della forza F al viaggio della forza E mantenere quella istessa proporzione, che ha la linea AC alla CB, cioè il peso E al peso F[71]. [But here it must be noted that, although the mobile E will have traversed the line AC in the same time that the other heavy body F will have descended in the same interval, nevertheless the heavy body E will not have deviated from the common center of heavy things more than the perpendicular CB; but, however, the heavy body F, descending perpendicularly, will have descended through a space equal to the entire line AC. And because heavy bodies do not resist transverse motions, unless they move away from the center of the Earth, and yet, since the mobile E is not raised more than the line CB throughout the motion along AC, while the other F lowered to perpendicular as the entire length AC, we will be able to rightly say that the journey of the force F to the journey of the force E maintains the same proportion that the line AC has to the CB, that is, the weight E to the weight F.]

The Galilean demonstration, which the Pisan scientist decided to use again over forty years later[72], confirms that Galileo conceives the relationship between the *momenti* of a heavy body along different inclinations as fully equivalent to that between the "vertical displacements" accomplished in equal times. And the law that establishes the proportion of the *momenti* of the same weight appears fully comprehensible only by starting from these principles. In the right-angled triangle

---

[71] *Ibid.*, pp. 185-186.

[72] Galileo decided to use this demonstration, of which he prepared a new draft, to confirm the supposition (bodies descending along different inclinations reach equal speeds at equal "vertical" distances from the starting point) premised on the theorems of *De motu naturaliter accelerato* from the Third Day of the *Discorsi*. See, *infra*, p. 299 (note 81) and *Discorsi e dimostrazioni matematiche*, edited by Carugo-Geymonat cit., pp. 782-783.

ABC (see fig. 10), the "partial" *momento* of a heavy body along AB, being able to neglect the lateral displacements, which do not require the application of appreciable forces, is made equal to AB. What will be the "total" *momento* of the same heavy body? It is easy to answer that the "total" *momento* will occur when its entire displacement, i.e., AB, is "vertical", that is, like BD. Hence, the proportionality of the *momenti* of the same weight to the spaces traversed in a given time:

"partial" *momento* for AB = BC, in other words = to the vertical descent carried out in the given time.

"total" *momento* for BC = AB = BD, in other words = to the vertical descent at the same time.

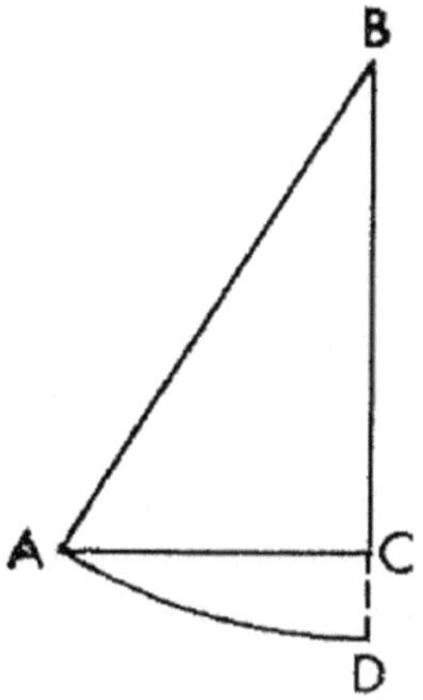

Figure 10

That is, again, *momento* for AB : *momento* for BC = BC : AB (BD).

With no greater difficulty, one can understand the *momento-speed* proportionality. By expressing the *momento* in terms of space traversed in equal times, the direct proportionality with speed is, in fact, automatically established.

Based on this reasoning, certainly present to Galileo – even if he preferred to "officially" propose a "geometric" demonstration of the law of *momenti* – it was difficult to ignore the dynamic implications of the inclined plane theorem and **223**
avoid the obligation to establish a relationship between *momento del discendere* [momento of descent] and speed of descent: by conceiving *momenti* in terms of spaces traversed in equal times, it seemed, in fact, that the relationship with speed was inevitably established. This, of course, is the case if the speeds of descent are uniform. In fact, the question would have appeared much more complicated if it had been admitted that, in descending, the heavy body exhibits a continuously accelerated speed. How could a constant *momento* produce a variable effect? Would it then be necessary, therefore, to consider the speed completely independent of the *momenti* of the bodies?

The uncertainties, the burdensome, uneasy silences, the rigorous confinement to a "static" scope, show that Galileo is aware of these difficulties and is working to overcome them. In this sense, the *Mecaniche* is an extremely important writing, testifying to a situation of profound crisis for our scientist and giving us the measure of the urgent need to renounce the model of uniform speeds of natural motions, in order to decisively address the problem of acceleration.

In the meantime, some progress has been made by Galileo, and not only in awareness. The Pisan scientist has, in fact, with very important terminological innovations, managed to simplify and make extremely versatile the concept – as used in the *De motu* – of the varying effectiveness of the weight according to the distances from the fulcrum (of the balance) of the point of suspension or according to the inclination of descent. The term *momento*, which makes its first appearance in the *Mecaniche*, is used to indicate this concept, of which we have emphasized a dual meaning: first, Archimedean, strictly "static", and the other substantially "dynamic" (the first corresponding to the geometric scheme of the balance, the second to the model of the inclined plane). In the *Mecaniche*, Galileo tries to take into consideration only the "static" value of *momento*, avoiding highlighting that the term also has a clear dynamic connotation (cause of motion and speed). However, both in combining the purely geometric Archimedean demonstration of the laws of *momenti* of the balance with that based on the principle of "virtual speeds", and in proposing, by analyzing the "screw", the laws of *momenti* of a heavy body along different inclinations, Galileo showed that he was very clear – even if he had no interest in emphasizing it – that *momento* guarantees an almost automatic transition from statics to dynamics, constituting a link between the two fields. The very fact that he chose *momento* (from *moveo* [to move]) to indicate a concept that, in the years of the *De motu*, he did not know how to indicate with a specific term, demonstrates that, even in the *Mecaniche*, dynamic concerns are crucial for Galileo.

224 We can, therefore, conclude that *momento* makes its appearance in the *Mecaniche* with the value of variable effectiveness of weight (in counterweighting, but also in descending "naturally"). The new term excellently serves to safeguard against the risk of confusion or unclear distinction between different concepts, which Galileo, due to the lack of adequate terminology, had repeatedly faced in the *De motu*. But why did Galileo choose this term?

It is difficult, if not impossible, to give a clear answer. It will be observed, first of all, that in Galileo's eyes, *momento* could hardly have appeared as a well-established "technical" term: as we have seen, there were in fact too few previous uses of the term with mechanical value. Nor, on the other hand, did he prefer it over other expressions – as has been claimed – by virtue of the fact that it was a "neutral" term, not compromised with previous conceptions, from which Galileo intended to distinguish himself[73]. *Momento* certainly wasn't neutral. The term – and Galileo was certainly aware of this – indicated, in fact, both in its etymology and in its most common use (the text and the main commentaries and expositions of Aristotle's

[73] This is the thesis supported by Altieri Biagi in the often cited *Galileo e la terminologia tecnico-scientifica* (p. 43).

Physics), the tendency, the *conatus* [endeavor] of bodies towards their "proper" place, and, consequently, the speed with which they would have moved towards that goal. Such tendency derived directly from weight. *Momento*, therefore, indicated the "instinct", the natural "desire" of the heavy body to fulfill itself: ultimately, a rather intangible cause of motion, an "occult" quality.

Likely, Galileo's attention was initially attracted precisely by these scarcely rational and geometric-mathematical values of *momento*. He must have recognized it as a term capable of effectively suggesting the natural tendency of heavy bodies to move towards the center of the universe. A fact of which the Pisan scientist had to take note, not being able to provide a satisfactory explanation. That Galileo felt *momento* as synonymous with "capacità [capacity]", "disposizione [disposition]", "inclinazione naturale [natural inclination]", appears, on the other hand, also from a passage in the *Mecaniche*, in which the term is used with this value:

> Ma se nella medesima circonferenza fusse applicata forza inanimata, la quale avesse momento di far impeto per tutti i versi, potria far l'effetto costituita in qual si voglia luogo di detta circonferenza[74]. [But if an inanimate force were applied to the same circumference, which had the momento to exert impeto in all directions, it could produce the effect constituted at any place of the said circumference.]

To indicate this same meaning, Galileo will also use "talento [talent]" in subsequent works, establishing, in one case, a direct relationship between *momento* and *talento*, a term that is certainly not quite "neutral"[75].

---

[74] *G.G.*, II, pp. 168-169.

[75] "*Simpl.* ...mi par che l'esperienza dovesse esser praticata non nell'elemento dell'aria, ma in un mezzo dove l'aria stessa potesse esercitare il suo *talento* del peso, se ella veramente ne possiede [It seems to me that the experiment should be conducted not in the element of air, but in a medium where the air itself could exert its *talent* of weight, if it truly possesses it].
Salv. ...l'operazione dunque... converrebbe farla nel vacuo, dove ogni grave esercita il suo *momento* senza diminuzione alcuna... [the operation therefore... would be best done in a vacuum, where every heavy body exerts its *momento* without any diminution…]" (*G.G.*, VIII, p. 125; the italics are ours). According to Altieri Biagi, Galileo made a subtle play here, placing a traditional term in Simplicio's mouth, compromised with magical-animistic conceptions ("talento [talent]") and making Salviati propose its replacement with a "mechanical" term that is absolutely "neutral" (*momento*). The interpretation suggested by Altieri Biagi is intriguing. It will be worthwhile, however, to observe that Galileo uses "talento [talent]" more than once, and without any ulterior motive, to express the tendency or propensity to move towards the center produced by heaviness, which, in general, he prefers to indicate with *momento*: "sì che essi non esercitino quello *puro e semplice talento* che depende dalla lor gravità [so that they do not exert that *pure and simple talent* that depends on their heaviness]" (*Lettera a Tolomeo Nozzolini*, *G.G.*, IV, p. 303; the italics are ours); "*Salv.* ...quel sasso, separato che e' sia dalla nave, si riduca alla sua naturalezza, e ritorni a esercitare *il puro e semplice suo natural talento* [*Salv.* …that stone, separated from the ship, is reduced to its naturalness, and returns to exerting *its pure and simple natural talent*]" (*Dialogo intorno ai due massimi sistemi*, *G.G.*, VII, p. 168; the italics are ours); "*Salv.* ...quel sasso che è in cima alla torre ha per suo primario istinto l'andare intorno al centro del suo tutto in ventiquattr'ore, e questo suo *natural talento* esercita egli eternamente [*Salv.* …that stone which is at the top of the tower has as its primary instinct the going about the center of its entirety in twenty-four hours, and it eternally exerts this *natural talent*]" (*ibid.*). The same *momento-talento* analogy was taken up by Nardi: "e perché siamo convenuti che tanto sia l'impeto, l'energia, il momento e il talento del mobile, quanta è la forza e la resistenza che basta a fermarlo… [and because we agreed that as much impeto, energy, momento and talent of the mobile, there is enough force and resistance to stop it…]" (*Scene*, Ms. Galil. 130, c. 276; on Nardi and his unpublished *Scene*, see *infra*, note 157). On "talento [talent]" and its relations with

225 Naturally, Galileo, in adopting it, made a notable effort to strip *momento* of this "occult" value, reducing it to a quality whose variation in intensity is geometrically (i.e., rationally) determinable. However, precisely in the first writings in which he used it, it appears evident that the use of *momento* remains strictly and rigidly connected to the idea of a "natural" inclination, produced by weight (thus, a consequence of its instinct). *Momento* is not yet, in other words, perceived as a "neutral" technical term, capable of expressing the variation of any type of force. In the *Mecaniche*, for example, he uses "momento del peso [momento of weight]" to indicate the variation of the tendency of a heavy body to descend along different inclinations, while "forza [force]" always indicates the weight that must be applied to make it ascend along the same inclinations. Galileo does not feel – and this is worth reflecting on – like using "momento del peso [momento of weight]" and "momento della forza [momento of force]" (which, from his point of view, coincide). This means that, while he feels *momento* is adequate for expressing the variation of the "natural" inclination of weight to descend, he does not consider that the term is suitable to indicate the variation of the "non-natural" inclination of force.

Subsequently, Galileo will free the term from these constraints, using it, in general and without any confusion, in expressions such as "momento della forza [momento of force]", "momento della resistenza [momento of resistance]". And yet, it's significant – and it tells us much about the motivations that led Galileo to choose precisely *momento* – that Galileo initially did not feel authorized to use our term with the value of variable effectiveness of any force, but only of the "natural" weight.

226 Nor should we overlook that *momento*, whose use in the Latin translations of the *Physics* and its main commentaries was known to Galileo at least as early as *De motu*, was introduced into the lexicon of the Pisan scientist only following the criticism of the concept of positive lightness and the rejection of admitting it as a "natural" propensity or quality of bodies. In other words, Galileo does not accept the "momentum gravium et levium [momentum of heaviness and lightness]" of the Aristotelian tradition: only heavy bodies are endowed with a "natural" instinct to move towards the center, conferred precisely by heaviness.

We cannot, therefore, properly speak of the term as either technical or "neutral". In Galileo's lexicon, *momento* will take on these characteristics only with time. It initially served Galileo, not only on a purely geometric level, as a device to subject the various static and dynamic effects of heaviness to rigorous measurement, but also because it suggested with its more common value that natural motion (and speed) derive from the effort of heavy bodies to return to the center. Galileo in the *Mecaniche* is still closely linked to this general, very non-relativistic view of movement. As he gradually distances himself from it (even if he will never completely abandon it), we will see the use of *momento* change (and become relative) in parallel.

Alongside this original, "philosophical" meaning of *momento*, we have seen another emerge, resulting from the effort of abstraction, that is, geometric "represen-

momento, cf. Gianni Mombello, *Les avatars de* "Talentum" cit., pp. 279-280.

tation". In the demonstrations of the laws of balance, through the "virtual speeds" approach, and in the inclined plane theorem, *momento* is always indicated (and measured) by the distance traveled in a given time. The geometer who, using the science of proportions, formulates such laws becomes progressively accustomed to conceiving the different *momenti* as the spaces traveled in given times. Since Galileo's analysis still substantially depends on the hypothesis of the uniform speed of natural motions, it is evident that the proportion between the *momenti* of the same heavy body along different inclinations can also be conceived as the proportion of its different speeds. However, it is necessary to reiterate that, on the level of geometric abstraction, *momento* is a distance and, only consequently, a speed. We will see that Galileo will later prefer to refer *momento* as an equivalent of speed, but he will conceive it, in other words, "represent" it, always as the greater or lesser distance traveled in equal times.

It remains to add some observations on the effects produced by Galileo's *momento* of the *Mecaniche*, which, even if it was not immediately printed, circulated widely in manuscript. The diffusion of the mechanical treatise was primarily ensured by the free French paraphrase prepared by Mersenne and published in 1634 in Paris. Father Mersenne showed quite a bit of confusion about *momento*. He reported, with some modifications – especially in recalling the parallelism with
ῥοπή – the definition of the term given by Galileo at the opening of the *Mecaniche*, **227**
transliterating it into *moment*[76]. But then he did not always adhere to this criterion, preferring to resort to *puissance* [power] and, more significantly, *mouvements des poids* [movements of weights] (for the Galilean *momento del peso* [momento of weight])[77]. Mersenne would later acknowledge the difficulty faced by the French reader regarding *momento*[78].

[76] "Le *moment* est l'inclination du mesme corps, lorsqu'elle n'est pas seulement considérée dans ledit corps, mais conjoinctement avec la situation qu'il a sur le bras d'un levier, ou d'une balance; et cette situation fait qu'il contrepese souvent à un plus grand poids, à raison de sa plus grande distance d'avec le centre de la balance. Car cette éloignement estant joint à la propre pesanteur du corps pesant, luy donne une plus forte inclination à descendre; de sort que cette inclination est composée de la pesanteur absolue du corps et de l'éloignement du centre de la balance, ou de l'appuy de levier. Nous appellons donc tousjours cette inclination composée, *moment*, qui répond au *ῥοπή* des Grecs [*Moment* is the inclination of the same body, when it is not only considered in the said body, but in conjunction with the position it has on the arm of a lever, or a balance; and this position means that it often counterweighs a greater weight, due to its greater distance from the center of the balance. For this distance being joined to the proper weight of the heavy body, gives it a stronger inclination to descend; so that this inclination is composed of the absolute weight of the body and the distance from the center of the balance, or from the fulcrum of the lever. We therefore always call this compound inclination, *moment*, which responds to the *ῥοπή* of the Greeks]" (*Les Mechaniques de Galilée*, ed. crit. cit., pp. 26-27).

[77] *Ibid*., pp. 33 ("la grande force qu'apporte la vistesse du mouvement, pour accroistre la *puissance* du mobile… [the great force brought by the speed of the movement, to increase the *power* of the mobile…]") and 32 ("Après avoir demonstré que les *mouvements des poids*… [After having demonstrated that the *movements of the weights*...]").

[78] See *infra*, note 168.

2. The "Hydrostatic" Momento

In the years immediately following 1600, Galileo worked hard to study the natural motion of falling and the acceleration that characterizes it, striving to attribute its behavior to mathematical laws. This is demonstrated by multiple documents, especially the letters to Guidobaldo and Sarpi, as well as Luca Valerio's response to a lost letter from Galileo, dated 1602, 1604 and 1609 respectively, as well as a number of the so-called *Frammenti* pertaining to the *Discorsi*, which are to be assigned to the first decade of the 17th century.

We will, for now, set aside these very important texts, which open the way to understanding the subsequent uses of *momento*. Our attention will focus, instead, on that group of writings, all dedicated to the analysis of hydrostatic questions, that Galileo wrote between 1611 and 1615 during a rather heated controversy he had with some Aristotelians[79]. These writings, in fact, document the extension of the concept and term of *momento* to new fields, just as we have seen them established in the *Mecaniche*.

From the point of view of our interest in the events of the term, this "leap" allows us to establish greater continuity in the exposition and to "conclude" the discussion of the main aspects of Galileo's early interests (hydrostatics and inclined plane-balances), which we have seen closely linked since the *De motu*.

228 It should be recalled that, in his early Latin treatise, Galileo attempted to establish a coherent and organic explanation of all types of movement based on the dual principle of hydrostatics and the balance. The attempts at revision to which the original draft was subjected demonstrate that Galileo discovered, at a certain point, the impossibility of combining the hydrostatic approach with that of the balance[80]. In the *Mecaniche*, in fact, he carefully avoided even alluding to a dynamic use of the balance model and its laws. Just as he avoided emphasizing the possible analogies between the balance model and hydrostatics.

During the summer of 1611, in Filippo Salviati's villa "Le Selve", near Florence, several meetings took place between professors of the Pisan University, gentlemen and literati of different orientations. During these meetings, an academic discussion began on the qualities of hot and cold, which, as we will see, soon transformed into a dispute on the causes of the buoyancy of heavy bodies in water. Galileo participated assiduously in these discussions, becoming the favored target of the polemical jabs from the Aristotelians.

From these controversies was born the *Discorso intorno alle cose che stanno in su l'acqua*, published in May 1612[81]. As in the case of the *Mecaniche*, we also

[79] All of these texts, including the writings of his opponents, were published by Favaro in the voluminous vol. IV of the Ed. Naz., entirely dedicated to the controversies on hydrostatics from the years 1611-1615.
[80] Cf. supra, p. 190.
[81] *Discorso al Serenissimo Don Cosimo II... Intorno alle cose che stanno in su l'acqua o che in quella si muovono...*, Appresso Cosimo Giunti, in Firenze 1612. Naturally, I use the text given by Favaro in the Ed.

retain a draft or, better yet, a first draft of the *Discorso*, in the form of a report of the controversy, written for Cosimo II[82]. The first printed edition was followed by a second, again in 1612, with some non-negligible modifications[83].

Also in this case, it will be convenient, therefore, to follow the progressive **229**
consolidation of the text through different phases, which, as we will see, present extremely significant variations, even with regards to *momento*. We will, however, omit, as it is not pertinent to our purpose, examination of the development of the discussions around the philosophical problem of whether heat causes rarefaction and cold condenses, limiting ourselves to examining the parts of these writings in which the phenomena of buoyancy are discussed in general and efforts are made to establish its laws.

In the *De motu*, Galileo, essentially following Archimedes, had established the laws that regulate the behavior of bodies in water. Premising that *gravitas* was understood as specific weight, he had demonstrated, in fact, that bodies with a specific weight equal to water are submerged, but do not descend in water; that bodies of higher specific weight sink; while, lastly, those lighter than water cannot "demergi tota [submerge entirely]". The demonstrations rest on the principle that in water – as in any other medium – a heavy body, brought to descend by natural inclination, is hindered by an equal volume of the medium. To overcome this resistance, it must, therefore, have a specific weight greater than that of the medium. If, in fact, the weight is equal, the heavy body will lose all inclination to descend, while, if it is less, the medium will "estruderà [extrude]" it, that is, force it to rise. We have also observed that, in the two chapters of the Latin treatise dedicated to the problems of buoyancy and sinking, Galileo does not use a particular term or expression to indicate "specific" weight, distinguishing it from "absolute" weight. Compared to the discussion of the balance, however, these two chapters of the *De motu* present less terminological confusion, given that there is no occasion to use the concept of the variable effectiveness of weight in the Archimedean hydrostatic model, which we have seen established, in the *Mecaniche*, with *momento*. "Minus

Naz. (vol. IV). For a detailed "history" of the controversy and disputes that took place in the presence of Maffeo Barberini (the future Urban VIII), see Antonio Favaro's *Avvertimento* (*G.G.*, IV, pp. 5-7); see also S. Drake, *Galileo Studies* cit., Ch. VIII (*The dispute over bodies in water*), pp. 159-176. For an analysis of the significance of the dispute on hydrostatics, in addition to the aforementioned essay by Drake, see R. Caverni, *Storia del metodo sperimentale* cit., VI, pp. 107 ff. and W. Shea, *La rivoluzione intellettuale di Galileo*, trad. it. cit., Ch. II (*L'idrostatica e l'uso normativo degli esperimenti*), pp. 31-71.

[82] This is the opinion of Favaro (*Avvertimento* cit., p. 8) who first published the draft of the *Discorso* (*G.G.*, IV, pp. 30 ff.). Unlike Favaro, Drake argued that it is not an incomplete first draft of the *Discorso*, but a finished work, written before the Grand Duke listened to Galileo's arguments on the issue at the center of the debate and, therefore, completely independent from the printed writing (*Galileo Studies* cit., p. 162).

[83] The new print differs in the title page from the previous one only by the addition of "Seconda Edizione [Second Edition]" and by Cosimo Giunti's foreword to the "Benigni Lettori [Benevolent Readers]", in which the publisher declares that he has resolved to print the *Discorso* again "per sodisfare a molti, che di Venezia, di Roma, e di altri luoghi mi chiedevano e mi chieggono con instanza il presente trattato [to satisfy many, who from Venice, Rome, and other places asked and are persistently asking me for this treatise]" (*G.G.*, IV, p. 63).

pondus non posse maius attollere [a lesser weight cannot lift a greater one]" is a universally valid principle in this type of hydrostatic conception, even though it does not always "hold" – as we have seen – in the balance model[84].

When, in the summer of 1611, Galileo, driven by the heat of the debate, proposed his own opinions on the behavior of bodies in water, he had to start – as was natural – precisely from the considerations made in the two "Archimedean" chapters of the *De motu* dedicated to hydrostatic questions.

This is demonstrated by the approach we find in the draft which constitutes the first Galilean writing relating to the controversy and which was, in all probability,
230 drawn up in September-October 1611[85]. In fact, the same principles from the *De motu* are sustained, while the terminology also appears substantially unchanged. Certain parts even constitute a translation of the two chapters of the Latin treatise. The concept of specific weight is still indicated with a periphrasis: "chiamonsi egualmente gravi quei corpi o quelle materie, delle quali moli eguali pesano egualmente [let us call equally heavy those bodies or materials of which equal bulks weigh equally]"[86]. *Momento* is not found either, but this is less surprising, because the whole discussion is based on the proportionality of equal volumes of materials that weigh equally or differently. It is, in other words, a treatment comparable, in the mechanical model, to the balance of equal arms, which can be addressed in terms of simple "weight", without resorting to the *momento del peso* [momento of weight].

The axioms premised to the proofs also correspond to those of the *De motu*:

> Che un peso maggiore non può esser sollevato ed alzato da un minore, *essendo pari tutte le altre circostanze*; o vero, che l'ordine naturale richiede che i corpi più gravi stiano sotto i men gravi, sotto i quali anderanno, non sendo impediti[87]. [That a greater weight cannot be lifted and raised by a lesser, *all other circumstances being the same*; or rather, that the natural order requires that the heavier bodies be under the less heavy, under which they will go, if not impeded.]

It is not difficult to recognize in this declaration the axiom of the *De motu*, not as it appears in the initial draft of *Liber I*, but rather according to the correction made in the subsequent revision: "id quod gravius est a minus gravi, *si cetera sint paria*, non posse attolli [that which is heavier cannot be lifted by that which is lighter, *if the other conditions are the same*]"[88]. The principle, exactly as in the *De motu*,

---

[84] Cf. *supra*, pp. 174 ff.

[85] This is taken from a letter of Galileo to Cigoli, dated 1 October 1611: "Sono in obligo di rispondere a due gratissime di V. S.; ma perché sono occupatissimo per finire una scrittura di 15 fogli, in proposito di certa contesa stata tra certi di questi filosofi Peripatetici e me questi giorni passati, la quale fo per il G. Duca e forse si stamperà, mi è forza esser brevissimo con lei… [I am obliged to respond to two very kind letters from Your Excellency; but because I am very busy finishing a 15-page writing regarding a certain dispute that took place between some of these Peripatetic philosophers and me in the past few days, which I am writing for the Grand Duke and perhaps will be printed, I am forced to be very brief with you…]" (*G.G.*, XI, pp. 213-214).

[86] *G.G.*, IV, p. 36.

[87] *Ibid.*, the italics are ours.

[88] *G.G.*, I, p. 348; the italics are ours. For the correction made by Galileo in the revision, cf. *supra*, pp. 189-190.

serves to establish all hydrostatic laws, given that Galileo's assumption is that, in descending, the heavy body displaces a quantity of water equal to its volume. The heavy body that sinks will, therefore, have a greater specific weight than the water.

From the revision and correction of this principle, and from the consequent need for a re-foundation, the *Discorso* and the introduction of *momento* arise, which until now had been unusable in Galilean hydrostatics.

The *Discorso*, like the *Mecaniche*, opens with a definition of terms whose necessity in the "progressione dimostrativa [demonstrative progression]" is energetically reiterated[89]. The first clarifications are related to heaviness "in specie [in species (i.e., specific)]" and "assoluta [absolute]" heaviness. We are faced with another important innovation in terminology: for the first time, in fact, Galileo uses a precise expression to distinguish specific weight from absolute weight.

Io, dunque, chiamo egualmente gravi in ispecie quelle materie, delle quali eguali moli pesano **231**
egualmente: come se, per esemplo, due palle, una di cera e l'altra d'alcun legno, eguali di mole, fussero ancora eguali in peso, diremmo quel tal legno e la cera essere in ispecie egualmente gravi.

Ma egualmente gravi di gravità assoluta chiamerò io due solidi li quali pesino egualmente, benché di mole fussero diseguali: come, per esemplo, una mole di piombo e una di legno, che pesino ciascheduna dieci libbre, dirò essere in gravità assoluta eguali, ancorché la mole del legno sia molto maggior di quella del piombo.

Più grave in specie chiamerò una materia che un'altra, della quale una mole eguale a una mole dell'altra peserà più: e così dirò il piombo esser più grave in ispecie dello stagno, perché, prese di loro due moli eguali, quella di piombo pesa più. Ma più grave assolutamente chiamerò io quel corpo di questo se quello peserà più di questo, senza aver rispetto alcuno di mole: e così un gran legno si dirà pesare assolutamente più d'una piccola mole di piombo, benché il piombo in ispecie sia più grave del legno. E lo stesso intendasi del men grave in ispecie e del men grave assolutamente[90].

[I, therefore, call equally heavy in species those materials of which equal bulks weigh equally: as if, for example, two balls, one of wax and the other of some wood, equal in bulk, are still equal in weight, we would say that such wood and wax are in species equally heavy.

But I will call two solids equally heavy in absolute heaviness that weigh equally, even though they are unequal in bulk: as, for example, a bulk of lead and one of wood, each weighing ten pounds, I will say are equal in absolute heaviness, even though the bulk of wood is much greater than that of lead.

I will call one material heavier in species than another, of which a bulk equal to a bulk of the other will weigh more: and thus I will say that lead is heavier in species than tin, because, in taking two equal bulks of them, that of lead weighs more. But I will call that more absolutely heavy than the other if it weighs more than the other, without any regard for its bulk: and thus a large piece of wood will be said to weigh absolutely more than a small bulk of lead, although lead is heavier in species than wood. And the same applies to the less heavy in species and the less absolutely heavy.]

Heaviness "in ispecie [in species]", "assoluta [absolute]" heaviness. Galileo belatedly adopted long-standing terms to indicate the specific weight and absolute

---

[89] "E perché così richiede la progressione dimostrativa, io definirò alcuni termini e poi esplicherò alcune proposizioni, delle quali come di cose vere e note, io possa servirmi a' miei propositi [And because the demonstrative progression requires this, I will define some terms and then explain some propositions, which I can use for my purposes as true and known things]" (*G.G.*, IV, p. 67).
[90] *Ibid.*, pp. 67-68.

weight of a body[91]. Apart from the distinction of the two "heavinesses", there is nothing new in Galileo's definition. The Pisan scientist still seems to be inspired by the Archimedean criterion adopted in the *De motu*: "Ciò fu sottilmente dimostrato da Archimede ne' libri Delle cose che stanno sopra l'acqua [This was subtly demonstrated by Archimedes in the books On Floating Bodies]"[92].

However, soon after, our author introduces a surprising innovation. He declares, in fact, that he now intends to address the problem of buoyancy in a different way than Archimedes.

> Io con metodo differente e con altri mezzi procurerò di concludere lo stesso, riducendo le cagioni di tali effetti a' principii più intrinsechi e immediati, ne' quali anco si scorgano le cause di qualche accidente ammirando e quasi incredibile[93]. [I will try to conclude the same with a different method and other means, reducing the causes of such effects to the most intrinsic and immediate principles, in which the causes of some astonishing and almost incredible accidents can be seen.]

232 The "intrinsic" principles from which Galileo proposes to start, principles that are no longer Archimedean, will provide the adequate interpretation of a phenomenon that, otherwise, could appear to be an insoluble paradox:

> qual sarebbe che una piccolissima quantità d'acqua potesse col suo lieve peso sollevare e sostenere un corpo solido, cento e mille volte più grave di lei[94]. [how could it be that a very small quantity of water could, with its slight weight, lift and support a solid body a hundred and thousand times heavier than itself.]

It is the first time that we find Galileo dissatisfied with an exclusively Archimedean approach. The fact is that the "paradox" does not seem to be explainable with the hydrostatic principles of the Syracusan, the same ones used by Galileo in the *De motu* and in the draft of the *Discorso*. The "hydrostatic paradox" indicates, in fact, a situation in which it is impossible to compare equal volumes of the displaced fluid and the immersed portion of the solid. Thus, in this case, the Archimedean principle of comparison between the "specific" weights of water and the solid appears inapplicable.

As we will see shortly, between the first draft of the *Discorso* and its final draft, Galileo had even discovered that the volume of water displaced by a body that

---

[91] For the traditional use of heaviness "in specie [in species]", cf. *supra*, pp. 196-197. It should be added that Giovanni Bardi, friend and student of Galileo, stated, in one of his works published in 1614, that *gravitas* per unit of volume is defined by the Pisan scientist "Cum Iordano, gravitatem in specie [with Jordanus, heaviness in species]" (*Eorum quae vehuntur in aquis experimenta...*, Roma 1614, cit. in M. Clagett, *La scienza della meccanica*, trad. it. cit., p. 182). Clagett derives from Bardi's testimony that Galileo would have known well the *de insidentibus in humidum*, Archimedeo, included by Tartaglia, without specifying the author, in the edition of *De ratione ponderis* (Venezia 1565, pp. 168-198), and that he attributed it to Jordanus. It is worth emphasizing that "cum Iordano" could simply mean "come Giordano [just like Jordanus]", not necessarily "seguendo, imitando Giordano [following, imitating Jordanus]". In any case, in the writings traditionally attributed to Jordanus, "gravitas in specie" is not found.

[92] *G.G.*, IV, p. 67.

[93] *Ibid.*

[94] *Ibid.*

immerses itself in it is always less than the volume of the immersed part of the solid. This means that the Archimedean procedure, which until that moment had appeared universal and necessary to Galileo, was no longer adequate to explain the phenomena.

The novelty was so profound as to require a radical reform of Galileo's investigation techniques of hydrostatics. We have clear evidence of this in the two "principles", derived from mechanics, which Galileo places alongside the definition of terms. We must not overlook that we are faced with the first occurrence of *momento* in a printed Galilean text. The first of these principles

è che pesi assolutamente eguali, mossi con eguale velocità, sono di forze e di momento eguali nel loro operare. Come, per esempio, due pesi d'assoluta gravità eguali, posti in bilancia di braccia eguali, restano in equilibrio, né s'inclina l'uno alzando l'altro: perché l'egualità delle distanze di ambedue dal centro, sopra il quale la bilancia viene sostenuta e circa il quale ella si muove, fa che tali pesi, movendosi essa bilancia, passerebbono nello stesso tempo spazii eguali, cioè si moverieno con eguali velocità... e perciò si fa l'equilibrio, e restano i momenti loro di virtù simili ed eguali[95]. [is that absolutely equal weights, moved with equal speed, have equal forces and momento in their operation. As, for example, two equal weights of absolute heaviness, placed on a balance of equal arms, remain in equilibrium, and neither one tilts in lifting the other: because the equality of the distances of both from the center, upon which the balance is supported and around which it moves, ensures that these weights, as the balance moves, would pass through equal spaces at the same time, that is, they would move with equal speeds... and therefore equilibrium is achieved, and their momenti remain of similar and equal virtue.]

The second mechanical principle assumed by Galileo is related to the case of the balance of unequal arms.

Il secondo principio è che il momento e la forza della gravità venga accresciuto dalla velocità del moto; sì che pesi assolutamente eguali, ma congiunti con velocità diseguali, sieno di forza, momento **233**
e virtù diseguale, e più potente il più veloce, secondo la proporzione della velocità sua alla velocità dell'altro. Di questo abbiamo accomodatissimo esemplo nella libra o stadera di braccia diseguali, nelle quali posti pesi assolutamente eguali, non premono e fanno forza egualmente, ma quello che è nella maggiore distanza dal centro, circa il quale la libra si muove, s'abbassa sollevando l'altro, ed è il moto di questo, che ascende, lento, e l'altro veloce: e tale è la forza e virtù che dalla velocità del moto vien conferita al mobile che la riceve, che ella può esquisitamente compensare altrettanto peso che all'altro mobile più tardo fosse accresciuto; sì che, se delle braccia della libra uno fosse dieci volte più lungo dell'altro, onde, nel muoversi la libra verso il suo centro, l'estremità di quello passasse dieci volte maggiore spazio che l'estremità di questo, un peso posto nella maggior distanza potrà sostenerne ed equilibrarne un altro dieci volte assolutamente più grave che non è egli: e ciò perché movendosi la stadera, il minor peso si moveria dieci volte più velocemente che l'altro maggiore[96]. [The second principle is that the momento and force of heaviness is increased by the speed of motion; such that absolutely equal weights, but combined with unequal speeds, are of unequal force, momento and virtue, the faster one being more powerful, according to the proportion of its speed to the speed of the other. We have a very good example of this in the scales or steelyard of unequal arms, in which, when absolutely equal weights are placed, they do not press and exert force equally, but the one which is at the greatest distance from the center, around which the scales move, is lowered in raising the other, and the motion of this one, which ascends, is slow, and the other fast: and such is the force and virtue that the speed of motion confers upon the mobile that receives it, that it can precisely compensate for the same weight that was added to the other slower mobile; so that if one of the arms of the scales were ten times longer

[95] *Ibid.*, p. 68.
[96] *Ibid.*, p. 68-69.

than the other, so that, in moving the scales towards its center, the extremity of the former would pass through ten times more space than the extremity of the latter, a weight placed in the greater distance will be able to support and balance another ten times more absolutely heavy than itself: and this is because, as the steelyard moves, the smaller weight will move ten times faster than the larger one.]

The two principles are well known to us, as Galileo used them extensively in the *Mecaniche*. However, in this presentation, the emphasis of the "velocità [speed]" with which the arms of the balance would move appears explicit. Of the two demonstrations provided in the *Mecaniche*, one rigorously static and the other based on the principle of "virtual speeds"[97], Galileo presents here exclusively the latter. *Momento*, thus defined, is reducible to the combined effect of the "absolute" weight and the speed, which – as already previously observed – is conceived as uniform speed, and expressed by the space traveled in a given time. It is known that the so-called "virtual speeds" approach was explicitly suggested in texts well known to Galileo, especially in pseudo-Aristotelian mechanics, as our author honestly admits:

Tal ragguagliamento tra la gravità e la velocità si ritrova in tutti gli strumenti meccanici, e fu considerato da Aristotele come principio nelle sue Questioni meccaniche: onde noi ancora possiamo prender per verissimo assunto che pesi assolutamente diseguali, alternatamente si contrappesano e si rendono di momenti eguali, ogni volta che le loro gravità con proporzione contraria rispondono alle velocità de' lor moti, cioè che quanto l'uno è men grave dell'altro, tanto sia in constituzione di muoversi più velocemente di quello[98]. [Such comparison between heaviness and speed is found in all mechanical instruments, and was considered by Aristotle as a principle in his Questions of Mechanics: so we can also take it as a very true assumption that absolutely unequal weights counterbalance each other and make each other equal in momenti, whenever their heavinesses correspond in inverse proportion to the speeds of their motions, that is, the less heavy one is than the other, the more capable it is of moving faster than the other.]

The laws of *momenti* of weights in the balance can be extended to problems of buoyancy. They allow us, rather, to conveniently resolve the paradoxes that
234 Archimedean hydrostatics proved incapable of resolving. But how can the laws of balance be applied to the analysis of hydrostatic questions? It will be sufficient – states Galileo – to concentrate attention on the "mutual operation" of solids and water. The solid, in fact, as it sinks, "viene discacciando l'acqua dal luogo dove egli successivamente subentra [displaces the water from the place where it subsequently enters]" and the water pushed away rises above the initial level, "al quale alzamento essa altresì, come corpo grave, per sua natura resiste [to which its raising also, as a heavy body, naturally resists]"[99]. As the solid submerges, it displaces an increasingly larger amount of water. The concept of *momento* and the laws of balance allow us to universally establish the behavior of bodies in water, that is, to establish whether they will float or sink. It will suffice, in fact,

conferire i momenti della resistenza dell'acqua all'essere alzata, co' momenti della gravità premente del

[97] Cf. *supra*, pp. 207 ff. It should be recalled that Galileo had proposed as a "demonstration" only the Archimedean one.
[98] *G.G.*, IV, p. 69.
[99] *Ibid.*, pp. 69-70.

solido: e se i momenti della resistenza dell'acqua pareggeranno i momenti del solido avanti la sua totale immersione, allora senza dubbio si farà l'equilibrio, né più oltre si tufferà il solido; ma se il momento del solido supererà sempre i momenti co' quali l'acqua scacciata va successivamente facendo resistenza, quello non solamente si sommergerà tutto sott'acqua, ma discenderà sino al fondo; ma se, finalmente, nel punto della total sommersione si farà l'agguagliamento tra i momenti del solido premente e dell'acqua resistente, allora si farà la quiete, e esso solido, in qualunque luogo dell'acqua, potrà indifferentemente fermarsi[100]. [compare the momenti of the water's resistance to being lifted, with the momenti of the pressing heaviness of the solid: and if the momenti of the resistance of the water equal the momenti of the solid before its total immersion, then undoubtedly equilibrium will occur, and the solid will not dive any further; but if the momento of the solid always exceeds the momenti with which the water driven out progressively offers resistance, it will not only submerge itself entirely under water, but will descend to the bottom; but if, lastly, at the point of total submergence, the momenti of the pressing solid and the resisting water achieve equality, then rest will occur, and the solid, at any place in the water, will be able to remain there indifferently.]

In Galilean hydrostatics, just as in Archimedean hydrostatics, we were accustomed until now to seeing the comparison proposed between the "gravità in ispecie [heaviness in species]" of bodies and water. Now every reference to specific weight has disappeared, and Galileo proposes to compare the *momenti* of the solid and of water, in which the *momenti* – as he noted above – are the product of the "assoluto [absolute]" weight and the speed.

The time has come to explain what has happened and to introduce, in Galileo's own words, the new fact that has made necessary a radical revision of the Archimedean approach used in the *De motu* and in the draft of the *Discorso*, as well as the adoption of the "virtual speed" model to explain hydrostatic equilibrium. Introducing the new fact, Galileo specifies that the comparison between the specific weight of water and solids on which hydrostatics (even Galilean hydrostatics) has been based is, in truth, if not false, at least not fully satisfactory:

tale comparazione potrebbe nel primo aspetto parere sufficiente per poter concludere e determinare, quali sieno i solidi che sopranuotino, e quali quelli che vanno in fondo, pronunziando che quelli sopranuotino che saranno men gravi in ispecie dell'acqua, e quelli vadano al fondo che in ispecie saranno più gravi[101]. [such comparison could, at first glance, seem sufficient to conclude and determine which solids will float, and which are those that go to the bottom, pronouncing that those that float will be less heavy in species than water, and those that go to the bottom will be heavier in species.]

This false impression, which occurs "at first glance", depends on the incorrect evaluation of the volume of water displaced by a submerging solid. In fact, it seems 235

che il solido nel sommergersi vada tuttavia alzando tant'acqua in mole, quanta è la parte della sua propria mole sommersa; per lo che impossibil sia che un solido men grave in ispecie dell'acqua si sommerga tutto, come impotente ad alzare un peso maggior del suo proprio, e tale sarebbe una mole d'acqua eguale alla mole sua propria; e parimenti parrà necessario che il solido più grave vada al fondo, come di forza soprabbondante ad alzare una mole d'acqua eguale alla propria, ma inferior di peso[102]. [that solid in submerging nevertheless raises as much water in bulk as is the portion of its own submerged bulk;

[100] *Ibid.*
[101] *Ibid.*
[102] *Ibid.*

therefore, it is impossible for a solid less heavy in species than water to submerge itself completely, as if it were unable to lift a weight greater than its own, and such would be a bulk of water equal to its own bulk; and likewise it will seem necessary for the heavier solid to go to the bottom, as if it has a superabundant force to raise a bulk of water equal to its own, but less in weight.]

This general principle of buoyancy – Galileo specifies – appears even more convincing as it is duly confirmed by experience, which demonstrates the impossibility for bodies of a specific weight lower than water to sink and, vice versa, for bodies of a higher specific weight to float. However – here's the novelty – "benché le conclusioni sian vere [although the conclusions are true]", this principle assigns "cagioni difettose [defective causes]". It is not true, in fact, that

il solido nel sommergersi sollevi e scacci mole d'acqua eguale alla sua propria sommersa, anzi l'acqua sollevata è sempre meno che la parte del solido ch'è sommersa, e tanto più, quanto il vaso, nel quale si contiene l'acqua, è più stretto: di modo che non repugna che un solido possa sommergersi tutto sott'acqua senza pure alzarne tanta, che in mole pareggi la decima o la ventesima parte della mole sua; sì come, all'incontro, piccolissima quantità d'acqua potrà sollevare una grandissima mole solida, ancorché tal solido pesasse assolutamente cento e più volte di essa acqua, tutta volta che la materia di tal solido sia in ispecie men grave dell'acqua; e così una grandissima trave, che, v.g., pesi 1000 libbre, potrà essere alzata e sostenuta da acqua che non ne pesi 50; e questo avverrà quando il momento dell'acqua venga compensato dalla velocità del suo moto[103]. [the solid in submerging raises and drives out a bulk of water equal to its own submersion, indeed the water raised is always less than the portion of the solid that is submerged, and even more so the narrower the vessel in which the water is contained: so it is not objectionable that a solid can submerge itself entirely under water without even raising so much water that in bulk it equals a tenth or twentieth part of its own bulk; just as, on the contrary, a very small quantity of water will be able to lift a very large solid bulk, even if that solid weighs absolutely a hundred or more times that of water, provided that the matter of this solid is less heavy in species than water; and so a very large beam, which, for example, weighs 1000 pounds, can be lifted and supported by water that does not weigh 50; and this will happen when the momento of the water is compensated by the speed of its motion.]

The volume of water displaced by the submerging solid is, therefore, always lesser than that of the immersed solid and lesser the narrower the container. We will not dwell on the demonstration that Galileo provides for this phenomenon[104]. What we are interested in emphasizing is that, having lost the possibility of comparing equal volumes, the Archimedean hydrostatic approach, entirely centered on the concept of specific weight and on the principle that the body in descending into the water displaces a quantity of fluid equal to the volume of its own part submerged, appears unusable.

On the other hand, the discovery that the volumes of the raised water and the
236 immersed solid are not automatically equal means that the comparison must be established not between "specific" weights, but between "absolute" weights[105]. It

[103] *Ibid.*, pp. 70-71.

[104] Cf. *ibid.*, pp. 71-72.

[105] Galileo is concerned, in any case, with establishing through a demonstration the relationship that exists between "specific" and "absolute" weights of bodies: "I pesi assoluti de' solidi hanno la proporzion composta delle proporzioni delle lor gravità in specie e delle lor moli [The absolute weights of solids have the compound proportion of the proportions of their heavinesses in species and of their bulks]"

is at this point, and in relation to the emergence of this awareness, that *momento* appears. Galileo can, in other words, return to proposing the hydrostatic-model analogy of the balance that he had to reluctantly renounce in the *De motu*, due to difficulties which now appear to have been overcome.

In the case proposed by the "hydrostatic paradox" that we referred to previously, if the bulk of water displaced is less than the volume of the immersed solid, the "cause" that makes buoyancy depend on the equality of the "specific" weights of the two volumes would be flawed. Buoyancy must, therefore, depend on the greater effectiveness with which the smaller bulk of water opposes the tendency of the larger bulk of the solid to descend. The concept of *momento del peso* [momento of weight] ("absolute" weight × speed), extensively tested already in the *Mecaniche*, and the laws of the balance of unequal arms offered a perfectly adequate model to explain the phenomenon: the volume of water displaced is less than that of the solid, but its "displacement" is greater, i.e., the speed is higher. Galileo had demonstrated in the *Mecaniche* that a minimum weight can balance even a very large one, provided that, at the same time, it moves through a space that stands to the distance covered by the other in an inverse ratio to their weights: that is, when the *momenti* of the two weights are equal:

> Accade dunque in questa operazione lo stesso a capello che nella stadera, nella quale un peso di due libbre ne contrappeserà un'altro di 200, tuttavolta che nel medesimo tempo quello si dovesse muovere per ispazio 100 volte maggiore che questo; il che accade quando l'un braccio della libra sia più cento volte lungo dell'altro[106]. [It thus happens in this operation the same as in the steelyard, in which a weight of two pounds will counterweight another of 200, provided that at the same time the former has to move through a space 100 times greater than the latter; which happens when one arm of the scales is more than a hundred times longer than the other.]

Having explained the hydrostatic paradox with the development of a new interpretative model, and using *momento*, Galileo can now formulate in new terms the laws that govern the behavior of bodies in water. There will be, in fact, buoyancy when the *momento* of the solid (the "absolute" weight of its immersed part × by the displacement) equals the *momento* of the resistance of the water ("absolute" weight of the water raised × displacement or speed). The sinking of the solid will occur when its *momento di discendere* [momento of descent] exceeds that of the resistance of the water. Whereas emergence will depend on the greater *momento* of the water compared to the *momento* of the solid.

The effectiveness of the new principles and their ability to offer a satisfactory explanation of every hydrostatic phenomenon is indicated by Galileo through a **237**
series of geometric demonstrations, all of notable interest, however, we will not focus on them, given that they do not bear any substantial enrichment to Galileo's concept of *momento*, nor to its use in the field of hydrostatics[107].

(*ibid.*, p. 74-75).

[106] *Ibid.*, p. 78.

[107] There are essentially five demonstrations, which establish just as many hydrostatic laws. In the first, Galileo proves that the volume of water raised by an immersing solid is less than the volume of the

Therefore, Galileo returned to energetically emphasizing the close analogy between hydrostatics and the balance model on which he had insisted in the initial drafting of the *De motu*. He was then forced to abandon it, both due to terminological difficulties and, above all, due to the recognized impossibility of attributing the comparison between specific weights, characteristic of Archimedean hydrostatics, to that between absolute weights, typical of the balance. The discovery of the hydrostatic paradox[108] and the consequent reform of the Archimedean principles, with the introduction of the "mechanical" concept of *momento*, allowed the Pisan scientist to revive the old hypothesis. In the analysis proposed by Galileo in the *Discorso*, the "specific" and the "absolute" weight of bodies perform a
238 complementary function[109]. Whether bodies will float or, conversely, sink, can in fact be determined, in general, by comparing their specific weight with that of

immersed part of the solid "e ad essa ha la medesima proporzione che la superficie dell'acqua circunfusa al solido alla medesima superficie circunfusa insieme con la base del solido [and has the same proportion to it as the surface of the water surrounding the solid to the same surface surrounding it together with the base of the solid]" (*ibid.*, pp. 71-72). In the second, it is demonstrated that, if a prismatic solid is immersed in a container full of water, "l'abbassamento dell'acqua all'alzamento del prisma avrà la medesima proporzione che l'una delle basi del prisma alla superficie dell'acqua circunfusa [the lowering of the water to the rising of the prism will have the same proportion as one of the bases of the prism to the surface of the water surrounding it]" (pp. 72-73). In the third, Galileo shows that a solid prism, less heavy "in species" than water, placed at the bottom of an empty container, "non resterà sotto, ma si solleverà, benché l'acqua circunfusa fosse pochissima e di gravità assoluta quanto si voglia inferiore alla gravità di esso prisma [will not remain below, but will rise, even though there were very little surrounding water and of as absolute heaviness as low as one might desire than the heaviness of the prism itself]" (pp. 73-74). In the fourth, the case of the prism is proposed again with a specific weight lower than water, placed at the bottom of an empty container; if water is poured in, the solid will remain immobile, "sin che l'acqua arrivi a tal parte dell'altezza di quello, alla quale tutta l'altezza del prisma abbia la medesima proporzione che la gravità in ispecie dell'acqua alla gravità in ispecie di esso solido [until the water reaches such a part of the height of the prism that the entire height of the prism has the same proportion as the heaviness in species of the water has to the heaviness in species of the solid itself]" (pp. 75-76). Lastly, in the fifth, Galileo demonstrates that the same solid, less heavy in species than water, immersed in containers of different sizes, one very large (that is, with a lot of water) and one very small (that is, with very little water), is raised, in both cases, when the condition posed in the previous proof is satisfied. In fact, "nel cacciamento e alzamento del solido [in the pushing and lifting of the solid]", the small amount of water in the small container exceeds "in velocità di moto [in speed of motion]" the large amount of water in the large container "di tanto di quanto ella vien superata da quella in quantità: onde i momenti loro in tale operazione sono ragguagliati [by as much as it is exceeded by that in quantity: hence their momenti in this operation are matched]" (pp. 76-77).

[108] On the "experimental" or, conversely, "intellectual" nature of this discovery, cf. the considerations of W. Shea, who believes that Galileo achieved this "non compiendo esperimenti, ma impiegando ragionamenti matematici [not by carrying out experiments, but by employing mathematical reasoning]" (*La rivoluzione intellettuale*, trad. it. cit., p. 40). It is also worth emphasizing how the renunciation of the hydrostatic-model analogy of the balance, in the first draft of the *Discorso*, and its re-proposal in the final draft, for the reasons illustrated above, demonstrate that the order of composition of the *De motu* suggested by Fredette is correct, while making Drake's different thesis unlikely (cf. *supra*, p. 190, note 42).

[109] In the *Discorso* there is more than one example of the complementary use of the "specific" and "absolute" weights of bodies.
Galileo also establishes the law that allows specific weights to be derived from absolute ones and vice versa: "I pesi assoluti de' solidi hanno la proporzion composta delle proporzioni delle lor gravità in specie e delle lor moli [The absolute weights of solids have the compound proportion of the proportions of their heavinesses in species and of their bulks]" (*G.G.*, IV, p. 74-75).

the medium. To understand, however, how a small quantity of water can lift an immense bulk (of course, with a lesser "specific" weight), it is necessary to use the concept of *momento* and the principle upon which it is based: a lesser "absolute" weight can overcome the resistance of one greater, when it moves faster.

It is no coincidence that the reunification of the two approaches automatically led to the revival of Galileo's old dynamic hypothesis, namely, the one according to which natural motions and their speeds depend on the difference in "specific" heaviness between the heavy bodies and the medium[110]. In the *Discorso*, we find more than one explicit mention of it, even though Galileo avoids addressing the issue directly. Concluding, for example, the discussion relating to the laws that govern the buoyancy of bodies in water, the Pisan scientist suggests that he still considers the interpretative model of natural motions developed in the *De motu* to be substantially valid, even though he avoids any reference to the accident of acceleration:

> Parmi di aver sin qui a bastanza dichiarata e aperta la strada alla contemplazione della vera, intrinseca e propria cagione de' diversi movimenti e della quiete de' diversi corpi solidi ne' diversi mezzi e in particolare nell'acqua, mostrando come in effetto il tutto depende dagli scambievoli eccessi della gravità de' mobili e de' mezzi[111]. [I seem to have thus far sufficiently declared and opened the pathway to the contemplation of the true, intrinsic and proper cause of the different movements and of the stillness of different solid bodies in different media, particularly in water, showing how in effect everything depends on the mutual excesses of the heaviness of the mobiles and the media.]

And later, to dispel any residual doubts regarding the survival, in 1612, of the old dynamic hypothesis, he, vigorously arguing against the Aristotelian conception of the speed of natural motions as proportional to the "absolute" weights of the mobiles, will solemnly proclaim that

> non la maggior gravità assoluta, ma la maggior gravità in specie, è cagione di velocità maggiore[112]. [not the greater absolute heaviness, but the greater heaviness in species, is the cause of greater speed.]

The *Discorso* did not fail to arouse controversy. It had barely been published when Arturo d'Elci, Superintendent of the University of Pisa, revealed his own *Considerazioni sopra il Discorso del Sig. Galileo Galilei*, in which he proclaimed to take up the defense of Aristotle's opinion[113]. D'Elci, who hid under the pseudonym **239**
of Accademico Incognito [Anonymous Academician][114], was followed in the same

---

[110] Cf. *supra*, pp. 170 ff.

[111] *G.G.*, IV, p. 79.

[112] *Ibid.*, p. 133.

[113] *Considerazioni sopra il Discorso del Sig. Galileo Galilei intorno alle cose che stanno in su l'acqua, o che in quella si muovono… fatte a difesa e dichiarazione dell'opinione d'Aristotele da Accademico Incognito*, Appresso Gio. Battista Boschetti e Giovanni Fontani, in Pisa 1612 (pubbl. in *G.G.*, IV, pp. 145-196). A Latin version of the *Considerazioni* must have circulated in manuscript form.

[114] The attribution of the writing to Arturo Pannocchieschi de' Conti d'Elci, Superintendent of the Pisan University, who, in the dedication to Maria Maddalena of Austria, declares himself a simple translator into the vernacular of the anonymous writing, is owed to Antonio Favaro and is based on quite convincing considerations and rather probative documents (cf. A. Favaro, *Concluzioni sull'Accademico*

year by Giorgio Coresio's *Operetta intorno al galleggiare*[115] and Lodovico delle Colombe's *Discorso Apologetico*[116]. Lastly, in 1613, Vincenzo di Grazia decided to add his own *Considerazioni* to the chorus of reproach raised by the Aristotelians against the *Discorso*[117]. The reprint of the *Discorso* should also be placed between Coresio's *Operetta* and Lodovico delle Colombe's *Discorso Apologetico*, suggested, as shown by a brief note from the publisher to the "benigni lettori [benevolent readers]", by the desire to satisfy "a molti, che di Venezia, di Roma e di altri luoghi [many, who from Venice, Rome and other places]" insistently requested Galileo's writing[118].

As we have already had the opportunity to point out, the second edition of the *Discorso* includes some additions which, although not substantially altering its physiognomy, are of some interest. Galileo carefully annotated the writings of the Accademico Incognito[119], thoroughly reviewed the manuscript in which Benedetto Castelli had collected Giorgio Coresio's most manifest errors[120], and, again in collaboration with Castelli, drafted and published, in 1615, the *Risposta*
240 *alle opposizioni del S. Lodovico delle Colombe e del S. Vincenzio di Grazia*[121]. The *Risposta* was published without the author's name, but with a dedication to Enea Piccolomini d'Aragona, signed by Benedetto Castelli, in which he referred to the *Risposta* as "mie presenti scritture [my current writings]"[122].

The additions introduced in the second edition of the *Discorso*, the annotations to his opponents' writings, and the *Risposta* all present interesting aspects for Galileo's conception of *momento*. It is precisely on these aspects that we will focus, setting aside the task of examining the substance of the controversy, in which

---

Incognito oppositore al Discorso di Galileo intorno alle cose che stanno in su l'acqua o che in quella si muovono, in "Bullettino di Bibliografia e di Storia delle Scienze Matematiche e Fisiche", XVIII (1885), pp. 321-326). In the *Galileo Studies* cit., Stillman Drake questioned the attribution to d'Elci, suggesting Flaminio Papazzoni, Professor of Philosophy in the Pisan University, as the author of the *Considerazioni* (pp. 170-172). It does not seem to me, however, that Drake's interesting observations can convince us to abandon the traditional attribution to d'Elci.

115 *Operetta intorno al galleggiare de' corpi solidi... di Giorgio Coresio, Lettore della lingua greca nel famosissimo Studio di Pisa*, Appresso Bartolommeo Sermartelli e fratelli, in Firenze 1612 (pubbl. in *G.G.*, IV, pp. 199-244).

116 *Discorso apologetico di Lodovico delle Colombe d'intorno al Discorso di Galileo Galilei... si come d'intorno all'aggiunte fatte dal medesimo Galileo nella Seconda Impressione*, Appresso il Pignoni, in Firenze 1612 (pubbl. in *G.G.*, IV, pp. 313-369).

117 *Considerazioni di M. Vincenzio di Grazia sopra il Discorso di Galileo Galilei....* presso Zanobi Pignoni, in Firenze 1613 (pubbl. in *G.G.*, IV, pp. 373-439).

118 Cf. *supra*, note 83.

119 Galileo's annotations are published by Favaro at their corresponding places in the *Considerazioni* to which they refer.

120 *Errori dei più manifesti commessi da Messer Giorgio Coresio... nella sua Operetta..., raccolti da D. Benedetto Castelli di Brescia* (*G.G.*, IV, pp. 247-285). At the foot of the page, Favaro includes Galileo's handwritten additions and corrections.

121 *Risposta alle opposizioni del S. Lodovico delle Colombe e del S. Vincenzio di Grazia contro al trattato del Sig. Galileo Galilei.... Nella quale si contengono molte considerazioni filosofiche remote dalle vulgate opinioni*, Appresso Cosimo Giunti, in Firenze 1615 (pubbl. in *G.G.*, IV, pp. 451-789). For the part played by Castelli and Galileo respectively, cf. the *Avvertimento* cit., by Antonio Favaro (*ibid.*, pp. 13-14).

122 *Ibid.*, p. 453. The dedication is dated 2 May 1615.

Galileo reaffirmed, without substantial modifications, though introducing some clarifications, the positions he had maintained in the *Discorso*.

Arturo d'Elci was the quickest to react. Along with a large number of objections, he also expressed some doubts about the legitimacy of Galileo's use of the *momento* term:

> Questa voce momento è latina e tolomeida, ma non usata nel preso significato dal volgare nostro moderno e meno dall'antico; poiché nel vocabolario copiosissimo ed esquisitissimo della Crusca non ve n'è esempio. Questo dico, non per attendere alla purità e proprietà della lingua, ma perché qui molto importa alla vera intelligenza e dichiarazione della materia proposta. Ma cosa di maggiore momento si è che l'Autore in questo luogo la forza confessa della resistenza e, poco di sotto, conclude esser manifesta la necessità di comparare insieme la gravità dell'acqua e de' solidi, senza più far menzione della sopradetta resistenza, la quale depende non poco dalla figura del corpo grave postole sopra[123]. [This word momento is Latin and Ptolemaic, but not used in the meaning it takes from our modern vernacular and less so from the ancient one; for there is no example of it in the very copious and exquisite vocabulary of the Crusca. I say this, not to concern the purity and propriety of the language, but because it is very important for the true understanding and declaration of the proposed matter. But what is more important is that the Author here admits to the force of resistance and, a little further on, concludes that it is evident to necessarily compare the heaviness of water and solids together, without further mentioning the aforementioned resistance, which depends significantly on the shape of the heavy body placed above.]

The good Aristotelian appears to be puzzled by *momento*, but he understood that it is a term that performs a central function in the discussion, that is, that "molto importa alla vera intelligenza e dichiarazione della materia [it matters a lot to the true understanding and declaration of the matter]". D'Elci knows that tradition attributes *περὶ ῥοπῶν* to Ptolemy, in other words, *de momentis*, which is lost. We have seen that in some versions of Eutocius's *Commentary* on the *Equilibrium of Planes*, which has been cited several times, reference was made to Ptolemy's *de momentis*[124]. The *Commentary* itself was probably the source for the Accademico Incognito. D'Elci intends to contest that in the vernacular, both modern and ancient, *momento* was ever used in the meaning assigned to it by Galileo. And the proof of the absolute lack of precedents would be provided by the *Dizionario della Crusca* [Crusca Dictionary], freshly printed[125], which did not record this use. In fact, the *Prima impressione* of the famous *Vocabolario* listed under *momento* only the meaning of "momento di 241
tempo [moment of time]" and the phraseological use ("cosa di momento [thing of moment (importance)]"[126]). However, it appears practically impossible to decipher

---

[123] *Ibid.*, p. 159.

[124] Cf. *supra*, Part One, Ch. III, paragr. 1.

[125] It was released in Venice in 1612.

[126] "MOMENTO. Brevissimo spazio di tempo. Latin. *momentum*. Zibald. And. Il punto si è dieci momenta, e 'l momento si è dodici, ecc. Petr. Son. 286. O giorno, o ora, o ultimo momento. E Son 243. In un momento ogni mio ben m'hai tolto. Filoc. lib. 6. 198. In momento riempié li vicini popoli dall'avvenuto male. L'usiamo anche per IMPORTANZA, Cosa di momento [MOMENTO. Very brief period of time. Latin. *Momentum*. Zibald. And. The point is ten momenta, and the momento is twelve, etc. Petr. Son. 286. Either day, or hour, or last momento. And Son. 243. In a momento, you have taken away all my good. Filoc. Lib. 6. 198. In a momento, he filled the neighboring peoples with the evil that had happened. We also use it for IMPORTANCE, thing of moment]" (*Vocabolario degli Accademici*

the meaning of the second part of d'Elci's accusation against Galileo's *momento*.

Galileo succinctly noted this paragraph, firmly denying that he had used *momento* with a meaning uncommon in the vernacular. Unfortunately, added the Pisan scientist, d'Elci immediately loses his bearings as soon as a mathematical term is used:

> Dal vulgare vostro non solamente non è usata questa voce *momento*, ma neanco niun'altra comunissima in tutte le matematiche[127]. [In your vernacular, not only is this term *momento* not used, but neither is any other term commonly found in all of mathematics.]

However, the Accademico Incognito's observation must have alerted him to a slight oversight in the *Discorso*: he had neglected to give an explicit definition of *momento*. He had done so in the *Mecaniche*, but the Paduan writing remained unpublished. It was logical, in a certain sense, that a term and concept like *momento*, truly central – d'Elci had understood this! – should be explicitly defined in the first printed work in which they were used. And Galileo remedied this oversight in the second edition of the *Discorso*, by introducing, immediately after the definitions of heaviness "in specie [i.e., specific]" and "absolute" heaviness, that of *momento*:

> Momento appresso i meccanici significa quella virtù, quella forza, quella efficacia, con la quale il motor muove e il mobile resiste; la qual virtù depende non solo dalla semplice gravità, ma dalla velocità del moto, dalle diverse inclinazioni degli spazii sopra i quali si fa il moto, perché più fa impeto un grave descendente in uno spazio molto declive che in un meno. Ed in somma, qualunque si sia la cagione di tal virtù, ella tuttavia ritien nome di momento. Né mi pareva che questo senso dovesse giugner nuovo nella nostra favella; perché, s'io non erro, mi par che noi assai frequentemente diciamo "Questo è ben negozio grave, ma l'altro è di poco momento" e "Noi consideriamo le cose leggiere, e trapassiamo quelle che son di momento": metafore stimere'io, tolte dalla meccanica[128]. [Momento in mechanics means that virtue, that force, that efficacy with which the motor moves and the mobile resists; which virtue depends not only on simple heaviness, but on the speed of the motion, on the different inclinations of the spaces over which the motion occurs, because a descending heavy body creates more impeto in a very steep space than in a lesser one. And in sum, whatever the cause of this virtue, it still retains the name of *momento*. Nor did it seem to me that this meaning should be new to our language; because, if I am not mistaken, it seems to me that we quite frequently say "This is indeed a serious (i.e., heavy and grave) matter, but the other is of little momento" and "We consider trivial things, and disregard those which are of momento": metaphors, I estimate, taken from mechanics.]

The "mechanics", just to avoid possible misunderstandings, means here, and often elsewhere, the "theoretical" mechanics, authors such as Archimedes, Commandino, Guidobaldo etc., not the "proti [mechanical technicians]" of the
242 Arsenal of Venice[129]. The second Galilean definition of *momento*, although presenting a quite different formulation, is substantially attributable to that of the *Mecaniche*. It will be observed, however, that the "natural propensity" has

della Crusca cit., 1612, p. 537).

[127] *G.G.*, IV, p. 158.

[128] *Ibid.*, p. 68.

[129] Of the hypothesis regarding the derivation of *momento* from the lexicon of practical mechanics supported by Altieri Biagi (*Galileo e la terminologia* cit., p. 48), I have already discussed it in my *A proposito di un errore dei traduttori di Vitruvio* cit. (pp. 87-88), to which I allow myself to refer.

disappeared and that, in general, the term is defined more “technically”. Even the role of heaviness appears blurred and *momento* is presented as the varying effectiveness not of heaviness, but of the mover of the mobile. This depends, in all likelihood, on the desire to give a more general definition, capable of allowing free use of *momento*. The definition in the *Mecaniche* spoke, in fact, of the “propensione naturale di andare al basso [natural propensity to go downwards]”, that is, it bound *momento* to descent. Galileo is preparing to broaden the scope of use for the term, extending it to include violent motions, impact and resistance, which can also involve a departure from the center.

It is also worth emphasizing that the definition in the *Discorso* is tailored to hydrostatic problems. Galileo, in fact, presents *momento* as “quella efficacia, con la quale il motor muove e ‘l mobile resiste [that efficacy with which the mover moves and the mobile resists]”. This comparison adequately expresses the “eccesso della gravità [excess of heaviness]” (the mover) of the bodies “sopra la gravità dell’acqua [over the heaviness of the water]” (the mobile), or vice versa, on which depends, in general, the sinking or emergence of the bodies.

Lastly, the conditions that produce the change in the effectiveness of *momento* are the same as those already indicated in the *Mecaniche*; but speed is clearly favored here, which was only implicit in the original definition. Also, in this we can see the intention to make the definition correspond to the specific use that is made of it in the text, where, in fact, *momento* is always presented as the effect produced by the combination of the weight with its speed (not of the weight with distance, as was predominantly in the *Mecaniche*).

Galileo added some comments to the definition, from which it is evident that the addition was suggested by the desire to escape the accusations of ambiguity directed at him by the Accademico Incognito. In fact, after defining it, Galileo, fearing that the addition could be understood as a tacit acknowledgement of the validity of the accusation made against him by d’Elci, took care to declare his surprise at the accusation of having assigned to *momento*, without warning the reader, a completely new meaning. “Nella nostra favella [In our language]” – Galileo thus asserts – expressions such as “negozio grave [serious (i.e., heavy and grave) business]”, “negozio di poco momento [business of little momento]” and similar are common, which appear to our scientist as “metafore tolte dalla meccanica [metaphors taken from mechanics]”.

We are faced with a very interesting statement. It will be noted, first of all, that Galileo does not consider himself the inventor of *momento*, but indicates, on the contrary, both in the definition and in this final observation, that it is used from “mechanics”. From what we have seen previously, we can conclude that Galileo’s assertion is well-founded. He himself probably derived it in some way from mechanics. Furthermore, by stating that the phraseological use of the term derived from the “mechanical” *momento*, he could refute d’Elci, and affirm that the **243**
“mechanical” *momento* is, at least implicitly, documented by the Vocabolario degli Accademici [Academic Vocabulary] (which recorded the phraseological use).

The analogy between the mechanical and phraseological use and, above all, the metaphorical derivation of the latter from the former raise many concerns. Indeed, to tell the truth, they seem decidedly unsustainable, as examples of phraseological use are recorded that are much older than those of the "mechanical" *momento.*

Despite its evident weakness, the Galilean interpretation had – as an extremely significant fact – enormous success. The *Seconda Impressione* [Second Print] of the *Vocabolario*, released from the presses in 1623, reserves us, in fact, a curious surprise. The entry "Momento", unchanged in the first part, presents an addition:

> Per quella forza, peso o violenza, che acquistano i corpi gravi nel muoversi naturalmente verso il lor centro, termine di meccanica. Onde, per metaf., Cosa di momento, cosa di importanza[130]. [For that force, weight or violence that heavy bodies acquire in moving naturally towards their center, term of mechanics. Hence, metaphorically, A matter of momento, a thing of importance.]

The good Academicians therefore not only recorded in the *Seconda Impressione* the use defined by Galileo in the *Discorso*, but also diligently subscribed to Galileo's proposal of a metaphorical derivation of the phraseological use of *momento* with a "mechanical" value. And the most curious thing is that the *Vocabolario* continued to precisely record this highly questionable derivation until the *Quinta Impressione* [Fifth Print], where, finally, any reference to the connection between the two meanings disappears[131]. It is truly a shame that notes for the *Seconda Impressione* have been lost, as well as a large part of the meeting minutes of the Academics. It is thus not known whether it was Galileo himself – as he was in fact an Academician – who suggested the addition, whether he had some friends propose it, or, more simply, whether the *Discorso* was among the works noted for the *Seconda Impressione.*

In Giorgio Coresio's *Operetta*, there are no observations relating to the use and meaning of *momento* that deserve examination. Lodovico delle Colombe, unlike the Accademico Incognito, did not dispute that the use of the term was legitimate, preferring to demonstrate that the laws of *momenti* did not provide an adequate explanation of the behavior of bodies in water. Delle Colombe, writing after the publication of the second edition of the *Discorso*, maliciously insinuated, however,
244 that the modifications introduced were proof of an unclear foundation. The reference to the addition of the definition of *momento* was evident:

> poi che vi siete messo per tal conto a ristamparlo e a levare e aggiugnere e dichiararvi, non vi essendo inteso, e per mutare in parte parere, come prudente, e parte per lasciarvi intendere, senza esserne da li avversari richiesto[132]. [then you have undertaken to reprint it in such a way and to remove and add and clarify yourself, not being understood, and partly to change your opinion, as prudent, and partly to make yourself understood, without being requested by your adversaries.]

Galileo dedicated a large part of the *Risposta* to the refutation of the unstable

---

[130] *Vocabolario... In questa seconda impressione riveduta e ampliata*, Iacopo Sarzina, Venezia 1623, p. 526.

[131] The *Terza* (Firenze 1691) and the *Quarta* (Firenze 1729-38) *Impressione* maintain, in fact, the metaphorical derivation of the phraseological use from the mechanical value. The two values are, however, presented as independent in the *Quinta Impressione* (Firenze 1910).

[132] *G.G.*, IV, p. 345.

edifice erected by Lodovico delle Colombe, demolishing with great ease the arguments on which it was based. As for the polemical jab regarding the additions introduced in the second edition of the *Discorso*, Galileo limited himself to observing that "dove niente niente si tocca qualche termine di geometria o si tratta alcuna dimostrazione con metodo matematico [where even the slightest term of geometry is touched upon or any demonstration is treated with a mathematical method]", delle Colombe would irremediably lose his bearings:

e quand'io credeva che egli, con l'occasione d'aver in mille propositi sentite replicar le medesime cose, dovesse aver capite almeno le definizion de' nomi, e intendesse quello che importa gravità assoluta, più o men grave in specie, quello che significhi momento e molt'altri termini dichiarati e usati dal Sig. Galileo, io mi trovo fortemente ingannato[133]. [and when I believed that, with the opportunity of having heard the same things repeated in a thousand arguments, he must have understood at least the definitions of the names, and understood what is meant by absolute heaviness, more or less heaviness in species, what momento meant and many other terms declared and used by Signor Galileo, I find myself greatly mistaken.]

Lastly, Galileo resolutely denied having reprinted the *Discorso* with the intention of making it clearer and more coherent. Nothing, in fact, has been deleted, compared to the first edition, and as regards the addition of the definition of *momento*:

Che egli si sia dichiarato per non si esser da sé medesimo inteso, come voi dite, è tanto falso, quanto è vero che voi non avete inteso delle venti parti una del suo trattato, benché egli molto bene si sia dichiarato[134]. [That he has declared that he did not himself understand it, as you say, is as false as it is true that you did not understand one of the twenty parts of his treatise, even though he declared himself very well.]

Even the last of the interlocutors to enter the fray, Vincenzo di Grazia, had something to object about the use of *momento* in the *Discorso* and about the definition introduced in the second edition. Di Grazia primarily contests the reduction of all movements to the effects of heaviness; that is, the exclusion of positive lightness as the cause of motion. Positive lightness – insisted the good Aristotelian professor – instead causes a natural inclination opposite to heaviness, as can also be proven with the authority of mathematicians:

La qual cosa non solo confronta con la natural filosofia, ma anche con le matematiche discipline,
quantunque repugni ad Archimede, quindi a poco vedremo se a ragione o a torto. Per ora, oltre al detto, **245**
siami lecito contro a un grandissimo matematico, qual fu Archimede, addurre l'autorità d'un più grande: e questi è l'ammirabile Tolommeo, nel libro che egli scrisse de' momenti, referito da Eutocio comentator del vostro Archimede; il qual libro, se per la voracità del tempo non si desiderasse, non solo per autorità servirebbe, ma ancora ragion gravissime e degne di Tolomeo in esso si scorgerebbono. Dice dunque Tolommeo che il genere del momento e dell'inclinazione alla gravità e alla leggerezza si estende; il che da noi con vivaci ragioni è stato provato[135]. [Which not only aligns with natural philosophy, but also with the mathematical disciplines, although it is repugnant to Archimedes, we will soon see whether rightly or wrongly. For now, other than what has been said, allow me to counter a great mathematician,

[133] *Ibid.*, p. 599.
[134] *Ibid.*, p. 603.
[135] *Ibid.*, p. 385.

Archimedes, by adducing the authority of an even greater one: and that is of the admirable Ptolemy, in the book he wrote On Momenti, referred to by Eutocius, commentator of your Archimedes; which book, if it were not lacking due to the voracity of the time, would serve not only as authority, but also for Ptolemy's very compelling and worthy arguments discerned in it. Ptolemy therefore says that the genus of momento and inclination extends to heaviness and lightness; which has been proven by us with vigorous arguments.]

Galileo distorted the original meaning of *momento*, using it to indicate only the inclination towards the center. The source of di Grazia, as in the case of the Accademico Incognito, is Eutocius's *Commentary*, in which the Archimedean position ("momentum omne gravitate produci [every momentum is produced by heaviness]") is associated to that of Plato and opposed to that of Aristotle and Ptolemy, who considered "momentum commune gravitatis et levitatis esse genus [momentum is a common genus of both heaviness and lightness]".

The same accusation is repeated a little further on, this time with explicit reference to the definition of *momento* added in the second edition of the *Discorso*:

Quanto alla nuova aggiunta, nella quale il Sig. Galileo dichiara quello che significhi *momento*, pare sia alquanto manchevole, non ci numerando una significazione ch' al suo discorso faceva di mestiero: e questa è che 'l momento denota quella potenzia e quella abilità naturale che hanno i mobili a esser mossi, sì come la gravità e la leggerezza al moto degli elementi; la qual significazione non solo è in uso appresso Aristotile e Platone, ma appresso i vostri meccanici, conciossiacosaché Eutocio ne' comenti de' libri d'Archimede si serva di questa significazione, dicendo: Il genere del momento, Aristotile e Tolomeo, che l'ha seguitato, dicono che non solo s'appartiene alla gravità (come vuol Platone), ma alla leggerezza ancora[136]. [As for the new addition, in which Signor Galileo declares what *momento* means, it seems to be somewhat lacking, as it does not include a meaning that was necessary for his discourse: and this is that momento denotes that power and that natural ability that mobiles have to be moved, just as heaviness and lightness do to the motion of the elements; which meaning is not only in use among Aristotle and Plato but also among your mechanicians, since Eutocius makes use of this meaning in the commentaries on the books of Archimedes, saying: The genus of momento, Aristotle and Ptolemy, who followed him, say that it not only belongs to heaviness (as Plato wanted), but also to lightness.]

Even to these criticisms, which were not particularly challenging, Galileo responded promptly. Using arguments substantially identical to those elaborated in the revision of the *De motu*, he denied that positive lightness, i.e., a *momentum levium*, could exist[137]. As for the accusation of having provided an incomplete definition of the term, omitting a meaning that would have been extremely important for the use of *momento* in the *Discorso*, Galileo's terse reply introduces – perhaps for the first time – his theory about the nature and value of the terms.

Dove io non dirò che il Sig. Grazia, o chi si sia altri, non possa chiamar momento tutto quello che piace a lui, essendo i nomi in arbitrio di ciascheduno; ma dirò bene che grande sproposito sarebbe stato
246 del Sig. Galileo il definirlo in tal maniera in questo luogo, non gli dovendo poi venir occasione di usarlo nel suo Discorso: e il Sig. Grazia, che per tal rispetto biasima il Sig. Galileo, doveva mostrar i luoghi particolari ne' quali momento venga usurpato in questo senso non definito dal Sig. Galileo; altramente

136 *Ibid.*, pp. 387-388.
137 *Ibid.*, pp. 704-705.

lo sproposito sarà tutto suo[138]. [Where I will not say that Signor Grazia, or anyone else, cannot call momento whatever he pleases, since names are at the discretion of each person; but I will certainly say it would have been a great mistake for Signor Galileo to define it in this way in this place, not then having any occasion to use it in his *Discorso*: and Signor Grazia, who criticizes Signor Galileo for this respect, ought to show the particular places in which *momento* is used in this sense not defined by Signor Galileo; otherwise the mistake will be all his.]

For the "imposizioni di nomi [impositions of names]" in Galileo, one can refer to the elegant and effective pages of Altieri Biagi, which clarify our scientist's attitude towards the relationship between language and reality[139]. What we are interested in emphasizing here is that this statement shows how the "neutralisation" process of *momento* has progressed considerably. Di Grazia's accusation tended to attack, in fact, the absence of the "propensione naturale [natural propensity]" in the definition, which was instead present – as we have seen – in the definition of the *Mecaniche*. Galileo admits that he could have defined it this way or in a thousand other ways, given that the terms do not necessarily pertain to things, to phenomena, and therefore are "in arbitrio di ciascheduno [at the discretion of each person]". Like his "matematici [mathematicians]", Galileo believes he has safeguarded himself by declaring the meaning of the terms.

Without wanting to diminish the scope and value of Galileo's "conventionalistic" attitude, it will be worth remembering that the "imposition" of *momento* in the *Mecaniche* was suggested – as we have previously emphasized – particularly from the more common traditional value of the term, the exclusion of which from Galileo's definition led di Grazia to protest. *Momento* has, however, now become, in Galileo's hands, a more flexible instrument, having lost many of the characteristics that initially made it usable only to indicate the tendency to move towards the center. It should be observed that, in the *Discorso*, Galileo contrasts the *momento* of the solid that presses on the surface of the water with the *momento* of the resistance of the water[140]. Certainly, in the years of the *Mecaniche*, our scientist would not have indicated this second force with the term *momento*, which is not an "istinto [instinct]", a "conato [striving]" towards the center.

---

138 *Ibid*., p. 707.

139 *Galileo e la terminologia* cit., Ch. III (*Le "imposizioni di nomi" di Galileo*), pp. 44 ff.

140 Cf., for example, *G.G.*, IV, p. 70.

3. The Momento of Resistance

From the discussions of 1611-1615 on hydrostatics, with a leap of over twenty years, even more daring than the previous one, we will move on to the examination of the *Giornata Seconda* [Second Day] of the *Discorsi e Dimostrazioni matematiche*, in which another significant field of application of the concept and term of *momento* is found, in the "mechanical" sense.

247 The *Giornata Seconda* presents the foundations of one of the two "nuove scienze [new sciences]" promised from the title of the *Discorsi*: the science of the resistance of solids to breaking. As can be seen from a letter from Galileo, dated 1609, to Don Antonio de' Medici, the demonstrations and discoveries published in 1638 dated back to many years earlier and were strictly dependent on the principles set out in the *Mecaniche*:

dopo il mio ritorno di Firenze sono stato occupato in alcune contemplazioni et in diverse esperienze attenenti al mio trattato delle meccaniche; nel qual ho speranza che la maggior parte saranno cose nuove, né da altri state tocche per addietro. Et pure ultimamente ho finito di ritrovare tutte le conclusioni, con le sue dimostrazioni, attenenti alle forze et resistenze de i legni di diverse lunghezze, grossezze et figure, et quanto sian più debili nel mezo che negli estremi, et quanto maggior peso sosterranno se quello sarà distribuito per tutto il legno che in un sol luogo, et qual figura doveria havere acciò fusse per tutto egualmente gagliardo[141]. [after my return from Florence, I have been occupied in some contemplations and in various experiments relating to my treatise on mechanics; in which I hope that the majority will be new things, not previously touched upon by others. And also recently I finished finding all the conclusions, with their demonstrations, relating to the forces and resistances of wood of different lengths, thicknesses and shapes, and how much weaker they are at the middle than at the extremes, and how much greater weight they will support if it is distributed throughout the wood than at a single place, and what shape it should have so that it is equally strong throughout.]

The problem of the resistance of solids had been posed in the *Giornata Prima* [First Day] of the *Discorsi*[142], but the digressions on the composition of the continuum and the wheel paradox had prevented it from being addressed. However, previously in the *Giornata Prima*, Galileo had specified the reasons why it is difficult to break material into minute parts, identifying them as two main phenomena:

l'uno de i quali è quella decantata repugnanza, che ha la natura all'ammettere il vacuo; per l'altro bisogna (non bastando questo del vacuo) introdur qualche glutine, visco o colla, che tenacemente colleghi le particole, delle quali esso corpo è composto[143]. [one of which is the vaunted repugnance that nature has to admit the vacuum; on the other hand, it is necessary (since this vacuum is not enough) to introduce

---

[141] *G.G.*, X, p. 229. Galileo returned to his interest in *de resistentia solidorum* during his forced exile in Siena, after the prohibition of the *Dialogo* (summer 1633), as demonstrated by the numerous letters on this subject that he exchanged with the Arrighetti, Aggiunti, Dino Peri and with others (cf. *G.G.*, XV, *passim*).
For matters relating to the printing of the *Discorsi* and for "textual" questions, please refer to Favaro's *Avvertimento* (*G.G.*, VIII, pp. 12-38).

[142] Cf. *ibid.*, pp. 50 ff.

[143] *Ibid.*, p. 59.

some glue, gum or paste, which tenaciously connects the particles of which the body is composed.]

The analysis and demonstrations proposed by the Pisan scientist, which examines the resistance of "geometric" bodies (cylindrical and prismatic solids), presents very close analogies with the procedures used in the *Mecaniche*, confirming the assertions contained in the aforementioned letter to Don Antonio de' Medici.

The discussion in the *Giornata Seconda* [Second Day] opens with the usual proposal of a general principle, valid for the lever: "la forza alla resistenza ha la proporzione contraria di quella che hanno le distanze tra 'l sostegno e le medesime **248**
forza e resistenza [the force to the resistance has the inverse proportion to that of the distances between the fulcrum and the same force and resistance]"[144]. Despite that Salviati explicitly recalls that this principle "nelle mecaniche si dimostra [is demonstrated in mechanics]", Sagredo invites him to repeat the demonstration, "già che questo principio è il fondamento di tutto quello che voi avete intenzione di volerci mostrare [since this principle is the foundation of everything that you intend to show us]"[145]. Sagredo's request is considered legitimate and worthy of being satisfied. Salviati therefore proposes the demonstration[146], which precisely corresponds to the "Archimedean" one of the *Mecaniche*, grounded in a strictly geometric approach.

Both in the statement of the principle (where distances from the fulcrum are mentioned, not of speed) and in the demonstration, there is no trace of the method of "virtual speeds", which Galileo employed in the *Mecaniche*, in a secondary approach[147], and in the *Discorso*, as a preferred procedure[148]. It cannot but arouse surprise that Galileo returned to the "Archimedean" demonstration, preferred in the *Mecaniche*, after having considered, on more than one indication in the entire series of writings dedicated to hydrostatics, the dynamic approach suggested by the pseudo-Aristotelian *Questions of Mechanics* as more effective and fruitful[149].

---

[144] *Ibid.*, p. 152.

[145] *Ibid.*

[146] *Ibid.*, pp. 152-153.

[147] Cf. *supra*, pp. 208-209. It should be recalled that Galileo introduced the theorem of "virtual speeds" in the *Mecaniche*, not as a true and proper demonstration, but rather as a "congruenza e probabilità [congruence and probability]".

[148] Cf. *supra*, pp. 232-233.

[149] Just as Salviati is about to propose the "Archimedean" demonstration of the law of *momenti* of the balance, Simplicio claims priority for Aristotle in the discovery of these laws: "Questo fu dimostrato da Aristotele, nelle sue Mecaniche, prima che d'ogni altro [This was demonstrated by Aristotle, in his Mechanics, before anyone else]". From Salviati's reply to this remark of the Aristotelian, it appears evident that Galileo, while recognizing the priority of the Stagirite (the Pisan scientist does not doubt that the *Questions of Mechanics* are his work), intends to specify that the Aristotelian demonstration is much less convincing than that proposed by the Syracusan scientist: "Voglio che gli concediamo il primato nel tempo; ma nella fermezza della dimostrazione parmi che se gli deva per grand'intervallo anteporre Archimede, da una sola proposizione del quale... dependono le ragioni non solo della leva, ma della maggior parte de gli altri strumenti meccanici [I want us to give him priority with regard to time; but in the firmness of the demonstration, it seems to me that Archimedes should be placed far before him, upon a single proposition of which... depend the arguments not only for the lever, but the majority of other mechanical instruments]" (*G.G.*, VIII, p. 152).

This return to a conception of *momento* as weight × distance (instead of weight × speed) deserves some consideration and clarification.

Before dedicating ourselves to this task, it is worth emphasizing that, to tell the truth, a brief reference to the principle of "virtual speeds" is found on a very important page of the *Giornata Seconda* of the *Discorsi*. Galileo intends to demonstrate what the relationship is between resistance to breaking (at the base of the prism) and the pulling force (at the other end) in a prismatic solid fixed in the wall and parallel to the horizon. The assumption premised for the demonstration,
249 which constitutes the "prima proposizione [first proposition]" of the treatise on the resistance of solids, is that "il momento della forza posta in C al momento della resistenza, che sta nella grossezza del prisma, cioè nell'attaccamento alla base BA con la sua contigua, ha la medesima proporzione che la lunghezza CB alla metà della BA [the momento of the force applied at C (cf. fig. 11) to the momento of the resistance, which lies in the thickness of the prism, that is, in the attachment to the base BA with its contiguous parts, has the same proportion as the length CB to half of the BA]"[150].

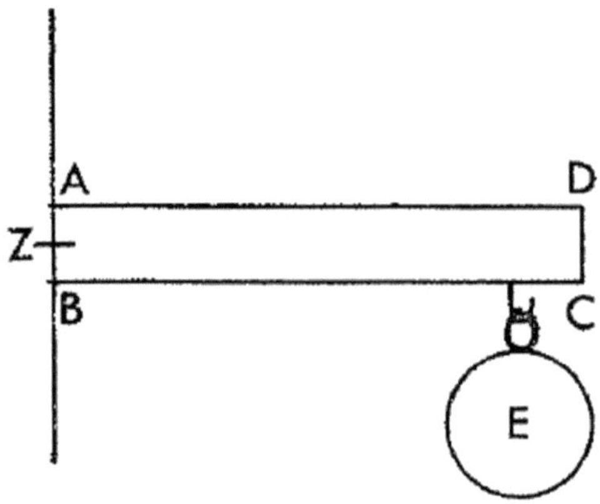

Figure 11

In other words, Galileo intends to state that the relationship between force and resistance is indicated by the ratio between the distances of the force and resistance from the fulcrum of the system. The example of the prismatic solid is, therefore, assimilated to that of the lever, in which the *momenti* of the weights applied to the ends are – as Galileo demonstrated in the *Mecaniche* – exactly proportional to the distances from the fulcrum. It follows that the solid can be broken more easily the more "la sua lunghezza eccederà la sua grossezza [its length exceeds its thickness]"[151].

---

150 *Ibid.*, p. 156.
151 *Ibid.*

To clarify further, the model of the solid AC, fixed in the wall, is reduced by Galileo to that of a lever BCA, with the fulcrum at C (cf. fig. 12), which is, in turn, equivalent to the lever with arms at right angles ZCA[152].

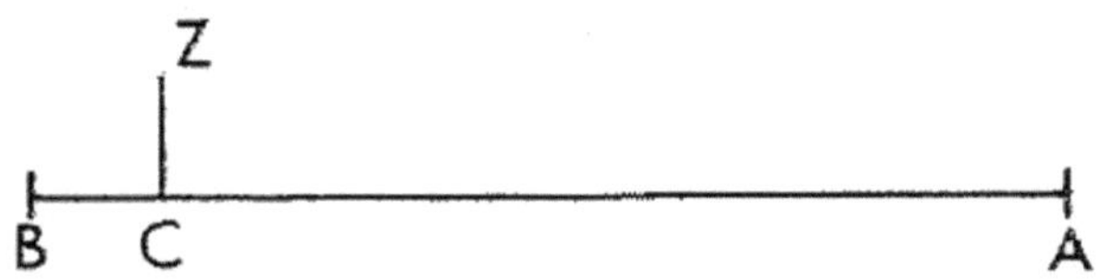

Figure 12

To further confirm the legitimacy of the assumption, Galileo reproposes the method of explanation used in the *Mecaniche* to determine the relationships between the different forces necessary to raise a weight along planes of various inclinations. It should be recalled that, in the Paduan treatise, Galileo had spoken of the "total" *momento* of the heavy body along the perpendicular, and, therefore, of the "maximum" force to make it rise vertically. Between the "total" *momento* and the "null" (when the heavy body lies on a plane perfectly parallel to the horizon), there are infinite "partial" *momenti.* Now the Pisan scientist explains that, even in the case of the beam fixed in the wall, there are "partial" or "absolute" resistances, **250**

[152] This is, essentially, the explanation that Galileo provided to Baliani, who had expressed the need for some clarification on this manner of determining the relationship between strength and resistance, in a letter to the Pisan scientist dated 1 July 1639 ("desidererei che V.S. havesse un tantino più dichiarato alla proposizione prima, che il momento della forza in C al momento della resistenza è come CD alla metà di DA [I would wish that your V.S. (Vostra Signoria, i.e., Your Excellency) would have stated a little more in the first proposition, that the momento of the force in C to the momento of resistance is as CD to half of DA]"; *G.G.*, XVIII, p. 70). Galileo's response, contained in the letter to Baliani dated 1 August of the same year, invited, in fact, to attribute the case of the cylindrical or prismatic beam fixed into the wall to the lever model: "nella quale operatione casca il medesimo accidente che avviene nel vette AB, il cui sostegno sia in C, dove posti nella minore distanza CB quantisivogliano pesi eguali, pendenti da distanze eguali, faranno la medesima resistenza alla forza messa in A, che se tutti i detti pesi ridotti in un solo, pendessero dal mezzo di BC [in such operation the same event occurs as what happens in the lever AB, whose fulcrum is at C, where placed at the shorter distance CB any desired number of equal weights, hanging from equal distances, will make the same resistance to the force placed at A, as if all the aforementioned weights, reduced to one, were hung from the middle of BC]" (*ibid.*, p. 77). Shortly thereafter (letter of 19 August), Baliani returned to specify that he did not intend to question the truth of the principle, but only to ask Galileo "se, per quanto sia vero, si dovessi domandarlo in una petizione [if, however true it is, it should be asked in a petition]" (*ibid.*, p. 87). Galileo responded promptly to Baliani's clarification (letter of 1 September), stating that he considered his own principle "come cosa per se stessa assai chiara [as something very clear in itself]", having been demonstrated, not only by Archimedes, but also by other mechanicians. The Pisan scientist then specified that the fact that the distance from the fulcrum of the resistance is perpendicular to that of the force does not alter the validity of the principle: "che poi, piegata ad angoli retti la minor distanza sopra la maggiore, trovi la forza il medesimo contrasto dalla resistenza, non mi pare che deva essere messo in dubbio, e tanto più che, se bene ho in memoria, credo che il Sig.r Guidobaldo nelle sue Mechaniche ponga questa medesima conclusione e che la dichiari assai a bastanza [that then, when the smaller distance is bent at right angles over the larger one, the force finds the same opposition from the resistance, does not seem to me to be in doubt, and all the more so since, if I recall correctly, I believe that Signor Guidobaldo in his Mechanics poses this same conclusion and states it very sufficiently]" (*ibid.*, pp. 94-95).

depending on the inclination of the applied force. As for "assoluta resistenza [absolute resistance]",

è quella che si fa per tirarlo per diritto, perché allora tanto è il moto del movente quanto quello del mosso[153]. [it is what is done to pull it straight, because then the motion of the mover is as much as that of the moved.]

Whoever attempts to tear the beam from the wall by applying a "per diritto [straight]" force would find that the force makes a "viaggio [journey]"[154] equal to that of the resistance. Galileo wants to say that, in this way, no advantage would be obtained, because, in an equal time, the force would move through a space equal to the resistance, that is, with the same speed.

The rapid, almost casual Galilean remark[155] constitutes a transparent and unequivocal reference to the principle of "virtual speeds". Galileo's reasoning,
251 to show the "partial" resistance of our solid to being torn, when the force is not applied "straight", but rather (see fig. 13) perpendicularly, consists in showing that, in case of breaking, the force makes, at the same time, a "viaggio [journey]" which stands to the "viaggio [journey]" of the resistance as the CC' arc is to the AA' arc, that is, it moves much faster.

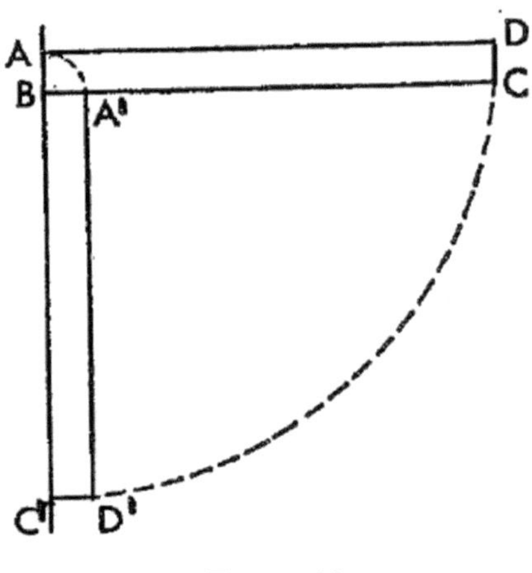

Figure 13

Galileo, however, avoids proposing such a demonstration, limiting himself to the quick and cryptic hint mentioned above.

Why, then, did Galileo explicitly avoid resorting to the method of proving the *momenti* of the lever offered by the theorem of "virtual speeds", even though he let this obvious hint slip through?

153 *G.G.*, VIII, p. 157.
154 It is the characteristic term used by Galileo in the *Mecaniche* to express the displacement of weight and force (cf. *G.G.*, II, p. 167).
155 The reference to the "virtual speed" theorem is inserted in parentheses and Galileo deliberately avoids developing it.

It will be observed, first of all, that, in the *Discorsi*, the theorem of "virtual speeds" is never used explicitly, whereas it is still used in the *Dialogo*[156]. Several important documents allow us to establish that Galileo was induced to abandon the dynamic approach due to extremely significant and precise reasons and difficulties, returning to the rigorously static method of proof of the laws of *momenti* of the balance.

In a page of the unpublished *Scene*[157] by Antonio Nardi, friend and correspondent of Galileo and member of the so-called Roman "triumvirate"[158], there is preserved an explicit declaration of the unsatisfactory use of the principle of "virtual speeds" to demonstrate the laws of *momenti* of the balance of unequal arms. According to Nardi, in fact,

> male si persuadono i Meccanici comunemente compensarsi in una bilancia di diseguali braccia, la velocità
> del moto con la grandezza del momento, onde cercano di render ragione perché questi pesi diseguali da **252**
> distanze reciprocamente diseguali pesino egualmente. Ma ciò non è in vero cagione dell'equilibrio, perché così discorrendo s'adduce di un effetto in atto una cagione in potenza[159]. [Mechanicians commonly persuade themselves poorly that the speed of motion is compensated by the magnitude of the momento in a scale of unequal arms, so it tries to explain why these unequal weights at reciprocally unequal distances weigh equally. But this is not truly the cause of equilibrium, because by reasoning in this way, a potential cause is adduced for an actual effect.]

That Nardi's objection was known to Galileo, and that our scientist considered it seriously, ultimately convincing himself of its validity, is demonstrated by one of the *Frammenti* relating to the *Discorsi*. The fragment, to which we intend to refer (there has been no shortage of controversy regarding its attribution to Galileo)[160],

---

[156] In the *Giornata Seconda* of the *Dialogo*, Salviati, to prove that "la velocità del mobile meno grave compensa la gravità del mobile più grave e meno veloce [the speed of the less heavy mobile compensates for the heaviness of the heavier and less fast mobile]", proposes the experiment "della stadera, nella quale voi vedrete il poco pesante romano allora poter sostenere ed equilibrare la gravissima balla, quando la sua lontananza dal centro, sopra il quale si sostiene e volgesi la stadera, sarà tanto maggiore dell'altra minor distanza dalla quale pende la balla, quanto il peso assoluto della balla è maggior di quel del romano. E di questo... altro non si vede poterne esser cagione che la disparità de i movimenti che e quella e questo far dovrebbero.... Il muoversi poi lo spazio di cento dita il romano, nel tempo che la balla si muove per un sol dito, è l'istesso che dire esser la velocità del moto del romano cento volte maggior della velocità del moto della balla [of the steelyard, in which you will see the little Roman weight [i.e., counterweight] then able to sustain and equilibrate the very heavy bale, when its distance from the center, above which the steelyard is supported and turns, will be as much greater than the other shorter distance from which the bale hangs, as the absolute weight of the bale is greater than that of the Roman. And of this... we cannot see anything else being the cause other than the disparity of the movements that this one and that one should make.... The Roman then moves the space of a hundred fingers, in the time that the bale moves by one finger, it is the same as saying that the speed of the motion of the Roman is a hundred times greater than the speed of the motion of the bale]" (*G.G.*, VII, pp. 241-242).

[157] The *Scene* by the Aretine, Antonio Nardi, are preserved in the Ms. Galileiano 130 of the B.N.F. It is a codex of over 1300 pages, which has remained unpublished until today, despite being a work of undeniable value. On Nardi, see Capone Braga, *Un filosofo dell'estremo Rinascimento* (*Antonio Nardi*), "Atti e Memorie della R. Accademia Petrarca di Sc., Lett. e Arti in Arezzo", n.s., V, fasc. 1, Arezzo 1925, pp. 36-135.

[158] The other "triumvirs" were Magiotti and Torricelli.

[159] Ms. Galil. 130, c. 661.

[160] It is published by Favaro in vol. VIII of the Ed. Naz. (pp. 438-440). Favaro attributes it without

handwritten by Vincenzo Viviani, is cast in dialogue form with the interlocutors Sagredo and Salviati. This suggests that Galileo considered including it in a possible second edition of the *Discorsi*, probably in the pages dedicated to the resistance of the materials that we are analyzing. The problem is introduced by Sagredo, who raises the same doubt that we have seen proposed by Nardi.

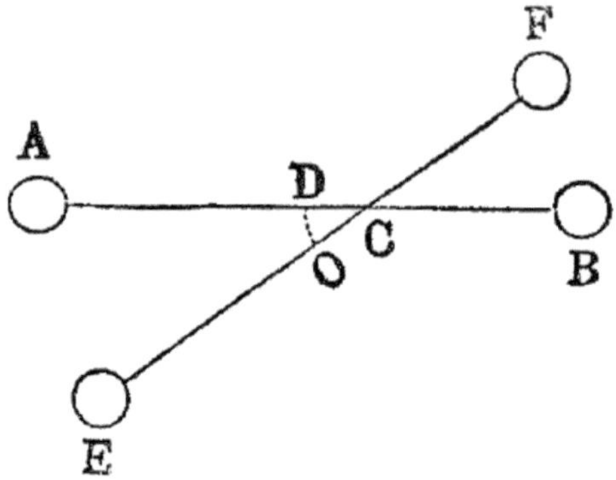

Figure 14

Sia sostenuta nel punto C la libra di braccia diseguali, AC maggiore, CB minore: cercarsi la cagione onde avvenga che, posti nell'estremità due pesi eguali A, B, la libra non resti in quiete ed equilibrio, ma inclini dalla parte del braccio maggiore, trasferendosi come in EF. La ragione che comunemente se ne assegna è perché la velocità del peso A, nello scendere, sarebbe maggiore della velocità del peso B, per essere la distanza CA maggiore della CB; onde il mobile A, quanto al peso eguale al B, lo supera quanto al momento della velocità, e però gli prevale, e scende sollevando l'altro. Dubitasi circa il valore
253 di tal ragione, la quale pare che non abbi forza di concludere: perché è ben vero che il momento di un grave si accresce, congiunto con velocità, sopra il momento di un grave eguale che sia costituito in quiete; ma che posti amendue in quiete..., quella maggioranza che non è, ma ancora ha da essere, possa produrre un effetto presente, ha qualche durezza nel potersi apprendere, ed io veramente ci sento difficoltà notabile[161]. [Let the scales of unequal arms, AC major, CB minor, be supported at point C (see fig. 14): to seek the cause of why, when two equal weights A, B are placed at the extremity, the scales

hesitation to Galileo, despite it being handwritten by Vincenzo Viviani, based on a note at the head of the fragment, also by Viviani, which reads: "Di questo ho l'originale [I have the original of this]". Favaro himself, in the footnote, correctly reports the existence, in Ms. Galil. 245 of the B.N.F. (cc. 18*r*-19*v*; "Raccolta di esperienze senz'ordine e di pensieri diversi di me Vincenzio Viviani, in diversi propositi sovvenutimi, intorno a materie meccaniche, fisiche, astronomiche, filosofiche ed altro [Collection of experiences without order and different thoughts by me, Vincenzio Viviani, on different subjects that occurred to me, around mechanical, physical, astronomical, philosophical and others]") of a note, handwritten by Viviani, in which he claims to be the inventor of the method of proving the law of *momenti* of the balance, contained in the dialogued fragment and to have shown it to a "Sig. Simplicio, Patrizio Anconitano [Signor Simplicio, Patrician of Ancona]" (Carlo Rinaldini?), in the presence of Cosimo Galilei. In this account, Viviani used practically the same words as in the dialogued fragment. According to Caverni, this note by Viviani demonstrates that the handwritten fragment is not by Galileo, but by Viviani, who should therefore be given the credit. Viviani himself would have tried, according to Caverni (whose preconceived anti-Galilean sentiments need no reminder), to hide his own discovery of the theorem, out of "zelo di salvare l'onore e la gloria del suo adorato Maestro [zeal to save the honor and glory of his adored Master]" (*Storia del metodo sperimentale* cit., IV, p. 167).

[161] *G.G.*, VIII, p. 438.

do not remain at rest and equilibrium, but incline towards the larger arm, transferring themselves like in EF. The reason commonly given is because the speed of weight A, in descending, would be greater than the speed of weight B, since the distance CA is greater than CB; therefore the mobile A, equal in weight to B, surpasses it in terms of momento of speed, and therefore prevails over it, and descends, lifting the other. There are doubts about the validity of this reason, which seems to have no strength to conclude: because it is very true that the momento of a heavy body increases, combined with speed, over the momento of an equal heavy body that is at rest; but when placed both at rest..., that excess, which does not yet exist but is still to come, that it can produce a present effect, is somewhat difficult to grasp, and I truly feel notable difficulty in it.]

In replying, Salviati acknowledges the validity of the doubt and states that he has devised a very simple way to bypass the difficulty, confirming the same principle without resorting to the questionable theorem of "virtual speeds". It involves considering the balance as a single heavy body which, when free, will move downwards, maintaining its center of heaviness along the vertical:

ma se voi a cotal moto opporrete un intoppo sotto il centro B, il moto si fermerà, restando la libra con i suoi pesi in equilibrio; ma se l'intoppo si metterà fuor del centro B, come sarebbe in C, tal intoppo non fermerà la bilancia, ma devierà il centro B dalla perpendicolare per la quale camminava, e lo farà scendere per l'arco DO[162]. [but if you oppose such a motion with an obstacle beneath the center of B, the motion will stop, leaving the scales with its weights in equilibrium; but if the obstacle is placed outside the center B, such as it would be at C, this obstacle will not stop the balance, but will deviate the center of B from the vertical along which it progressed, and will make it descend along the arc DO.]

In this way, one could return to explicitly declaring the effect of speed on *momento*.

We cannot assign a precise date to this fragment, but it is certain that it was conceived after the publication of the *Discorsi* (the dialogued form itself demonstrates this)[163]. In the years between the publication of the *Dialogo* and that of the *Discorsi*, Galileo had thus become convinced of the weakness of the principle of "virtual speeds" as a method of proving the laws of *momenti*. Unable to provide a more effective foundation, he decided to relinquish its use in the *Discorsi*, striving, contemporaneously, to devise a different demonstration. However, convinced that the theorem of "virtual speeds" allusively explained the variation of the *momento* of the weight well, Galileo resorted to it exceptionally – and without stating it explicitly – in the first proposition of the *Giornata Seconda*, to clarify how the **254**
resistance to the breaking of the prismatic solid varies according to the inclination along which the force is applied.

---

162 *G.G.*, VIII, p. 439.

163 It is probable that Galileo arrived at the new demonstration as a consequence of the solicitations received from the exchange of letters with Nardi and Torricelli, in the summer of 1641, regarding the necessity to better establish the supposition premised on the *De motu naturaliter accelerato*. See Nardi's letters to Galileo of 10 August and 7 September 1641 (*G.G.*, XVIII, pp. 342 and 350), by Torricelli to Galileo on 17 August 1641 (*ibid.*, p. 345) and of Galileo to Torricelli on 27 September 1641 (*ibid.*, XVIII, p. 359). The "mechanical" demonstration that Galileo planned to add to confirm the supposition, and of which he prepared a draft dialogue (see it in *G.G.*, VIII, pp. 214-219), essentially repeated a speculation of the *Mecaniche* in which he had used the principle of "virtual speeds".

Having clarified the reasons that led Galileo to avoid resorting to the theorem of "virtual speeds", we can now examine other interesting aspects of the *Giornata Seconda* of the *Discorsi*.

Following the declaration of the mechanical principle of the lever, Galileo precedes the actual demonstrations with an important remark. In fact, it is a proper declaration of terms, which concerns, once again, but in a new way, our *momento*:

avanti che passiamo più oltre, devo metter in considerazione come queste forze, resistenze, momenti, figure, etc., si possono considerar in astratto e separate dalla materia, ed anco in concreto e congiunte con la materia; ed in questo modo quelli accidenti che converranno alle figure considerate come immateriali, riceveranno alcune modificazioni mentre gli aggiugneremo la materia, ed in consequenza la gravità[164]. [as we proceed further, I must take into consideration how these forces, resistances, momenti, figures, etc., can be considered abstractly and separated from matter, and also concretely and joined with matter; and in this way, those accidents, that will apply to the figures considered as immaterial, will receive some modifications when we add matter, and consequently heaviness, to them.]

Let us consider, for example, the case of a lever AB, with a fulcrum at C (see fig. 15), which operates to lift the body D.

Figure 15

It is evident, from the above principle of the lever, that the force at B will balance the resistance of the weight D, "se il suo momento al momento di esso D abbia la medesima proporzione che ha la distanza AC alla distanza CB [if its momento to the momento of D has the same proportion that the distance AC has to the distance CB]"[165]. This rule will be considered valid when not taking

in considerazione altri momenti che quelli della semplice forza in B e della resistenza in D, quasi che l'istessa leva fusse immateriale e senza gravità: ma se noi metteremo in conto la gravità ancora dello
255 strumento stesso della leva, la quale sarà talor di legno e tal volta anco di ferro, è manifesto che, alla forza in B aggiunto il peso della leva, altererà la proporzione, la quale converrà pronunziare sotto altri termini. E però, prima che passar più oltre, è necessario che noi convenghiamo in por distinzione tra queste due maniere di considerare, chiamando un *prendere assolutamente* quello quando intenderemo lo strumento preso in astratto, cioè separato dalla gravità della propria materia; ma congiugnendo con le figure semplici ed assolute la materia, con la gravità ancora, nomineremo le figure congiunte con la

[164] *G.G.*, VIII, p. 154.
[165] *Ibid.*

materia *momento* o *forza composta*[166]. [into consideration other momenti than those of the simple force at B and the resistance at D, almost as if the lever itself were immaterial and without heaviness: but if we also take into account the heaviness of the instrument of the lever itself, which is sometimes made of wood and sometimes also of iron, it is clear that, by adding the weight of the lever to the force at B, the proportion will alter, which must then be expressed under other terms. And therefore, before going any further, it is necessary that we agree on a distinction between these two ways of consideration, calling it an *absolute consideration* when we understand the instrument taken in the abstract, that is, separated from the heaviness of its own matter; but by joining matter with simple and absolute figures, with heaviness as well, we will call the figures joined with matter *momento* or *compound force*.]

For the first time, in Galileo's mechanical speculations, we see the need for a clear distinction between "abstract" geometric consideration and the analysis of phenomena in which material bodies intervene. This is a remark that is not reducible to Galileo's frequent, characteristic reminder of the need to "sottrarre gli impedimenti [remove the impediments]" due to matter. The new science of the resistance of material can be founded, using the same principles, on two different levels. It should be recalled that Archimedes had assigned "heaviness" to planes, in other words, to two-dimensional geometric quantities, and that Galileo had done the same for solids in the *Theoremata* and *Mecaniche*[167], without feeling the need to distinguish the "astratto [abstract]" *momento* from the "composto [compound]" *momento*.

Now something has changed. In studying the resistance of materials, Galileo frequently uses the case of a beam, cylindrical or prismatic, fixed in a wall and parallel to the horizontal plane. He attempts to establish how far such a beam can be extended before breaking right at the base. In this model, as the beam is lengthened, there is an increase in the distance, in other words, in the *momento* of the force compared to the *momento* of the resistance, which remains unchanged. There is also, at the same time, an increase in the "absolute" weight of the beam. To calculate the effect of these two factors in relation to the resistance to being broken at the base, it will therefore be necessary not to limit ourselves to the "geometric" consideration of distances, but to include the more "physical" aspect of weight. The two factors must be distinguished in order to calculate their combined effect, to obtain that "compound" *momento* which allows the laws of resistance of geometric solids to be extended to real bodies.

Of the necessity to combine the two *momenti* and of the way in which the compound must be carried out, there is a very clear example in the demonstration by which Galileo establishes, in a prism, or in a cylinder fixed in a wall, parallel to the horizon, the *momenti* of "dirompente [disruptive]" force with respect to the **256**
resistance to being broken, as the solid is lengthened.

Let the prism AD be initially fixed in the wall, at AF (see fig. 16), and subsequently extended to C.

[166] *Ibid.*, pp. 154-155; the italics are ours.

[167] In the *Mecaniche*, for example, Galileo had never taken into consideration the weight of the balance beam or the lever; cf. for example, the pages dedicated to the examination of the steelyard and the "lieva [lever]" (*G.G.*, II, pp. 165-167).

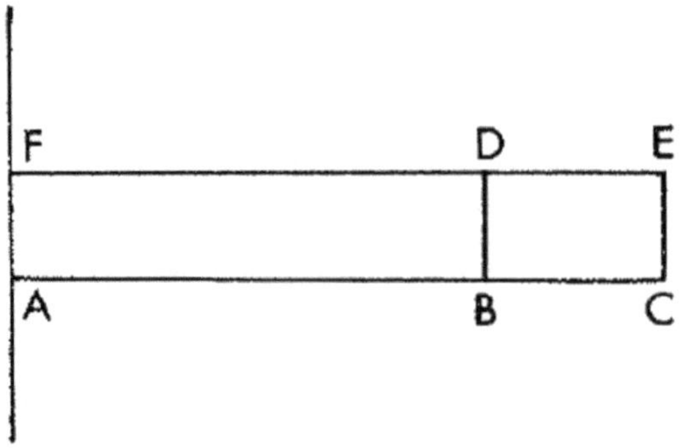

Figure 16

È manifesto che l’allungamento della leva AB sino in C cresce per sé solo, cioè assolutamente preso, il momento della forza premente contro alla resistenza dello staccamento e rottura da farsi in A secondo la proporzione di CA a BA: ma oltre a questo, il peso aggiunto del solido BE al peso del solido AB cresce il momento della gravità premente secondo la proporzione del prisma AE al prisma AB, la qual proporzione è la medesima della lunghezza AC alla AB: adunque è manifesto che congiunti i due accrescimenti delle lunghezze e della gravità, il momento composto di amendue è in doppia proporzione di qualunque di esse. Concludasi per tanto, i momenti delle forze de i prismi e cilindri egualmente grossi, ma disegualmente lunghi, esser tra di loro in duplicata proporzione di quella delle lor lunghezze, cioè esser come i quadrati delle lunghezze[168]. [It is clear that the extension of the lever AB up to C by itself alone, i.e., absolutely taken, increases the momento of the pressing force against the resistance of the detachment and breakage to be made at A according to the proportion of CA to BA: but in addition to this, the added weight of the solid BE to the weight of the solid AB increases the momento of the pressing heaviness according to the proportion of the prism AE to the prism AB, which proportion is the same as the length AC to AB: therefore it is clear that when the two increases in length and in heaviness are combined, the momento composed of both is in double proportion to either of them. Therefore, let it be concluded that the momenti of forces of equally thick, but unequally long, prisms and cylinders are in double proportion to each other in proportion to that of their lengths, that is, they are as the squares of the lengths.]

On the “physical” level, that is, when we refer not to geometric quantities, but to real bodies, the calculation of the *momenti* of force must therefore be conducted differently. These *momenti* will stand to each other not as the distances from the support, but as their squares. Galileo’s “compound” *momento* is, in reality, the “physical” momento compared to the “simple” one of Archimedean mechanics, which is strictly geometric.

---

[168] *G.G.*, VIII, p. 159. It is worth emphasizing that the use of *momento*, in this new meaning, caused Father Mersenne some confusion. In translating the present passage of the *Discorsi*, the Frenchman does not feel comfortable maintaining *moment* and at least correctly warns the reader: “Mais quand on considere la pesanteur du levier, il est certain que la raison change, parce qu’il en resulte une force composee de la force H jointe à la pesanteur du levier CH. Or il faut remarquer que je me sers de la diction, *force* ou *puissance*, au lieu du *moment* de Galilee, parce que nous n’avons point d’autre diction [But when we consider the weight of the lever, it is certain that the ratio changes, because it results in a force composed of the force H joined to the weight of the lever CH. Now it should be noted that I use the diction, *force* or *power*, instead of Galileo’s *moment*, because we have no other diction]” (*Les nouvelles pensées de Galilée*, par M. Mersenne. Édition critique par P. Costabel et M. P. Lerner, 2 voll., Vrin, Paris 1973, I, p. 118; see also the editors’ note, II, p. 217).

We will not dwell on the long series of propositions of the *Giornata Seconda* in which, using the "compound" *momento* and the principle of the lever, Galileo 257
establishes some fundamental laws *de resistentia solidorum* [of the resistance of solids], determining the proportions, under different conditions, of the *momenti* of forces and resistances.

Instead, we will focus on a proposition, as it is of particular interest from our point of view, in which Galileo demonstrates that

dei cilindri e prismi simili i momenti composti, cioè risultanti dalle lor gravità e dalle loro lunghezze, che sono come leve, hanno tra di loro proporzione sesquialtera di quella che hanno le resistenze delle medesime lor basi[169]. [of similar cylinders and prisms, the compound momenti, those which result from their heavinesses and their lengths, which are like levers, have between them a sesquialteral proportion (i.e., 3:2) of that of the resistances of their same bases.]

Galileo speaks of "similar" cylinders and prisms, in which the ratio between the base diameter and height is constant. The demonstration presupposes the Galilean assumption which we have previously discussed: the resistance is "rappresentata [represented]" by the base circle of the solid fixed in the wall, while the "compound" *momento* of the force is indicated by the product of the distance from the fulcrum and the weight of the cylinder or prism.

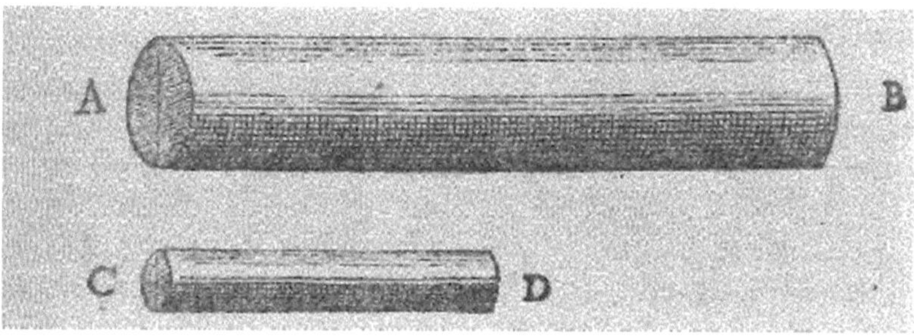

Figure 17

In the two similar cylinders AB, CD (see fig. 17), the ratio between the distances from the fulcrum of the forces and the resistances are the same. Therefore, as Galileo affirms, the "total" (that is, "compound") *momento* of the cylinder AB will be to that of the cylinder CD simply as the heaviness of AB to the heaviness of CD, that is, as the cylinder AB to the cylinder CD. This is equivalent to saying that the compound *momenti* are as the cubes of the base diameters. But the resistances of the two cylinders are as the bases, that is, as the square of the diameters of the bases. It can, therefore, be concluded that, in similar cylinders, while the resistances increase as the squares of the base diameters, the "compound" *momenti* of the forces

[169] *Ibid.*, p. 162.

increase as the cubes of the same diameters. Therefore, as Galileo affirms,

> i momenti de i cilindri son in sesquialtera proporzione delle resistenze delle basi loro[170]. [the momenti of the cylinders are in sesquialteral proportion to the resistances of their bases.]

258 From Galileo's demonstration, it follows that, by multiplying the dimensions of a cylindrical or prismatic solid with a base fixed in a vertical wall by any coefficient, the *momento* of resistance to being broken will increase in a much smaller proportion than the "compound" *momento* of the force. This means that similar solids of larger dimensions are more fragile.

Simplicio admits his surprise:

> Questa proposizione mi è veramente giunta non solamente nuova, ma inaspettata, e nel primo aspetto assai remota dal giudizio che io ne averei conietturalmente fatto: imperò che, essendo tali figure in tutto il restante simili, arei tenuto per fermo che anco i momenti loro verso le proprie resistenze avessero ritenuta la medesima proporzione[171]. [This proposition has truly come to me not only as new, but unexpected, and in the first aspect, very remote from the judgment that I would have conjecturally made of it: because, since these figures were similar in all other respects, I would have considered it certain that their momenti towards their own resistances also had the same proportion.]

From the considerations on the non-homogeneous increase of the *momenti* of the force in relation to those of the resistances, Galileo derives the explanation of a rule followed not only by craft, but also by nature: larger machines break down more easily than small ones:

> Or vegghino come dalle cose sin qui dimostrate apertamente si raccoglie l'impossibilità del poter non solamente l'arte, ma la natura stessa, crescer le sue macchine a vastità immensa: sì che impossibil sarebbe fabbricar navilii, palazzi o templi vastissimi, li cui remi, antenne, travamenti, catene di ferro, ed in somma le altre lor parti, consistessero.... Dal che è manifesto che chi volesse mantener in un vastissimo gigante le proporzioni che hanno le membra in un uomo ordinario, bisognerebbe o trovar materia molto più dura e resistente, per formarne l'ossa, o vero ammettere che la robustezza sua fusse a proporzione assai fiacca che ne gli uomini di statura mediocre; altrimente, crescendogli a smisurata altezza, si vedrebbono dal proprio peso opprimere e cadere. Dove che, all'incontro, si vede, nel diminuire i corpi non si diminuir con la medesima proporzione le forze, anzi ne i minimi crescer la gagliardia con proporzion maggiore[172]. [Now let us see how, from the things demonstrated up to now, it is clearly gathered that it is impossible not only in craft, but in nature itself, of being able to grow machines to an immense vastness: so that it would be impossible to build vessels, palaces, or vast temples whose oars, masts, beams, iron chains, and in short all their other parts, would hold together.... From this, it is clear that anyone who wanted to maintain in a vast giant the proportions that the limbs have in an ordinary man would have to either find much harder and more resistant material to form his bones, or indeed admit that his strength was in proportion much weaker than in men of moderate stature; otherwise, as they grow to an enormous height, they will be crushed and fall by their own weight. Whereas, on the contrary, it is seen that, in diminishing bodies, the forces do not decrease in the same proportion, but rather in the smallest ones the strength increases in a greater proportion.]

The study of the *momenti* of resistance of solids, of the "compound" *momenti*

---

[170] *Ibid.*, p. 164.
[171] *Ibid.*
[172] *Ibid.*, pp. 169-170.

of forces and the use of the laws of the lever, as well as the wise use of the science of proportions, therefore allows us to reveal the mechanical reasons for the very construction criteria followed by nature, that is, dictated by God, the perfect geometer. What greater ambition is there for an "entirely new" science? And what greater success for the "mechanical" *momento*?

With the *Giornata Seconda* of the *Discorsi*, the examination of the use of *momento* in the sense that we have seen emerge in the *Mecaniche* can be said to be concluded. At this point, *momento* has undergone a notable expansion of use, compared to the very limited functions assigned to it in the early Paduan treatise. *Momento* is no longer exclusively related to weight, nor does it necessarily indicate a downward tendency. There is *momento* of force, *momento* of resistance, "absolute" **259**
momento, "partial" *momento*, *momento* of heaviness, *momento* of speed, etc. The initial close connection of the term with the natural tendency, the intrinsic propensity towards the center now appears to have faded. *Momento* has now effectively become a "neutral" technical term, that is, without any compromises with previous conceptions. Its emancipation – if it is legitimate to speak of emancipation – was however – as we have seen – gradual and not automatic, and entailed a notable change in Galileo's way of dealing with questions of statics and mechanical motions.

We will now leave what we have defined as the "mechanical" *momento* to move on to carefully analyze the gradual emergence of a different use of the term, characterized in a more distinctly "dynamic" sense, whose relationship with the "mechanical" *momento* will appear quite interesting and peculiar.

In doing so, we will change the scene and go back to the Paduan years, focusing our attention on the effort that Galileo made to explain and attribute the natural motion of fall and acceleration to precise laws, attempting to start from the same principles used in the analysis of the static problems. In many of the pages that Galileo dedicated to solving this problem – truly a cross borne by scientists for centuries – we will see what intelligent use he was able to make, once again, of *momento*. We will return to the "mechanical" use of *momento* later, to illustrate the final reworking that Galileo proposed, in light of the awareness that a more precise understanding of the "symptoms" of acceleration had given him.

Chapter III

FROM "MECHANICAL" MOMENTO TO "MOMENTUM VELOCITATIS"

In introducing the problem of acceleration in the *De motu*, Galileo had specified how problematic it was to establish "causa cur naturalis motus in fine augeatur [the cause of why natural motion increases at the end]"[1]. Acceleration, in essence, was then presented as an accidental [i.e., incidental] phenomenon, a consequence, in every "natural" motion, of the inevitable pre-existence of a violent state. This initial "violence", by conferring an accidental lightness to the heavy body, prevented it from descending with a *uniform* speed proportional to the ratio between its own specific weight and that of the medium. Once the accidental lightness was "consumed", the heavy body would then continue its fall with uniform motion[2]. 261

At the time in which he wrote the pages on acceleration in the Latin treatise, Galileo was already convinced that heavy bodies of the same material but of different bulks [i.e., volumes] descend with the same speed[3]. He had, in the end, developed an ingenious theory to justify the observable phenomenon of the greater initial speed with which less heavy bodies (of the same species) fall: having stored a greater amount of lightness (i.e., having undergone a more intense "violence"), heavier bodies are "slowed down" to a greater extent[4], in the initial phase of the fall.

Following the pages of the *De motu*, assigned to 1590, for many years, we have no Galilean documents that allow us to establish with absolute certainty whether the Pisan scientist remained faithful to this conception of accelerated fall, or whether, instead, he walked away from it. If, in the *Mecaniche*, he in fact developed and made some parts of the *De motu* more terminologically coherent, he carefully avoided any temptation of dynamic extrapolation. We have, however, pointed out that, precisely in the chapter of the Paduan manual dedicated to the "screw" in the examination of the inclined plane, to which this instrument is essentially reduced, Galileo had
resumed the demonstration from the *De motu*, reproposing it in such a way that an 262
important consequence could hardly have been missed: the *momento* of a heavy

[1] *G.G.*, I, p. 315.
[2] *Ibid.*, pp. 318-319.
[3] *Ibid.*, pp. 262-273.
[4] *Ibid.*, pp. 333-337.

P. Galluzzi, *Momento: Galilean Studies*, Archimedes 78,
https://doi.org/10.1007/978-3-032-35049-7_7

body progressively decreases as the lever arm, leaving the horizontal position, rotates around the fulcrum, describing a quadrant of a circle; the *momento* becomes null once the perpendicular position with respect to the other arm is reached[5].

There remains certain evidence that Galileo, after having brought the *Mecaniche* to the level of elaboration corresponding to the draft published by Favaro in the Edizione Nazionale, continued to work on some parts of the treatise, focusing his attention primarily on the dynamic implications of the mechanical theorems. This very important unification effort can be placed in Padua, in the years from 1600 to 1609.

It is explicitly demonstrated by the letter that the Pisan scientist addressed to Vinta during the repatriation negotiations that most of the fruits of this attempt were gathered by Galileo before returning to Tuscany[6]. In it, Galileo reminded the Grand Duke's Secretary that his merits were not limited to the discovery of the telescope and the Medicean Planets. He emphasized, in fact, that among his practically completed works were "tre libri *De motu locali*, scienza interamente nuova, non havendo alcun altro, né antico, né moderno, scoperto alcuno dei moltissimi sintomi ammirandi che io dimostro essere ne i movimenti naturali et ne i violenti [three books *De motu locali* (i.e., on local motion), an entirely new science, as no one else, neither ancient nor modern, has discovered any of the many admirable symptoms that I demonstrate to exist in natural and violent movements]"[7]. The title and the internal partition were destined to remain unchanged. And certainly – as we will soon see – most of the fundamental *de motu locali* propositions were, in 1609, already demonstrated.

However, there remained major foundational problems, problems from which Galileo would never be able to free himself completely. On the verge of setting out for Florence, where previously unthinkable honors and rewards awaited him, Galileo had already "exhausted" the attempt to found his own science *de motu* [i.e., science of motion] based on static principles.

In this endeavor, on which we will now focus to reconstruct its main stages, Galileo's exploration of the concept of *momento* held extraordinary importance: to see if it could provide the fulcrum/anchor for the unification of statics and dynamics, which Galileo had shown he was aiming for since the years of the *De motu*.

The first document that informs us of Galileo's effort to further develop the *Mecaniche* after 1600 is the letter that our scientist wrote to Guidobaldo on
263 29 November 1602[8]. Unfortunately, this is an isolated document, as neither a previous letter from Galileo nor the response from Guidobaldo, who had raised some doubts, has reached us. Galileo strives to convince his authoritative friend of the truth of the "proposizione de i moti fatti in tempi eguali nella medesima

[5] *G.G.*, II, pp. 181-183; cf. *supra*, Part Two, Ch. II, pp. 215-216.

[6] *G.G.*, X, pp. 348-356; letter dated 7 May 1610, from Padua.

[7] *Ibid.*, pp. 351-352.

[8] *Ibid.*, pp. 97-100; the letter, sent from Padua, is preserved in a modern copy. This is a very important document to which all scholars of the Galilean conception of movement have dedicated considerable attention. For this reason, it is considered superfluous to give bibliographical indications, referring instead to the works mentioned in the previous chapter.

quarta del cerchio [proposition of the motions occurring in equal times in the same quarter of the circle]"[9]. And he proposes to the incredulous correspondent "l'esperienza con che mi sono principalmente chiarito di tal verità [the experiment with which I have principally clarified this truth]"[10]. Galileo therefore does not have a "certain demonstration" of the isochronism of pendulums of equal length, but only experimental proof. He invites Guidobaldo to repeat the experiment, which consists of observing two pendulums of equal length, however displaced at different angles. They will therefore describe noticeably different arcs. If we take the halves of these oscillations, we will observe the isochronism of the two motions along different arcs of the same quadrant of the circle[11].

Having dispelled the doubts expressed by Guidobaldo, Galileo adds valuable information on some discoveries made recently (i.e., after the previous letter to the Marquis dal Monte), "non meno inopinabili [no less unexpected]" than the isochronism of pendulums of equal length. We thus learn that Galileo demonstrated:

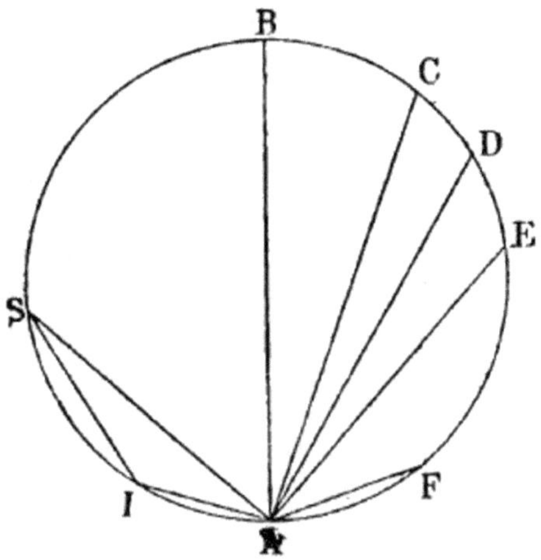

Figure 18

mobili eguali cadere in tempi uguali per la perpendicolare BA e per piani inclinati secondo le linee CA, DA, EA, FA; sicché, partendosi nell'istesso momento dalli punti B, C, D, E, F, arriveranno in uno stesso momento al termine A, e sia la linea FA piccola quant'esser si voglia[12]. [equal mobiles fall in equal times along the perpendicular BA (see fig. 18) and along inclined planes along the lines CA, DA, EA, FA; so that, starting at the same moment from the points B, C, D, E, F, they will arrive at the terminus A at the same moment, and the line FA can be as small as one desires.]

264

[9] *Ibid.*, p. 97.
[10] *Ibid.*, p. 98.
[11] *Ibid.*, pp. 98-99.
[12] *Ibid.*, p. 99.

Yet “più inopinabile [more unexpected]” is the subsequent demonstration obtained by Galileo:

essendo la linea SA non maggiore della corda di una quarta, e le linee SI, IA *utcumque*, più presto fa il medesimo mobile il viaggio SIA, partendosi da S, che il viaggio solo IA, partendosi da I[13]. [since the line SA (see fig. 18) is not greater than the chord of a quarter, and the lines SI, IA *howsoever*, the same mobile makes the journey SIA, starting from S, faster than the journey IA alone, starting from I.]

The concluding addition is of considerable interest. Galileo, in fact, admits that he obtained these results by demonstrating “senza trasgredire i termini meccanici [without transgressing the mechanical terms (i.e., principles)]” and that he was, however, in a serious quandary because of his inability to demonstrate “come gli archi SIA e IA siano passati in tempi eguali [how the arcs SIA and IA pass in equal times]”[14].

Galileo will never achieve a “demonstration” of the isochronism of pendular motions (i.e., along the arcs), nor will he ever go beyond the level of proof based on the observation of the identical times with which the same pendulum performs different “reciprocazioni [reciprocations]”[15]. However, it is probable that Galileo initially tried to demonstrate the isochronism of the motions of the same pendulum along different arcs by resorting to the concept of *momento della gravità* [momento of heaviness]. Having experimentally observed the isochronism of pendulums of equal length, Galileo intuitively thought that it could be explained by focusing attention on the continuous decrease in *momento della gravità* at the individual points of the circle’s quadrant from top to bottom, as observed in the *Mecaniche*, in the theorem of inclined planes[16]. But the difficulties encountered in this type of analysis were not insignificant. So, unable to demonstrate isochronism directly, he tried to circumvent the obstacle. This is indicated by the first of the two discoveries communicated to Guidobaldo: equal mobiles descend along all the chords drawn from any point of the circumference to the foot of the perpendicular diameter in equal times, the maximum and the minimum, “piccola quant’esser si voglia [however small it may be]”. This proposition, which we will see demonstrated in the *Discorsi* as *Theorema VI*[17], originally derived from the observed isochronism of pendulums of equal length, was, in the end, used by Galileo as the foundation of this phenomenon. Guidobaldo had justified his incredulity by emphasizing the improbability that, “pigliandosi una quarta lunga cento miglia, due mobili eguali
265 possino passarla, uno tutta e l’altra un palmo solo in tempi eguali [in taking a quarter

[13] *Ibid.*, p. 100.

[14] *Ibid.*

[15] Galileo did not have sufficient mathematical tools to allow him to demonstrate the isochronism of the motions along the arcs described by the pendular oscillations, that is, the free descents of the same weight along different arcs of the same quadrant.

[16] *G.G.*, II, pp. 181-182.

[17] “Si a puncto sublimi vel imo circuli ad horizontem erecti ducantur quaelibet plana usque ad circumferentiam inclinata, tempora descensuum per ipsa erunt aequalia [If planes are drawn from any point, whether at the highest or lowest part of the circle, to the horizon, up to the inclined circumference, the times of descent will be equal]” (*G.G.*, VIII, p. 221).

(of a circle) one hundred miles long, two equal mobiles could traverse, one the entirety and the other only a span (i.e., handbreadth), in equal times]"[18]. Galileo responds to Guidobaldo's doubt by proposing motion not along arcs of circles, but rather along inclined planes, as fully equivalent examples[19]. That is, he suggests that there is no difference between the two models: a circle can indeed be considered a regular polygon with infinitely many sides (this argument, absent from the letter to Guidobaldo, will be presented later)[20]: it is, therefore, reasonable to rely upon the demonstration of the isochronism of the chords for the assertion of the arcs, which he is unable to demonstrate independently.

In the letter to Guidobaldo, Galileo claims to have the "demonstration" of the isochronism of the concurrent motions along the chords to the foot of the perpendicular diameter. The Pisan scientist, however, does not communicate the demonstration to his correspondent, limiting himself to informing him that he has not transgressed "i termini meccanici [the mechanical terms (i.e., principles)]". It is not, however, impossible to reconstruct the "mechanical" demonstration of 1602. *Theorema VI* of the *Discorsi* presents, in fact, a twofold demonstration of the isochronism of the motions along the chords of the quadrant. The first is based on purely geometric methods[21]. To this, Galileo adds a proof "ex mechanicis [from mechanics]"[22], attributable to the inclined plane theorem used in the *De motu* and the *Mecaniche*, and based on the law of *momenti*. Of the "ex mechanicis" demonstration, Galileo provides three different versions. In all of them, although the techniques of proof vary, the foundation is constituted by the use of *ponderis momentum* [momentum of weight] and by the law which, on the inclined plane, states that the partial *momento* of the weight along the inclined plane is to its total *momento* (i.e., along the perpendicular) as the perpendicular is to the incline. Through a geometric analysis, the case of concurrent chords at the lowest point of a circumference is, in fact, reduced to this foundation[23].

---

[18] *G.G.*, X, p. 99.

[19] "Se consideriamo che può esser un piano tanto poco declive, qual saria quello della superficie di un fiume che lentissimamente si muovesse, che in esso non haverà camminato un mobile naturalmente più d'un palmo nel tempo che un altro sopra un piano molto inclinato (ovvero congiunto con grandissimo impeto ricevuto, anco sopra una piccola inclinazione) haverà passato cento miglia, né questa proposizione ha seco per avventura più inverisimilitudine di quello che si habbia che i triangoli tra le medesime parallele et in basi eguali siano sempre eguali, potendone fare uno brevissimo e l'altro lungo mille miglia [If we consider that there can be such a plane so slightly inclined, such as that of the surface of a river which moves very slowly, that on it no mobile will have moved more naturally than a handbreadth in the time than another on a very inclined plane (or endowed with a very great received impeto, even over a small inclination) will have traversed a hundred miles, nor does this proposition carry with it perhaps more implausibility than the fact that triangles between the same parallels and on equal bases are always equal, even though one can be made very short and the other a thousand miles long]" (*G.G.*, II, p. 99).

[20] *G.G.*, VIII, p. 95.

[21] *Ibid.*, p. 221, *Theorema VI, Propositio VI.*

[22] "dem aliter demonstratur ex mechanicis: nempe... mobile temporibus aequalibus pertransire CA, DA [it is otherwise demonstrated from mechanics: namely... that the mobile traverses CA and DA in equal times]" (*ibid.*).

[23] Cf. the three equivalent versions of the demonstration based on the laws of *ponderis momentum*, which Galileo claims to derive "ex elementis mechanicis", i.e., from the *Mecaniche*. (*ibid.*, pp. 222-223).

266 In November of 1602, the demonstration that Galileo declared to Guidobaldo that he possessed – based, in fact, according to our author’s explicit statement, on mechanical principles – must not have been very different from the one included, many years later, in the *Discorsi*. However, a handwritten fragment could preserve the demonstration, exactly as Galileo conceived it at the time of the letter to Guidobaldo.

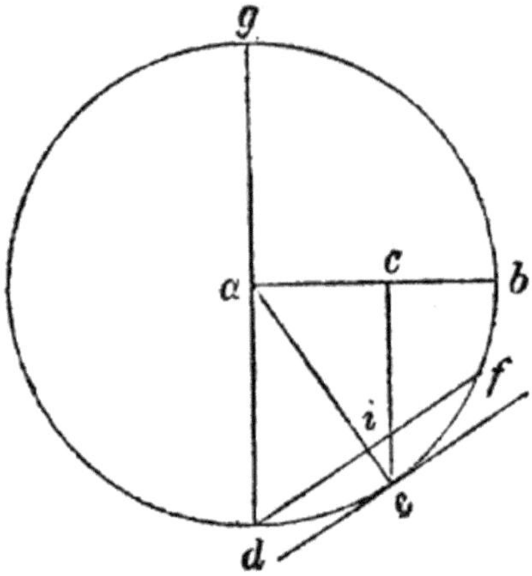

Figure 19

Sit *gd* erecta ad orizontem, *df* vero inclinata: dico eodem tempore fieri motum ex *g* in *d* et ex *f* in *d*.

Momentum enim super *fd* est idem ac super contingente in *e*, quae ipsi *fd* esset parallela; ergo momentum super *fd* ad totale momentum erit ut *ca* ad *ab*, idest *ae*: verum ut *ca* ad *ae*, ita *id* ad *da* et dupla *fd* ad duplam *dg*; ergo momentum super *fd* ad totale momentum, scilicet per *gd*, est ut *fd* ad *gd*: ergo eodem tempore fiet motus per *fd* et *gd*[24].

[Let *gd* (see fig. 19) be perpendicular to the horizon, and *df* inclined: I say that the motion from *g* to *d* and from *f* to *d* occurs in the same time.

For the momentum on *fd* is the same as on the tangent at *e*, which would be parallel to *fd*: therefore, the momentum on *fd* to the total momentum will be as *ca* is to *ab*, that is, *ae*: indeed, as *ca* is to *ae*, so is

---

24 *Ibid.*, p. 378; the theorem is found on page 151*r* of Ms. 72. The technique used in the demonstration attested by the fragment is different from those used in the three versions of the “ex mechanicis” proof of the *Discorsi*. The fragment is to be considered the oldest mechanical proof, among those that have come down to us, of the isochronism of motions along the chords concurrent to the foot of the perpendicular diameter. The similarities in the demonstrative technique and in the figure with the inclined plane theorem in the *Mecaniche* lead us to place its drafting at the very beginning of the 17th century. Of the “mechanical” proof of the *Discorsi* there are, again in Ms. 72, two handwritten drafts of Galileo, on pages 172*r* and 160*r*, respectively. These constitute intermediate elaborations between our fragment and the draft published in ‘38. The drafts were not published among Favaro’s *Frammenti*. In fact, he limited himself to giving the variations compared to the definitive text in the critical apparatus (*ibid.*, pp. 221-222). The text of the draft of page 160*r* can be read, reconstructed as an autonomous entity, in W. Wisan, *The new science of motion* cit., p. 164. Wisan also suggests a close relationship between the letter to Guidobaldo and the theorem of folio 151*r*.

id to *da*, and the double of *fd* to the double of *dg*; therefore, the momentum on *fd* to the total momentum, namely through *gd*, is as *fd* is to *gd*: thus, the motion through *fd* and *gd* will occur in the same time.]

The handwritten fragment demonstrates that the times along the perpendicular diameter and along the chord are equal, given that the *momenti* of the same mobile **267**
are in the same ratio as the diameter to the chord. The dynamic principle employed in the *De motu* and carefully avoided in the *Mecaniche* is implied here: the speeds are in proportion to the *momenti*. In the case where the *momenti* are as the distances, as occurs in our fragment, it is obvious that the motions will occur in equal times. In the terminology and in the demonstration schema, we find strong analogies with the theorem from the *Mecaniche*:[25] the "total" *momentum*, along the perpendicular, the use of representing *momenti* as distances, the schema of the quadrant of a circle and the assimilation of the *momento* along the chord to the *momento* along the parallel tangent to the circle, etc. However, the implicit, but very evident, *momento-speed* proportionality indicates that the Pisan scientist has resumed work on the dynamic hypothesis used in the *De motu*. There is no precise element that allows us to establish whether Galileo is thinking, in these theorems, of accelerated or uniform motions; although the reappearance of the theorem from the *De motu*, previously abandoned, could lead us to suppose that Galileo wanted to verify its validity in the hypothesis of accelerated descents.

The second discovery, even more "inopinabile [unexpected]", mentioned by Galileo to Guidobaldo consisted – as we have seen – in the demonstration of the more rapid descent of the body along the path SI, IA, starting from S, than along the path IA, starting from I (see fig. 18). In essence – as demonstrated by the *Scholium* placed at the end of the demonstration of this phenomenon in the *Discorsi*[26] – this is equivalent to affirming the principle that, in a quadrant of a circle, the movements of descent for an identical difference in height are all the more rapid, the longer the path that takes place approximates the arc contained in this difference in height; the path along the arc will descend more rapidly[27].

Even in this demonstration, Galileo had proceeded without departing from mechanical terms. It refers, without a shadow of a doubt, to *Theorema XXII* of the *Discorsi*, dedicated precisely to the proof of our principle[28]. But reconstructing Galileo's "mechanical" demonstration, at the time of the letter to Guidobaldo, is **268**

[25] Cf. *supra*, p. 215.

[26] *G.G.*, VIII, pp. 263-264. This is the *Scholium* to *Theorema XXII*.

[27] This entails – and Galileo makes this clearly evident – that, in the circle, the quickest descent does not occur along the shortest path: "Ex his quae demonstrata sunt colligi posse videtur, lationem omnium velocissimam ex termino ad terminum non per brevissimam lineam, nempe per rectam, sed per circuli portionem fieri [From what has been demonstrated, it seems to be inferred that the fastest transfer from one point to another occurs not along the shortest line, namely the straight line, but along a portion of the circle]" (*ibid.*, p. 263). The *Scholium* intends to propose a non-demonstrable, but intuitively convincing consequence of *Theorema XXII*, specifically that which identifies the arc as the fastest descent path. On Galileo's *Scholium* and its significance, cf. W. Wisan, *The new science of motion* cit., pp. 184-187.

[28] "Si in circulo ad horizontem erecto ab imo puncto elevetur planum non maiorem subtendens circumferentiam quadrante, a terminis cuius duo alia plana ad quodlibet circumferentiae punctum inflectantur, descensus in planis ambobus inflexis breviori tempore absolvetur, quam in solo priori plano elevato, vel quam in altero tantum ex illis duobus, nempe in inferiori [If, in a circle with an

much more difficult, given that in the *Discorsi* it is proven with purely geometric procedures, and Galileo does not mention a different way of demonstrating it "ex mechanicis". Of the fragments relating to *Theorema XXII* of the *Discorsi* that we have preserved, only one, handwritten, very obscure and incomplete, records a trace of the original mechanical approach:

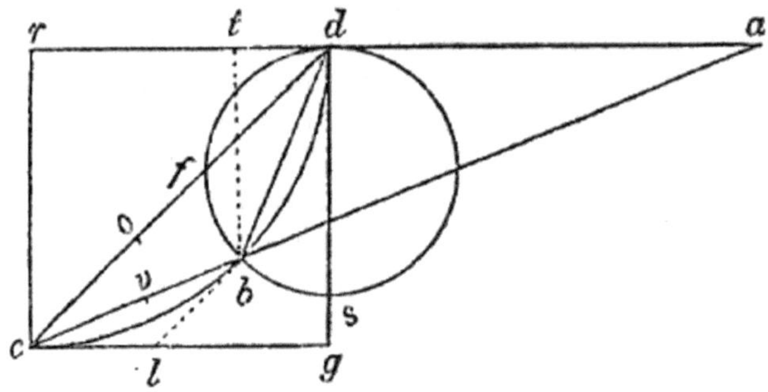

Figure 20

accipiatur media inter *rc*, *bt*, cui aequalis ponatur *do*, erit reliqua *co* eadem quae invenitur per mediam inter *cd*, *df*, quae est eadem *do*.

Considera, momentum in singulis circumferentiae quadrantis punctis imminui pro ratione accessus puncti perpendicularis, ut *t*, ad centrum.

ut *bc* ad *cd*, sic *cd*...

momentum super plano *dc* ad totale momentum est ut linea *tr* ad *rd*, ducta *lb* aequidistante *cd*[29].

[let the mean be taken between *rc*, *bt* (see fig. 20), to which *do* is set equal, and the remaining *co* is the same as what is found through the mean between *cd* and *df*, which is the same as *do*.

Consider, the momentum in the individual points of the quadrant's circumference diminishing in proportion to the approach of the perpendicular point, as *t* to the center.

---

upright horizon, a plane is elevated from the lowest point, subtending no greater arc than a quadrant of the circumference, and from its two endpoints, two other planes are inclined to any point on the circumference, the descent on both the inclined planes will be completed in a shorter time than on the first elevated plane alone, or than on just one of those two, namely the lower one]" (*G.G.*, VIII, pp. 261-262). On the demonstrative procedures used by Galileo in *Theorema XXII*, cf. W. Wisan, *The new science of motion* cit., pp. 179-184.

[29] Ms. 72, c. 131*r*, publ. in *G.G.*, VIII, p. 417. The first paragraph is deleted, probably by Galileo himself. The fragment is composed of four parts, corresponding to the paragraphs, which were almost certainly not written contemporaneously. On the significance of the fragment, cf. Wisan's observations (*The new science of motion* cit., pp. 176-177), who is uncertain whether to assign the theorem of page 131*r* to the search for a proof for isochronism, or for *Theorema XXII*. It should be emphasized, however, that the figure used in the fragment presents clear analogies with that of *Theorema XXII*, although it does not correspond exactly, while it appears different from the one used in the demonstration of the isochronism of the chords (*Theorema IV*).

as *bc* to *cd*, so *cd*...

the momentum over the plane *dc* to the total momentum is as the line *tr* to *rd*, drawn *lb* equidistant from *cd*.]

The figure and the brief notes clearly allude to the problem of the descent times along the arc *cd*, starting from *c*, and along the arcs *db*, *bc*, starting from *d*. The circle *dfb*, which we will also find in the demonstration of the *Discorsi*, serves, in all probability, to affirm the isochronism of the descents along *df*, *db*; that is, the *momento* over *db* is to the *momento* over *df* as *db* to *df*. It is, however, impossible to reconstruct the way in which Galileo intended to achieve the demonstration by working on the proportion of *momenti*. In the *Discorsi*, in fact, the theorem is proven with the use of the proportional [i.e., geometric] mean, which follows from the law **269**
that places the proportionality of the spaces traveled to the square of the times[30]. Our fragment also opens with a reference to the proportional mean theorem, but the two lines containing it are struck out.

In this fragment, Galileo shows that he is still operating in the conceptual orbit of the *Mecaniche*. In fact, it recalls the continuous decrease of *momento*, as the heavy body descends along the quadrant perpendicular to the horizontal (a phenomenon observed for the first time in the Paduan writing); furthermore, it measures this *momento* with the segment determined on the horizontal radius of the quadrant (which represents the "total" *momento*) from the projection of the point of the quadrant through which the tangent passes parallel to the chord (a procedure explicitly used in the Paduan treatise and already outlined in the *De motu*).

If we are unable to reconstruct Galileo's "mechanical" demonstration, we must not overlook the indecipherable entry of the older mechanical approach to the problem of fastest descent, through identical differences in height of the quadrant, along paths closest to the trajectory of the corresponding arc (even if longer), which is closely connected to the awareness of the continuous decrease in *momento*. It had by then become impossible to treat the downward motions as uniform movements. At the time of the letter to Guidobaldo, Galileo had already acknowledged that acceleration was a necessary and natural reality. And he tried to learn how to master it. In the peremptory *Considera* [Consider] of the fragment, a new understanding seems to be reflected. The motion of descent does not have a constant speed and therefore cannot be attributed to the effect of an invariable *momento*, as the Pisan scientist had done in the *De motu* and, with considerably greater caution, in the *Mecaniche*. The dynamic extrapolation of static principles needed to be reconsidered in terms of acceleration. In the reflection of the Pisan scientist, a new phase begins,

[30] This, at least, is the general opinion (cf., for example, W. Wisan, *The new science of motion* cit., p. 189). However, it should be recalled that, in an interesting essay, W. C. Humphreys supported a different thesis. He has, in fact, sought to demonstrate that Galileo reached the theorem of the proportional mean (which would become *Corollarium II* of *Theorema II* of the *Discorsi*), before the discovery of the law of fall and independently of it. From the proportional mean theorem, the Pisan scientist would later derive the law of fall (*Galileo, falling bodies and inclined planes*, "The British Journal for the History of Science", III, 1967, pp. 224-244.).

which will soon bear important fruits. Less than two years after sending this letter to Guidobaldo, Galileo informed his Servite friend Paolo Sarpi that he had long possessed the law of falling bodies and was searching for an evident principle to establish as its foundation.

Prompted by Sarpi to express his opinion on certain questions regarding the equivalence between the effects of natural and violent motion and about the
270 consequences of applying the same "virtù [virtue]" to two different bodies[31], Galileo replied on 16 October 1604. This is a central document. It is known, in fact, that the law that establishes the proportionality of the spaces traveled to the square of the times in natural fall is stated for the first time.

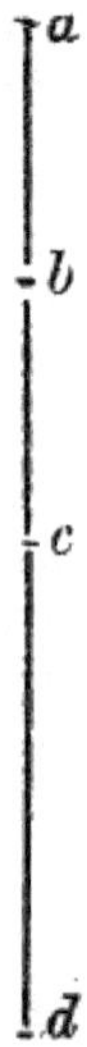

Figure 21

Riflettendo circa le cose del moto, nelle quali, per dimostrare li accidenti da me osservati, mi mancava principio totalmente indubitabile da poter porlo per assioma, mi son ridotto ad una proposizione la quale ha molto del naturale et dell'evidente; et questa supposta dimostro poi il resto, cioè gli spazzii passati dal moto naturale esser in proporzione doppia dei tempi, et per conseguenza gli spazii passati in tempi eguali esser come i numeri impari *ab unitate*, et le altre cose. Et il principio è questo: che il mobile naturale vadia crescendo di velocità con quella proportione che si discosta dal principio del suo moto; come, v.g., cadendo il grave dal termine *a* per la linea *abcd*, suppongo che il grado di velocità che hebbe in *b* esser come la distanza *ca* alla distanza *ba*, et così conseguentemente in *d* haver grado di velocità maggiore che in *c* secondo che la distanza *da* è maggior della *ca*[32]. [Reflecting on matters concerning motion, in which, in order to demonstrate the phenomena I had observed, I lacked a wholly indubitable principle that I could posit as an axiom, I have arrived at a proposition which has much of the natural and the evident about it; and, this assumed, I then demonstrate the rest, that is, the spaces passed by

[31] Sarpi's letter to Galileo was sent from Venice on 9 October 1604; see it in *G.G.*, X, p. 114.
[32] *Ibid.*, p. 115.

natural motion are in duplicate proportion to the times, and consequently that the spaces passed in equal times are as the odd numbers *ab unitate* [from unity], and the other things. And the principle is this: that the natural mobile increases in speed in proportion as it deviates from the beginning of its motion; as, for example, when the heavy body falls from the terminus *a* (see fig. 21) along the line *abcd*, I suppose that the degree of speed it has in *b* is as the distance *ca* to the distance *ba*, and thus consequently, it will have a degree of speed greater at *d* than at *c*, since the distance *da* is greater than *ca*.]

It is pointless to insist – since it is well known – on the fact that the formulation of the correct law of fall in the letter in question was accompanied, curiously, by the declaration of a false general principle: the speed increases "con quella proportione [with that proportion]" with which the heavy body "si discosta dal principio del suo moto [deviates from the beginning of its motion]"[33]. Acceleration

[33] As is known, Galileo will admit in the *Discorsi* that he had, at a certain point, considered the direct proportionality acceleration-distance from rest to be plausible. In fact, regarding the erroneous principle, he makes Salviati say: "Il nostro medesimo Autore non mi negò, quando io glielo proposi, d'esser egli ancora stato per qualche tempo nella medesima fallacia [Our very own Author did not deny to me, when I proposed it to him, that he had been for some time in the same fallacy]" (*G.G.*, VIII, p. 203). Furthermore, Salviati adduces as the reason for the rejection of such proportionality the absurd consequence that derives from it: "Quando le velocità hanno la medesima proporzione che gli spazii passati o da passarsi, tali spazii vengon passati in tempi eguali; se dunque le velocità con le quali il cadente passò lo spazio di quattro braccia, furon doppie delle velocità con le quali passò le due prime braccia (si come lo spazio è doppio dello spazio), adunque i tempi di tali passaggi sono eguali: ma passare il medesimo mobile le quattro braccia e le due nell'istesso tempo, non può aver luogo fuor che nel moto istantaneo… [When the speeds have the same proportion as the spaces traversed or to be traversed, such spaces are traversed in equal times; if, therefore, the speeds with which the falling object passed the space of four arm's lengths were double the speeds with which it passed the first two arm's lengths (just as the space is double the space), then the times of those passes are equal: but for the same mobile to pass the four arm's lengths and the two in the same time, this can only happen in instantaneous motion…]" (*ibid.*, pp. 203-204). This cryptic justification by Galileo has confused interpreters quite a bit. I. B. Cohen, in an essay that appeared in "Isis", in 1956, believed he could affirm that Galileo's reasoning depended on the illicit application of the Merton College theorem, valid for accelerations proportional to time, to the case of a motion accelerated according to distance from the starting point. Cohen obtained, among other things, from his own reconstruction the evidence of a direct influence of 14th century kinematics on Galileo (*Galileo's rejection of the possibility of velocity changing uniformly with respect to distances*, "Isis", XLVII, 1956, pp. 231-235). A. Rupert Hall promptly replied to Cohen's essay, reproposing Galileo's reasoning on a geometric basis, which was undoubtedly more suitable than the algebraic techniques used by Cohen to reconstruct the procedure of the Pisan scientist. Hall, however, believed it necessary to confirm, in essence, Cohen's judgment: the passage from the *Discorsi* reveals the illegitimate recourse to the mean speed theorem (*Galileo's fallacy*, "Isis", XLIX, 1958, pp. 342-344). In the same issue of "Isis", Cohen's response to Hall's observations (*ibid.*, pp. 344-345) and a letter from Stillman Drake was also published, in which he reviewed Cohen's first paper. Drake, in essence, showed that he did not consider the hypothesis advanced by Cohen unacceptable (*ibid.*, pp. 345-346). Subsequently, however, Stillman Drake revisited the issue with a lengthy essay. He argued that the use of the plural "velocità" ("le velocità con le quali il cadente passò lo spazio di quattro braccia [the speeds with which the falling object passed the space of four arm's lengths…]") indicates that Galileo was thinking of accelerated motion. Based on this, after having reviewed the French, English and German translations of Galileo's passage in the *Discorsi* and having shown that, in all, "le velocità" had been rendered in the singular, Drake concluded that Galileo's reasoning does not at all presuppose the incorrect use of the Mertonian mean speed theorem. Here is how Drake believed the passage in question should be read: "If all the infinite instantaneous velocities occurring in actual accelerated fall from rest over any space, say one of four arms, were the respective doubles of all the infinite velocities occurring over the first half, or two arms, of the same fall, then no difference in the times of fall could be accounted for. But we observe a difference in the times; hence proportionality of speed to space traversed from rest cannot govern the

271 is, in fact, as Galileo himself eventually recognized, directly proportional to time, not to the distance of fall. We will focus later on the nature and significance – which is very important, also for the history of Galileo's *momento* – of this singular misunderstanding. We would like now, however, to emphasize that Galileo, in soliciting his friend's opinion, added that he also intended, by using the proposed principle, to reunify natural and violent motion, demonstrating that, in both cases, the proportions of speeds remain equal.

> Imperò che se il proietto vien gettato dal termine *d* al termine *a*, è manifesto che nel punto *d* ha grado di impeto potente a spingerlo fino al termine *a*, et non più; et quando il medesimo proietto è in *c*, è chiaro che è congiunto con grado di impeto potente a spingerlo sino al medesimo termine *a*; et parimente il grado d'impeto in *b* basta per spingerlo in *a*: onde è manifesto, l'impeto nei punti *d*, *c*, *b*, andar decrescendo secondo le proporzioni delle linee *da*, *ca*, *ba*; onde, se secondo le medesime va nella caduta naturale acquistando gradi di velocità, è vero quanto ho detto et creduto sin qui[34]. [Because if the projectile is thrown from terminus *d* to terminus *a*, it is evident that at point *d* it has a degree of impeto capable of pushing it up to terminus *a*, and no further; and when the same projectile is at *c*, it is clear that it is conjoined with a degree of impeto capable of pushing it to the same terminus *a*; and likewise the degree of impeto at *b* is enough to push it to *a*: from which it is evident that the impeto at the points *d*, *c*, *b* is decreasing according to the proportions of the lines *da*, *ca*, *ba*; hence, if in natural descent it acquires degrees of speed in the same manner, then what I have said and believed up to now is true.]

Galileo's letter to Sarpi provides essentially three pieces of information. Firstly,
272 it demonstrates that the assumed principle (i.e., the direct proportionality between acceleration and distance from the starting point) serves to explain "accidents" already previously observed. This means that Galileo has long harbored the belief that, in natural motion, acceleration increases according to a very precise proportion; this belief – as we have seen – was already clear in the letter to Guidobaldo of 1602.

Secondly, the proposed principle allows for a unitary and analogical interpretation of the naturally accelerated motions and the progressively "ritardati [slowed]" upward violent motions of projectiles. The tendency towards unification, albeit in a different manner, had already appeared evident in the *De motu*.

Thirdly, the principle confirmed our scientist in what he has "detto e creduto sin qui [said and believed up till now]". That is, as we will see shortly, on the possibility of establishing a close connection between static principles and science *de motu* [i.e., science of motion][35].

The opening lines of the letter to Sarpi reveal the need, felt by our scientist, to postulate a general theory that justifies the progressive increase in speed in falling motion.

In the *De motu*, the general principle was identified in the violence experienced by the heavy body before starting its descent. Here, that hypothesis is abandoned

---

fall of actual bodies (*Uniform acceleration, space and time. Galileo Gleanings XIX*, "British Journal for the History of Science", V, 1970, pp. 21-43; the passage cited is on p. 36). Maurice A. Finocchiaro declared himself opposed to Drake's reconstruction (*Vires acquirit eundo: the passage where Galileo renounces space-acceleration and causal investigation*, "Physis", XIV, 1972, fasc. 2, pp. 125-145).

[34] *G.G.*, X, pp. 115-116.

[35] *G.G.*, I, pp. 321-323.

and replaced with another, which does not involve the reduction of acceleration to an accidental phenomenon (unlike what occurred in the Pisan writing).

In the letter to Sarpi, Galileo does not clarify whether he affirms the principle and the law of fall as invariably valid for all bodies, whatever their size and kind. However, the general reference to "spazzii passati dal moto naturale [spaces passed by natural motion]" seems to indicate that our scientist has already reached full awareness of the theoretical equivalence of all free falls. This means that the hydrostatic model used in the *De motu* had essentially been removed, although there was no shortage of attempts to propose it again later.

Lastly, of particular importance from our perspective is a reflection on the specific significance of the principle set forth by Galileo in the letter to Sarpi. It should not escape us that, in fact, in substance, it recovers, for speed, the proportion for *momento* stated in the *De motu* and, with greater clarity, in the *Mecaniche*. The latter, in fact, was measured by the vertical displacement of the heavy body. The "indubitable" foundation assumed by Galileo in 1604 consists, therefore, in the simple extension to the acceleration of natural motion of a principle already used in mechanical speculations. This helps us quite a bit to understand the meaning of Galileo's reflections during this period and sheds much light on – as we will soon see – the genesis of the curious error committed by rendering the acceleration proportional to the distance from the starting point. Galileo was, in fact, strongly conditioned by the static model and the "mechanical" concept of *momento*.

Temporarily setting aside the incorrect principle, it remains to be established **273**
how Galileo arrived at the correct formulation of the law of fall (i.e., affirming the proportionality of the spaces covered from rest to the square of times). This is a very delicate and extremely important issue. Addressing it in depth would require an entire chapter, even just to review the different hypotheses that have been put forward by the numerous scholars who have focused on it. The problem has, in fact, exercised extraordinary interest, both for its intrinsic value and because it was seen as a central "node" for the understanding of Galileo's scientific conception. Determining whether Galileo arrived at the law on the basis of purely intellectual procedures, whether he instead achieved it by drawing upon and developing the efforts of some skilled predecessors, or whether, lastly, he achieved success by virtue of simple observation and his ability to conceive and carry out experiments, has significance and meaning for many in capturing the characteristic aspect of that complex phenomenon known as the scientific revolution. The debate on this topic has carefully examined the different hypotheses, especially emphasizing – thanks mainly to Duhem – the close analogy between Galileo's method of proving the law of fall and the geometric procedures of Oresme, the Mertonians and the logic of the *calculationes* to determine the *uniformiter difformis* [uniformly difformed] motion. However, no one has ever been able to provide concrete proof of the influence exerted by such widespread and famous theories on the Pisan scientist[36]. Nor **274**

[36] Duhem's thesis, supported in the *Troisième Série* of his famous *Études sur Leonard de Vinci* cit. (especially pp. 570 ff.), is well known. Galilean physics would, in general, be the heir of the Buridanian theory of *impetus*, while Galileo would have arrived at the discovery of the law of fall by using Oresme's

could more convincing or definitive evidence be provided by the proponents of "deductivism" of geometric-mathematical character in Galileo[37]. Recently, the debate has concentrated – primarily thanks to Stillman Drake, who first raised the

reflections on movement and the so-called Merton College theorem to measure uniform acceleration in terms of average speed. The influence of the French scientist's thesis has been enormous even though, from the outset, there has been no shortage of opponents (see the very recent reconsideration of the entire issue in the essay by S. Caroti, *Nicole Oresme precursore di Galileo e di Descartes?*, "Rivista Critica di Storia della Filosofia", 1977, fasc. 1, pp. 11-23, and fasc. 4, pp. 414-436). Recently, the dependence of Galileo's dynamic considerations on 14th century sources was vigorously re-proposed by M. Clagett (*The science of mechanics*, trad. it. cit., pp. 282, 441; see primarily, p. 371, where Clagett states: "è difficile, secondo me, dubitare del fatto che Galileo abbia tratto sia il teorema sia gli elementi essenziali della sua dimostrazione dalla tradizione medievale oxoniense-parigina, anche se le sue fonti non ci sono note... [it is, in my opinion, difficult to doubt the fact that Galileo drew both the theorem and the essential elements of its demonstration from the medieval Oxford-Parisian tradition, even though his sources are not known to us...]"). Stillman Drake instead declared himself, on several occasions, decidedly opposed to the influence of the "medievals", emphasizing that the proportionality between acceleration and distance from rest, initially proposed by Galileo, is not compatible with the speculations of Oresme and the Mertonians regarding *uniformiter difformis* motion. They are, in fact, founded, as is known, on the direct proportionality acceleration-time (*Galileo's 1604 fragment on falling bodies. Galileo Gleanings* XVIII, in "The British Journal for the History of Science", 1969, n. 16, pp. 350 ff.; *Galileo and the Merton Rule*. Paper read at the History of Science IV, Annual Meeting, Dec. 1969, Washington; *Galileo's discovery of the law of free fall*, "Scientific American", vol. 228, 1973, n. 5, p. 75; and *Galileo's work on free fall in 1604*, in "Physis", XVI, 1974, fasc. 4, pp. 309-310). Particularly on the basis of the observation that no documents survive attesting to Galileo's knowledge of the Merton College theorem and its use (as is well known, the reduction of the accelerated speed to a uniform "mean" speed will, however, be carried out in the *Discorsi* and Galileo will even present it as *Theorema I*), today we are inclined to exclude that Galileo was subjected to a decisive influence by the authors of the 14th century prior to the letter to Sarpi and to admit, instead, as plausible the hypothesis that he used Merton's theorem and the analytical procedures characteristic of the logic of the *calculationes* in a later period, between 1606 and 1609, to base the examination of accelerated motion according to the proportion of time. Wisan, in particular, pointed out that analogies with the technical language and with the representational schema (the right-angled triangle) of terministic logic begin to appear more evident in the fragments dating back to the letter to Sarpi (*The new science of motion* cit., pp. 201, 204-207, 226 etc.). It should be recalled that, among other things, perhaps the most characteristic term of terministic logic, *gradus* [degree, step, level], appears for the first time in Galileo's lexicon precisely in the letter to Sarpi. C. J. T. Lewis revisited the issue of the influence exerted by the *calculatores* on Galileo, concluding that the hypothesis of Galileo's knowledge of the terministic texts is extremely unlikely (*The fortune of Richard Swineshead in the time of Galileo*, "Annals of Science", 33, 1976, pp. 561-564). W. A. Wallace recently dedicated an essay to the relationship between Galileo and the Parisian physicists (*Galileo Galilei and the "Doctores Parisienses"*, in R. E. Butts and J. C. Pitt ed., *New Perspectives on Galileo* (Dordrecht 1978, pp. 87-139)); Wallace founded his own investigation on the *Juvenilia*.

37 Reference to Koyré and his *Studi galileiani* is obligatory (see, primarily, pp. 82-84, of the trad. it. cit.). It is worth remembering, however, that S. Drake, in the aforementioned *Galileo's 1604 fragment*, from 1969, did not rule out the possibility that the Pisan scientist could have arrived at the law of fall by working on numerical series (cf. p. 349). Later, as we will see, the North American scholar changed his opinion.

issue[38], Winifred Wisan[39] and Ronald Naylor[40], who discussed it – on some pages **275**

[38] A detailed examination of the various hypotheses advanced by Stillman Drake in the numerous essays dedicated to Galileo's law of fall over the last few years would require space that a note cannot allow. We will therefore limit ourselves to outlining the main conclusions reached by the North American scholar. Drake initially stated that Galileo would have reached the belief that, in a natural fall, the spaces traveled are as the square of the times, based on experimental observations of descents along moderately inclined planes (*Galileo's 1604 fragment* cit., pp. 348-349). Subsequently, he became convinced that the experiment did not play a decisive role in the discovery; Galileo would have arrived at it through the incidental use, in page 152*r* of Ms. 72 (only partially published by Favaro), of two square numbers (4 and 9) to indicate the distance traveled by "immaginati [imagined]" natural motions with different speeds (*Galileo's discovery* cit., p. 89; page 152*r* is reproduced on p. 84). The importance of the experiment – this time for the determination of the laws of motion of projectiles – was claimed again by Drake based on the reconstruction of an experimental test carried out by Galileo, most certainly in Padua, and recorded on page 166*v* of the Ms. 72 (*Galileo's experimental confirmation of horizontal inertia: unpublished manuscripts*, "Isis", LXIV, 1973, pp. 291-305). Lastly, Stillman Drake returned to the problem of the law of fall, admitting that he was induced to revise his opinions as a result of a thorough examination of other pages of Ms. 72, not published by Favaro. In essence, he believed he had found evidence of an experiment carried out by Galileo in some numerical values in the column recorded on page 107*v*. Given that the calculations of folio 107 precede, according to Drake, the discovery of the law of fall, the latter would have been obtained by Galileo on an experimental basis (*Galileo's work on free fall* cit., pp. 312 ff.).

[39] Wisan's important essay, cited several times, is primarily dedicated to examining the mathematical-geometric procedures used by Galileo in the analysis of movement. For this reason, Wisan does not pay special attention to the problem of how Galileo reached the law which, in free descent, places spaces directly proportional to the square of times. Moreover, Wisan's essay precedes Drake's recent work (*Galileo's work on free fall* cit.), in which the experiment is claimed to have a decisive role in the discovery of the law of fall. The American scholar therefore limits herself to reviewing the different hypotheses put forward by Drake in the essay cited above, published in "Scientific American", of which she shows several aspects to be unconvincing, if not downright contradictory. Wisan's way of interpreting the meaning of the figure and notes on sheet 152*r* is substantially different (cf. *The new science of motion* cit., pp. 210-215; in particular, for the judgment on the Drake hypothesis, cf. note 9, p. 214). Wisan is also credited for reporting a fragment, previously overlooked, in which the correct law of fall (acceleration-time proportionality) is deduced from the supposition of the parabolic trajectory of the projectiles (pp. 227-229; the fragment in question is found on page 91*v* of Ms. 72).

[40] A radical critique of Drake's works is contained in a recent work by Ronald Naylor. He emphasizes the frequent shifts in the American scholar's position, the numerous unfounded statements (on watermarks, for example), and the many distortions applied in the reading of the documents (*Galileo's theory of motion: Processes of conceptual change in the period 1604-1610*, "Annals of Science", XXXIV, 1977, pp. 365-392; for the critical examination of the interpretation of pages 152*r* and 107*v* of Ms. 72 by Drake, see pp. 367-377). Naylor also carefully reviews numerous pages of Wisan's essay. He appears to share many of her observations, even though he puts forward a different interpretation of the notes and the figure on page 152r. In essence, Naylor carefully disputes Drake's claim that page 107 preserves proof of an experiment that was decisive for the discovery of the law of fall. He argues, however, that page 91*v* (reported by Wisan) and 116*v* (brought to attention by Drake) indicate clear traces of experimental evidence that could have helped Galileo to reach the correct law of fall. On the significance of the experiment on page 116*v* and on the well-founded possibility that Galileo conceived it not to experimentally confirm the principle of inertia (as supported by Drake in *Galileo's experimental confirmation* cit.), but rather to verify the law of fall, see the previous essay by Naylor, *Galileo and the problem of free fall* ("British Journal for the History of Science", VII, 1974, n. 76, pp. 105-134), which also presents the results of the repetition of the experiment. In a lengthy, argumentative review of Wisan's cited essay, Naylor objected to the American scholar that she had underestimated the role of the experiment in the genesis of the *De motu locali*, insisting too much on the examination of logical-mathematical procedures, in the attempt to reconstruct a unified account of Galilean thoughts between 1602 and 1610, which appear to us, so to speak, chopped up in the mass of the *Frammenti* of

of Ms. 72, in which there remains compelling evidence of experiments carried out by Galileo to establish the law of fall.

276 If we must withhold judgment regarding the manner in which Galileo originally arrived at the law of fall[41], there still remain more than plenty of indications of the effort made by the Pisan scientist to construct, starting from static principles, a new conception of movement. This effort, in which the mathematical law of fall should undoubtedly be included as its most fruitful achievement, appears evident not only from the letter to Guidobaldo but also from countless other testimonies, especially from some of the *Frammenti* pertaining to the *Discorsi*.

Before moving on to the examination of these fragments and other fundamental epistolary documents, it should be emphasized that, in the letter to Sarpi, there is no mention of *momento*. For the first time, expressions appear such as "grado di velocità [degree of speed]", which indicates the speed reached, in natural motion, at a given distance from the *terminus a quo* [the starting point], and "grado di impeto [degree of impeto]", which, however, expresses the same concept in the case of the upward motion of projectiles. These are phrases, especially the first, with which we will soon become familiar. For now, it is enough to point out that "impeto", until now always felt to be very close to "velocità [speed]", appears here more characteristically linked to speed impressed with violence. "Grado [Degree]" – it is worth noting – is a term very dear to the logic of the *calculationes*, to express the *intensio* [intensity] of forms. *Gradus velocitatis* [degree of speed], in Heytesbury like in Bradwardine, indicates a certain level of "estensione [extension]" of motion. It is always *gradus* that occurs in medieval discussions on accelerated motion[42]. With the exception of the justifiably suspected *Juvenilia*, *grado*/*gradus* in Galileo's writings *never* appears before the letter to Sarpi – it is a fact that no one has previously observed and on which it would be worth reflecting – not even in the *De motu*, where, on various occasions, our scientist compares acceleration to alterative motions[43].

Of the *Frammenti* relating to the *Discorsi* that have reached us, the oldest that records the erroneous supposition of the direct proportionality of acceleration-distance from the starting point is likely the one preserved on page 128 (*r.* and *v.*)

Ms. 72 ("Annali dell'Istituto e Museo di Storia della Scienza di Firenze", I, 1976, fasc. 2, pp. 88-97). Touching on the general themes posed by Naylor, Wisan responded promptly, denying that she had overlooked the evidence of Galileo's experimental engagement, but admitting, in essence, that she considered 1) that the experiment carried out primarily had a function of verification and 2) that Galileo himself does not consider experimental verification capable of providing an acceptable foundation for the *de motu* theorems (*Mathematics and experiment in Galileo's science of motion*, "Annali dell'Istituto e Museo di Storia della Scienza di Firenze", II, 1977, fasc. 2, pp. 149-160).

[41] Not inappropriately, it seems to me, W. Wisan has emphasized the practical impossibility of determining "precisely how Galileo came to think of the times-squared law" (*Mathematics and experiment* cit., p. 154).

[42] Cf. M. Clagett, *La scienza della meccanica*, trad. it. cit., Ch. IV (*L'emergere della cinematica di Merton College*), pp. 229-282.

[43] *G.G.*, I, pp. 334-337. Undoubtedly, this is a singular fact, which seems to instinctively suggest an influence of school texts on Galileo in Padua.

of Ms. Galil. 72. The fragment is handwritten.

Io suppongo (e forse potrò dimostrarlo) che il grave cadente naturalmente vada continuamente accrescendo la sua velocità secondo che accresce la distanza dal termine onde si partì: come v.g., partendosi il grave dal punto *a* e cadendo per la linea *ab*, suppongo che il grado di velocità nel punto *d* sia tanto maggiore che il grado di velocità in *c*, quanto la distanza *da* è maggiore della *ca*, e così il grado **277**
di velocità in *e* esser al grado di velocità in *d* come *ea* a *da*, e così in ogni punto della linea *ab* trovarsi con gradi di velocità proporzionali alle distanze de i medesimi punti dal termine *a*. Questo principio mi par molto naturale, e che risponda a tutte le esperienze che veggiamo negli strumenti e machine che operano percotendo, dove il percuziente fa tanto maggiore effetto, quanto da più grande altezza casca: e supposto questo principio, dimostrerò il resto[44]. [I suppose (and perhaps I will be able to demonstrate it) that the naturally falling body continuously increases its speed as the distance from the terminus from which it begins increases: as shown, starting from point *a* (see fig. 22) and falling along the line *ab*, I suppose that the degree of speed at point *d* is as much greater than the degree of speed at *c*, as the distance *da* is greater than *ca*, and thus the degree of speed at *e* is to the degree of speed at *d* as *ea* is to *da*, and so in every point of the line *ab* there are degrees of speed proportional to the distances of the same points from the terminus *a*. This principle seems very natural to me, and it corresponds to all the experiences that we see in instruments and machines that operate by striking, where the percussive object has a greater effect the more it falls from a greater height: and assuming this principle, I will demonstrate the remainder.]

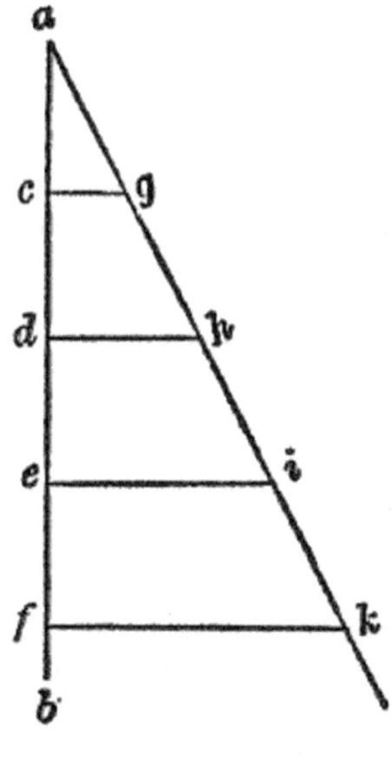

Figure 22

We are, therefore, at the same level of understanding as the letter to Sarpi, and, once again, we see the expression "grado di velocità [degree of speed]"

[44] G.G., VIII, pp. 373-374. The fact that the note is written in the vernacular, unlike the vast majority of the other *Frammenti*, suggests that Galileo intended it for private use. According to Stillman Drake, who assumed a close relationship between the letter to Sarpi and the vernacular fragment, the latter constitutes the explanation that Galileo provided for the incorrect principle and the law of fall, shortly after the letter of 16 October 1604 (*Galileo's 1604 fragment* cit., pp. 343-344).

appear. Galileo does not have the demonstration of the assumed principle, but he considers it very natural and compliant with "esperienze [experience or experiment]". Just in illustrating its conformity to experience, Galileo introduces an important clarification: the principle corresponds to the indications provided by mechanical science, where we observe the effect of the impact of a heavy body being doubled when the "viaggio [journey]" of the mobile is doubled[45].

Here, then, is yet another confirmation that Galileo has the *Mecaniche* continually in mind and that he attempts to develop his own dynamic investigation based, as much as possible, on the same principles used in the Paduan treatise. The erroneous supposition thus appears, with even greater clarity, as the consequence of this effort towards unification.

The "resto [remainder]" that Galileo promises to demonstrate, and which he had neglected to specify in the letter to Sarpi, is the series of implications connected to the supposition of direct proportionality between acceleration and departure from the starting point. Take the case of a heavy body that descends freely from *a* to *b* (see fig. 22) and, taking *ak* arbitrarily inclined, draw the parallels *cg*, *dh*, *ei*, *fk*, which will be between them like the spaces *ac*, *ad*, *ae* etc. The speeds at *c*, *d*, *e* etc., will, therefore, be among themselves as *cg*, *dh*, *ei* etc.,

vanno dunque continuamente crescendo i gradi di velocità in tutti i punti della linea *af* secondo
278 l'incremento delle parallele tirate da tutti i medesimi punti. In oltre, perché la velocità con la quale il mobile è venuto da *a* in *d* è composta da tutti i gradi di velocità auti in tutti i punti della linea *ad*, e la velocità con che ha passato la linea *ac* è composta di tutti i gradi di velocità che ha auti in tutti i punti della linea *ac*, adunque la velocità con che ha passata la linea *ad*, alla velocità con che ha passata la linea *ac*, ha quella proporzione che hanno tutte le linee parallele tirate da tutti i punti della linea [*ad* sino alla *ah*, a tutte le parallele tirate da tutti i punti della linea][†] *ac* sino alla *ag*: e questa proporzione è quella che ha il triangolo *adh* al triangolo *acg*, ciò è il □° *ad* al □° *ac*. Adunque la velocità con che si è passata la linea [*ad*, alla velocità con che si è passata la linea][‡] *ac*, ha doppia proporzione di quella che ha *da* a *ca*. E perché la velocità alla velocità ha contraria proporzione di quella che ha il tempo al tempo (imperò che il medesimo è crescere la velocità che sciemare il tempo), adunque il tempo del moto in *ad* al tempo del moto in *ac* ha subduplicata proporzione di quella che ha la distanza *ad* alla distanza *ac*. Le distanze dunque dal principio del moto sono come i quadrati de i tempi, e, dividendo, gli spazii passati in tempi eguali sono come i numeri impari *ab unitate*: che risponde a quello che ho sempre detto e con esperienze osservato; e così tutti i veri si rispondono[46]. [the degrees of speed at all points of the line *af* are therefore continually increasing according to the increase in the parallels drawn from all the same points. Moreover, because the speed with which the mobile came from *a* to *d* is composed of all the degrees of speed acquired at all the points of the line *ad*, and the speed with which it traversed the line *ac* is composed of all the degrees of speed acquired at all the points of the line *ac*, therefore the speed with which the line *ad* is traversed, to the speed with which the line *ac* is traversed, has that proportion that all the parallel lines drawn from all the points of the line [(*ad* up to *ah*, to all the parallels drawn from all the points of the line)] *ac* up to *ag*: and this proportion is the one that the triangle *adh* has to the triangle *acg*, that is the square on *ad* to the square on *ac*. Therefore the speed with which the line [(*ad* is traversed, to the speed with which the line)] *ac* is traversed has duplicate proportion to that of *da* to *ca* [i.e., the speeds are as the squares of the distances]. And because speed to speed has the contrary

[45] *G.G.*, II, pp. 185-186. The expression is used several times to express the path of force in the *Mecaniche*.

[†] This was omitted from Galluzzi's recitation, now restored.

[‡] This was omitted from Galluzzi's recitation, now restored; see also [281] for the intended recitation.

[46] *G.G.*, VIII, pp. 373-374.

proportion to that of time to time (since increasing speed is the same as decreasing time), therefore the time of motion along *ad* to the time of motion along *ac* has a subduplicate proportion of that which the distance *ad* has to the distance *ac* [i.e., the times are as the square roots of the distances]. The distances therefore from the beginning of the motion are as the squares of the times, and, dividing, the spaces traversed in equal times are as the odd numbers *ab unitate* [from unity]: which corresponds to what I have always said and observed through experience; and thus all the truths correspond to one another.]

This demonstration must have been known to Galileo since the letter to Sarpi. The figure used and some of the procedures used are the same as those proposed later in the *Dialogo* and in the *Discorsi*. We are faced with a text that presents, in embryonic form, all the subsequent developments of Galileo's conception of natural movement, but in which significant conceptual errors and representational limitations are evident.

The procedure suggested by Galileo requires no explanation, as it is so crystal clear. It is worth stopping, however, to emphasize some undeclared suppositions that he uses and which he will later try, albeit confusingly, to justify. The degrees of speed are depicted by the parallels to the base of the triangle. It is evident that each of these degrees is greater than the previous one and corresponds (indicating the descent with the line *ab*) to a precise distance from the *terminus a quo* [starting point]. Then Galileo adds that the "velocità [speed]" with which the mobile descends from *a* to *c* is composed of all the degrees of speed reached along *ac*, that is, by the set of infinite parallels to the base drawn by the infinite points that constitute *ab*. But these infinite parallels constitute the triangle *acg*, just as the infinite parallels, drawn by the infinite points of *ad*, constitute the triangle *adh*. Therefore the speeds along *ac* and *ad* are as the triangles *acg* and *adh*, that is, as the squares of the distances.

Now, it is evident that, in this type of demonstration, Galileo made use of a whole series of concepts and suppositions relating to the composition of geometric quantities, even suggesting that the proportionality between two lines or between two surfaces can be resolved into the proportionality of the points or lines (both infinite) from which they are constituted. We will later see that Galileo recognized the need to find an acceptable justification for this type of procedure, and at that time, **279**
we will evaluate its true meaning[47]. For now, it is enough to have indicated that this type of analysis is certainly the same that Galileo used at the time of the letter to Sarpi, that is, in 1604. The analogy with the epistolary document, in case it needs to be reinforced, is further suggested by the end of our fragment, where once again we see the close analogy proposed between the increase in speed in the natural fall and the decrease in speed in violent vertical motion. The same parallelism can be observed on a terminological level. Just as in the letter to Sarpi, *momento* is absent, "grado di velocità [degree of speed]" is frequent, with the connotation indicated above, and a constant alternation between "grado di velocità [degree of speed]" (natural fall) and "impeto" (violent motion). The only difference is that Galileo does not use "grado di impeto [degree of impeto]" here, but simply "impeto".

Historians have often focused their attention on our fragment. An intense debate

[47] Cf., *infra*, Ch. V.

has also taken place in recent years regarding the meaning of Galileo's reflections made there. Or rather, it can be said that the discussion is still ongoing. The fragment makes the statements contained in the letter to Sarpi more explicit. And it offers an example of how Galileo was able to deduce from an incorrect principle (the increase in speed proportional to distance[†]) a correct conclusion: the distances traveled are as the square of the times. In essence, many have pointed out that the "crucial" phase of Galileo's demonstrative progress is assigning a "contraria [contrary]" proportion to the relationship between the times and the speeds. For Koyré, this is explained by the fact that the Pisan scientist – here's his error – would have established the inverse proportionality between speed and times *for different distances*, whereas it is only valid in the case in which the paths are equal[48]. In truth, Koyré's interpretation does not explain at all how Galileo could arrive at the correct conclusion that distances are as the square of the times. The fact is – as has been observed – that, in our context, Galileo makes an extremely peculiar, and certainly unacceptable, use of the "contraria [contrary]" proportion. Initially, he seems to, in fact, present it as an inverse relationship (the antecedent is to the consequent as the consequent is to the antecedent); but then he deduces a different conclusion. Making [speeds] V_1 and V_2 respective to the distances *af* and *ak* (see fig. 22)[‡], Galileo's reasoning can be schematized in the following way:

$$V_1 : V_2 = af^2 : ak^2$$

Therefore, invoking the "contraria [inverse]" proportionality,

$$t_1 : t_2 = \sqrt{af} : \sqrt{ak}$$

280 In other words, it seems that Galileo intends the relationship of the times as "contrario [contrary]" to that of the speeds (which are to one another as the squares of the distances), as it is directly proportional to the square roots of the distances. This is – obviously – an untenable assertion[49], of which, however, it is more important

[†] The parenthetical here originally recited "...the increase in speed proportional to *time*," which appears to be in error since this would recite the correct relationship for natural fall. Based on the context of the discussion, in which the incorrect principle is desired, the translator believes *distance* was intended.

[48] "È vero, senza dubbio, che i rapporti delle velocità sono in ragione inversa di quelli dei tempi; a condizione che il termine di comparazione, ovverosia lo spazio percorso, sia lo stesso e non, come nel nostro caso, differente [It is true, without a doubt, that the ratios of the speeds are inversely proportional to those of the times; provided that the comparison term, i.e., the distance travelled, is the same and not, as in our case, different]" (*Studi galileiani*, trad. it. cit., p. 103).

[‡] It is unclear why distance *ak* is chosen for this example since line *ak* is not a distance traversed by the heavy body in free fall. It is also unclear why the chosen distances *af* and *ak* differ from the example distances provided by Galileo, namely *ac* and *ad*. Thus, either the text contains a single error: that point *k* was mistaken as being located on line *ab* such that *ak* is a distance traversed; or a duplicate error: *af* and *ak* should be *ac* and *ad*. The latter appears more plausible since it would fall in line with Galileo's earlier demonstration.

[49] The first to denounce Koyré's curious misunderstanding was A. Rupert Hall, who emphasized the mathematical error committed by Galileo in asserting that, if the speeds are like the squares of the distances, then the distances will be as the squares of the times (*Another Galilean error*, "Isis", L., 1959,

to grasp the significance rather than to denounce the error[50] or try to justify it[51].

Let's start from the beginning. It will be useful to specify that, in our fragment, **281**
Galileo uses the term "velocità" with two different meanings. Right at the beginning, in fact, he declares that the "velocità" increases "secondo che accresce la distanza dal termine onde si parti [as the distance from the terminus from which it begins]". Here, "velocità" means "grado di velocità [degree of speed]", as Galileo himself defines it immediately afterwards: "in ogni punto della linea *ab* trovarsi con gradi di

n. 161, pp. 261-262).

[50] This is the case of Hall's essay, cited in the previous note.

[51] One cannot overlook the attempt at a comprehensive reconstruction of the meaning of the entire fragment and its inclusion in Galileo's reflections on naturally accelerated motion carried out in several essays by Stillman Drake. First of all, Drake tried to show that Galileo's "contraria proporzione [contrary proportion]" should not be confused with what we mean by "inverse proportionality". That is, the Pisan scientist would have been referring here not to relationships between pairs of termini, but to comparisons between infinite series of termini of the same quantity. He would have, in essence, observed that if the speeds are as the square roots of the spaces traversed from rest, the times are as the squares of the spaces. And speaking of "contraria proporzione [contrary proportion]" between speeds and times, Galileo would have intended to indicate the special relationship that exists between two infinite series (speeds and times) of the same "quantity" (the distances from rest), one according to the square roots, the other according to the squares. Therefore, according to Drake, it would not be an error by Galileo, but a correct statement, misunderstood by historians, who would not have reconstructed the genuine meaning of the expression "contraria proporzione" (*Galileo's 1604 fragment* cit., especially pp. 352-356). The weakest point of Drake's reasoning consists in the fact that, according to the North American scholar, Galileo started, in the demonstration of our fragment, from the awareness that the distances from rest increase, in equal times, as the series of odd numbers (*ibid.*, p. 354), which is precisely what the fragment seeks to demonstrate. It is worth adding, as evidence of the impracticality of the interpretation suggested by Drake, that Galileo uses the expression "contraria proporzione" on other occasions, always with the unequivocal meaning of "inverse proportion"; see, for example, a passage from the *Giornata Sesta* [Sixth Day] of the *Discorsi*: "...e così in tutte le altre proporzioni, quanto una cacciata fatta dal medesimo percuziente è più breve, tanto, per l'opposito, con *proporzione contraria* vi si ricerca, per far l'istesso, gravità maggiore di peso morto premente [...and so, in all other proportions, the shorter a thrust made by the same striker is, conversely, with a *contrary proportion*, a greater heaviness of pressing dead weight is required to do the same thing]" (*G.G.*, VIII, p. 340; the italics are ours). In a subsequent essay, Drake revisited our fragment to support an even more radical thesis. In essence, Galileo would never really have considered the so-called erroneous distance-acceleration proportionality as a viable hypothesis. From the first speculations, the Pisan scientist would have been driven by the belief that acceleration is proportional to time, but for a certain period (corresponding to the letter to Sarpi and to our fragment), he would have expressed himself in an equivocal manner, in which historians would have erroneously recognized the statement of acceleration-distance proportionality from rest. Drake's frankly paradoxical thesis is based on a distinction between the different meanings with which Galileo uses "velocità". In our fragment, at the opening where it is stated that speed increases according to the distance from rest, Galileo would have intended to indicate, not the speed, but the square of the speed. According to Drake, this conclusion is suggested by the analogy introduced by Galileo himself between an increase in speed and an increase in the effectiveness of mechanical instruments and percussion. Since, as Drake states, the effect of percussion and mechanical instruments "is governed not by speed, but by its square... the effect does not support a distance proportionality assumption in the sense that we have always imputed to Galileo's words" (*Galileo's discovery* cit., p. 92; cf. also, again by Drake, *Galileo's work on free fall* cit., pp. 319-321). Drake's operation leaves quite a bit of perplexity, such is the distortion of the literal meaning of the fragment, which is otherwise very clear, that accepting it entails. The American scholar's interpretation appeared unsatisfactory to many. In particular, Wisan (*The new science of motion* cit., p. 210) and Naylor have spoken out against it, who also contested against Drake that no traces of Galileo's demonstrations or calculations are found which use the incorrect proportionality distance from rest-acceleration (*Galileo's theory of motion* cit., p. 385).

velocità proporzionali alle distanze de i medesimi punti dal termine *a* [in every point of the line *ab* there are degrees of speed proportional to the distances of the same points from the terminus *a*]". The assumed principle indicates, therefore, that the degree of speed, not the speed *tout court*, increases proportionally to the distance.

In the subsequent fragment, Galileo introduces "velocità" again, but with a different meaning: "adunque la velocità con che si è passata la linea *ad* alla velocità con che si è passata la linea *ac* ha doppia proporzione di quella che ha *da* a *ca* [therefore the speed with which the line *ad* (see fig. 22) is traversed to the speed with which the line *ac* is traversed has duplicate proportion to that of *da* to *ca*]". This "velocità" is, in fact, the "overall" speed, which is not declared proportional to the distances, but rather to the square root of the distances. The distinction between the two "velocità" is important, because it allows us to understand how Galileo was able to state, in the same fragment, first that speeds are proportional to distances and, a few lines later, that they are proportional to the square roots of distances[52].

As, then, for the justification he gives to assert the contrary proportionality between "mean" speeds and times, it should be noted that in it he used a consideration that, from his point of view, was legitimate only for instantaneous speeds: "e perché la velocità alla velocità ha contraria proporzione di quella che ha il tempo al tempo (imperò che il medesimo è crescere la velocità che sciemare il tempo)... [and because speed to speed has the contrary proportion to that of the times (since increasing speed is the same as decreasing time)...]". If, in naturally accelerated motion, two points or minimal equal portions of space are taken along *ab*, the speeds (which, for such minimum distances, can be assumed uniform) with which they will be traversed are inverse to the times. This is presumably the meaning of Galileo's statement. Naturally, it remains to be justified how he could erroneously extend the inverse proportionality between instantaneous speeds and times to the relationship
282 between "overall" (or "mean") speeds and times, deducing, moreover, from this untenable premise the correct law of fall: the spaces from rest, in natural motion, are as the square of the times. He probably knew that such a relationship existed between spaces and times and did not pay too much attention to the subtleties in the effort to demonstrate it[53].

The fragment in the vernacular on which we have focused upon is of considerable significance. The paralogisms and the mathematical and conceptual errors committed by Galileo tell us a great deal, in fact, about the directions and characteristics of his reflections on movement around 1604.

The most evident limitations that can be seen concern the geometric "representation" of the quantities examined; these include terminological difficulties that are no less apparent.

---

[52] The transition from the direct proportionality of distance from rest-acceleration to a speed-square proportionality of distances depends, obviously, on the assimilation of the "overall" speed to the area of the triangle (i.e., to a two-dimensional element), while the degree of speed remains, like the distance, represented in the figure by a segment (i.e., by a one-dimensional element).

[53] This is what Humphreys, among others, also claimed (*Galileo, falling bodies and inclined planes* cit., p. 231).

The depiction used warrants some commentary. If *ab* indicates the distance of fall (see fig. 22), *cg* and *db* will be the degrees of speed – progressively increasing – with which equal portions of distance are traversed (let’s say all the infinite points of *ab*). It follows, necessarily, that the points of *ab* will be traversed in progressively shorter times and such that they are in inverse proportion to each other in terms of the degrees of speed (which evidently, for such infinitesimal distances, can be considered uniform). Now – reasoning in Galilean terms – the “inverse” proportionality between instantaneous speeds along equal points of *af* and the times taken to traverse them with such speeds, becomes a “direct” proportionality, when the overall speeds for two distances are compared (i.e., the sums of all the progressively increasing instantaneous speeds for the individual points of these distances) and the times taken to traverse them (i.e., the sums of the times in which the equal points of the two distances were passed). But the Pisan scientist did not make this consideration, which appears elementary to us, and instead maintained, even for the overall speed, the inverse proportionality with time.

The error – which did not, however, prevent him from establishing the correct proportion between spaces and times – depended, first of all, on the inadequacy of the representation. In fact, in the figure used by Galileo, there is no “indice” of time, while for the space of fall, the overall speeds and the degrees of speed, there are precise indices. In short, Galileo was unable to “visualize” the behavior of time, as he would do later, by modifying the representation schema. Thus, it may have escaped him that time (taken as a whole) is directly, not inversely, proportional to the total speed, even though it is true that – in free fall accelerated proportionally to the distance from rest – along minimum equal spaces of fall, at distances different from rest, the speeds are inversely proportional to the times.

The error – as well as the inadequacy of the representation – was facilitated by **283**
terminological confusion. In essence, Galileo does not conceptually distinguish clearly the two speeds (degree of speed and “mean” speed), and this lack of distinction, facilitated by the use of the same term to indicate them, prevents him from properly realizing that the inverse speed-times proportionality cannot be extended to the ratio between overall speeds and overall times, with which different distances are traversed *starting from rest*.

We are faced with another betrayal of the language which – as we will soon see – will produce serious repercussions to *momento* as well. The fragment in the vernacular reveals the enormous difficulties that Galileo initially had to face in the attempt to establish the relationships between space, time and speed on a geometric-mathematical level, in the hypothesis of a continuously accelerated natural motion. The emphasis on the increase in the speed of this motion in proportion to the distance from the starting point was evidently suggested both by the analogy with the motions of mechanical instruments and by the static *momento*. Certainly – as Koyré observed – it also responded to the need for “geometrization” strongly felt by Galileo[54]: time appeared to our scientist much

[54] *Studi galileiani*, trad. it. cit., p. 101.

less easily geometrizable than space. It is not surprising, therefore, that he initially sought to anchor speed and its rate of change to distance rather than time.

The 1604 fragment in the vernacular, as well as the letter to Sarpi, concluded, however, with the statement of the correct proportion between distances traveled and times: $d_1 : d_2 = t_1^{(2)} : t_2^{(2)}$. This could, therefore, for a certain number of years – at least until 1608, according to prevailing opinion – allow Galileo to demonstrate a considerable number of theorems[55]. The erroneous principle will be definitively abandoned when, after achieving a precise clarification of the terminological difficulties and developing – as we will soon see – a representation schema in which each of the quantities compared can be easily and conveniently visualized, it will appear incompatible with the laws that Galileo had deduced from it[56].

284 We will not subject this complex development to a detailed analysis, which would require the space of an entire volume. Instead, we will limit ourselves to following the aspects that most clearly document Galileo's awareness of the terminological inadequacy and the limits in geometric schematization, as well as the attempts he made to resolve them. The introduction of *momentum* with a meaning different from the one examined so far can, in fact, be fully understood only from this perspective.

"Così tutti i veri si rispondono [Thus all the truths correspond to one another]", as Galileo triumphantly concluded in the vernacular fragment; "è vero quanto ho detto e creduto fin qui [what I have said and believed up to now is true]", he declared in the letter to Sarpi. What did our scientist mean with these words? He evidently believed, by assuming the incorrect principle, that he had brought mechanical motions, natural falling motions and violent motions back to a substantial unity of conception and treatment, in all of which the speed (and, therefore, every other effect) varies with

---

[55] Cf., for example, W. Wisan, *The new science of motion* cit., pp. 207-222.

[56] Of the fragment based on the erroneous supposition, we have also retained, again in Ms. 72, a slightly modified Latin version (*G.G.*, VIII, p. 383). The fragment is by Guiducci's hand and is canceled by two crossed lines. This is, in all likelihood, the draft that Galileo intended to include in the *De motu locali*, designed in Latin. The Pisan scientist, in the initial part, uses *velocitas* with the dual meaning of instantaneous speed and overall speed. He again proposes the proportionality of (overall) speeds to the squares of distances. As in the vernacular fragment, although in a much clearer manner, he states that the increase in speed entails, for equal spaces, a proportional decrease in time. In this part, *velocitas* is replaced by *acceleratio*: "pro ratione incrementi accelerationis tempora quibus motus ipsi fiunt debent imminui [in proportion to the increase in acceleration, the times in which the motions themselves occur must decrease]". The greater terminological precision, obtained with the use of *acceleratio* [acceleration], does not prevent Galileo from mistakenly deriving – as in the fragment – the proportion between overall times and overall speeds based on the relationship between instantaneous speeds and times taken to traverse equal spaces with these speeds, at different distances from rest. This time, however, instead of concluding that the spaces of fall are as the squares of the times, Galileo states that the times are to each other as the first distance to the mean proportional between the first and second distances. This is a different formulation of the previous relationship, to which it can be attributed. The figure used in the Latin fragment is substantially equivalent to the one used in the vernacular version. The only difference is the presence of a segment, alongside the triangle, with the indication *S*. In any case, an index of time remains absent. According to Wisan, the Latin version precedes the vernacular version. I, on the contrary, particularly given his use of the mean proportional, am inclined to postdate the Latin fragment.

the distance from rest[57]. For a few more years, Galileo would strive to grasp the relationships between statics and dynamics more firmly, eventually lending the term and concept of *momento* to the developing science of movement, which had played a central role in the Paduan treatise on mechanical principles.

Once more we must rely on the *Frammenti* to reconstruct the line of thought and investigation pursued by Galileo in these years. An incomplete handwritten fragment, again from Ms. 72, presents a Latin note presumably based still on the erroneous supposition. The depiction technique used is identical to that of the vernacular fragment:

Si in linea naturalis descensus a principio lationis sumantur duae distantiae inaequales, momenta velocitatis cum quibus mobile permeat illas distantias sunt inter se in duplicata proportione ipsarum distantiarum.

Sit in linea naturalis descensus *ab*, in qua ex principio lationis *a* sumantur duae distantiae *ac*, *ad*: **285**
dico momenta velocitatis cum quibus mobile permeat *ac* esse in duplicata proportione distantiarum *ad*, *ac*. Ponatur linea *ae* ad *ab* quemlibet angulum continens...[58].

[If, in the line of natural descent, two unequal distances are taken from the starting point of motion, the momenta of speed with which the mobile traverses those distances are to one another in the duplicate proportion of the distances themselves.

Let there be in the line of natural descent *ab* (see fig. 23), in which, from the starting point of motion *a*, two distances *ac* and *ad* are taken: I say that the momenta of speed with which the mobile traverses *ac* are in duplicate proportion of the distances *ad* and *ac*. Let the line *ae* be drawn forming any angle with *ab*...]

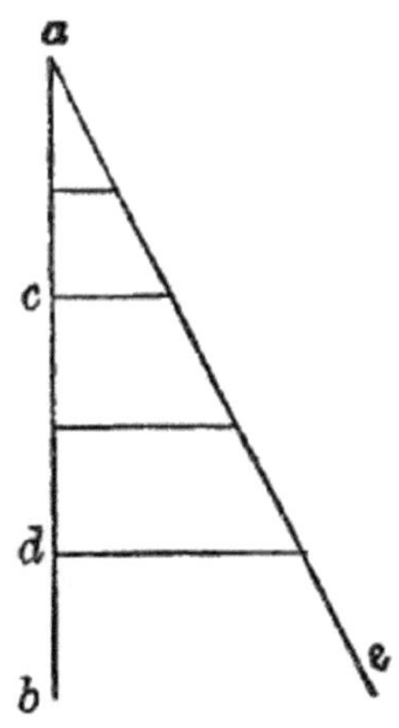

Figure 23

The law is not new. That the overall speeds are as the squares of the distances

[57] "Videatis, ergo, quam bene vera inter se congruant [See, therefore, how well the true things agree with each other]", Galileo had declared, in the *De motu* (*G.G.*, I, p. 322), after having demonstrated the close analogy between natural motions and violent upward motions.
[58] *G.G.*, VIII, p. 380; the note is on folio 179*v*.

traversed had already been stated in the vernacular and Latin fragments previously examined, both based on the erroneous supposition. However, here we encounter the expression *momentum velocitatis* for the first time. It clearly appears equivalent to "grado di velocità [degree of speed]" or "velocità istantanea [instantaneous speed]", used by Galileo in the vernacular fragment and in the letter to Sarpi. In this way, Galileo manages to clarify more effectively that the overall speed is the sum of a series of instantaneous speeds, which he defines as *momenta velocitatis*. Why a new term? Does it, then, have a particular meaning?

The fragment is incomplete and appears covered by two strips of paper, which bear other handwritten notes by Galileo. According to Favaro, the "due strisce di carta furono incollate l'una sopra l'altra sul *tergo* della carta 179, probabilmente da Galileo stesso [two strips of paper were glued one on top of the other on the *back* of folio 179, probably by Galileo himself]"[59]. On the strips are recorded the two laws of the momenti of heavy bodies along variously inclined planes, which are well known to us since the *De motu*. We only recall the statement, since the demonstrations correspond to those already illustrated previously:

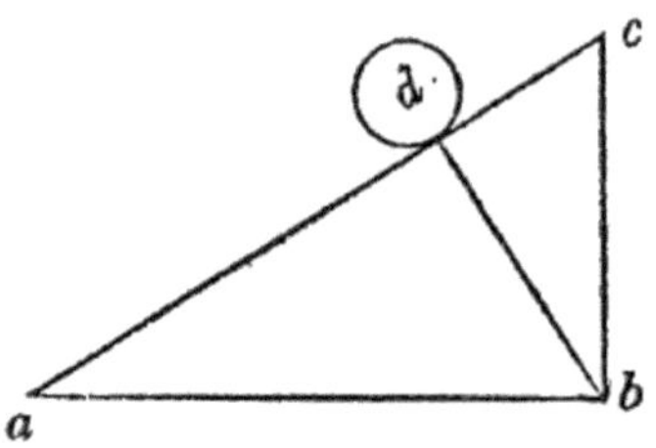

Figure 24

Momenta gravitatis eiusdem mobilis super plano inclinato et in perpendiculo permutatim respondent longitudini et elevationi eiusdem plani.... Id autem in mecanicis probatum est. [The momenta of heaviness of the same mobile on an inclined plane and in the perpendicular (see fig. 24) respectively correspond to the length and elevation of the same plane.... This has been proven in mechanics.]

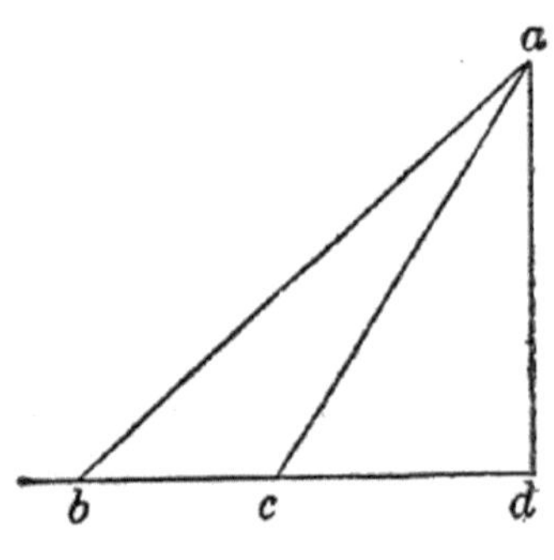

Figure 25

[59] *Ibid.*, p. 377.

Momenta gravitatis eiusdem mobilis super diversas planorum inclinationes habent inter se permutatim eandem rationem quam eorumdem planorum longitudines, dum eidem elevationi respondeant...[60]. [The momenta of heaviness of the same mobile on different inclinations of planes (see fig. 25) have between them a reciprocal ratio the same as that of the lengths of those planes, while corresponding to the same elevation...] **286**

It is a curious fact, but one whose meaning can be guessed. Galileo repeats, first, that, in a natural fall, the overall or mean speed is proportional to the square of the distances from rest. In doing so, he introduces the expression *momentum velocitatis* – this is the first time we have encountered it – instead of *gradus velocitatis*, or *velocitas*, which he had used before. Then, he leaves his considerations incomplete and covers the note with two strips of paper, which record the laws of *momenta gravitatis* along the inclined planes and the perpendicular of equal elevation.

If, as it seems, Galileo himself concealed the first fragment, it is difficult to escape the impression that the reason lies in his awareness that proposing the proportion between spaces and speed suggested in the vernacular fragment was no longer a possibility. The introduction of *momentum velocitatis* seems to anticipate Galileo's decision to make a new attempt at analysis, even more strictly dependent on mechanical principles. It was a matter – as we will see – of courageously exploring another path: could the analysis of natural motion be attributed to the *momento della gravità* [momento of heaviness] and its laws? The operation was undoubtedly daring and far from simple. To base the analysis of naturally accelerated motion, composed of progressively increasing *momenta velocitatis*, on the *momento della gravità* meant reconsidering all of mechanics in terms of acceleration. A constant force (heaviness) would, therefore, have generated not a constant effect (uniform speed), but rather continuously increasing *momenti di gravità* [momenti of heaviness], from which proportional accelerations (*momenta velocitatis*) would have arisen. The *momenti* of the heavy body, and therefore its degrees of speed, would then have increased as the mobile moved away, in its free descent, from its starting point.

That Galileo, at a certain point, considered this plan possible, that he made a **287**
notable effort to realize it and that in this attempt the central goal was to suggest a very close connection between *momentum gravitatis* and *momentum velocitatis*, is demonstrated by a certain number of fragments and some epistolary documents[61].

---

[60] *Ibid.*, p. 376.

[61] This is what Wisan recognized (*The new science of motion* cit., pp. 220-222), according to whom this new approach, consisting in the use of the dynamic postulate of the *De motu* and the theorem of inclined planes from the *Mecaniche*, should be placed, essentially, at the end of the Paduan period. Previously (in the letter to Sarpi and in the vernacular fragment, for example), Galileo attempted to base his treatise on exclusively geometric-mathematical procedures, which appear in some way attributable to the kinematics of the 14th century. According to Wisan, the return to an attempt at a mechanical foundation, after 1604, would have been suggested to Galileo by the difficulties encountered in the attempt to correctly deduce the proportionality of spaces to the square of times. I, on the contrary, believe that, from the letter to Guidobaldo up to Valerio's response (1609), Galileo devoted himself with substantial continuity to the same attempt: to base the science *de motu naturaliter accelerato* on the same principles used in the *Mecaniche*. In this, my investigation confirms what Thomas Settle argued in his unpublished dissertation. The examination of the fragments, which Settle had not carried out, allows, in fact, to emphasize the substantial continuity of Galileo's reflections from the *Mecaniche* to the letter

This means – as it is worth emphasizing – that the Pisan scientist consciously coined the expression *momentum velocitatis* (of which no trace is found before 1604 and whose introduction is, presumably, to be assigned to the two-year period 1606-1608) precisely on the model of the *momento della gravità* [momento of heaviness] (or *del peso* [of weight]) so frequently used in mechanics. *Momentum velocitatis* thus came to express, in Galileo's intentions during these years, the peak speed (or final speed) corresponding to a specific *momento della gravità* reached by the body in free descent at a certain distance from rest. The overall speed would have resulted from the integration of all the different *momenta velocitatis*. The increase
288 in *momento della gravità* and, therefore, *della velocità* is still conceived by Galileo in terms of distance from rest. We will see, however, that the Pisan scientist will soon reach the new understanding of the direct proportionality between *momenta velocitatis* and time of fall from rest. But, by that point, the relationships between the "static" *momento* and the "dynamic" *momento* will have already undergone a sharp attenuation, a clear prelude to the subsequent, radical separation of the two concepts.

It should be recalled that Galileo had shown, since the *De motu*, that he harbored the belief that speed varies proportionally to the variation in the effectiveness of heaviness (i.e., its *momenti*). It was, therefore, a matter of returning enthusiastically to work on an old project, to be approached, however, with new methods of analysis, imposed by the awareness that acceleration is a natural phenomenon.

It is still the notes of Ms. 72 that we must rely on to search for traces of this subsequently abandoned foundation attempt. A handwritten fragment, on folio 172*v*, presents a demonstration with which Galileo attempts to establish a relationship between the *momento della gravità* and accelerated speed.

From point *d* (see fig. 26), draw two inclined planes *ad*, *bd* from the horizon. Let *de* be perpendicular to *bd* and let the triangle *bde* be circumscribed by a circle, whose perpendicular diameter is *eb*. Lastly, let's draw the straight line *fb*, parallel to *ad*:

---

of Valerio and to illustrate, on a documentary basis, the emergence of *velocitatis momentum* and its connotation as term and concept derived from the "mechanical" *momento*. In summary, for a long time, in Galileo's eyes, extremely blurred boundaries existed between statics and dynamics (cf. Th. Settle, *Galilean Science* cit., especially Ch. V, pp. 201 ff.). It is worth emphasizing that Maurice Clavelin, in his well-known work on Galileo, in a group of very important pages, also expressed the conviction that the "mechanical" notion of the *momento del discendere* [momento of descent] (or *della gravità* [of heaviness]) deduced from the inclined plane theorem of the *Mecaniche*, seemed at a certain point to the Pisan scientist capable of providing the interpretative model for the way the natural driving force (the *gravità* [heaviness]) acts, justifying the acceleration of free falling motions. However, Clavelin, who based his work exclusively on the text of the *Discorsi*, neglecting the *Frammenti* published by Favaro, believed that this attempt should be assigned to the last years of Galileo's activity, after the publication of the *Discorsi*. In his opinion, the mechanical theorem added at the suggestion of Viviani would unequivocally attest to this (cf. *G.G.*, VIII, pp. 214-219). In essence, instead of seeing it as an original project, subsequently abandoned due to a series of difficulties, on which we will focus later, Clavelin considered the use of *momento del discendere* to establish the naturally accelerated motion as the fruit of a final attempt (cf. M. Clavelin, *La philosophie naturelle de Galilée* cit., pp. 352-359 and 372-373).

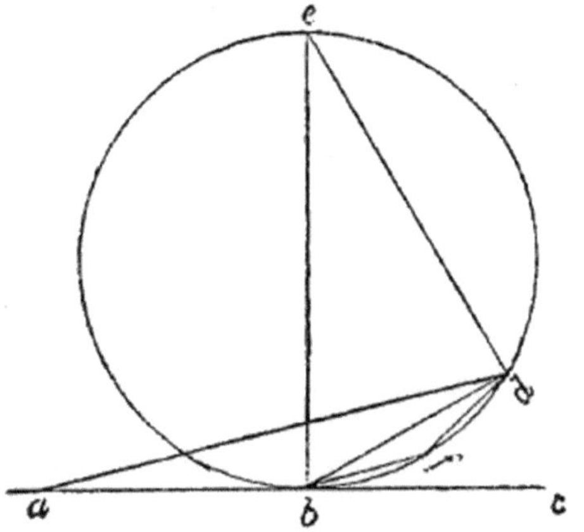

Figure 26

Patet, tarditatem per *fb* esse consimilem tarditati per *da*; quia vero tempore eodem movetur mobile per *db* et *fb*, patet velocitates per *db* ad velocitates per *fb* esse ut *db* ad *fb*, ita ut semper iisdem temporibus duo mobilia ex punctis *d*, *f*, venientia, linearum *db*, *fb* partes integris lineis *db*, *fb* proportione respondentes peregerint. Cum vero angulus *bfd* in portione angulo *dba* ad tangentem sit aequalis, angulus vero *dbf* alterno *bda*, aequiangula erunt triangula *bfd*, *abd*, et ut *bd* ad *bf*, ita *ad* ad *db*: ergo ut *ad* ad *db*, ita **289**
velocitas per *db* ad velocitatem per *da*, et, ex opposito, tarditas per *da* ad tarditatem per *db*[62]. [It is evident that the slowness along *fb* is similar to the slowness along *da*; and because the mobile moves through *db* and *fb* in the same time, it is evident that the speeds along *db* to the speeds along *fb* are as *db* is to *fb*, so that, always in the same times, two mobiles starting from points *d* and *f* traverse portions of the lines *db* and *fb* proportional to the entire lengths of *db* and *fb*. Since the angle *bfd* in the segment is equal to the angle *dba* with respect to the tangent, and the angle *dbf* alternates with *bda*, the triangles *bfd* and *abd* will be equiangular, and as *bd* is to *bf*, so is *ad* to *db*: therefore, as *ad* is to *db*, so is the speed along *db* to the speed along *da*, and conversely, the slowness along *da* to the slowness along *db*.]

What Galileo wants to demonstrate here, using the law of isochronism of chords converging at the lowest point of the circumference, is that along inclined planes of identical elevation, in which – as was demonstrated in the *Mecaniche* – the *momenti della gravità* are inverse to the lengths of the planes, a heavy weight will traverse, in equal times, distances that are always inversely proportional to the lengths of the planes. This is equivalent to saying that speed is proportional to *momento*.

The significant fact is not in the reassertion of the dynamic principle already proposed in the *De motu* (speeds are proportional to the effectiveness of heaviness); here, in fact, Galileo explicitly refers to not uniform, but rather accelerated, descents. He speaks of the ratio of “velocitates per *db* ad velocitates per *fb* [speeds along *db* to speeds along *fb*]”. The use of the plural leaves no doubt on this point.

The meaning of the operation carried out by Galileo begins to appear clear. Even when accelerated descents are examined outside the mechanical schema of uniform motions, the proportions of the speeds of such motions can be attributed to the proportions of the respective *momenti della gravità*. The spaces traversed by a

[62] *G.G.*, VIII, p. 378.

heavy body starting from rest, in equal times, along planes of different lengths but of the same elevation, are always inversely related to the length of the planes. The speeds will therefore also be in the same relationship and will be proportional to the *momenti della gravità*. This is Galileo's reasoning, which conceals – as we will see shortly – quite a few pitfalls. *Velocitas* here obviously means mean speed, while the plural *velocitates* indicates the collection of speed degrees (later he will say *momenta velocitatis*), all distinct.

The proportionality between overall speeds and *momenti della gravità* in natural motion evidently implies – this suggestion is very clear – that the proportion between any two degrees of speed, reached on the two inclined planes of identical elevation, *always* corresponds to the proportion of the *momenti della gravità* from which they are generated. This means that the *momento della gravità* increases constantly from the beginning of the motion[63].

What remains to be clarified – and it is no small detail – is the way in which Galileo conceives, in this fragment, the constant increase of the *momenti*
290 *della gravità* and, therefore, of speeds. Did he think that the *momenti* increased proportionally with the flow of time? Or did he instead believe that the increase was still proportional to the perpendicular distance from rest?

Let us set this problem aside for a moment to present the last passage of the fragment, which assists in its solution. Galileo is perfectly aware that the transfer of the laws of "mechanical" *momento* to the analysis of accelerated motions, however suggestive, cannot be presented as a "certain demonstration". For this reason, after having suggested it, he proposes it correctly, not as a law, but in the form of a supposition or principle, assuming which the rest can be rigorously demonstrated.

> Si hoc sumatur, reliqua demonstrari possent. Ponatur, igitur, augeri vel imminui motus velocitatem secundum proportionem qua augentur vel minuuntur gravitatis momenta; et cum constet, eiusdem mobilis momenta gravitatis super plano *bd* ad momenta super plano *da* esse ut longitudo *da* ad longitudinem *db*, idcirco velocitatem per *db* ad velocitatem per *da* esse ut *ad* ad *db*.[64] [If this is assumed, the remaining things could be demonstrated. Let it be supposed, therefore, that the speed of motion increases or decreases in proportion to the way the momenta of heaviness increase or decrease; and since it is established that the momenta of heaviness of the same mobile on the plane *bd* (see fig. 26) compared to the momenta on the plane *da* are as the length of *da* to the length of *db*, therefore the speed along *db* is to the speed along *da* as *ad* is to *db*.]

The supposition is very clear and of great interest. The use of the plural *gravitatis momenta* [momenta of heaviness] for a motion along the same plane suggests, without any possibility of misunderstanding, that Galileo restructured the constant "mechanical" *momento* in terms of acceleration.

---

[63] This is also what Wisan maintains, who sees in our theorem both the attempt to return to the dynamic postulate of the *De motu*, the effort to reinterpret it in light of the principle that the *momento della gravità* is continuously increasing, and to relate to each "grado [degree]" of the *momento della gravità* a proportional *grado di velocità* [degree of speed]. Wisan does not explicitly attempt to date the fragment, but emphasizes its close proximity to Valerio's response to Galileo's lost letter (*The new science of motion* cit., pp. 222-224).

[64] *G.G.*, VIII, p. 379.

What, then, is the rate that marks the increase in speed in natural motion? Unfortunately, the fragment doesn't tell us, at least not explicitly. In the last supposition, Galileo limits himself to stating that the speed increases according to the same proportion with which the *momenti della gravità* [momenta of heaviness] increase or decrease. Nothing more.

As vague as it may appear, this is a valuable indication. All the previous mechanical reflections of the Pisan scientist, in fact, tend to suggest his belief in the direct proportionality between the *momenti della gravità* [momenti of heaviness] and the perpendicular distance from rest (the "viaggio [journey]" of the force, as he had stated in the *Mecaniche*).

It seems, therefore, that our fragment must still be inscribed within this conceptual framework, which corresponds, moreover, to what has already been observed in the vernacular fragment, certainly from 1604, in which – not by chance – the analogy between natural motion and the behavior of "instruments" and "machines" was emphasized.[65]

By advancing the supposition of a constant increase in the *momento della gravità* [momento of heaviness] which corresponds to continuous and proportional increases in the *momenta velocitatis* [momenta of speed] (the expression is not used in the fragment, but it is conceived precisely to make it understandable and to legitimize its use), Galileo posed, therefore, a principle not substantially different from that **291**
stated in the letter to Sarpi.

However, the new formulation expressed in a much clearer way the nature of the attempt to derive the laws of natural motion from "mechanical" concepts and principles. Moreover – and this is perhaps the most significant point – it seems that the new principle served Galileo primarily to develop a precise analysis of the laws of the natural motions *of the same heavy body* along planes of different lengths, but of identical elevation above the horizontal. As we will see shortly, the Pisan scientist will only use it in this context, which is the same as in the *De motu* and the *Mecaniche*, while there remains no trace of attempts to use it in the analysis of the laws of free fall of heavy bodies.

Between 1604 and 1609, Galileo must have committed himself deeply to this analysis, as evidenced by the numerous manuscript notes related to this aspect that we have preserved. Evidently, Galileo considered for a long time this different formulation of the principle as perfectly equivalent to the supposition put forward in the letter to Sarpi, but much more suitable for guiding the analysis of motions along inclined planes.

The persistence, in our fragment, of the schema based on the direct proportionality between the increase in speed and distance from rest is confirmed by the absence of any reference to fall times. Galileo suggests, in fact, the relationship of the overall speeds and degrees of speed exclusively with the length of the planes. Furthermore, in the last supposition, having established the relationship between increase in speed and increase of *momento*, the Pisan scientist recalls the law of

---

[65] Cf. *supra*, p. 277.

the *Mecaniche* to establish the proportion between different *momenti*. If he had conceived at the time that the increase in *momento* is proportional to the "duration" of the descent (which is marked, that is, by the passage of time) he would hardly have neglected to clearly emphasize it. In fact, this would have been precisely the most significant and distinctive aspect of the supposition.

However, it was precisely this new supposition and the more explicit attempt to derive the laws of naturally accelerated motion from the "mechanical" *momento* that must have led Galileo to engage in careful reflection, in order to give some coherence and foundation to the notion of speed and its relation with not only space, but especially with time.

Already in the now examined fragment, there are evident traces of uncertainties and aporias in the way of conceiving speed. In fact, Galileo employs both *velocitas* [speed] and *tarditas* [slowness] (he did the same in the *De motu*). The *velocitas* [speed] is declared to be inversely proportional to the lengths of the planes, i.e., directly proportional to the *momenta gravitatis* [momenta of heaviness]. For *tarditas* [slowness], however, exactly the opposite happens. The use of the pair of opposite terms betrays the lack of a precise and satisfactory concept of "speed" and the strong, instinctive tendency to consider it as an independent quantity, not as a variable relationship between spaces traveled and times taken.

292 In other notes of Ms. 72, we observe Galileo's effort to develop a precise definition of "velocità". This attempt, also of relevant importance for the investigation of *momento*, appears particularly evident in a handwritten note, in which we once again find *momentum velocitatis* [momento of speed] employed. The fragment, rather than solving it, declares the difficulty.

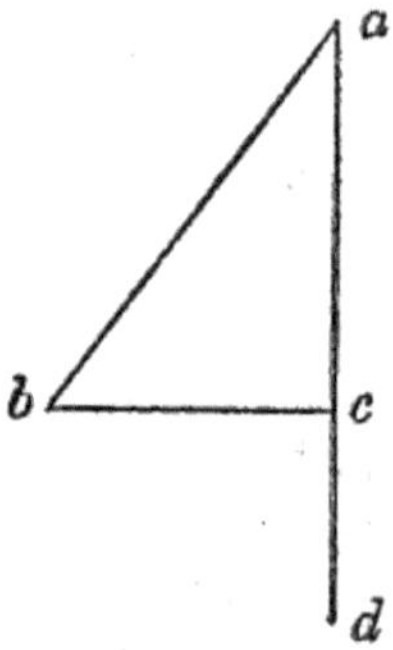

Figure 27

Mirandum. Numquid motus per perpendiculum *ad* velocior sit quam per inclinationem *ab*? Videtur esse; nam aequalia spatia citius conficiuntur per *ad*: attamen videtur etiam non esse; nam, ducta orizontali *bc*, tempus per *ab* ad tempus per *ac* est ut *ab* ad *ac*; ergo eadem momenta velocitatis per *ab* et per *ac*: est enim una eademque velocitas illa quae, temporibus inaequalibus, spatia transit inaequalia, eandem quam tempore rationem habentia[66]. [Amazing. Could the motion along the perpendicular *ad* (see fig.

[66] Ms. Galil. 72, c. 164*v*; publ. in *G.G.*, VIII, p. 375.

27) be faster than along the incline *ab*? It seems so; for equal spaces are traversed more quickly along *ad*: however, it also seems not to be so; for, with the horizontal *bc* drawn, the time along *ab* to the time along *ac* is as *ab* to *ac*; therefore, the momenta of speed along *ab* and *ac* are the same: for it is one and the same speed that, in unequal times, traverses unequal spaces that bear the same ratio to each other as the times do.]

On the one hand, namely, if equal distances from rest are taken along the incline and the perpendicular, the time of fall along the perpendicular is shorter. It can, therefore, be said that the speed is greater along the perpendicular. On the other hand, the times along *ab* and *ac* are in the same ratio – as stated in the previously examined fragment – as *ab* to *ac*. Here, therefore, the speeds, or rather the *momenta velocitatis*, can be considered equal in the perpendicular and the incline, given that the heavy body overcomes with these speeds, in different times, spaces proportional to the times.

The fragment preserves for us the idea of the paradox deriving from a dual way of considering the proportion of speeds along different inclinations of identical elevation above the horizon: either as a proportion of the times necessary to travel equal distances, or as a proportion between unequal times and distances, but always proportional two by two.

The paradox will be proposed again by Galileo in the *Giornata Prima* [First Day] of the *Dialogo*. The Pisan scientist will point out that a heavy body can be considered faster than another (Galileo refers to accelerated motions) when it travels, in the same time, a greater distance, or the same distance in less time. Equal speeds are then declared to be those of two bodies that cover the same distance in the same time[67]. Sagredo adds, lastly, another "more universal" definition of equal speeds: "quando gli spazii passati hanno la medesima proporzione che i tempi ne' quali sono passati [when the spaces traversed have the same proportion as the times in which they were traversed]"[68]. This definition exactly coincides with the second one proposed in the *Mirandum* fragment. Derived from an Archimedean theorem relating to uniform **293**
motions[69], Galileo, after proposing its use for accelerated motions as well, included it in the *Discorsi* as *Theorema I* of the *De motu aequabili*[70].

The connection between the *Mirandum* fragment and the previous one is all too evident. Here, however, *velocitas* and *velocitates* have been replaced by *momenta velocitatis*. The sum of all degrees of speed – this is the meaning of Galileo's statement – is equal both on the perpendicular and on the incline. This obviously

---

[67] *G.G.*, VII, pp. 47-48.

[68] *Ibid.*, p. 48.

[69] Cf. Proposition I of the *Spirals*: "Se un punto viene mosso con velocità costante su una linea e su questa si prendono due segmenti, essi avranno tra loro lo stesso rapporto che hanno i tempi nei quali il punto li percorre [If a point is moved with constant speed on a line and two segments are taken on this, they will have the same relationship between them as the times in which the point traverses through them]" (*Opere di Archimede*, trad. it. cit., p. 322).

[70] "Si mobile aequabiliter latum eademque cum velocitate duo pertranseat spatia, tempora lationum erunt inter se ut spatia peracta [If a mobile moves uniformly and with the same speed across two distances, the times of the motions will be in the same ratio as the distances traveled]" (*G.G.*, VIII, p. 192).

means that, at the foot of the perpendicular and the incline, the speeds must be equal.

Another testimony to the remarkable care Galileo dedicated to the attempt to base a precise analysis of accelerated descents along inclined planes on the static concept of *momento* is offered by another handwritten note, probably preceding the *Mirandum*. Galileo shows some doubt about the legitimacy of using the inclined plane theorem of the *De motu* and the *Mecaniche* (for uniform motions) to interpret accelerated descents along the same planes.

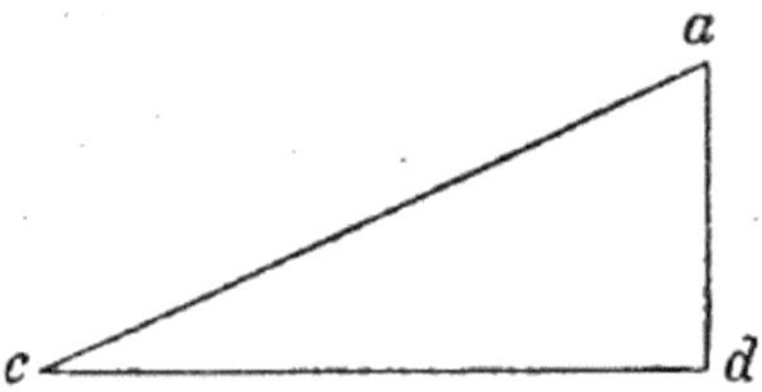

Figure 28

Velocitates mobilium quae inaequali momento incipiunt motum, sunt semper inter se in eadem proportione ac si aequabili motu progrederentur: ut, verbi gratia, mobile per *ac* incipit motum cum momento ad momentum per *ad* ut *da* ad *ac*; si aequabili motu progrederetur, tempus per *ac* ad tempus per *ad* esset ut *ac* ad *ad*, quod in accelerato dubito quidem; et ideo demonstra...[71]. [The speeds of mobiles that begin motion with an unequal momento are always in the same proportion to each other as if they were progressing with uniform motion: for example, a mobile along *ac* (see fig. 28) begins motion with a momento in proportion to the momento along *ad* as *da* is to *ac*; if it progresses with uniform motion, the time along *ac* to the time along *ad* would be in the ratio of *ac* to *ad*, which is doubtful for the accelerated case; and therefore demonstrate...]

The doubt concerns the legitimacy of the extension of laws established for uniform motions to accelerated descents. The *inaequale momentum* to which the
294 Pisan scientist refers is obviously the disposition to descend more or less quickly along the perpendicular and the incline (but always with uniform speed) which he discussed in the *De motu* and in the *Mecaniche*. In such a case, it is obvious that the speeds will always remain in the same proportion as the initial *momenti*. Now – here is the problem to be demonstrated – Galileo asks whether, at equal vertical distances from rest, the degrees of speed reached will always be in proportion to the initial *momenti*. Which he mistakenly believes is the same as asking whether the

[71] The note, in the hand of Arrighetti, is written on the *recto* [right] of page 177 of Ms. 72 (cf. *G.G.*, VIII, p. 386 and note 2). The passage is canceled by two crossed pen strokes. Our note is followed by another, which does not seem to be linked to the previous one: "Aliter sic: tempus per *ac* ad tempus per *ab*, ex praecedentibus, est ut linea *ac* ad linea *ab*; sed etiam ad tempus *ad* habet eandem rationem, cum *ab* sit media inter *ac*, *ad*; ergo tempora *ad*, *ab* erunt aequalia [Alternatively, thus: the time through *ac* to the time through *ab*, from the preceding (arguments), is as the line *ac* to the line *ab*; but also to the time *ad*, it has the same ratio, since *ab* is the mean between *ac* and *ad*; therefore, the times *ad* and *ab* will be equal]" (*ibid.*, see figure at the bottom of the page).

times of accelerated descent will remain, even in the case of accelerated motions, in the proportion of the distances travelled[72].

This note would seem to be written before the two previously examined ones as suggested by the fact that Galileo questions the possibility of considering, in accelerated motion, the relationship between the times along the incline and the perpendicular to be identical to that between the lengths of the incline and the perpendicular. In the two previous fragments, he indeed takes this relationship for granted. In an earlier note, he had also specified that, in the time in which the heavy body descends along the entire perpendicular, it passes a certain distance on the incline. Between this distance and the incline, the perpendicular is the mean proportional[73].

The most interesting aspect of the doubtful note, unfortunately incomplete, consists, however, in the evidence it provides of Galileo's verification regarding the "validity" of the static principles when confronted with the reality of acceleration. It constitutes perhaps the most direct document of this attempt, whose importance for understanding the functions assigned to *momento* by Galileo is decisive.

Galileo's doubt about the possibility of legitimately using the laws of "mechanical" *momento* for the analysis of accelerated descents along inclined planes of identical elevation must have soon vanished. In fact, we retain a long, very important handwritten note in which the operation is finally declared explicitly. Galileo also makes a serious effort to justify it. It can be said that the fragment in question actually constitutes the most detailed attempt at a "static" foundation of the science *de motu* by Galileo that has come down to us.

It is an organic fragment, which starts from the erroneous supposition of the letter to Sarpi, aimed at attempting a definitive arrangement of the space/time/speed relationships. The operation that we have seen progressively emerge in numerous notes now appears in its clearest form. Galileo states, firstly, what he intends to
demonstrate: the times of descent of the same heavy body along the perpendicular **295**
and the incline of equal elevation "erunt inter se ut plani inclinati et perpendiculi longitudines [will be to each other as the lengths of the inclined plane and the perpendicular]"[74]. What must be demonstrated, having drawn the triangle *abc* (see fig. 29), is that the time of motion along *ab* to that along *ac* = *ab* : *ac*. This is, indeed, the relationship between times and spaces, the legitimacy of which, in the case of accelerated descents, Galileo had shown some doubt in the previous fragment. Let's look at Galileo's demonstration:

Cum enim assumptum sit, in naturali descensu velocitatis momenta eadem semper reperiri in punctis aequaliter ab orizonte distantibus iuxta perpendiculares distantias, constat quod, producta linea orizontali

[72] It has rightly been observed that the proportionality of the descent times along the perpendicular and the incline to the lengths of the planes (assuming that the descent speeds – as Galileo does here – are inversely proportional to the lengths of the planes), holds *exclusively* in the case of accelerated descents. The statement is, in fact, not sustainable on the level of uniform descents (cf. W. Wisan, *The new science of motion* cit., p. 201).

[73] Cf. *supra*, note 56.

[74] The fragment is located on p. 179*v* of Ms. 72 (*G.G.*, VIII, pp. 387-388).

am, quae ipsi *bc* erit parallela, sumptisque in perpendiculo *ab* quotcumque punctis *e, g, i, l*, et per ipsa ductis parallelis orizonti *ed, gf, ih, lk*, erit mobilis per *ab* momentum seu gradus velocitatis in puncto *e* idem cum gradu velocitatis lati per *ac* in puncto *d*, cum punctorum *e, d* eadem sit distantia perpendicularis ab orizonte *am*: et similiter concludetur, in punctis *f, g* idem esse velocitatis momentum, et rursus in punctis *h, i*, et *k, l* et *c, b*[75]. [Since it has been assumed that, in natural descent, the momenti of speed are always found to be the same at points equally distant from the horizontal according to perpendicular distances, it is evident that, by extending the horizontal line *am*, which will be parallel to *bc*, and selecting any points *e, g, i, l* on the perpendicular *ab*, and drawing through them parallels to the horizontal, namely *ed, gf, ih, lk*, the momento or degree of speed of the mobile along *ab* at point *e* will be the same as the degree of speed attained along *ac* at point *d*, since the points *e, d* are at the same perpendicular distance from the horizontal *am*: and similarly, it will be concluded that at points *f, g*, the momento of speed is the same, and likewise at points *h, i*, and *k, l*, and *c, b*.]

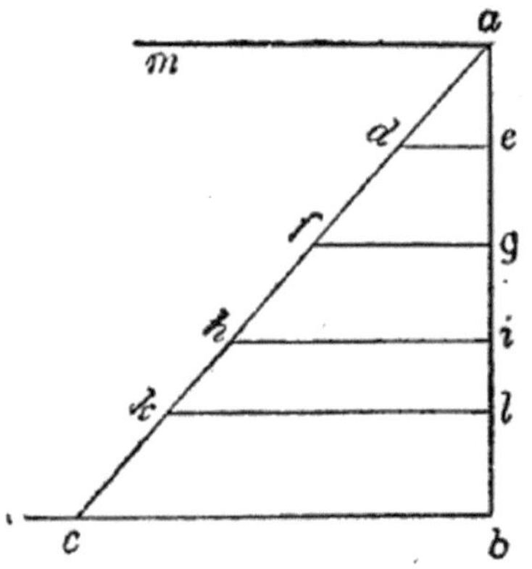

Figure 29

296 The equality of the degrees of speed reached at points on the perpendicular and the incline equally distant "vertically" from rest (which is the same as saying at equal distances from the horizontal plane) is presented by Galileo as a supposition ("cum assumptum sit [since it has been assumed]"). In fact, this coincides with the supposition, already advanced in a previously examined note, of the proportionality between increase in speed and the growth in *momenta gravitatis*. Since the *momenta gravitatis* of the same mobile are measured in relation to the perpendicular

[75] From *eadem* to *distantias* is written in the margin, presumably replacing a previously written passage, which appears underlined: "continue augeri secundum rationem elongationis perpendicularis a linea orizontali, in qua fuit lationis initium [to increase according to the ratio of the perpendicular's elongation from the horizontal line, where the motion began]" (cf. the critical apparatus, *ibid.*, p. 388). In essence, Galileo would subsequently replace the declaration of the incorrect supposition of acceleration-distance from rest with the correct one of the equality of degrees of speed at equal vertical distances from rest. It is worth reflecting, however, that Galileo did not explicitly delete the passage containing the erroneous supposition. This may lead one to think that the variant did not appear to him as an alternative, but rather as a different formulation of the same principle. In favor of the neither necessary nor probable alternative nature of the marginal insertion, Wisan argued and appropriately pointed out that, in the copy of the Galilean theorem, written by Arrighetti, which is preserved on p. 88 of Ms. 72, the two passages coexist next to each other (cf. *The new science of motion*, p. 218).

distances from rest, they will coincide at equal distances on the incline and on the perpendicular, therefore generating identical speeds.

Except that now the Pisan scientist, referring to the increase in speed, no longer uses *velocitas*, as used in the previous supposition, but rather *velocitatis momenta* [speeds' momenta]. The expression, obviously, more clearly evoked the close connection with the analysis of the proportional increase of the *momenta gravitatis* and revealed the derivation of the principle from mechanical theorems. Furthermore, for the first time, the parallelism *momentum seu gradus velocitatis* [momento or degree of speed] is suggested, which we will so frequently find in the *Discorsi*.

That the presumed supposition actually appears to Galileo to be a consequence of the acceleration-distance from rest proportionality is clarified by the following passage, of which interpreters have expressed different opinions on its meaning:

(Et quia velocitas semper intenditur pro ratione elongationis a termino *a*, constat, in latione *ab* tot esse velocitatis gradus seu momenta diversa, quot sunt in eadem linea *ab* puncta magis ac magis a termino *a* distantia; quibus totidem in linea *ac* respondent et per parallelas lineas determinantur, in quibus iidem sunt gradus velocitatis)[76]. [(And because speed is always increased in proportion to the distance from terminus *a*, it is evident that, in the motion along *ab*, there are as many different degrees or momenta of speed as there are points on the same line *ab* at progressively greater distances from point *a*; and for each of these, there correspond the same number in the line *ac*, determined by parallel lines, in which the degrees of speed are the same).]

The acceleration-perpendicular distance from rest proportionality constitutes the foundation of the "supposition", which we will find advanced as a Postulate in the *Discorsi*. Or rather, the supposition is actually a simple consequence of it. The **297**
passage containing these statements is underlined by the Pisan scientist, certainly – despite contrary opinions – to highlight its extreme importance[77]. This is, in fact, a decisive step, of which we must carefully evaluate every nuance. Before examining the basis of reasoning of which Galileo attempts to establish that the times of descent along *ab* and *ac* are to each other as *ab* is to *ac*, it will therefore be appropriate to open a parenthesis.

First of all, it must be emphasized that the framework in which the principle is used is that of the comparison between descents along the inclined plane and the perpendicular of identical elevation; furthermore, the reference is to the descents *of the same mobile*. The context is, therefore, different from that of the letter to Sarpi,

[76] The passage, as in the case examined in the previous note, is underlined by Galileo. Furthermore, it is included in round brackets. This was generally interpreted as a clear sign of suppression. Cf., in this sense, in addition to the opinion of Favaro himself, who separated it from the text of the theorem, relegating it to the critical apparatus, the opinion, substantially in agreement, of Geymonat-Carugo, who even declare that "questo tratto venne poi cancellato [this passage was then canceled]" (*Discorsi e Dimostrazioni* cit., p. 874). Stillman Drake, on the other hand, believes that the marginal note should be recognized as a substitute proposal, but shows that he is convinced that the long passage was underlined by Galileo, not to cancel it, but to highlight it (*Uniform acceleration, space and time* cit., p. 24). Wisan believes that the passage should be considered an integral part of the text of the fragment. In fact, even in Guiducci's copy, it appears in the text, not underlined and without brackets (*The new science of motion* cit., p. 218).

[77] Cf. the previous note.

in which reference is made exclusively to the free descent of heavy bodies. Nor are there any documents that allow us to affirm that Galileo did not perceive any difference between the two models.

Galileo tries to adapt the laws of *momenti* developed in the *Mecaniche* to the actual accelerated descents along inclined planes. In doing so, he uses the principle established in the Paduan treatise, in which *momento* was considered the generator of uniform motions: the effectiveness of *momento* is measured in proportion to the vertical displacement[78]. In all probability, Galileo took a cue from this reference, once he had ascertained that the speed is accelerated (i.e., that the *momento* increases continuously), to affirm that the degrees of speed reached on the incline and on the perpendicular are proportional to each other as the vertical distances from rest. Which means recognizing, as Galileo also does in our fragment, that, at equal vertical distances from rest, the degrees of speed will be identical.

The deduction of this principle seems to strengthen Galileo in the validity of the erroneous supposition. In all likelihood – however peculiar it may seem – Galileo, for some time, did not reflect on the fact that the equality of the degrees of speed at identical vertical distances from rest, along the incline and the perpendicular, occurs, in naturally accelerated motion, *even when starting from the supposition that speed increases proportionally with the passage of time*. In other words, it seems that Galileo, bound by the theorem of inclined planes of the *Mecaniche* and the laws of *momenti della gravità*, believed that the principle of equality of degrees of speed could only be affirmed on the basis of the erroneous supposition of 1604. This depended primarily on the strength of the analogy with the "mechanical" concept of *momento*, which he had relied on in trying to found the laws of accelerated descents
298 along inclined planes. However, the reference to the model of inclined planes, that used in the *De motu* and the *Mecaniche*, must have been no less influential.

In this model, and in Galileo's representational figure (the right-angled triangle), the intersecting points of equal vertical distances from rest with horizontal parallels on the incline and on the perpendicular are easily visible. The relationship of direct proportionality between the increase in distance from rest and the increase in speed is therefore instinctively suggested – albeit erroneously.

In essence – at least, this is our hypothesis – Galileo would have inadvertently gone from the legitimate assertion that, at equal vertical distances from rest, identical degrees of speed are reached on the incline and on the perpendicular, to the erroneous belief that the acceleration is proportional to the vertical distance from rest. That is, for a certain number of years he would have considered the two statements as fully equivalent. In reality, he must have seen the first as a simple consequence of the second.

The use of the notion of "vertical displacement" from rest, derived from the *Mecaniche*, allowed Galileo, in fact, to consider two distances, along the

[78] It should be recalled that, to justify this assertion, Galileo had adduced the principle of the indifference of heavy bodies to lateral displacement: "i corpi gravi non fanno resistenza a i moti trasversali, se non in quanto in essi vengono a discostarsi dal centro dalla Terra… [heavy bodies do not resist transverse motions, except insofar as they move away from the center of the Earth…]" (*G.G.*, II, p. 186).

perpendicular and the incline, as *substantially* equal, although they were actually different. This seemed to make it immediately justifiable that, after equal "vertical" descents, the same degrees of speed were always reached, regardless of the inclination.

Although, limiting the analysis to free fall, the equality of the degrees of speed at equal distances from rest could also seem to Galileo an obvious consequence of the acceleration-time proportionality, conversely, the derivation of this principle from the correct supposition appeared difficult when the problem had become that of comparing the descents of the same heavy body along the incline and the perpendicular of equal elevation; even assuming – as Galileo does in our note – that the times of the descents of the heavy body along the incline and the perpendicular stand in proportion as the incline with the perpendicular. So difficult, in fact, that Galileo would not succeed in doing so during the considerable number of years that span from the discovery of the correct speed-time proportionality (i.e., presumably, between 1606 and 1608) to the publication of the *Discorsi*. In the last of his major works, in fact, Galileo will suppress the close connection, established in our fragment, between the erroneous supposition and the equality of the degrees of speed at equal vertical distances from rest. But, not knowing how to deduce this principle from the correct supposition, he will introduce it immediately after the definition of naturally accelerated motion, asking that it be granted to him as
a Postulate[79]. The only confirmation he will be able to provide is, in fact, the **299**
experiment with the interrupted pendulum[80].

It was mainly thanks to Viviani that, a couple of years after the publication of the *Discorsi*, Galileo finally came into possession of a theorem capable of demonstrating, starting from the acceleration-time proportionality, the equality of the degrees of speed at the base of the incline and the perpendicular[81].

If these considerations are correct, it is clear that some important facts follow. Galileo used the proportionality of acceleration to the distance from rest for many

[79] Salviati, whose role is to introduce it, strongly emphasizes that the Academician requests to be granted "un solo principio [a single principle]": "Accipio gradus velocitatis eiusdem mobilis super diversas planorum inclinationes acquisitos tunc esse aequales, cum eorumdem planorum elevationes aequales sint [let the degrees of speed acquired by the same mobile on different inclinations of planes be equal when the elevations of those same planes are equal]" (*G.G.*, VIII, p. 205).

[80] *Ibid.*, pp. 205-207.

[81] As is known, Galileo, at Viviani's suggestion, considered preparing a possible new edition of the *Discorsi*, an integration intended to confirm, with a demonstration derived from the theorem of the inclined plane of the *Mecaniche*, that, at equal vertical distances from rest, the degrees of speed reached by the same body on the incline and on the perpendicular are equal. In this demonstration, Galileo explicitly started from the supposition that "vada crescendo la velocità o la quantità dell'impeto, con la proporzione del tempo [the speed or quantity of impeto increases in proportion to time]" (*ibid.*, VIII, p. 218). The drafting of the integration, certainly inspired and approved by Galileo, was curated by Viviani himself (cf. Favaro's *Avvertimento* to the *Discorsi*, *ibid.*, pp. 23-24). Galileo gave news of the discovered demonstration, and of his intention to include it in a new edition of the *Discorsi*, to Benedetto Castelli in a letter dated 3 December 1639 (*G.G.*, XVIII, pp. 125-126). The addition was first published in the *Opere di Galileo Galilei* (2 voll. Per gli Heredi del Dozza, in Bologna 1656, I, p. 105), edited by Rinaldini and Viviani. Favaro included the mechanical demonstration into the text of the *Discorsi* (cf. *G.G.*, VIII, pp. 214-219).

years, especially in the analysis of the descents of the same heavy body along planes of different inclination but of equal elevation. He had to arrive at the discovery of the correct acceleration-time proportionality through the recognition of errors committed in deducing from the erroneous principle, in the case of free fall, the relationships between spaces, times and speed, and/or with the aid of experimental observations from which he became convinced that speed does not increase with distance. Finally, if – as it seems – Galileo used the incorrect and correct suppositions simultaneously for some time, it cannot be ruled out that he remained bound to the former, out of a sort of necessity, in the analysis of descents along inclined planes of equal height (in fact, he did not know how to deduce the equality of degrees of speed from the acceleration-time proportionality), while already using the second to interpret the acceleration in the free fall of heavy bodies.

These are indications of considerable weight, which would require precise verification within the bulk of the *Frammenti attinenti ai Discorsi*. A need that, obviously, cannot be satisfied here. For now, it is enough to have demonstrated with the utmost clarity that the inclined plane theorem of the *De motu* and the *Mecaniche*, as well as the concept of *momento della gravità*, which had played a decisive role, were used by Galileo for several years, at least between 1602 and 1609, to serve as the basis for a rigorous treatment of accelerated motions. The mechanical model,
300 originally conceived as the framework for uniform motions, strongly suggested the erroneous supposition of 1604. The reformulation of the concept of *momento della gravità* in terms of constant acceleration has also determined the appearance – as we have already seen – of *momentum velocitatis* alongside *gradus velocitatis*.

Having closed this indispensable parenthesis, we can return to the Galilean fragment to observe how, starting from the erroneous supposition and the principle of the equality of the degrees of speed at equal vertical distances from rest, Galileo demonstrated that the times of descent along the incline and the perpendicular are in proportion as the incline to the perpendicular. The passage is indeed of considerable interest, not only for the analytical procedures employed, but also because it clarifies the meaning attributed by Galileo to the expression *momentum velocitatis*.

Sunt igitur in linea *ab* quasi innumera quaedam spaciola, quibus multitudine quidem aequalia, et bina sumpta secundum eandem rationem respondentia, alia signantur in *ac* per lineas innumeras parallelas ex punctis lineae *ab* extensas (intercepta enim spacia *ad*, *df*, *fh*, etc. ad spatia *ae*, *eg*, *gi* etc. respondent singula singulis secundum rationem *ac* ad *ab*); suntque in singulis binis sibi respondentibus iidem velocitatis gradus. Ergo, ex praecedenti, tempora omnia simul sumpta lationum omnium per *ab*, ad tempora omnia similiter accepta lationum omnium per *ac*, eandem habebunt rationem quam spacia omnia lineae *ab* ad spacia omnia lineae *ac*; hoc autem idem est, ac tempus casus per *ab* ad tempus casus per *ac* esse ut linea *ab* ad *ac*: quod erat demonstrandum[82]. [Therefore, on the line *ab* (see fig. 29), there are, as it were, an innumerable number of small spaces, which, in terms of quantity, are equal, and in pairs taken according to the same ratio, others are marked in *ac* by innumerable parallel lines extended from the points of the line *ab* (for the intersecting spaces *ad*, *df*, *fh*, etc. correspond to the spaces *ae*, *eg*, *gi*, etc. respectively, according to the ratio of *ac* to *ab*); and in each of the pairs corresponding to each other, the same degrees of speed are found. Therefore, from the preceding, the total times taken for all the motions along *ab* will have the same ratio to the total times taken for all the motions along *ac* as the

[82] *Ibid.*, VIII, p. 383.

total spaces of the line *ab* to the total spaces of the line *ac*; and this is the same as saying that the time of the fall along *ab* is to the time of the fall along *ac* as the line *ab* is to *ac*: which was to be demonstrated.]

The Pisan scientist emphasizes, therefore, that on the perpendicular *ab* there are "quasi innumera spaciola [nearly innumerable small spaces]", which correspond to the same number of small spaces, but not of the same dimensions, on *ac*. Each *spaciolum* [small space] of *ab* is to the corresponding one of *ac* as *ab* is to *ac*. Now, since we know that, at the points intersected by the horizontal parallels to the incline and the perpendicular, the degrees or *momenti* of speed are equal, to travel through two corresponding *spaciola* [small spaces], proportional to *ab*, *ac*, the heavy body will take times which are also proportional to the perpendicular and the incline. The sum of the times along the two descents will, therefore, be as the sum of the *spaciola*. In this way, Galileo believed he had demonstrated that the times along the perpendicular and the incline are in the same proportion as the perpendicular is to the incline.

Galileo's reasoning corresponds – as has been observed – to that proposed in the paradox of the *Mirandum* fragment regarding the two different ways of understanding speed: the speeds along the perpendicular and the incline can be considered equal since "gli spazii passati hanno la medesima proporzione che i tempi ne' quali sono passati [the traversed spaces have the same proportion as the times in which they were traversed]"[83].

We will focus later on the important analysis of the *quasi innumera spaciola* **301**
in proportional pairs along the perpendicular and the incline[84]. What we want to emphasize now is the use of *momentum velocitatis* and the meaning of the expression. We have already mentioned the correlation with *gradus*, a term dear to the logic of the *calculationes*, with which it will frequently recur in the *Discorsi*. Here the *momenta velocitatis* increases constantly according to the *elongatio perpendicularis* [perpendicular elongation] of the heavy body from the *terminus a quo* [starting point]. Speed, like weight – Galileo seems to assert – is articulated into *momenti*; and its maximum *momento* is on the perpendicular. Evidently Galileo considers the *momenti* or degrees of speed along the corresponding pairs of infinitesimal *spaciolas* in which the incline and perpendicular are divided, as instantaneous motions in which the speed is *uniform*. This unstated supposition essentially legitimizes the stated equality of the proportions between descent times and distances travelled. Later, it will be clearer that *momento* served Galileo to reduce the accelerated speed to a series of progressively increasing instantaneous uniform speeds[85]. *Momento* is presented here as a real measure of speed, which is once again treated as an independent quantity, not as a relationship between space and time.

If the *momenti della velocità* are identical on the perpendicular and on the incline, their "duration", however, is not equal. On the incline, each *momento* "survives", so to speak, longer than its corresponding one on the perpendicular (since the distance

---

[83] Cf. *supra*, p. 292.

[84] Cf. *infra*, pp. 360-362.

[85] Cf. *infra*, pp. 365-366.

to traverse is greater), before a new, more intense *momentum velocitatis* is produced. As for the subsequent *momenti* along the perpendicular (or the incline), they increase with each vertical distance from rest. This means that, when taking successive equal vertical displacements, the time it takes for the mobile to complete them tends to constantly decrease. The progressive increase in the *momenta velocitatis* of the same motion is, therefore, conceived by Galileo in terms of an increasingly shorter time necessary to carry out successive equal vertical displacements. Soon, once the correct proportionality of acceleration to time is established, we will see that the constant increase in *momenta velocitatis* will be expressed instead (and from then on without exception) as the ever greater space traversed in equal times.

The fragment we have been analyzing constitutes, in all likelihood, the most advanced point reached by Galileo in the attempt to closely link the erroneous supposition with the equality of *gradi* or *momenti di velocità* at equal vertical distances from rest. Even from a terminological point of view, as well as from the perspective of conceptual analysis, the demonstration of page 179*v* can be considered the most mature development that Galileo reached, inspired primarily by
302 the analogy with the theorem of inclined planes of the *De motu* and the *Mecaniche*, his desire to reform it in light of the understanding of accelerated descents, and his attempt to use the "static" concept of *momento*, appropriately restructured, in this operation.

That Galileo, at a certain point, succeeded in clearly isolating some general principles from which everything else could be deduced; that these principles were the same ones that we saw gradually and laboriously emerge from the previously examined manuscript notes; that on these principles he intended to base a rigorous treatment of the descents of the same heavy body along different inclinations of identical height; and, lastly, that in them, the analogy with the theorem of the *De motu* and the *Mecaniche* appeared more than evident, as well as the attempt at an analysis based on the concept of *momento della gravità* and its laws; all this appears confirmed – and is a testimony of enormous importance – by an epistolary document. It also allows us to assert that Galileo's reflection on these themes was still active in 1609, that is, at the end of his Padua period.

The notes we have examined do not bear dates. Although serious attempts have recently been made to determine material and conceptual analysis tools that allow at least approximate dating[86], isolating a precise intellectual path from this

[86] In this direction, the most substantial and significant effort is certainly that made by Wisan in the important essay cited several times. Although using various tools (watermarks, handwriting, careful examination of terminological changes, etc.), the American scholar considered the preferred dating method to be that consisting in the careful inspection of the geometric-mathematical procedures used by Galileo, in search of a logical succession that coincides with the chronological one (cf. *The new science of motion* cit., especially pp. 127-132). Numerous references to possible chronological successions of the Galilean fragments of Ms. 72 can be found in almost all of Drake's works cited in this chapter. Naylor dedicated special attention to the "techniques and problems in the interpretation of the Galilean manuscript" in his *Galileo's theory of motion* cit. (pp. 365-367), primarily to limit the excessive reliance placed by Drake in datings based on watermarks. Naylor also discussed Wisan's dating method (cf. the review cited in *The new science of motion*). It is worth recalling that the first scholar to face the thankless task of establishing a chronological succession in the bulk of fragments relating to the *Discorsi*

bulk of notes remains extremely difficult and risky. Although we are aware of these difficulties and do not feel confident suggesting that the order in which we examined the above fragments corresponds, in substance, to the chronological succession of their writing, we nevertheless believe that we can affirm, without fear of contradiction, that the demonstrations and the theorems that we have emphasized can reasonably be assigned, all, to the years between 1604 and 1609. Among them, namely, the letter to Sarpi and the valuable, albeit confused, response from Luca Valerio to an unfortunately lost letter from Galileo. This last document, in fact, allows us to ascertain with absolute certainty that, at that date, Galileo was
still seriously working on a mechanical foundation for his treatise on naturally **303**
accelerated motion.

On June 5th of 1609, Galileo wrote from Padua to Luca Valerio to inquire about the reasonableness and legitimacy of two principles, which he proposed to use as the foundations of his science *de motu naturaliter accelerato*. Valerio's response, although somewhat unclear, at least in the first part, allows us to reconstruct the two principles that were dear to the Pisan scientist[87].

Valerio replies, firstly, that, for a "scienza di mezzo [intermediate science]", such as geometry[88], the proposed principles are to be considered "very clear". However, to strengthen its credibility, it will be useful to recall the metaphysical principle, indisputable in Valerio's opinion, which places the effect subordinate to the cause: "poiché la virtù causale, in quanto tale, alla quantità dell'effetto si riferisce, et con la quantità dell'effetto la quantità della causa misuriamo, si in quanto all'estensione et intensione, come alla perfettione et nobiltà [since the causal virtue, as such, refers to the quantity of the effect, and with the quantity of the effect we measure the quantity of the cause, both in terms of extension and intension, as well as in perfection and nobility]".

Given that the multiplication of the virtue of a sufficient cause automatically entails the multiplication of its effect, Valerio believes he can move on to the examination of the two Galilean petitions:

> Dunque, se l'impeto o inclinatione della gravità del corpo A sopra il piano inclinato all'orizonte secondo l'angolo B si supponga esser doppio dell'impeto della gravità del medesimo A sopra il piano inclinato all'orizonte secondo l'angolo C, maggiore dell'angolo B; et tali due diversi impeti nascono dalla gravità di A, limitata verso la produttione dell'impeto diversamente per le diverse inclinationi de' detti piani, si vede, per immediata conseguenza, che la velocità del moto naturale di A sopra il piano meno inclinato, sarà doppia della velocità del moto del medesimo A sopra quell'altro piano più inclinato. Dunque, il vigore della causa immediata della doppia velocità, ch'è l'impeto o inclinatione alla doppia velocità, dovea essere doppia dell'inclinatione alla meza velocità, secondo la maggior inclinatione dell'altro piano[89]. [Therefore, if the impeto or inclination of the heaviness of the body A on the plane inclined to the horizontal according to the angle B is supposed to be double the impeto of the heaviness

was Caverni, in his *Storia del metodo sperimentale* cit. (IV, pp. 266-436; for the dating techniques used by Caverni, based primarily on calligraphic examination, cf. *The new science of motion* cit., p. 109).

[87] Valerio's letter was sent from Rome on 18 July 1609 (cf. *G.G.*, X, pp. 248-249). Valerio replied to a letter from Galileo "delli 5 giugno [of June 5th]".

[88] According to the traditional peripatetic classification.

[89] *G.G.*, X, pp. 248-249.

of the same A above the plane inclined to the horizontal according to the angle C, greater than angle B; and these two different impeti arise from the heaviness of A, constrained differently in the production of the impeto by the different inclinations of the said planes, it is seen, as an immediate consequence, that the speed of the natural motion of A on the less inclined plane will be double the speed of motion of the same A on the other, more inclined plane. Therefore, the strength of the immediate cause of the double speed, which is the impeto or inclination to the double speed, must have been double the inclination to the half speed, according to the greater inclination of the other plane.]

The meaning of the quoted passage is far from clear. Evidently, Valerio only partially understood Galileo's thoughts and, in recounting them, he fell into glaring errors.

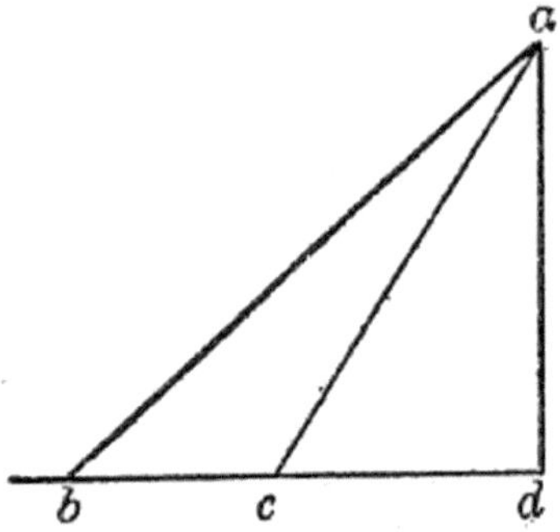

Figure 30

The figure relating to the statements contained in the quoted passage is missing from the letter, and the opinions regarding the way in which it should be conceived are not in agreement. Unlike other scholars, I am convinced that Galileo, in the lost letter to which Valerio was replying, used the same diagram used in many of
304 the fragments examined above: a right triangle (see fig. 25) from whose vertex a segment is drawn to any point of the base leg[90]. In other words, the same figure that is found in the fragment pasted on the reverse of folio 179[91] (see fig. 30). It is worth noting that, in this figure, the angles of elevation are marked respectively *b*

[90] Favaro, "per agevolare l'intelligenza del testo [to facilitate the understanding of the text]", proposes a different figure; in it, the two inclined planes do not appear to be of identical elevation (cf. *ibid.*, p. 248, note 1). Even the reconstruction of the figure proposed by Caverni (*Storia del metodo sperimentale* cit., IV, pp. 355-356) is, in essence, attributable to that of Favaro. Wisan, if I am not mistaken, suggests a more complex solution. In his opinion, in fact, Galileo would have reasoned in terms of planes of equal elevation, even though he used a figure in which the planes are of different heights above the horizon. Galileo would, in fact, have intended to ask Valerio for confirmation of the general principle which places *momento* (or effective weight) of the heavy body along an inclined plane proportional to its speed of descent ("the impeto"). According to Wisan, the fact that the "impeti" are inversely proportional to the lengths of the planes *of identical elevation* is a simple consequence of this principle (*The new science of motion* cit., pp. 224-225).

[91] *G.G.*, VIII, p. 376 (cf. *supra*, p. 286); I'm referring to the figure at the bottom of the page. Wisan herself recognizes the close connection between the discussion of the first principle in Valerio's letter and the ideas "obscurely expressed on folio 179*v*" (*The new science of motion* cit., p. 225).

(the smaller) and *c* (the larger), as in Valerio's letter, who, without a doubt, had the figure contained in Galileo's letter in front of him. Furthermore, the vertex (and, therefore, the heavy body that descends from the vertex) is indicated with *a*.

If our reconstruction is correct, Galileo was therefore referring to planes of different lengths, but of equal elevation – Valerio does not say this, but neither does he deny it. Once this schema is accepted, Valerio's passage appears comprehensible and the Galilean principle emerges clearly. In essence, due to a trivial error, Valerio would have declared the impetus along the plane of lesser inclination to the horizon, i.e., *ab*, "double" compared to that of greater inclination, i.e., *ac* (the assertion can be – obviously – accepted only by reversing the relationship: *ac* will be double). **305**
Furthermore, he evidently believed that the quantity of the "impetus" of the same heavy body along two planes of different inclination on the horizon, but of equal elevation, could be indicated by the ratio of the angles of elevation (i.e., according to one of the traditional methods)[92] just as well as by the inverse ratio of the lengths of the planes themselves (as Galileo expressed in his letter). Since he was more familiar with the relationship between the angles of inclination and considered it equivalent, Valerio used it in his response.

Lastly, it seems – this is, at least, the impression suggested in the reading of the quoted passage – that Valerio did not understand that Galileo was referring to accelerated, non-uniform, descents. Most likely, the Pisan scientist had failed to explain it with the necessary clarity, mistakenly convinced that the use of the plural "le velocità" for the same descent would be sufficient to prevent the correspondent from being misled. As it is conceived, Valerio's confirmation appears, in fact, to refer to the dynamic principle used by Galileo in the *De motu* for uniform descents along inclined planes of equal height. Conversely, the Pisan scientist had certainly presented Valerio with another principle – whose similarity with that of the *De motu* – from which it derived – was so close that the diligent correspondent ended up confusing them.

The principle proposed by Galileo was, in all likelihood, the same one that we saw put forward, as a supposition, in one of the notes examined in the previous pages and on which we, not by chance, focused:

> Ponatur igitur augeri vel imminui motus velocitatem secundum proportionem qua augentur vel minuuntur gravitatis momenta[93]. [Let it therefore be assumed that the speed of motion increases or decreases in proportion to the increase or decrease of the momenta of heaviness.]

We have already discussed the meaning of the principle and the role assigned to it by Galileo previously[94]. It is, therefore, unnecessary to focus on it. It will suffice to point out that, in 1609, Galileo was still seriously engaged in the project of a mechanical foundation of the science of motion, and that in this foundation, the concept of *momento* had a decisive role.

---

[92] For example, it was the method used by Cardano (cf. Caverni, *Storia del metodo sperimentale* cit., pp. 232-233).
[93] *G.G.*, VIII, p. 379.
[94] Cf. *supra*, p. 290.

The fact that *momento* is absent in Valerio's letter should not be misleading. From his response, we can, indeed, reconstruct with some approximation the terms and expressions used by Galileo in the letter of June 5th. Valerio compares "l'impeto o inclinatione della gravità [the impetus or inclination of heaviness]" to the "impeto o inclinatione della velocità [impetus or inclination of speed]". In all likelihood, Galileo expressed himself in terms of "impeto o momento della gravità [impeto or
306 momento of heaviness]" and "impeto o momento della velocità [impeto or momento of speed]". Valerio, who was not familiar with these terms, especially *momento*, although he understood their meaning, replied using characteristic expressions from his own lexicon[95].

Fortunately, it is much simpler to reconstruct, from Valerio's words, the second principle presented to him by Galileo.

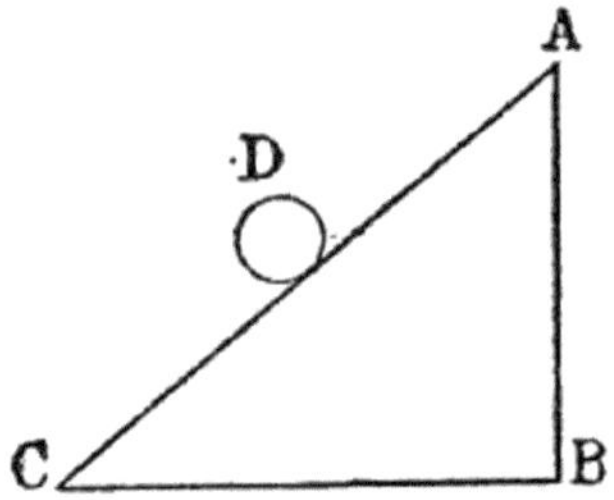

Figure 31

La seconda supposizione non mi si rende men chiara della prima: per ciò ch'essendo il moto del corpo grave D, mosso per l'AC all'orizonte BC, composto da due moti retti, l'uno per una parallela alla BC, mobile verso la BC, et l'altro per una perpendicolare all'orizonte, essa ancora mobile, cosa chiara è che quando D sarà in C, havrà acquistato tanto impeto o inclinatione a velocemente muoversi, ch'è la quantità dell'effetto (in quanto effetto, dico, di quella parte del moto composto che si fa per la perpendicolar mobile, eguale alla stabile AB), quanto havrebbe acquistato se D si fusse mosso per la sola perpendicolare AB: et ciò dico in vigore del sopradetto principio metafisico[96]. [The second supposition is no less clear to me than the first: since the motion of the heavy body D (see fig. 31), moved through AC to the horizontal BC, is composed of two rectilinear motions, one parallel to the BC, mobile towards BC, and the other perpendicular to the horizontal, which is again mobile, what is clear is that when D is at C, it will have acquired as much impeto or inclination to move quickly, which is the quantity of the effect (as an effect, I mean, of that part of the compound motion that occurs along the mobile perpendicular, equal to the steady AB), as much as it would have acquired if D had moved along the perpendicular AB alone: and I say this by virtue of the aforementioned metaphysical principle.]

It is not difficult to recognize in Galileo's "second supposition" the now familiar one that places the *gradi* or *momenti di velocità* reached at equal vertical distances

[95] It is worth emphasizing that *momento* never occurs in Valerio's *De centro gravitatis solidorum libri tres* (Typis Bartholomaei Bonfadini, Romae 1604).

[96] *G.G.*, X, p. 249. The figure corresponds perfectly to the second one drawn on the strips applied on the reverse side of folio 179. This seems to confirm our reconstruction of the missing figure, derived from the same note (cf. *supra*, note 90). It is worth specifying that the second figure is not, unlike the first, conjectural, as Wisan mistakenly believes (*The new science of motion* cit., p. 225, note 20), but is found in Valerio's handwritten letter (cf. Ms. Galil. 88, c. 97*v*).

from rest as equal. It is, therefore, the same principle that we saw recalled by Galileo in the last of the demonstrations examined: "cum enim assumptum sit in naturali descensu velocitatis momenta eadem semper reperiri in punctis aequaliter ab horizonte distantibus iuxta perpendiculares distantias… [since it has been assumed that in natural descent the momenta of speed are always found to be the same at points equally distant from the horizontal along perpendicular distances…]"[97].

The justification of the principle provided by Galileo to Valerio must have consisted, as in the demonstration cited above, in emphasizing the notion of "vertical" distance from rest, dependent on the principle of the indifference of heavy bodies to the motion along planes parallel to the horizontal stated in the *Mecaniche*. It is on the basis of these same considerations that Valerio, in fact, replied that he considered it evident, and that it conformed to the metaphysical principle stated initially, that at the foot of the perpendicular and the incline "l'impeto o inclinazione a velocemente muoversi [the impetus or inclination to move quickly]" (Galileo would have said "il grado o momento della velocità") will be equal.

We have already emphasized the close relationship that, at a certain point, **307**
Galileo must have established between the erroneous supposition and what would become the Postulate in the *De motu naturaliter accelerato*. Valerio's response does not help us in the slightest in establishing whether Galileo was still convinced of the erroneous supposition, or whether he had, instead, already achieved the understanding that acceleration increases with time. It can only be observed that he presented the Postulate to Valerio as an independent supposition, without explicitly suggesting its analogy with the erroneous acceleration-to-distance proportionality, as he had done instead in the aforementioned demonstration. The desire to assert the independent worth of the Postulate – as in the *Discorsi* – might indicate that Galileo was already aware of the falsity of the principle stated in the letter to Sarpi.

Valerio's letter thus offers further proof of Galileo's attempt to derive dynamics from the same principles on which he had based his treatise on statics. And it allows us to affirm, as we have already seen in numerous fragments and as documented by some letters[98], that, in the second decade in Padua, Galileo worked on the theorems and concepts of the *Mecaniche*, seeking to found upon them, with the adequate corrections imposed by the discovery of the naturalness of acceleration, his science *de motu*.

After the strictly static use of *momento* in the Paduan treatise, the Pisan scientist thought, for several years, that it could constitute the link, so to speak, between statics and dynamics. As a consequence of this operation, we have seen the notion and expression of *momentum velocitatis* appear alongside *momento della gravità*. In the notes examined so far, *momentum velocitatis* appears to be a direct consequence, "effect", of that of heaviness. Furthermore, it has come to be configured as a uniform speed, continuously increasing (according to the distance from rest), with which the mobile travels through successive equal spaces in progressively shorter times.

---

[97] *G.G.*, VIII, p. 389 (cf. *supra*, pp. 295-296).

[98] In addition to the ones already mentioned, see the letter – also very well known – from Galileo to Don Antonio de' Medici, dated 11 February 1609 (*G.G.*, X, pp. 228-230).

The story of *momentum velocitatis* as we have illustrated it so far is only a prelude. We will soon see that the original close connection with *momento della gravità* will first be attenuated and then even cleverly hidden. *Momentum velocitatis* will, at this point, take on a completely independent physiognomy. But this presupposes Galileo's abandonment of the attempt at a mechanical foundation for the science of motion, in other words, the renunciation of explaining accelerated natural motion as a consequence or, in some way, an effect of heaviness.

CHAPTER IV

THEORIES OF ACCELERATION

Our analysis has led us to the end of Galileo's Padua period. It is certain that, 309
prior to making his way back to Florence with his astronomical discoveries, the Pisan scientist had already used the correct supposition of the proportionality between acceleration and duration of the fall as the foundation of the science *De motu naturaliter accelerato* [On naturally accelerated motion]. It seems, however – as we have already emphasized – that, for some time, he continued to employ the incorrect principle as well. The discovery of the telescope and the sensational consequences produced by its astronomical use not only produced a cultural upheaval of the time, but also forced Galileo to make an abrupt and sudden shift in interests as well as in life. In his early Florentine years, he presumably must have dedicated very little attention to the project of the three *De motu locali* [On local motion] books, of which he had already lucidly written in the letter to Vinta of 7 May 1610. The almost total absence of references to the *de motu* questions in Galileo's correspondence between 1610 and 1615 authoritatively confirms this turning point, imposed not by free choice, but by unforeseeable circumstances.

Around 1615, Galileo resumed work on the *De motu locali*[1], drafting most of the theorems, which were published with a quick final revision in 1638. We will see the belief in the possibility of basing the science of movement, in all its aspects (natural fall, violent motions, projectiles, percussive impacts, etc.), on the same
principles on which he had built his statics (that is, as far as our specific interest, in 310
the possibility of basing the *momento della velocità* on the *momento della gravità*), gradually fade. At the end of this process, the difficult and demanding attempt, to

[1] The renewed interest in the science of movement was undoubtedly facilitated by the anti-Copernican decree of 1616, which forced Galileo to at least attenuate, if not to abandon, the attention he had until then devoted to astronomical questions. Moreover, with 1615, the controversy over buoyancy came to an end, the final phase of which, as we will see, did not leave out interesting ideas relating to the science *de motu*. Also in '15, Galileo communicated to Giovan Battista Baliani, who went to visit him, some speculations *de motu naturaliter accelerato* (*G.G.*, XIII, pp. 348-349; the letter is from 1627, but it seems that Baliani refers to an oral communication, which could have only occurred in 1615). Galileo's return to work on *De motu locali* around 1615, after his astronomical period, is recognized by all scholars (cf. for example, S. Drake, *Galileo Studies* cit., pp. 226 ff., who assigns some of the *Frammenti attinenti ai Discorsi*, a certain number of which are by Guiducci's hand).

P. Galluzzi, *Momento: Galilean Studies*, Archimedes 78,
https://doi.org/10.1007/978-3-032-35049-7_8

which Galileo had dedicated himself for many years, will leave practically no trace in his printed works.

The reasons that led Galileo to make, after careful consideration, a final, conscious separation of the two *momenti* are not difficult to identify. The difficulties Galileo encountered in carrying out his unification project were, in fact, predominantly of a general nature. In the "mechanical" schema, ultimately based on the balance model (to which Galileo had also attributed the analysis of *momenti* of a heavy body along variously inclined planes)[2], the weight of the heavy body performed an essential function. *Momento* was, essentially – as we have seen – the result of a composition of the distance from the fulcrum of the system with the weight of the heavy body.

In the absence of precise documents, it is difficult to establish with acceptable approximation when Galileo developed the conviction that the speed of heavy bodies, in free fall from rest, increases according to a precise proportion, which is absolutely identical for all bodies. At that point, however, the hypothesis of deriving the laws of naturally accelerated motion from static principles must have appeared to him in all its considerable problematic nature. From this new perspective, "weight", fundamental for the "mechanical" *momento*, effectively became – unless through some clever stratagem – a negligible element in the science *De motu*. How could he, then, propose as the foundation of his own conception of naturally accelerated motion, in which the speed increases from rest in the same manner for all bodies, of any size and weight, a concept that assigned a discriminating function to weight?

Galileo must have acknowledged, presumably in the years between the *Discorsi* and the *Dialogo*, that statics constituted a field of investigation substantially independent of dynamics and that static principles did not lend themselves to serve as a foundation for the construction of the science of movement.

Galileo did not reach this conclusion directly, nor did he reach it without making attempts to "mend", with a general theory, the unpleasant rift that had opened between mechanics and the science of movement, that is, between *momento della gravità* and *momento della velocità*.

In reality, the Pisan scientist never abandoned his deep conviction that heaviness was the essential causal factor of motion. Even his final reflections, communicated or dictated to his students, attest to the stubborn persistence with which he maintained
311 this perspective, which was the traditional one that he had adopted since the *De motu*, albeit modifying it with Archimedean hydrostatic principles. Eventually, unable to offer a satisfactory justification, he had to give up. He decided, then, to suppress in the *Discorsi* any explicit reference to the "causes" of natural motion, limiting himself to describing the "symptoms" and contenting himself with declaring that, fortunately, these symptoms fully corresponded to the actual unfolding of the natural motions[3].

---

[2] It should be recalled that the inclined plane theorem of the *De motu* and the *Mecaniche* attributed the *momento* of the heavy body along a certain inclination to the case of an identical weight applied at a certain distance from the fulcrum of the balance.

[3] See the reasoning prefaced to the *De motu naturaliter accelerato*, in which this attitude is declared:

It is from these bases – how contradictory and uncertain! – that what took shape in Galileo was what many have recognized as one of the most characteristic – if not the most characteristic – traits of the modern scientist: disinterest in causes (and general theories), replaced by a careful recognition of phenomena in order to grasp their patterns and laws, and then translate them into the objective language of mathematics and geometry[4]. In truth, it would be all too easy to show that Galileo **312**
was led to this "modern" attitude almost exclusively by a strong and painful sense of frustration. For the Pisan scientist, it was not a freely meditated choice, but rather a necessary fallback, which must have caused him quite a bit of suffering. The hesitation in publishing a work that about thirty years earlier was already in an advanced stage of development must be attributed, in part, also to the uncertainty of someone who, being unable to indicate the causes of phenomena whose laws he described, felt himself illegitimately evading a fundamental issue and failing in his precise duty as a scholar.

"Quamvis enim aliquam lationis speciem ex arbitrio confingere, et consequentes eius passiones contemplari, non sit inconveniens (ita, enim, qui helicas aut conchoides lineas ex motibus quibusdam exortas, licet talibus non utatur natura, sibi finxerunt earum symptomata ex suppositione demonstrarunt cum laude), tamen, quandoquidem quadam accelerationis specie gravium descendentium utitur natura, eorundem speculari passiones decrevimus, si eam quam allaturi sumus de nostro motu accelerato definitionem, cum essentia motus naturaliter accelerati congruere contigerit. Quod tandem, post diuturnas mentis agitationes, repperisse confidimus [Although it is not inappropriate to fabricate some kind of species of motion at will, and to contemplate the consequent passions, (for indeed, those who have invented the helices or conchoidal lines derived from certain motions, although nature does not use such motions, have demonstrated their symptoms with commendation), nevertheless, since nature uses a certain kind of acceleration for descending heavy bodies, we have decided to examine their passions, provided that the definition we are about to present of our accelerated motion should coincide with the essence of naturally accelerated motion. Which, at last, after long mental deliberations, we confidently believe we have discovered]" (*G.G.*, VIII, p. 197).

[4] This is a common trait that is difficult to overlook. It undoubtedly originated from the attitude towards Galilean science on the part of the neo-Kantians. Above all, Cassirer recognized Galileo as the first scientist aware of the absurdity of the attempt to establish objective causal relationships, owing to the impossibility of drawing on the reality of things. For this reason, he would have limited himself to trying to establish geometric-mathematical relationships between phenomena, without claiming to assert their objective existence. Galileo would have, in short, intuited "quel concetto critico di esperienza formulato da Kant nel principio, in base al quale l'esperienza è possibile soltanto attraverso la rappresentazione di una relazione necessaria delle percezioni [that critical concept of experience formulated by Kant in the principle, according to which experience is possible only through the representation of a necessary relationship of perceptions]" (E. Cassirer, *Storia della Filosofia Moderna*, trad. it. di A. Pasquinelli, 4 voll., Einaudi, Torino 1961, vol. I, pp. 421-461; the passage cited is at p. 440). With undoubtedly greater caution, Koyré also contributed to endorsing this image of the scientist. He would have, in essence, pursued a project that was independent of identifying the causes of motion and acceleration. The supreme form of causality of motion was identified by the Pisan scientist in "time", that is, in a purely subjective relationship (*Studi galileiani*, trad. it. cit., pp. 100-105 and 157-158). However, it did not escape Koyré's attention that the temptation to recognize heaviness as the causal factor of movement always remained very strong for Galileo (*ibid.*, pp. 246-262). Today, in light of more recent studies, and on the basis of a more careful reading of Galileo's work, supporting these theses appears impossible. This does not, of course, mean disregarding the great innovation introduced by Galileo, despite a thousand contradictions and driven by necessity, deciding to publish a treatise in which, formally, he avoided explaining the causes of the phenomenon studied. Conscious of introducing an unusual practice, he, precisely on the opening page of the *De motu naturaliter accelerato*, not coincidentally referenced the previous Archimedean *Spirals* (G.G., VIII, p. 197).

Thus, the emphasis on the strict static-dynamic analogy that we focused on in the previous chapter gradually disappears. Galileo will still resort to it on some occasions, either to confirm principles for which he lacks demonstration (such as the supposition introduced in the *De motu naturaliter accelerato*)[5], or to add further

[5] The pendulum experiment, whose oscillations are artificially interrupted, was proposed by Galileo to make plausible the postulate of the equality of the degrees of speed of the same mobile at the base of the perpendicular and the incline of identical elevation. The interruption of the oscillatory motion, by modifying the trajectory of the pendulum ball, does not, however, prevent it from rising to the same elevation above the horizon that it would have reached if it had continued freely (i.e., the same from which it started). The force that drives the ball is defined by Galileo in terms of "mechanical" *momento*, a concept that inevitably refers to that of the heaviness of the ball (*G.G.*, VIII, pp. 205-207). We have already observed that, in 1639, Galileo, at Viviani's request, considered inserting, in a possible new edition of the *Discorsi*, after the *Scholium* to the first two *Theoremata* (cf. *supra*, Part Two, Chapter III, note 81), a "mechanical" theorem capable of confirmatory demonstration. In essence, Galileo, after the publication of the *Discorsi*, had succeeded in deducing from the acceleration-time proportionality the equality of the degrees of speed of the same mobile at the base of the perpendicular and the incline of equal elevation. The demonstration, drawn up by Viviani but certainly inspired and corrected by Galileo, is quite interesting, even from a terminological point of view. It is based on *momento della gravità* and its laws, set out forty years earlier in the *Mecaniche*. Taking inclined planes and perpendiculars of equal elevation (see fig. 32), Galileo proceeds as follows:

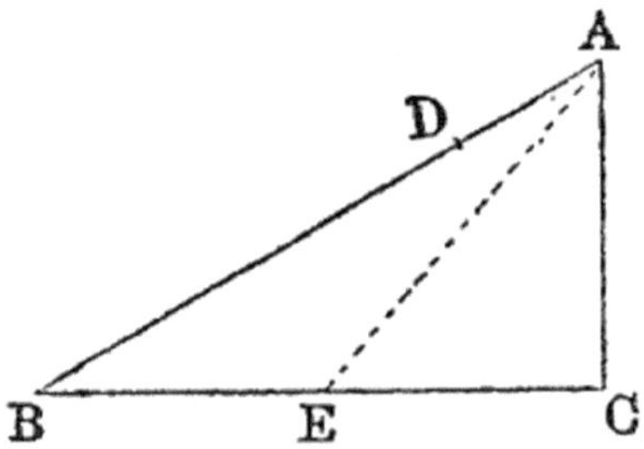

Figure 32

"e perché, come poco fa si è concluso, l'impeto d'un mobile per la perpendicolare AC all'impeto del medesimo per l'inclinata AB, sta come AB ad AC, prendasi nell'inclinata AB, la AD, terza proporzionale delle AB, AC: l'impeto dunque per AC all'impeto per la AB, cioè per la AD, sta come la AC alla AD; e perciò il mobile nell'istesso tempo che passerebbe lo spazio perpendicolare AC, passerà ancora lo spazio AD nell'inclinata AB (essendo i momenti come gli spazi), ed il grado di velocità in C al grado di velocità in D averà la medesima proporzione della AC alla AD. Ma il grado di velocità in B al medesimo grado in D sta come il tempo per AB al tempo per AD, per la definizione del moto accelerato, ed il tempo per AB al tempo per AD sta come la medesima AC, media tra le BA, AD, alla AD, per l'ultimo corollario della seconda proposizione; adunque i gradi in B ed in C al grado in D hanno la medesima proporzione della AC alla AD, e però sono eguali; che è il teorema che intesi dimostrare [and because, as was concluded a little while ago, the impeto of a mobile along the perpendicular AC to the impeto of the same through the inclined AB, is as AB to AC, let AD be taken on the incline AB, the third proportional of AB, AC: the impeto therefore for AC to the impeto for AB, in other words, for AD, is as AC to AD; and therefore the mobile in the same time that the perpendicular space AC is traversed, will also traverse the space AD on the incline AB (the momenti being as the spaces), and the degree of speed at C to the degree of speed at D will have the same proportion as AC to AD. But the degree of speed at B to the same degree at D is as the time for AB to the time for AD, by the definition of accelerated motion, and the time for AB to the time for AD is as the same AC, the mean between BA, AD, to AD, according to the last corollary of the second proposition; therefore the degrees at B and C to the degree at D have the same proportion

evidence to propositions demonstrated through purely geometric-mathematical 313
procedures (this is the case with *Theorema VI*)[6]. The Pisan scientist will thus show that he does not consider the *de motu naturaliter accelerato* evidence obtained through the use of mechanical demonstrations based on the concept of *momento della gravità* to be indisputably conclusive.

Momento della gravità will, therefore, no longer have a relevant role in Galileo's conception of natural motion. In the *De motu locali*, the expression occurs only a few times, and its relationship with *momentum velocitatis* is never emphasized. The abandonment of using the "static" *momento* did not, however, lead to the abandonment of *velocitatis momentum*, to which, on the contrary, Galileo progressively assigned an increasingly important function. After recognizing the substantial "specificity" of dynamics, the development of the concept of *momentum velocitatis* will follow a path that is completely independent of the "mechanical" concept of *momento* and in no way attributable to it. However, *momento della velocità* will continue to suggest to the reader, automatically, a substantial continuity between mechanical and dynamic principles and speculations, a connection in which Galileo, despite lacking a general theory that would allow him to foundationalize it conveniently, continued to believe until the end.

Before focusing on the evolution of *momentum velocitatis*, following the
abandonment of the "mechanical" foundation, it will be useful to try to focus, 314
at least in broad terms, on the attitude (or, rather, attitudes) that Galileo took on, from the return to Florence until the publication of the *Discorsi*, towards the "cause" of natural motion and the acceleration that characterizes it. Following the various interpretations that he developed of this perplexing phenomenon, trying to grasp its meaning and emphasizing its connection with the parallel elaboration of certain theorems *de motu naturaliter accelerato* is an indispensable preliminary operation to even be able to reconstruct the events of *velocitatis momentum*.

The intense reflection on the composition of the continuum and on a whole series of infinitesimal concepts – which we will examine in the next chapter – together with the effort to develop a general theory of natural acceleration capable of conveniently justifying all phenomena, constitute, in fact, the most eloquent

as AC to AD, and therefore are equal; which is the theorem I intended to demonstrate]" (*ibid.*, p. 218). The demonstration is very clear and requires no comment. Our scientist establishes a close relationship between the "mechanical" *momento* and the accelerated speed. Having specified that these are descents *of the same heavy body*, there are no problems. However, the theorem, like the postulate, despite being presented with this limitation, is in fact used by Galileo in an absolute sense. And Galileo knows he is not authorized to do so. Here, then, Galileo, with the clear intention of muddying the waters, tries to conceal the derivation of the postulate from the laws of *momento della gravità* (i.e., the "mechanical" foundation of naturally accelerated motion) by using terms different from the usual *momento della gravità* and *momento della velocità*. Substituting, in fact, the first with *impeto*, and the second with *grado di velocità*. The meaning of the operation is all too obvious. Galileo does not intend to point out the relationship between the two *momenti*. It would, indeed, be like admitting a dependence of dynamics on static principles, which he has no arguments to support. "Impeto" is for Galileo a less compromising term than *momento*. Moreover, it has been used since the *Mecaniche* to indicate speed. It therefore appears capable of suggesting less traumatically the effect of heaviness on accelerated descents. In the entire demonstration, *momento* appears only once, and is related to distance, not speed.

[6] *Ibid.*, pp. 221-223.

and evident proofs of the desperate attempts that Galileo made, before deciding to abandon it, to make the mechanical approach plausible, of the awareness with which he dedicated himself to the project of "refounding" science *de motu*, and, lastly, of the central function that *momentum velocitatis* served.

We have already focused on the theory developed by Galileo in the *De motu* to explain the increase in speed in falling motion. In fact, he had expunged acceleration from the category of natural phenomena, reducing it to a pure accident. At the same time, he had identified the specific weight of the mobile as the direct cause of the (uniform) speed of descent. This implied that bodies of the same species, whatever their specific weight, would all move with the same speed[7]. This, in essence, was the framework of Galileo's reflections at the time of the drafting of the Latin treatise. We know that the decision to abandon its completion must have depended, however, to a considerable extent, on the difficulty of sustaining the theory of "accidental" acceleration itself.

It is obvious that it would be extremely important for us to be able to follow the course of thought of the Pisan scientist, after the setback of the *De motu*, to see in which direction it oriented in the effort to justify acceleration and propose a convincing cause for it. Unfortunately, the available documents do not allow this need to be met. In fact, between the 1590s and the 1630s, precise indications are missing regarding the way in which Galileo approached acceleration. Indeed, neither the *Mecaniche*, the bulk of the *Frammenti*, nor the works preceding the *Dialogo* offer conclusive ideas in this sense. The fragments allow us to affirm that Galileo, starting from the early years of the 17th century, made a serious effort to give accelerated motions a rigorous geometric-mathematical treatment. We can also emphasize that, in this attempt, he initially accentuated the analogies with the treatment
315 of mechanical motions (the erroneous supposition of the acceleration-distance proportionality and the use of *momento della gravità*). In this way, he arrived at a certain point to suppose that the speed increases proportionally to the growth of the *momenti della gravità*, suggesting that *momento* increases constantly starting from rest. Acceleration would therefore appear to be an effect of *intentio gravitatis* [intensification of heaviness].

In modern terms, Galileo would have, in other words, understood that a constant force generates a continuously accelerated speed. If the indication is clear, it should not, however, be overlooked that the fragment that preserves it is very explicit in emphasizing the frame of reference: descents *of the same mobile* along differently inclined planes, but of identical elevation[8]. Galileo's supposition therefore implies that when the mobile changed (that is, in the case of a different *momento della*

---

[7] Cf. *supra*, pp. 180 ff.

[8] "Ponatur igitur augeri vel imminui motus velocitatem secundum proportionem qua augentur vel minuuntur gravitatis momenta; et cum constet, *eiusdem mobilis* momenta gravitatis super plano *db* ad momenta super plano *da* esse ut longitudo da ad longitudinem *db*… [Let it therefore be assumed that the speed of motion increases or decreases according to the proportion in which the momenta of heaviness increase or decrease; and since it is established that the momenta of heaviness *of the same mobile* on plane *db* are to the momenta on the plane *da* as the length of *da* is to the length of *db*…]" (*G.G.*, VIII, p. 379; the italics are ours; cf. also *supra*, pp. 290-291).

gravità), different speeds of descent occur. Starting from this supposition, in short, it cannot be maintained that the acceleration from rest is the same for all heavy bodies.

The letter to Sarpi is much less clear on this point. The law that Galileo presents there would seem to be valid for all heavy bodies (Galileo, in describing it, speaks of a "mobile naturale [natural mobile]" and limits himself to specifying the degrees of speed that such a mobile reaches at different distances from rest, without mentioning the differences that the choice of mobiles of different weights would entail for the "proportion" of speed)[9]. In the other fragments that we have stressed, there are no indications that betray the way in which Galileo conceived the cause of natural motion and its constant acceleration. The letter to Sarpi, however, leads us to think that, since 1604, our scientist was convinced that the speed of any heavy body increases, starting from rest, according to an identical proportion. The persistence of the schema of inclined planes along which *the same mobile* descends indicates, however, that he was absolutely unable to justify the equality of the speeds for all mobiles (in reality, as we will see, he will never succeed)[10].

The controversies on buoyancy, in which Galileo was involved between 1612 and 1615, preserve some valuable evidence of the survival, in essence, of the "Archimedean" explanation of natural motions set out in the *De motu*. In the *Discorsi*, the Pisan scientist, in fact, reiterates twice that the speed of mobiles
depends on the relationship between their specific heaviness and that of the medium **316**
in which they move[11]. This obviously implied that, among bodies of different specific weight, there will be differences in speed within the same medium. Now, in the *Discorsi*, Galileo does not mention the nature of this "speed" in the slightest; in other words, he does not specify whether it is uniform or accelerated. Obviously, in 1612, Galileo knew very well that descents were accelerated. If he avoided clarifying it, it was because the hydrostatic theory – as it had been shown in the *De motu* – could not be used, neither to explain naturally accelerated descents, nor (much less) the equality of all descents.

The hydrostatic theory had been restated – again, avoiding any connection to the problem of acceleration – also in the *Risposta* of 1615. In the last of the writings related to the controversy on buoyancy, Galileo was, to tell the truth, decidedly more daring, ultimately recognizing that in a vacuum even heavy bodies of different "specific" weights would descend with identical speeds. Of this assertion, he had not provided a true demonstration, limiting himself to declaring that he had based it on an observation: the natural falls of two heavy bodies of different specific weight in different media occur at speeds whose difference tends to appear all the more

---

[9] *G.G.*, X, p. 115.

[10] Again, in the Postulate preceding the *De motu naturaliter accelerato* of the *Discorsi*, he will be forced to express himself in terms *of the same mobile*: "Accipio gradus velocitatis eiusdem mobilis super diversas planorum inclinationes acquisitos tunc esse aequales, cum eorundem planorum elevationes aequales sint [I assume that the degrees of speed of the same mobile, acquired over different inclinations of planes, are equal when the elevations of those planes are equal]" (*G.G.*, VIII, p. 205).

[11] Cf. *supra*, p. 238.

imperceptible the less dense the medium becomes:

sì che attenuandosi ancora più il mezzo, ed in consequenza facendosi tal differenza di velocità minore, molto probabilmente si può concludere che nel vacuo tutti i corpi si moverebbono con la medesima prestezza[12]. [so that by attenuating the medium even more, and consequently making this difference in speed smaller, it can very likely be concluded that in a vacuum all bodies would move with the same swiftness.]

This, it seems to me, is a clear indication of the conviction that Galileo had developed regarding the equality in the speed of descent of all bodies. Furthermore, Galileo, on the same page of the *Risposta*, took the opportunity to also state that the descents of mobiles of different species "nell'aria saranno tanto simili, che non ci sarà a pena un centesimo di differenza [in the air will be so similar, that there will barely be a hundredth of a difference]"[13]. The bodies – as Galileo is now certain – fall – as experiment attests – all with the same speed.

It remains to be clarified how Galileo could reconcile the explanation of natural motion through the hydrostatic theory (conceived to account for uniform motions) with constant acceleration, equal for all bodies. In all likelihood, the hydrostatic theory served excellently to justify the different speeds observed in the descents of heavy bodies of different species in dense media such as water (i.e., the typical context of the *Discorsi* and the *Risposta*). On the other hand, he knew that, in air and – even more so – in a vacuum, the speeds of descent are all equal and accelerated.
317 The cause of the equal descents could not, therefore, be attributed to the different specific weight, nor could it explain (nor hypothesize – as had been done in one of the previously examined fragments – that it constantly increases) the uniform and constant acceleration for all bodies from rest.

Some irrepressible experimental certainties, therefore, posed serious foundational questions for the science *de motu*. The contrast, never conveniently resolved, between these two essential poles (experimental certainties and theoretical justifications) clearly emerges even from Galileo's subsequent reflections on the causes of natural motion and uniform, constant acceleration.

The hydrostatic theory was not, in fact, abandoned. We find it proposed again in the *Dialogo*, in terms substantially no different from the *Risposta*. This time, however, it is introduced immediately after the explicit declaration of uniform and constant natural acceleration for all heavy bodies[14]. The reference to the hydrostatic theory of the *Dialogo* is very brief, and Salviati refers to the exposition of "certa esperienza [certain experience]" and the "ferma dimostrazione [firm demonstration]" on which theory is supported[15].

The demonstration is found precisely in the *Prima Giornata* [First Day] of the *Discorsi*, where Salviati contests the Aristotelian thesis, supported by Simplicio, that the speed of fall is proportional to the weight of heavy bodies. The proof consists of

[12] *G.G.*, IV, p. 659.
[13] *Ibid.*
[14] *G.G.*, VII, p. 248.
[15] *Ibid.*, p. 250.

a *reductio ad absurdum*. Assuming that, of two bodies equal in specific heaviness but different in absolute weight, the heavier one moves faster, the question arises what will happen if the two bodies are joined in such a way as to form a single body. On the one hand, one is led to say that the compound will descend with a speed slower than that of the heavier body, being, so to speak, "braked" by the slower speed of the lighter one. On the other hand, since the compound is a heavier body, it should move with a greater speed than the heavier of the components. "Vedete dunque come dal suppor che 'l mobile più grave si muova più velocemente del men grave, io vi concludo, il più grave muoversi men velocemente [You see therefore that from the supposition that the heavier body moves faster than the less heavy one, I conclude that the heavier body moves less quickly]"[16]. The demonstration – as promised by Salviati in the *Dialogo* – is followed by "certain experience". This corresponds exactly to that proposed in the *Risposta* of 1615: the observation indicates that the differences in speed between heavy bodies of different absolute and specific weights tend to decrease as the motions occur in thinner media; in a vacuum, therefore, "tutti i mobili si moverebbe con i medesimi gradi di velocità [all mobiles would move with the same degrees of speed]"[17].

To these reasonings, Galileo added some clarifications, from which we can draw **318**
valuable indications regarding his way of conceiving the cause of accelerations in natural motions from rest, uniform for all heavy bodies:

> Dico per tanto che un corpo grave ha da natura intrinseco principio di muoversi verso 'l comun centro de i gravi, cioè del nostro globo terrestre, con movimento continuamente accelerato, ed accelerato sempre egualmente, cioè che in tempi eguali si fanno aggiunte eguali di nuovi momenti e gradi di velocità[18]. [I, therefore, say that a heavy body has an intrinsic natural principle of moving towards the common center of heavy bodies, that is, of our terrestrial globe, with continuously accelerated movement, and always equally accelerated, in other words, that in equal times, equal additions of new momenti and degrees of speed are made.]

The fact that the speed of fall from rest of all heavy bodies of any specific weight increases in the same way is, therefore, a consequence of a precise law of nature. Heaviness is not one of the causes of this phenomenon. This is an important fact on which we will shortly focus.

Even in a long chapter of the *Postille* to Antonio Rocco's *Esercitazioni Filosofiche*, Galileo had presented, with even greater energy and clarity, the fully equivalent principle of a natural constitution, which imposed equal speeds of fall on all heavy bodies. Galileo had even spoken of "velocità innata [innate speed]":

> passando col discorso più oltre, mi formai un assioma da non esser revocato in dubio da nessuno, e supposi qualsivoglia corpo grave descendente aver nel suo moto gradi di velocità, dalla natura limitati ed in maniera prefissi, che il volergli alterare, col crescergli la velocità o diminuirgliela, non si potesse fare senza usargli violenza per ritardargli o concitargli il detto suo limitato corso naturale[19]. [going further

[16] *G.G.*, VIII, p. 108.
[17] *Ibid.*, p. 118
[18] *Ibid.*
[19] *G.G.*, VII, p. 731.

with the discussion, I formed an axiom that should not be called into doubt by anyone, and I supposed that any descending heavy body would have degrees of speed in its motion, limited by nature and in a predetermined manner, such that to alter it by increasing or decreasing its speed, could not be done without using violence to delay or hasten its limited natural course.]

The *momenti della velocità* would be acquired by all mobiles in natural motion according to the same proportions: "le innate velocità da tutti i mobili dovere essere nel vacuo tra di loro tutte simili ed eguali, crescenti però continuamente in duplicata proporzione de' tempi [the innate speeds of all mobiles must be all similar and equal in a vacuum, continually increasing in duplicate proportion to the times]"[20]. The innateness of speeds made it possible, just as well as the law of nature, to escape the obligation to justify why all bodies fall at the same speed.

The natural statute is valid, of course, when any impediments are excluded. But, in human experiences, motion always takes place in a medium. This medium – dense or thin – nevertheless opposes "con resistenza or minore ed or maggiore, secondo che lentamente o velocemente ei deve aprirsi per dar il tramite al mobile [with sometimes less or greater resistance, depending on whether it must open slowly or quickly to give passage to the mobile]"[21]. There is, therefore, "nel mezzo, accrescimento di resistenza [in the medium, an increase in resistance]", not because – as Galileo adds – "la sua essenza [its essence]" changes, but because the speed with which it must open and move laterally to let the falling body pass through changes. From the principle of the equality of all accelerated motions in the natural
319 fall from rest, under ideal conditions (i.e., in a vacuum), one can easily deduce the proportions of the speeds of mobiles of different specific weights in various media. This objective will be achieved if we pay attention to

quanto la gravità del mezzo detrae alla gravità del mobile, la qual gravità è lo strumento col quale il mobile si fa strada, rispingendo le parti del mezzo alle bande, operazione che non accade nel mezzo vacuo e che però differenza nissuna si ha da attendere dalla diversa gravità; e perché è manifesto il mezzo detrarre alla gravità del corpo da lui contenuto quant'è il peso d'altrettanto della sua materia, scemando con tale proporzione le velocità dei mobili, che nel mezzo non resistente sarebbero (come si è supposto) eguali, aremo l'intento[22]. [how much the heaviness of the medium detracts from the heaviness of the mobile, which heaviness is the instrument with which the mobile makes its way, pushing aside the parts of the medium to the sides, an operation that does not occur in the vacuum medium and, thus, no difference is to be expected by a different heaviness; and because it is evident that the medium detracts from the heaviness of the body the weight of its matter of an equally contained amount, decreasing in such proportion the speeds of the mobiles, which in the non-resisting medium would be (as has been supposed) equal, accomplish our intent.]

The proposed calculation of the speeds of mobiles of different specific heaviness in different media requires no comment, it is so crystal clear. It corresponds to the hydrostatic interpretation already presented several times previously by Galileo.

However, at this point, Galileo had, for the first time, clarified how the hydrostatic approach, based on the comparison between specific weights of the

[20] *Ibid.*, pp. 742-743.
[21] *G.G.*, VIII, p. 119.
[22] *Ibid.*

mobile and the medium, could coexist with the established equality of acceleration from rest for all bodies, whatever their specific heaviness.

Precisely in justifying the equality of accelerated descents in a vacuum (whereas the speeds follow the hydrostatic rule for a filled medium), Galileo ended up assigning to heaviness a rather particular and significantly reductive function, one that is worth focusing our attention on. Heaviness has, in fact, become the "tool" that the heavy body has at its disposal to carve its way through the medium. In a vacuum, lacking any resistance to overcome, the operation of heaviness is null. Differences in heaviness cannot, therefore, produce any difference in speed. It is worth recalling that the conception of heaviness as the instrument that the mobile uses to overcome the resistance of the medium had been developed by Themistius in the *Paraphrase* to Aristotle's *Physics*, in a paragraph to which Galileo had dedicated a page in the *De motu* for its refutation. Themistius's thesis was that, in a vacuum, the speeds (uniform, of course) of heavy bodies of any weight were all the same[23].

The serious limitation of the role of heaviness must have been imposed on Galileo by his inability, despite numerous attempts, to use it as the foundation of
his science *de motu naturaliter accelerato*. Acceleration is thus detached and made **320**
independent of heaviness. The speed of all mobiles in free fall increases constantly, by virtue not of their absolute or specific weight but of a natural law or constitution promulgated by God. This makes it possible to circumvent the obstacle posed by the inability of the Pisan scientist to reconcile heaviness as the cause of natural motion with the equality of acceleration from rest for all heavy bodies[24].

We have thus arrived at a conception of heaviness that is vastly distant from the one we saw in the *Mecaniche*[25]. This illustrates the extent of the gap that separates Galileo's original dynamic conception from the positions he was forced to assume after the failure of his attempt at a mechanical foundation of naturally accelerated motion. Once again, clear evidence leads us to maintain that Galileo, if he ever arrived at it[26], was forced to adopt the supposedly "modern" attitude, consisting of limiting himself to the geometric-mathematical description of natural motions, by

---

[23] *Paraphrasis* cit., p. 139 (cf. *supra*, Part One, Ch. III, p. 99). The refutation of Themistius's thesis by Galileo, in which he maintained that in a vacuum the speed of descent would have been proportional to the heaviness of the bodies, is found in the *De motu* (*G.G.*, I, pp. 294 and 412; cf. *supra*, Part Two, p. 172, note 73). It is curious that, after more than half a century, Galileo conceded and made Themistius's thesis his own. Aristotle had, as is known, maintained that in a vacuum, lacking any opposition to descent, the excess of heaviness with respect to the resistance would be infinite. The mobile would, therefore, have had infinite speed: that is, it would have moved instantaneously.

[24] Already in the *Postille* to Rocco's *Esercitazioni*, Galileo had stated that heaviness has no effect on the speed of descent: "per quello che depende dalla gravità, tutti i mobili, quanto si vogliano disuguali in grandezza, si muoveranno con la medesima velocità; ma se *ab extra* si interporrà qualche ostacolo, sì come sempre di necessità si interpone, allora la regola, per altro sicura, della gravità vien perturbata tal volta e più che sommamente alterata [for what depends on heaviness, all mobiles, no matter how unequal in size, will move with the same speed; but if *ab extra* [externally] some obstacle is interposed, as it always necessarily is, then the rule, otherwise certain, of heaviness is sometimes disturbed and more than greatly altered]" (*G.G.*, VII, p. 734).

[25] *G.G.*, II, p. 159.

[26] Let us not forget that the Galilean conception of movement ultimately presupposes a divine decree.

the insurmountable difficulties encountered in the attempt to give the science *de motu* a non-contradictory foundation.

With Galileo, we have now become accustomed to not considering statements of principle as definitive, and to expect timely corrections and denials. And, in fact, in *Giornata Terza* [Day Three] of the *Discorsi*, we find a suggested interpretation of acceleration in which heaviness once again plays an important role. The new proposal is, however, to be considered of lesser authority, since it is presented by Sagredo. Salviati, Galileo's official spokesperson, defends it from Simplicio's objections, but, in fact, avoids explicitly endorsing it.

Sagredo's intervention takes place during the discussion relating to the principle outlined on the opening page of the *De motu naturaliter accelerato*: a heavy body, which begins to descend starting from rest, reaches a certain *momento di velocità* only after having passed through the infinite *momenta velocitatis* between that *momento* and rest; the same happens for violent upward motions[27]. The objections of Simplicio and Sagredo emphasize how, in order to pass through an infinite number of degrees of speed, an infinite amount of time seems necessary. Salviati
321 replies that this would happen if the mobile remained for some time in each of the individual degrees of speed, "ma egli vi passa solamente, senza dimorarvi oltre a un instante [but it only passes through, without remaining there for more than an instant]"[28].

At this point, Sagredo speaks again to inform his interlocutors that Salviati's[†] explanation has suggested to him some considerations, from which one could derive an "assai congrua [very congruous]" justification of the question "agitata tra i filosofi [agitated among philosophers]" regarding the cause of acceleration in natural motion. If, indeed, the speed of the heavy body pushed upwards violently progressively decreases, passing through the infinite *momenti of* smaller *speeds*, until it reaches rest, this means that there is a parallel decrease in the "virtù impressa [impressed virtue/force]", of the opposite sign compared to the heaviness with which it was thrown. Therefore, when the "virtù impressa" and "gravità" of the mobile have reached equilibrium,

resta il mobile di più salire e passa per lo stato della quiete, nel quale l'impeto impresso non è altramente annichilato, ma solo consumatosi quell'eccesso che pur dianzi aveva sopra la gravità del mobile, per lo quale, prevalendogli, la spigneva in su, continuandosi poi la diminuzione di questo impeto straniero, ed in conseguenza cominciando il vantaggio ad esser dalla parte della gravità, comincia altresì la scesa, ma lenta per il contrasto della virtù impressa, buona parte della quale rimane ancora nel mobile; ma perché ella pur va continuamente diminuendosi, venendo sempre con maggior proporzione superata dalla gravità, quindi nasce la continua accelerazione del moto[29]. [the mobile ceases to climb further and passes through the state of rest, in which the impressed impeto is not otherwise annihilated, but only that excess which it had previously over the heaviness of the mobile has been consumed, by which, prevailing over it, it was pushing it upwards, then the decrease of this foreign impeto continues, and

---

[27] *G.G.*, VIII, pp. 198-200.
[28] *Ibid.*, p. 201.
[†] The text originally indicated "Sagredo"; however, Sagredo is actually responding to Salviati's explanation.
[29] *Ibid.*

consequently the advantage begins to be on the side of heaviness, the descent also begins, but slowly due to the opposition of the impressed force, a good part of which still remains in the mobile; but because it continues to decrease continuously, being increasingly overcome by heaviness, thus the continuous acceleration of the motion arises.]

Salviati remains extraneous to the polemical exchange between Simplicio and Sagredo following the exposition of the interpretative criterion. Then he intervenes to assert energetically the inappropriateness of discussing the cause of natural acceleration, "intorno alla quale da varii filosofi varie sentenzie sono state prodotte [about which various opinions have been produced by various philosophers]"[30]. Salviati invites our attention to the "passioni [passions]" of an accelerated motion (whatever the cause may be), in which the speed increases according to "la continuazion del tempo [the continuation of time]"[31].

Sagredo's proposal – as has been observed – coincides exactly with the explanation of acceleration provided by Galileo in the *De motu.* Acceleration appears to be an "accident", produced not directly by heaviness (which would generate uniform motions) but by the progressive prevalence of heaviness over the "impressed force" of the opposite sign that kept the mobile at rest. Obviously, once the impressed force is consumed, the mobile will proceed with uniform speed. It is strange that Galileo – even without involving Salviati – thought it appropriate to propose a conception of accelerated motion as non-natural motion[32]. And no less **322**
curious is the fact that he returns to giving a central and causal role to heaviness in the natural fall, which shortly before he had reduced to an instrument of the mobile.

Sagredo, who presents the explanation of acceleration based on the relationship between heaviness and impressed force, knows very well that acceleration is the

---

[30] *Ibid.*, pp. 201-202.

[31] *Ibid.*, p. 202.

[32] It is worth recalling the theory, presented by Galileo in the *Dialogo*, of the semicircular trajectory described by a heavy body pushed by its own heaviness with accelerated speed towards the center of the Earth and simultaneously transported from east to west by the diurnal motion of the Earth. Galileo had attempted to demonstrate that during this fall, the heavy body would travel equal distances in equal times. That is, it would have moved along the trajectory resulting from the composition of the two motions with uniform motion. "Talché – aveva aggiunto trionfalmente – noi venghiamo liberi di ricercar nuove cause di accelerazione o di altri moti, poiché il mobile, tanto stando sulla torre, quanto scendendone, sempre si muove nel modo medesimo, cioè circolarmente, con la medesima velocità. Or ditemi che vi pare di questa mia bizzarria [Thus – he added triumphantly – we are free to seek new causes of acceleration or other motions, since the mobile, whether standing on the tower or descending from it, always moves in the same manner, that is, circularly, with the same speed. Now tell me, what do you think of this bizarreness of mine]" (*G.G.*, VII, p. 192). The "bizarreness", which Galileo substantially retracted not long after (cf. letter to Carcavi, who had conveyed his objections to him, dated 5 June 1637, in *G.G.*, XVII, p. 81), allowed – just like the explanation proposed by Sagredo in the Discorsi – for the exclusion of acceleration from "real" motions within the Copernican cosmos (acceleration would, in fact, only occur if the Earth were immobile). On the "bizarreness" of the *Dialogo*, cf. the classic work by A. Koyré, *A documentary history of the problem of fall from Kepler to Newton*, "Transactions of the American Philosophical Society", New Series, vol. 45, part 4, Philadelphia 1955; see also S. Drake, *Semicircular fall in the Dialogue, Galileo Gleanings XVI*, "Physis", X (1968), pp. 89-100; lastly, let me also refer to my *Galileo contro Copernico. Il dibattito sulla prova "galileiana" di G. B. Riccioli alla luce di nuovi documenti*, "Annali dell'Istituto e Museo di Storia della Scienza di Firenze", II (1977), fasc. 2, pp. 87-148.

same for all bodies starting from rest. However, he avoids indicating how his theory can justify this phenomenon. In truth, the solution was not difficult. In fact, it was enough to suppose that the ratio between "virtù impressa" and "gravità" was always constant. To keep two mobiles of different heaviness at rest, since heaviness tends to make them descend, it is necessary to impress on them two forces of opposite sign, proportional to each other in the ratio of their heaviness. It could thus appear logical that, once the fall had begun, the greater heaviness, having to overcome the greater impressed force, would produce degrees of speed identical to those generated by the lesser heaviness, hindered by a proportionately lesser opposing force.

The fact that the Pisan scientist lingered at some point on this type of reflection seems to be confirmed by some documents that are not directly Galileo's. They also indicate with strong probability (suggested by numerous other pieces of evidence) that Galileo, in the last years of his life, developed a third interpretation of acceleration and its equality for all heavy bodies. With this last justification, Galileo could assign the *momento della gravità* a causal role in natural motion, simultaneously managing to "save" the equality of the accelerated speeds of all mobiles, whatever their weight. At this point, the Pisan scientist had to return
323 to viewing with well-founded confidence the possibility of subjecting statics and dynamics to a unitary treatment.

The documents to which I intend to refer are some letters in which Evangelista Torricelli, at the beginning of 1645, defended a Galilean principle, which he had employed in his own *De motu*, from the attack of Marino Mersenne, to whom it had not seemed convincing. "Supponis cum Galilaeo velocitates in diversis planorum inclinationibus ita esse ut sunt momenta, quando fuerit eadem moles [You assume – the French Minim[†] had written to him – with Galileo that the speeds on different inclined planes are in the same ratio as the momenta, when the bulks are the same]"[33]. And he asked whether Torricelli was able to provide a demonstration "ab experientia independentem [independent of experience]". Torricelli responded promptly, emphasizing, first of all, that the proportionality *momenti/speed* was indisputable and proposed a *ratio physica* [physical principle]. In fact, he invites us to consider two equal spheres, which lie on two different planes. Assuming that the *momento* of a sphere on one of the two planes is double that of the other,

> quis non concedat et velocitatem ad velocitatem esse duplam? Moles a me supponuntur aequales eiusdemque materiae; virtus vero quae impellit alteram molem dupla demonstratur virtutis alterius, ergo si dupla virtus impellens est, et subiecta aequalia, duplam procul dubio velocitatem efficiet[34]. [who does not concede that the one speed is double the other? The bulks are assumed by me to be equal and of the same material; the force, however, that drives one bulk is shown to be double that of the other, therefore, if the driving force is double and the subjects are equal, it will undoubtedly produce double the speed.]

The physical argument proposed by Torricelli refers to the dynamic postulate

[†] Mersenne was a member of the Catholic Order of Minims.

[33] Edizione Nazionale delle *Opere dei Discepoli di Galileo. Carteggio*, Vol. I (1642-1648), a cura di P. Galluzzi e M. Torrini, Giunti-Barbèra, Firenze 1975, p. 180. Mersenne's letter is dated 10 January 1645.

[34] *Ibid.*, p. 182. Torricelli's response, undated, is to be assigned to the end of January 1645.

used by Galileo in the *De motu* and should be seen as part of the attempt to base statics and dynamics on identical foundations, on which we have previously examined at great length.

As for acceleration, the most interesting clues are found in the following passage. Torricelli, in fact, anticipating Mersenne's inevitable objection, took care to clarify how the proportion of speeds would follow that of the *momenti* even in the case of descents along the same plane, as occurs in free fall. Applying to the free fall the same rule of *momenti* along the inclined plane, two bodies, of which one weighs twice as much as the other (i.e., one has double the *momento* of the other), would have descended one at double the speed of the other. This, obviously, would have clashed with Galileo's assertion according to which "omnia gravia, cuiuscunque molis, ponderis et figurae sint, libera demissa a loco absque impedimentis eadem velocitate feruntur deorsum [all heavy bodies, of whatever bulk, weight, and shape, when freely dropped from a place without impediments, move downward with the same speed]"[35]. The justification proposed by Torricelli is based on the recognition **324**
that the *momento* along the same plane is always the same for all heavy bodies of any kind and weight:

Tantum enim est in unoquolibet corpore virtutis moventis quantum est et materiae; exempl. gr.: in uncia auri atque in uncia coerae tantundem est et materiae et virtutis moventis, licet cera appareat multo maiorem locum occupare, propterea dum quiescent pariter gravitant et manifeste aequalitatem virtutum indicant. Quando vero moventur, aurum praecedet (sed longe minus quam pro ratione specierum gravitatis) ipsam coeram, quod quidem accidit quia cum virtutes aequales sint in utraque materia, si altera cum maiori mole ambienti medii, altera vero cum minori luctari debet. Quando vero consideramus duas sphaeras eiusdem materiae, sed alteram minus unciae, alteram vero decem librarum, aequaliter hae descendunt in eodem plano, quia in utraque sphaera virtutes illae arcanae, licet inaequales sint, inter se eandem habent rationem quam resistentiae, hoc est corpora ipsa movenda. Vel si mavis hoc modo: virtus minor ad minus pondus a se movendum eandem habet rationem quam virtus maior ad maius pondus a se movendum[36]. [For in each body there is as much moving force as there is matter; for example: in an ounce of gold and in an ounce of wax, there is the same amount of matter and moving force, although the wax appears to occupy a much larger space. Therefore, while at rest, they are equally heavy and clearly indicate an equality of forces. However, when they move, gold will precede the wax itself (but by far less than in proportion to the specific heaviness of the materials). This happens because, while the forces are equal in both materials, one must struggle against a greater bulk of the surrounding medium, and the other against a lesser bulk. However, when we consider two spheres of the same material, one weighing less than an ounce, the other weighing ten pounds, they descend equally on the same plane. This is because in each sphere, those hidden forces, though unequal, have the same ratio to each other as the resistances, that is, the bodies themselves that are to be moved. Or, if you prefer, a lesser force in moving a smaller weight has the same ratio to that weight as a larger force has in moving a larger

[35] *Ibid*.

[36] *Ibid*. That Torricelli intended to emphasize without hesitation the equality of the *momenti della gravità* of all heavy bodies along the same plane is indicated by a subsequent passage in the same letter: "Mirum ergo non sit metalla, lapides, ligna etc. tam in descensu libero, quam in eodem plano collocata pene eadem velocitate descendere, cum singula aequale momentum, hoc est aequalem sibi ipsis virtutem moventem habeant [Let it not be surprising, therefore, that metals, stones, wood, etc., descend with almost the same speed, both in free fall and when placed on the same plane, since each has an equal momentum, that is, an equal moving force relative to itself]" (*ibid.*).

weight.]

Although impeccable in formally explaining the equality of the speeds of fall from rest for all heavy bodies, Torricelli did not, to tell the truth, specify why these speeds are constantly accelerated. We know – as we will focus on this aspect later – that the Faentine scientist understood that a constant *momento* generates an accelerated speed[37]. It was therefore sufficient for him to demonstrate that all heavy bodies, at the instant in which they begin to descend along the same plane, have the same *momento*, from which increasing degrees of speed will therefore be generated according to an identical proportion[38].

The justification given by Torricelli is substantially different from that presented by Sagredo in the *Discorsi*. In fact, acceleration appears here as a phenomenon that is neither transitory nor accidental, but rather permanent and natural. However, the interpretation given by the Faentine scientist of the equality of all speeds along the same plane was – as a justificatory mechanism – similar to the one implicit in
325 Sagredo's proposal, which was based on the recognition of the constant relationship between heaviness and impressed force. In any case, it too returned to assigning to heaviness a decisive role and a precise causal function, both in natural fall and in acceleration.

Torricelli's hypothesis had already been proposed in 1633 by Baliani[39]. Galileo himself, after the publication of the *Discorsi*, probably adopted it as his own or, at the very least, took it into serious consideration. It was at this point – not before – and based on this hypothesis, that he must have developed the conviction that the constant action of the *momento della gravità* generates uniformly accelerated motions[40]. In

---

[37] Cf. *infra*, pp. 404-405.

[38] I discuss the significance of the Torricelli-Mersenne discussion regarding the legitimacy of extending to speed the proportions between the *momenti della gravità* in a recent essay, to which I take the liberty of referring for further indications. (*Evangelista Torricelli*: *Concezione della matematica e segreto degli occhiali*, in "Annali dell'Istituto e Museo di Storia della Scienza di Firenze", I, 1976, fasc. 1, pp. 71-95).

[39] Giovanbattista Baliani, in the preface to the *De motu gravium* (Genuae 1646, pp. 5-7) recounts having carried out experiments from the Rocca di Savona by simultaneously dropping, first, cannonballs of the same material but of different weight, and then, spheres of different material (lead and wax). He was able to observe that, in the first case, the two balls touched the ground simultaneously, while in the second, the lead sphere preceded the wax one, but by a space absolutely not proportional to the difference in weight. On this basis, it seemed natural to him to formulate a principle that corresponds exactly to that proposed by Torricelli in the *De motu*: "Porro, cum ex experimentis satis superque liqueret in naturali motu gravium proportionem gravitatum communiter creditam non servari, in eam descendi sententiam ut arbitrarer fortasse gravitatem se habere ut agens materiam vero, seu mavis materiale corpus, ut passum, et proinde gravia moveri iuxta proportionem gravitatis ad materiam, et ubi sine impedimento naturaliter perpendiculari motu ferantur moveri aequaliter quia ubi plus est gravitatis plus pariter sit materiae, seu materialis gravitatis [Furthermore, since it was clearly evident from the experiments that the commonly accepted proportion of heaviness is not maintained in the natural motion of heavy bodies, he came to the opinion that perhaps heaviness behaves as the agent and matter, or if you prefer, the material body, as the recipient, and therefore heavy bodies move according to the proportion of heaviness to matter, and where, without impediment, they are naturally carried in a perpendicular motion, they move equally because where there is more heaviness, there is equally more matter, or material heaviness]" (cit. from Caverni, *Storia del metodo sperimentale* cit., IV, p. 278).

[40] For a precise review of the main theories from antiquity up to Gassendi regarding the way of conceiving

this way, he could attempt to present acceleration as a natural phenomenon, equal **326**
for all bodies, without giving up on heaviness as the cause of motion. Furthermore, the *momento della gravità* could once again be proposed as a fundamental element for the science *de motu*: in short, in the light of the new principles, the long-pursued unification of statics and dynamics once again appeared not impossible. Various elements show that Galileo, at a certain point, returned to orient himself in this direction. First, the reflections of Torricelli, who was at Galileo's side in his last years and gathered his last recommendations[41]. Then, a fragment in Viviani's hand (Galileo was blind by this time)[42]. Lastly, some of the considerations on

---

the action of the driving force (of which the acceleration produced by the application of a constant force constitutes one case), cf., of Duhem, the still useful *De l'accélération produite par une force constante* cit. Duhem assigns the first formulation of this theory to the conception of "impeto" by Alessandro Piccolomini, whose footsteps Julius Caesar Scaliger and Giovanni Battista Benedetti would have followed (pp. 881-887). Conversely, Galileo would have remained traditional on this point: a constant force generates uniform motion (p. 898). For his part, Koyré emphasized Galileo's manifest and reiterated declarations of esteem and admiration for Gilbert and his magnetic theories. According to Koyré, Galileo, at a certain point, would have considered plausible a physical theory of gravity based on magnetic attraction. In essence, the Pisan scientist would have glimpsed in the *De magnete* the possibility of justifying the acceleration of natural motion: "il corpo che cade accelera il proprio movimento perché, in ciascun istante seguente, subisce la medesima azione istantanea - l'attrazione della Terra [the falling body accelerates its movement because, in each subsequent instant, it undergoes the same instantaneous action – the attraction of the Earth]" (*Studi galileiani*, trad. it. cit., p. 105). It could be added that, by conceiving the attraction as proportional to the heaviness of the object (i.e., to the resistance), Gilbert's conception could also have been useful to Galileo to explain the equality of acceleration from rest for all bodies. However, there is a lack of documents that grant a basis for arguing that Galileo perceives this possibility in Gilbert's text. In fact, he never used the magnetic theory to base his own conception of naturally accelerated motion (Koyré also recognizes this; cf. p. 105, note 47). And when, in a set of pages of the *Discorsi*, he applauds the bold and intelligent work of the English scientist, he will show that he appreciated above all the essentially Copernican spirit by which it was animated (*G.G.*, VII, pp. 431 ff.). Clavelin also states that Galileo reached, at a certain point, the idea that the action of a constant force generates a continuously increasing speed, emphasizing, however, that the Pisan scientist reached this result by distinguishing two functions in heaviness. In one, he would have identified the cause of motion (*fonction motrice*), in the other, the weight (*fonction gravifique*). These two functions would have corresponded, respectively, to the *momento della gravità* and the *momento della velocità* (*La philosophie naturelle de Galilée* cit., pp. 362-363).

[41] In addition to the discussion with Mersenne, it is necessary to take into account the fact that heaviness is presented as the cause of naturally accelerated motion in the *Lezioni Accademiche* on the force of percussion (cf. *infra*, pp. 404-405).

[42] "...Surto su l'ancore, si dessero le vele al vento e si sarpasse: dove la forza del vento non subitamente imprime la massima velocità alla nave, ma successivamente e con tempo, avvenga che nel principio la trovò immota, e poi di mano in mano opera sopra il mobile continuamente affetto di maggiore velocità. Né doviamo porre alcuna differenza tra gl'impulsi dati per intervalli e quello che vien conferito con forza continuata, perché, sì come tra gl'impulsi interrotti nessuna varietà si deve considerare se talvolta in 10 minuti di tempo si dieno 20 vogate o 30 o 100 o 1000, così né anche può cader alcuna alterazione tra quelli e l'impulso continuato, non sendo questo altro che una frequentissima moltitudine di spinte, cioè di infinite, dentro all'istesso tempo. Non basta, dunque, che il mobile, il mezzo e la facoltà sieno sempre l'istesse, a fare l'introduzione di una tanta celerità; ma vi vuole, partendosi il mobile dalla quiete, una successione di tempo. In simil guisa penso io che proceda il negozio ne i mobili naturali, partendosi dalla quiete, dove da qualche impedimento erano ritenuti; perché, il mezzo sia sempre l'istesso, l'istesso il mobile e l'istessa la gravità movente, tutta via essa gravità su 'l principio opera sopra un mobile non abituato di moto alcuno, ma poi successivamente va operando sopra mobile affetto di velocità; onde, operando essa virtù nel modo istesso, muove più, perché accresce moto sopra mobile che ella ritrova in

percussion[43]. Unfortunately, advanced age and blindness prevented Galileo from finding the strength to return to the *Discorsi* for a radical refoundation. It will fall upon Torricelli to rewrite the *De motu locali*, developing these latest directions of the Master and adding what his fertile mathematical genius had allowed him to refine. It is no coincidence that, in Torricelli's treatise, the mechanical derivation of the laws *de motu naturaliter accelerato*, carefully hidden by Galileo in the *Discorsi*, will be boldly reintroduced with crystal clear evidence[44].

327 Having reached this point, the question remains whether it is possible to identify a criterion for reading Galileo's different attitudes[45] towards acceleration that allows

---

moto [...Having lifted the anchor, the sails were set to the wind, and set out: where the force of the wind does not immediately impart maximum speed to the ship, but progressively and with time, it happens that in the beginning it finds it motionless, and then little by little it acts on the mobile, continuously affected with greater speed. Nor must we place any difference between the impulses given at intervals and that which is given with continuous force, because, just as no variety must be considered between the interrupted impulses, whether 20 strokes or 30 or 100 or 1000 are given in 10 minutes of time, so neither can any alteration occur between those and the continuous impulse, since this is nothing other than a very frequent multitude of impulses, that is, infinite ones, within the same time. It is not enough, therefore, that the mobile, the medium, and the faculty are always the same, to make the introduction of so great a swiftness; but, starting from rest, it requires a succession of time. In a similar way, I think that the matter proceeds with natural mobiles, starting from rest, where they were held back by some impediment; because the medium is always the same, the mobile is the same and the moving heaviness is the same, this same heaviness initially operates on a mobile unaccustomed to any motion, but then gradually continues acting upon a mobile affected by speed; therefore, by operating this virtue in the same way, it moves more, because it increases motion over the mobile that it finds in motion]" (*G.G.*, VIII, pp. 441-442; the note is handwritten by Viviani, who, at the bottom, noted: "di questo ho l'originale [of this, I have the original]").

43 Cf. *infra*, Ch. VII.

44 The *De motu* by Torricelli, divided into two books (I: *de motu gravium naturaliter accelerato*; II: *de motu proiectorum*) was published in the *Opera Geometrica*, Typis Amatoris Massae, Florentiae 1644. For the significance of the *De motu*, cf. my E. Torricelli. *Concezione della matematica* cit., pp. 73 ff.

45 It will be worth adding, for completeness, that, in the *Giornata Prima* [First Day] of the *Discorsi*, Galileo suggests, albeit in a very particular context, an interpretation of the acceleration of fall that is still different from the previous ones. Preparing to illustrate the famous "Platonic myth", Salviati proposed the way in which the Lincean Academician conceived of acceleration. In essence, this explanation suggested that the continuous increase in speed could be an effect of approaching the natural place. This was one of the traditional justifications, which should not be given excessive weight. In all likelihood, Galileo formulated it, because it served very well to support "the Platonic myth": "Ogni corpo costituito per qualsivoglia causa in istato di quiete, ma che per sua natura sia mobile, posto che sia in libertà, si moverà, tutta volta però ch'egli abbia da natura inclinazione a qualche luogo particolare; ché, quando e' fusse indifferente a tutti, resterebbe nella sua quiete.... Dall'aver questa inclinazione ne nasce necessariamente che egli, nel suo moto, si anderà continuamente accelerando; e cominciando con moto tardissimo, non acquisterà grado alcuno di velocità che prima e' non sia passato per tutti i gradi di velocità minori, o vogliamo dire di tardità maggiori.... Ora, questa accelerazion di moto non si farà se non quando il mobile nel muoversi acquista; né altro è l'acquisto suo, se non l'avvicinarsi al luogo desiderato, cioè dove l'inclinazion naturale lo tira... [Every body constituted in a state of rest for any reason, but is naturally mobile, provided that it is placed freely, will move, whenever it has a natural inclination to some particular place; because, if it were indifferent to all, it would remain at rest.... From having this inclination, it necessarily follows that, in its motion, it will continually accelerate; and starting with a very slow motion, it will not acquire any degree of speed that has not previously passed through all the degrees of lesser speeds, or we may say of greater slowness.... Now, this acceleration of motion will not occur unless the mobile acquires movement; and its acquisition is nothing other than approaching the desired place, that is, where its natural inclination draws it...]" (*G.G.*, VII, p. 44). Of course, this

us to understand them and refer each of them to the precise needs felt by the Pisan scientist at the same time or in different situations and periods.

We have seen, in essence, that Galileo proposed two distinct solutions (the third, "Torricellian," one will be formulated after the publication of the *Discorsi*, and we will evaluate it better later). The first made the acceleration of fall and its uniformity depend on a natural constitution (in its final analysis, a law of creation), removing any causal influence from heaviness. The second left a precise causal function to heaviness, but ended up relegating acceleration to an accidental, transitory phenomenon, produced by the "consumarsi [consumption]" of a "virtù impressa [impressed force]" of opposite sign. These two interpretations cannot be chronologically ordered, although the second certainly has much older origins. In fact, it dates back to the *De motu*. However, they were both represented in the *Discorsi*, albeit with the distinction observed above.

In the first interpretation, the effort to found a science of movement completely detached from mechanical principles, entrusting to a metaphysical principle the task of clearing away all contradictions, should be acknowledged. The *De motu locali* – as we will see better shortly – presupposes this type of foundation – declared several times – in which heaviness appears to be a negligible element, as it indeed occurs in the definitions and theorems *de motu naturaliter accelerato*. The *momenta velocitatis* of the *Discorsi* are, in fact, gained by the heavy bodies, by all the heavy **328**
bodies, by an inflexible order of nature, and have no connection either with heaviness or with its *momenti*.

But Galileo – despite the best intentions – cannot abandon, at least on the analogical level, the mechanical theorems and the *momento della gravità*. The second interpretation of acceleration and its uniformity thus allows us to bring mechanics and the science of movement back to the same principles: heaviness and its *momenti*.

The presence of both these interpretations in the *Discorsi* can certainly be traced back to similar conditioning. Although he had been convinced by now of the necessity of keeping mechanical proofs distinct from mathematical-geometrical ones in the analysis of natural motion, Galileo made a fundamental supposition for most of the theorems in the *De motu naturaliter accelerato*. The close analogy of the supposition with the principles employed in the mechanical inclined plane theorem was clear. It coincided, in fact, with the second of the foundations on whose legitimacy Galileo had invited, in 1609, Luca Valerio to express his opinion[46].

> Accipio gradus velocitatis, eiusdem mobilis super diversas planorum inclinationes acquisitos tunc aequales, cum eorumdem planorum elevationes aequales sint. [I take the degrees of speed, acquired by the same mobile on different inclinations of planes, to be then equal when the elevations of those same planes are equal.]

The supposition – as it is worth emphasizing – is proposed by Salviati, just two

---

interpretation did not account for why the acceleration from rest was the same for all heavy bodies.

[46] Cf. *supra*, pp. 306-307.

pages after the presentation of Sagredo's interpretation of acceleration[47]. This is likely not a coincidence. Preparing to propose a principle that he could only make plausible by emphasizing its analogy with mechanical principles and making use of heaviness and its *momenti*, our scientist felt the need to preface it with the revival of the old theory of the *De motu* with calculated ambiguity, in which heaviness had central importance. It cannot be ruled out that our scientist, very skilled in careful dosages, thought of having Sagredo propose the old interpretation of acceleration, without either rejecting it or explicitly accepting it, to somehow justify the use of a principle of evident mechanical derivation as a Postulate for the treatise *de motu naturaliter accelerato*. The other interpretation, the only one officially provided by Galileo in the *Discorsi*, through the official spokesperson, seemed, in fact, to rigorously exclude the use of the "mechanical" *momento* to demonstrate theorems *de motu naturaliter accelerato*.

Behind the complex operations to establish, first, the close connection between
329 the two *momenti*, then, to separate them clearly and, lastly, to make the combined use of the two approaches somehow plausible, there were serious problems of the general foundation of the science of movement, different ways of interpreting the nature of acceleration and its patterns, insurmountable difficulties in proposing a cause of natural motion capable of univocally accounting for all phenomena. Without full awareness of Galileo's puzzlement, uncertainties and shifts in the face of these general problems, it is not easy to follow and understand the complicated evolutions of *momento* from the *Mecaniche* to the *Discorsi* and beyond.

---

[47] *G.G.*, VIII, p. 205.

CHAPTER V

THE INFINITESIMAL FOUNDATION

After the forced abandonment of his interests in the *de motu* theorems following 331
his astronomical discoveries, the arguments that ensued, and his return to Florence, Galileo could not have resumed serious study of natural motion and its laws prior to 1611-1615. Then again, his prevailing interest in astronomical questions returned to distract him until the years of the *Dialogo*, in which he anticipated on several occasions some of the main conclusions of the science *de motu*. Finally, starting from 1634, until the publication of '38, the Pisan scientist dedicated himself to drafting the *Discorsi*, obviously concentrating his attention on the *De motu locali* theorems.

We have already discussed how the demonstration of a substantial core of theorems included by Galileo in the *Discorsi* can, in all probability, be assigned to the Paduan period. The further development of the old theorems, the irreversible decision to favor the correct acceleration-time supposition, as well as, probably, the definitive abandonment of the "mechanical" foundation (i.e., the clarification of the *momento della gravità–momento della velocità* relationship) are to be placed in the first Florentine phase of dynamic interests (around 1616-1618)[1]. The *Dialogo* indicates that, between 1629 and 1632, Galileo had made further progress, better specifying a project with now definitive characteristics[2]. In fact, quite a few documents suggest that, when Galileo finally dedicated himself to the drafting of the *Giornata Terza* and *Quarta* [Third and Fourth Days] of the *Discorsi*, he essentially limited himself to a less meticulous revision of the theorems, almost all of which had already been written previously[3].

[1] After his return to Florence, Galileo was engaged not only in astronomical observations and writings, but also – as we have seen – in the long and intense controversy on buoyancy.

[2] Many pages of the *Dialogo* are dedicated to the exposition of issues relating to movement, especially in the *Giornata Seconda* [Second Day] (cf., for example, *G.G.*, VII, pp. 253 ff.).

[3] Wisan has demonstrated that some of the published *Theoremata* present, even at the drafting level, the same form in which they had been conceived many years before. That, in finally deciding to publish the *De motu locali*, Galileo started from material mostly developed earlier, is demonstrated, among other things, by the letter that the Pisan scientist wrote to Diodati on 6 December 1636, admitting to his friend with much candor that he was unable to easily understand the demonstrations that he himself had obtained as a young man: "provo quanto la vecchiaia tolga di vivezza e di velocità agli spiriti, mentre

P. Galluzzi, *Momento: Galilean Studies*, Archimedes 78,
https://doi.org/10.1007/978-3-032-35049-7_9

332 Attempting to closely follow the evolution of Galileo's science *de motu* after returning to Florence is not an easy task. The dating of the *Frammenti* that preserve the first drafts of the theorems later included in the *Discorsi* becomes, in fact, anything but easy. The abandonment of the correct supposition no longer allows us to have a discriminating element for dating. Not even the watermarks bring any benefit. The handwriting itself tends to stabilize[4]. What remains are the indications suggested by the terminology used: an indication too vague to be relied upon to establish a chronological sequence. In essence, the only practicable path is to make an effort to trace a logical thread in the mass of fragments, which can lead to establishing a sequence, at least approximately.

This investigation, although interesting, remains extraneous to our work. In fact, doing so would contribute very little to clarifying the history of Galileo's *momento*. Instead, we will turn our attention to another problem, far more interesting from our point of view. Having resolved the question of the causal foundation of naturally accelerated motion, removing it from the effect of heaviness and appealing to a rigid law of nature, Galileo could believe that he had circumvented one of the most serious obstacles encountered up to that point. This entailed, of course, the abandonment of the long-pursued project of unitary treatment of mechanical and natural motions. Made independent of *momentum gravitatis*, *momentum velocitatis* would, from here on, follow its own path in full autonomy.

It is precisely at this point that new foundational problems must have become evident to Galileo. These concerned, fundamentally, the techniques of geometric representation used in the demonstrations and a certain type of infinitesimal analysis used to justify the laws of naturally accelerated motion. The foundation of *momentum velocitatis* therefore required new efforts. And it is precisely to these efforts – quite significant – that we will now turn our attention.

It is legitimate to think that, before returning to Florence, Galileo, along with the correct law of fall, had now specified the geometric "schema" capable of conveniently indicating, in naturally accelerated motion, the relationships between time, space and speed.

This schema must have been the same one that we will find used in the fundamental demonstrations of the *Discorsi* (the right-triangle). On the other hand,
333 we have already seen that a similar technique of representation had also been used by Galileo in the Paduan demonstrations based on the erroneous supposition of the acceleration-distance from rest proportionality[5]. After discovering that acceleration is proportional to time, the Pisan scientist had to limit himself to reforming this schema, that is, by making the perpendicular leg indicate time instead of space. The

duro fatica ad intendere non poche delle cose nell'età più fresca ritrovate e dimostrate da me [I'm proof how old age takes away liveliness and speed from the spirit, while I struggle to understand many of the things discovered and demonstrated by me in my younger, fresher age]" (*G.G.*, XVI, p. 524). For the composition of the *Discorsi*, please refer to the always very useful *Avvertimento* by Favaro (*G.G.*, VIII, pp. 11-38).

[4] Cf. W. Wisan, *The new science of motion* cit., pp. 229-230.

[5] Cf., for example, the Latin fragment published by Favaro, which we previously examined (*G.G.*, VIII, p. 383). Naturally, in that figure, the perpendicular leg indicates distance, not time.

parallels at the base could, however, continue to "depict" the "instantaneous speeds", which, more or less in the same period of years, Galileo had become accustomed to indicating with *gradus* or *momenta velocitatis*.

Now, even after having chosen to limit himself to a purely geometric-mathematical exposition of the naturally accelerated motion according to the proportion of time, Galileo intended to firmly maintain – and he reiterated this several times – that this motion corresponded adequately to that of real bodies. The laws that he deduced from the proposed principles were, therefore, to be considered true and proper laws of nature (absent, of course, any material impediments). For the Pisan scientist, this meant the need to make every possible effort to develop techniques for the geometric representation of falling motion and types of logical and conceptual analysis that could plausibly justify the reference to "nature" and the behavior of real bodies.

In this attempt, several difficulties were encountered. Galileo probably ended up realizing the impossibility of overcoming them all. He therefore preferred to bring to light the *De motu naturaliter accelerato* without explicitly presenting the multiple foundational problems that he had tried to resolve. He limited himself – as we will see – to "mending" the relationship of his own geometric-mathematical construction with the reality of natural motions in an extrinsic and ambiguous manner. He had, however, thought long and hard about these problems, and in his writings, abundant and significant traces of such speculations remain.

The definition and the first two theorems of the *De motu naturaliter accelerato* presuppose a whole series of undeclared principial assumptions, each of no small complexity. As we will see, the demonstrations are based on a very precise relationship between the adopted geometric schema and a particular (and far from convincing) conception of the composition of continuous magnitudes and the nature of the infinite. In some hints included in the definition and theorems, Galileo shows that he is aware of the non-negligible implications of the infinitesimal level. We know that he was aware of it from the beginning and that he tried for a long time to find a solution to the problem. He must not have been completely satisfied since he preferred to carefully separate the justifications elaborated for the *de motu* theorems, avoiding emphasizing that they served to ground the analysis conducted in the Latin **334**
treatise and to make its reference to reality plausible. There is, in fact, a very close relationship between the geometric-mathematical analysis of accelerated motion and the pages of the *Dialogo*, the *Postille* to Rocco and the *Discorsi* in which Galileo addresses the aporias of infinity and the composition of the continuum. It could be said, without fear of exaggeration, that Galileo laid them out intentionally, intending, namely, to offer some justification – if not a first foundation – to the techniques used in the interpretation of accelerated movement. To fully understand the story of *momentum velocitatis*, after the suppression of any reference to heaviness and its *momenti*, it is essential to reconstruct the general meaning of the new founding effort made by Galileo. For this reason, after having indicated, in a group of theorems of the *Giornata Terza* [Third Day] of the *Discorsi*, the operation of a series of relevant undeclared suppositions, we will try to show in what manner Galileo had previously

tried to support their plausibility.

Of the page that opens, in the *Discorsi*, the *De motu naturaliter accelerato* discussion and concludes with its definition, we preserve a first draft, written in Galileo's own handwriting. Antonio Favaro mistakenly believed he could date it 1604, due to the close similarity, in his view, that the draft bore to the letter to Sarpi[6]. In truth, Galileo used the correct acceleration-time supposition, which – as is known – he acquired only later. The important document cannot, therefore, be assigned to a date prior to the two-year period 1606-1608. Most likely, however, it was drafted later; perhaps, as has been argued, between 1629 and 1630[7].

We are faced with something much more complex than a simple definition. Before making it explicit, Galileo proposes, in fact, a whole series of methodological and general considerations – or better yet, principles. The motion that he intends to define – Galileo immediately declares this – is not an imaginary or hypothetical motion, but rather real and fully corresponding to actual natural movement[8]. The
335 diligent observation of nature and of the very simple and easy means that it ordinarily employs in the production of phenomena has led our scientist, almost by the hand, to the definition of accelerated motion, that is, a motion with which a moving body progressively acquires "nova velocitatis incrementa [new increments of speed]". On

---

[6] In fact, the Paduan scholar included it in vol. II of the Ed. Naz. (see the *Avvertimento*, in *G.G.*, II, pp. 259-260; the draft is published on pp. 261-266). It is worth emphasizing that the writing is preserved in Ms. Galil. 71 of the B.N.F., which also contains the *De motu.* According to Fredette, it was Galileo himself who placed the draft near the *De motu* (unpublished dissertation cit.). The definition of naturally accelerated motion with which the Latin draft ends is substantially the same as that included in the *Discorsi*, from which it differs, not too much, only in terms of the drafting: "Motum uniformiter, seu equabiliter, acceleratum dico illum, cuius momenta, seu gradus celeritatis a discessu ex quiete augentur iuxta ipsiusmet temporis incrementum a primo instanti lationis [I say that the uniformly, or equally, accelerated motion is that whose momenta, or degrees of swiftness, increase from rest according to the very increment of time from the first instant of motion]" (*G.G.*, II, p. 266). For the definition in the *Discorsi*, see *G.G.*, VIII, p. 205.

[7] The proposal to bring the date of composition towards the preparatory years of the *Dialogo* was supported primarily by Wisan (*The new science of motion* cit., pp. 276-278, especially note 4).

[8] "Quamvis enim aliquam lationis speciem ex arbitrio confingere, et consequentes eius passiones contemplari, non sit inconveniens (ita, enim, qui helicas aut conchoides lineas ex motibus quibusdam exortas, licet talibus non utatur natura, sibi finxerunt, earum symptomata ex suppositione demonstrarunt cum laude), tamen quandoquidem quadam accelerationis specie in suis quibusdam motibus, gravium scilicet descendentium, utitur natura. Eorundem speculari passiones decrevimus, si eam, quam allaturi sumus de nostro motu accelerato definitionem, cum motus naturaliter accelerati essentia congruere contigerit. Quod tandem, post diuturnas mentis agitationes, reperiisse confidimus; ea potissimum ducti ratione, quia symptomatis, deinceps a nobis demonstratis, apprime respondere atque congruere videntur ea, quae naturalia experimenta sensui repraesentant [Although it is not inappropriate to invent some form of motion arbitrarily and contemplate its consequent passions (for indeed, those who have invented helices or conchoids as lines arising from certain motions, although such motions are not used by nature, have demonstrated their symptoms from assumption with praise), nevertheless, since nature uses a certain species of acceleration in some of its motions, namely in the descending motions of heavy bodies, we have decided to observe the passions of the same, if the definition we are about to present of our accelerated motion happens to align with the essence of naturally accelerated motion. Which, finally, after long deliberations, we confidently believe we have discovered; chiefly guided by this reasoning, because the symptoms, subsequently demonstrated by us, seem to correspond closely and align with those which natural experiments represent to the senses]" (*G.G.*, II, p. 261). In the reference to the "fictions" of spirals and conchoids, one must obviously recognize an explicit reference to Archimedes.

closer inspection, in fact, "nullum additamentum, nullum incrementum simplicius inveniemus quam illud quod semper eodem modo superaddit [we will find no addition, no increment simpler than that which always adds in the same manner]"[9]. This unrivaled simplicity is clearly evident when one reflects on the "maximam temporis atque motus affinitatem [greatest affinity between time and motion]". If, in fact, it is correct to define uniform motion as the motion with which the same distances are covered in equal times, we will just as easily admit that in accelerated motion, in equal times[10], there is the addition of equal "celeritatis incrementa [increments of swiftness]". It will, therefore, be permissible to conceive "motum illum uniformiter atque eodem modo continuo acceleratum esse, dum temporibus quibuscunque aequalibus aequalia ei superaddantur celeritatis additamenta [that motion is uniformly and in the same manner continuously accelerated, as equal additions of speed are added to it in any equal times]".

In practice, taking as many equal small quantities of time as one wishes

> a primo instanti in quo mobile recedit a quiete et descensum aggreditur, celeritatis gradus in prima cum secunda temporis particula acquisitus, duplus sit gradus quem acquisivit mobile in prima particula; gradus itidem quem obtinet in tribus temporibus, triplus; quem in 4, quadruplus eiusdem gradus primi temporis; adeo ut, si mobile lationem suam continuaret iuxta velocitatis gradum seu momentum in prima temporis particula acquisitum, motumque suum deinceps aequabiliter cum tali velocitate extenderet, talis latio duplo esset tardior ea, quam iuxta gradum velocitatis in secundo tempore acquisitae obtineret.
>
> Apparet, proinde, a recta ratione absonum nequaquam esse, si accipiamus intensionem velocitatis fieri iuxta temporis extensionem[11].
>
> [from the first instant in which the mobile departs from rest and begins its descent, the degree of swiftness acquired in the first time interval together with the second, is double the degree acquired by the mobile in the first interval; the degree likewise obtained in three time intervals is triple; that obtained in four, quadruple of the same degree of the first time interval; so that, if the mobile were to continue its motion according to the degree of speed or momentum acquired in the first time interval, and extend its motion thereafter uniformly with such speed, such a motion would be twice as slow as the one it would obtain according to the speed degree acquired in the second time interval.
>
> It appears, therefore, that it is by no means absurd according to sound reasoning if we accept that the intensity of speed increases according to the extension of time.]

Here ends the section that will be published later, with very few corrections, none of them significant, in the *Discorsi*[12]. This is an extremely interesting passage, **336**
as well as for the rich terminology that it presents in just a few lines. The concept of the increase in speed according to the passage of time is indicated with the following expressions: *velocitatis incrementa*, *velocitatis additamenta*, *celeritatis incrementa*, *celeritatis gradus*, *velocitatis gradus seu momentum*. This fact should not be surprising. The text that Galileo was writing was the first, in the *Discorsi*, in which

---

[9] *Ibid*., pp. 261-262.

[10] The draft is preceded by the title "Liber Secundus, in quo agitur de motu accelerato [Book Two, in which the subject of accelerated motion is discussed]". Evidently the *Liber Primus* was intended to be dedicated (as it will be in the *Discorsi*) to the treatment of uniform motion, the definition of which is presupposed here. For the definition of the uniform motion of the *Discorsi*, see *G.G.*, VIII, p. 191.

[11] *G.G.*, II, pp. 262-263.

[12] *G.G.*, VIII, pp. 197-198.

velocitatis momentum would be used, emphasizing its analogy with *velocitatis gradus*. The attention paid to preparing the reader to understand the technical term should not, therefore, be considered useless or excessive, as the concept is introduced with easily understandable formulas. The lexical progression, in fact, is clearly directed from the "common" to the "technical": *incrementum, additamentum, gradus, gradus seu momentum*. This type of reading is confirmed by the fact that, *gradus* and *momentum velocitatis*, once introduced, are the only terms that recur in the remainder of the draft[13].

It is worth emphasizing that the cited passage is also of terminological interest from another point of view, specifically in the last portion, in which Galileo uses language that reveals, all too clearly, scholastic influences. In addition to *gradus*, the Pisan scientist introduces *intensio* to suggest the continuous growth of speed, while *extensio* is assigned the task of expressing the proportional passage of time[14].

As for the content, Galileo alludes to the substantial difficulties he encountered prior to reaching the valid solution of the acceleration-time proportionality ("post diuturnas mentis agitationes [after prolonged mental deliberations]") and strongly underscores the analogies between movement and time.

The Latin draft continues for several more pages. Galileo introduces a series of considerations to defend himself from what he admits is a natural objection. In the *Discorsi*, this section is translated into the vernacular for the dialogue in a substantially faithful manner, and it is up to Sagredo to raise the objection. The problem lies in the fact that, from the close analogy between motion and time suggested by Galileo, a paradoxical consequence seems to arise. In fact,

> si a primo termino lationis ex quiete fit deinceps perpetua novae celeritatis additio iuxta rationem legemque eandem, secundum quam temporis discursus a primo instanti nova perpetuo suscipit additamenta, considerandum occurrit quod, sicuti post primum instans non est assignare tempus aliquod tam breve, quin aliud brevius atque brevius inter hoc et primum instans non mediet, ita post relictam quietem in latione non poterit assignari gradus adeo exiguus velocitatis, seu tam magnus tarditatis, quin in altero adhuc illo tardiori mobile descendens constitutum antea non fuerit; cumque tarditas
> **337** in infinitum augeri, aut velocitas imminui, possit, fateri oportebit, mobile aliquando tam immensum obtinuisse tarditatis momentum ut, cum eo latum, vel integri anni curriculo ne spatium transversi digiti pertransisset[15]. [if from the first instant of motion from rest, a continuous addition of new speed occurs according to the same ratio and law by which the passage of time from the first instant perpetually receives new additions, it must be considered that, just as after the first instant, no time can be assigned so brief that another, shorter time does not intervene between it and the first instant, so, after leaving rest in motion, no degree of speed, however small, or of slowness, however great, can be assigned without the mobile previously being in an even slower state; and since slowness can increase infinitely, or speed decrease, it must be admitted that the mobile at some point acquired such an immense momentum of slowness that, with it, in the course of an entire year, it would not have traversed the space of a transverse finger.]

---

[13] To tell the truth, there is still an example of both *incrementum* and *additamentum*, but both terms refer to time, not speed.

[14] We have previously examined the problem of the influence of the logic of the *calculationes* on Galileo; cf. *supra*, Part Two, Ch. III, p. 273 and note 36.

[15] *G.G.*, II, p. 263; cf. also VIII, pp. 198-199.

The meaning of the objection is clear: if the *momenta velocitatis* grow according to the passage of time, we must conceive that to each instant of time there corresponds a degree of speed. But the instants of time are infinite, so the *momenta velocitatis* will also be infinite. In other words, if between even the smallest degree of speed and rest we must admit infinitely many smaller degrees, we get the impression of reaching an infinite *tarditatis momentum* [momentum of slowness] and the movement appears to vanish. In essence, it was – it is unnecessary to emphasize this, as it is so evident – an ancient argument to deny the reality of the movement. And Galileo could draw from a vast amount of literature the reasons to counter it.

The response is instead entrusted to an observation that is, among other things, hardly convincing[16]. Galileo, in fact, invites us to consider that a heavy object exerts a minimal effect on another when it is simply placed on top of it. The effect increases when the simple placement is replaced by an impact at a certain speed (greater speed = greater impact). The speed is determined by the distance from which the weight is dropped, which can be infinitesimally small: for example, equal to the thickness of a sheet of paper. The placement and the impact produced by falling from a height corresponding to the thickness of a sheet of paper, or from greater distances, effectively demonstrate, according to Galileo, with their different effects, that a mobile[17]

ex quiete naturaliter descendens, per omnes tarditatis gradus facere transitum, in nullo tamen commorari, sed, ad instantium temporis successionem, novas maioresque semper acquirere celeritatis gradus[18]. [naturally descending from rest, passes through all degrees of slowness, yet does not remain in any, but, with the successive instants of time, continually acquires new and greater degrees of swiftness.]

It is peculiar that he proposed this proof. Moreover, the acquisition of *momenta velocitatis* appears to be proportional to the distance from rest, not to *time*. Galileo must have later realized the weakness of the argument. In fact, he proposes it again
through Salviati in the *Discorsi*, but he added another consideration, explicitly **338**
reducing the importance of the mechanical proof[19].

In the Latin draft, Galileo also addressed the problem of the increase in speed along planes of different inclination, attempting to show, with an appropriate geometric demonstration, that there is no plane of such minimal inclination that,

---

[16] “Quod profecto mirum, seu potius absurdum videtur: verumtamen, licet primo intuitu mirum, falsum tamen non esse neque absurdum, experientia, qualibet demonstratione haud infirmior, quemlibet admonere potest [Which indeed seems strange, or rather absurd: nevertheless, although it seems strange at first glance, it is neither false nor absurd, as experience, which is no less reliable than any demonstration, can instruct anyone]” (*G.G.*, II, p. 263).

[17] *Ibid.*, pp. 263-264.

[18] *Ibid.*, p. 264. Galileo added that he could confirm this principle with numerous other experiments, “quas in meis Mechanicis Quaestionibus, tamquam loco convenientiori repono [which in my Questions on Mechanics, I place in a more suitable place]”.

[19] *G.G.*, VIII, pp. 199-201. Let us see how the Pisan scientist, after having reproposed the experiment, advanced the new consideration, based on pure reasoning: “Ma senza ridursi a tale esperienza (che senza dubbio è concludentissima), mi pare che non sia difficile co ‘l semplice discorso penetrare una tal verità [But without being limited to this experiment (which is undoubtedly very conclusive), it seems to me that it is not difficult to penetrate such a truth with simple discourse]” (*ibid.*, p. 200).

descending along it, a heavy body does not reach, even after a very long time, a *grado* or *momento di velocità* equal to that which, in a shorter time, it gains by descending perpendicularly.

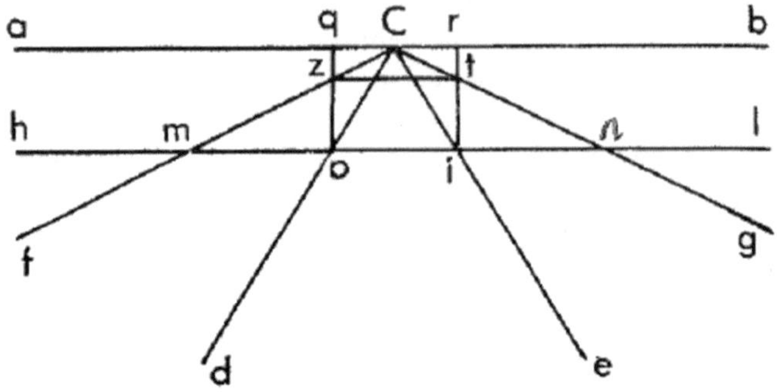

Figure 33

The geometric schema used to illustrate this phenomenon is extremely interesting. From point C (see fig. 33) on the parallel to the horizontal *ab*, pairs of rays *cf*, *cg* and *cd*, *ce* emanate and form, at C, two angles that are respectively obtuse and acute. These angles represent the inclinations of planes to the horizontal along which the motion occurs: the largest angle corresponds to the most inclined plane. The line *hl*, a moving parallel to *ab*, through the different distances from *ab* that it reaches by moving away with uniform motion, represents the flow of time (*temporis effluxum*). The degrees of speed reached at a given instant during the descents from C along the two inclinations correspond to the segments of the parallel *hl* included between the pairs of rays forming, at C, the two angles *fcg* and *dce*. It is evident that, by construction, at equal distances of *hl* from *ab*, that is, after equal times, the degree of speed reached along the more inclined plane above the horizontal is always greater than that achieved along the less inclined plane. However, if we take a degree of speed as large as we want along the plane of greater inclination, we will *always* have, in the plane of lesser inclination, an equal (or lesser) degree of speed, even though it will take a longer time to reach it (that is, *hl* will have to be further away from *ab*). The same can be repeated for any small degree of speed taken on the less
339 inclined plane: an equal (or smaller) one will *always* be found on the other plane, and therefore achieved in less time (that is, with *hl* being less distant from *ab*)[20].

In essence – as this is what Galileo intended to point out – the infinite *gradi* or *momenti di velocità* achieved in the motion along the plane of less inclination (that is, the set of infinite parallels to the base of the triangle *oci*) have all been reached, one by one, during the descent along the more inclined plane, even if in a shorter time.

To better "visualize" the meaning of Galileo's statement, it will suffice to drop perpendiculars from *o* and from *i* onto *ab*. They will intersect *fc* and *cg* at *z* and

[20] *G.G.*, II, pp. 264-265.

t, respectively. This serves to show that the *momento di velocità oi*, reached at a certain temporal "distance" from the beginning of the motion along the less inclined plane, was also achieved previously on the plane of greater inclination: *zt* is in fact equal to *oi*. The same construction can be repeated for each of the parallels to the base of the triangle *oic*. Galileo can therefore conclude:

> Hinc lineas omnes quotquot in triangulo *oci*, ipsi *oi* parallelas, intelligere quis potest (quae infinitae sunt), fuisse quoque comprehensas a triangulo *mcn*, in descensu lineae *hl*, sed tempore breviori[21]. [Hence, one can understand that all the lines, as many as there are in the triangle *oci*, which are parallel to *oi* (and these are infinite), were also encompassed by the triangle *mcn*, in the descent of the line *hl*, but in a shorter time.]

The presumed demonstration serves Galileo to argue that the law of natural acceleration is strict and without exceptions: before reaching any *momentum velocitatis* starting from rest, one must pass through all the infinite lesser *momenti di velocità* (or greater slowness). There is no possibility of skipping even one, as demonstrated by the comparison between motions that reach the same *momenti di velocità* in significantly different times (in the figure, it will be sufficient to bring one of the two angles closer to 180° and make the other tend towards 0°)[22]. According to Galileo, only the almost instantaneous motion of light perhaps eludes this criterion[23].

The analysis technique and the conclusions reached are very important. It is, in fact, evident that these pages were responding to a clear need for foundation. The legitimacy of the procedures employed by the Pisan scientist in the demonstration rests on a series of suppositions not explicitly declared, relating to the composition of continuous magnitudes and to the comparison between infinite series. A particularly problematic statement by Galileo appeared: the infinite parallels (*momenta velocitatis*) included in the triangle *oci* correspond one by one to the infinite parallels at the base of the triangle *zct*. This was a conclusion that was not **340**
easy to digest. The triangle zct is, in fact, smaller than *oic* (it has the same base but a smaller height); how could it, then, contain the infinite parallels to the base of the triangle *oic*?

To confirm the difficulty, it must be said that Galileo dropped the alleged demonstration, of which no trace remains in the *Discorsi*[24]. It does, however,

---

[21] *Ibid.*, p. 265.

[22] *Ibid.*, pp. 265-266.

[23] "Numquid eiusmodi sit luminis expansio, quod certe loco uno ac terminato generatur, et tempore eodem locis omnibus late circumquaque patentibus reperitur? [Is such the expansion of light, which certainly is generated in one place and at a defined point, and is found at the same time in all places widely spread around?]" (*ibid.*, p. 266).

[24] It is, however, worth noting that Galileo proposed the demonstration to Pierre Carcavi, who had informed him of Fermat's reluctance to admit that, in natural motion from rest, before reaching a certain degree of speed, one must pass through all the infinite smaller degrees of speed. In fact, for Fermat, it seemed that in this manner the movement would end up disappearing. The demonstration communicated to Carcavi is, however, modified compared to the version in the Latin draft. Galileo has, in fact, suppressed the comparison between the infinite parallels of the two unequal triangles. In fact, by limiting himself to a single triangle, he shows that a certain degree is preceded by all the smaller

indicate an effort, even though unsuccessful, to arrive at a foundation of the infinitesimal techniques indispensable for the geometric analysis of naturally accelerated motion according to the proportion of the passage of time.

The demonstration in the Latin fragment also shows several limitations in the representation technique. The figure, in fact, does not record the spaces of fall. In short, the limitation that we observed in the geometric schematization at the time of the erroneous supposition appears reversed. Before, the figure indicated space and degrees of speed, but not time[25]. Now, the introduction of time has come at the cost of excluding space.

We mentioned that the demonstration of the Latin fragment was later dropped. This type of comparison between infinite series appeared – as we will see shortly – untenable at the time when Galileo returned to the draft for the final version. Moreover, by that time, Galileo had already developed a different technique of geometric representation, with precise and proportionate indices for time, space and speed.

Let us now move on to the *Giornata Terza* [Third Day] of the *Discorsi*. Having defined naturally accelerated motion and advanced the postulate of the equality of degrees of speed at equal vertical distances from rest[26], Galileo introduces the first two theorems. Both present very important problems, both from the perspective of infinitesimal analysis and that of the technique of representation.

Let's start with the statement of *Theorema I*:

> Tempus in quo aliquod spatium a mobili conficitur latione ex quiete uniformiter accelerata, est
> **341** aequale tempori in quo idem spatium conficeretur ab eodem mobili motu aequabili delato, cuius velocitatis gradus subduplus sit ad summum et ultimum gradum velocitatis prioris motus uniformiter accelerati[27]. [The time in which a certain space is covered by a mobile from rest with uniform acceleration is equal to the time in which the same space would be covered by the same mobile in uniform motion, the speed of which is halved compared to the maximum and final speed of the previous uniformly accelerated motion.]

The demonstration allows us to attribute the case of a motion uniformly accelerated according to the proportion of time to that of a uniform motion with a speed corresponding to the "mean" *grado* or *momento* of the acceleration. We will not dwell on the meaning of the *Theorema*[28], but will instead focus our attention on the proof techniques employed by Galileo.

> Repraesentetur per extensionem AB tempus in quo a mobili latione uniformiter accelerata ex quiete in C conficiatur spatium CD; graduum autem velocitatis adauctae in instantibus temporis AB maximus et ultimus repraesentetur per EB, utcunque super AB constitutam; iunctaque AE, lineae

---

infinities, represented by all the parallels included between the degree and the vertex of the triangle. The modification introduced in the letter to Carcavi of 5 June 1637 proves, therefore, unequivocally that Galileo was aware of the unsustainability of the comparison between the two triangles (*G.G.*, XVII, pp. 92-93).

[25] Cf. *supra*, p. 282.

[26] *G.G.*, VIII, p. 205.

[27] *Ibid.*, p. 208.

[28] Cf. *infra*, pp. 363 ff.

omnes ex singulis punctis lineae AB ipsi BE aequidistanter actae, crescentes velocitatis gradus post instans A repraesentabunt. Divisa deinde BE bifariam in F, ductisque parallelis FG, AG ipsis BA, BF, parallelogrammum AGFB erit constitutum, triangulo AEB aequale, dividens suo latere GF bifariam AE in I: quodsi parallelae trianguli AEB usque ad IG extendantur, habebimus aggregatum parallelarum omnium in quadrilatero contentarum aequalem aggregatui comprehensarum in triangulo AEB; quae enim sunt in triangulo IEF, pares sunt cum contentis in triangulo GIA; eae vero quae habentur in trapezio AIFB, communes sunt. Cumque singulis et omnibus instantibus temporis AB respondeant singula et omnia puncta lineae AB, ex quibus actae parallelae in triangulo AEB comprehensae crescentes gradus velocitatis adauctae repraesentant, parallelae vero intra parallelogrammum contentae totidem gradus velocitatis non adauctae, sed aequalibus, itidem repraesentent; apparet, totidem velocitatis momenta absumpta esse in motu accelerato iuxta crescentes parallelas parallelogrammi GB: quod enim momentorum deficit in prima motus accelerati medietate (deficiunt enim momenta per parallelas trianguli AGI repraesentata), reficitur a momentis per parallelas trianguli IEF repraesentatis. Patet, igitur, aequalia futura esse spatia tempore eodem a duobus mobilibus peracta, quorum unum motu ex quiete uniformiter accelerato moveatur, alterum vero motu aequabili iuxta momentum subduplum momenti maximi velocitatis accelerati motus; quod erat intentum[29]. [Let the time in which the space CD is completed by a mobile with uniformly accelerated motion from rest be represented by the extension AB (see fig. 34); the greatest and final degrees of increased speed at the time instants in AB shall be represented by EB, whatever it may be set over AB; and the line AE, with all lines drawn from each point of the line AB parallel to BE, will represent the increasing degrees of speed after the instant A. Then divide BE into two parts at F, and by drawing parallels FG, AG to BA, BF, a parallelogram AGFB will be formed, equal to triangle AEB, dividing its side GF into two equal parts at I: and if the parallels of triangle AEB are extended up to IG, we will have the sum of all the parallels contained in the quadrilateral equal to the sum of those contained in triangle AEB; for those in triangle IEF are equal to those in triangle GIA; and those found in the trapezoid AIFB are common. And since each and every instant of time AB corresponds to each and every point of line AB, from which the parallel lines drawn in triangle AEB represent the increasing degrees of accelerated speed, while the parallels contained within the parallelogram represent an equal number of speed degrees, but not increasing ones, it is evident that the same number of momenta of speed are consumed in the accelerated motion along the increasing parallels of parallelogram GB: for what is lacking in the first half of the accelerated motion (for the momenta represented by the parallels of triangle AGI are deficient), is replenished by the momenta represented by the parallels of triangle IEF. It is evident, therefore, that the spaces traveled in the same time by two mobiles will be equal, one moving with a motion uniformly accelerated from rest, the other moving uniformly according to a momentum half the maximum momentum of speed of the accelerated motion; which was the intended result.]

Let us first focus our attention on the figure proposed by Galileo (see fig. 34). It will be immediately observed that, this time, all three quantities to be related are explicitly indicated: the space is, in fact, represented laterally by the straight **342**
line CD. The analogies between the demonstrative techniques used in the Latin draft and in *Theorema I* appear considerable. The comparison is indeed established, even in this case, at the level of infinite sets of lines parallel to the base. We do not find any explicit suppositions, but it is evident that Galileo is fully aware of the complexity of the operation he is carrying out. The aggregate of parallels contained in the parallelogram GABF is compared with the aggregate of parallels contained in the triangle EAB. Furthermore, it is stated that the two infinite aggregates stand in relation to each other as the area of the parallelogram to that of the triangle. Moreover, since AB represents the flow of time (from A to B), Galileo adds that each point of the infinite constituents of AB corresponds to *one and only one* of the

---

[29] *G.G.*, VIII, pp. 208-209.

infinite perpendiculars to AB, which are precisely the *gradi* or *momenti di velocità*. Thus, each of the infinite instants of time corresponds to a specific degree of speed in the naturally accelerated motion.

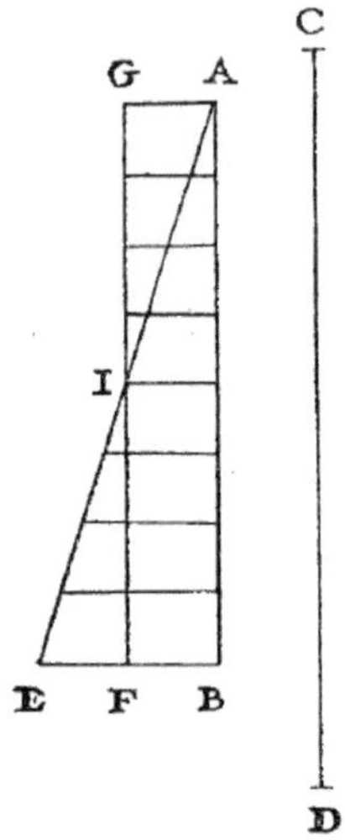

Figure 34

Theorema I demonstrates that Galileo did not give up using methods of infinitesimal analysis to establish the proportionality between the constant and uniform growth of speed in natural motion and the passage of time. It is worth noting, however, that his audacity has, so to speak, been tempered compared to the demonstration attempted in the Latin fragment. The comparison between infinite aggregates of parallels seems to be made more plausible here by the equality of the areas that contain such aggregates. Similarly, the correspondence between instants of time and *momenta* or *gradus velocitatis* appears convincingly suggested by the "representation" of the former with the infinite points of AB and of the latter with the infinite perpendiculars to each of such points.

Even the relations between time and speed with the spaces traversed, which the figure fails to present explicitly (the space is indicated by CD[†] but this representation makes it seem extraneous), are guaranteed by a reasoning that can be reduced to the comparison between infinite series. Each *momentum velocitatis* of uniform motion, in a given time, will make the mobile travel an infinitesimal amount of space that is always the same. The sum of the infinitesimal spaces traversed with infinite equal *momenti di velocità*, in infinite equal instants of time, in uniform motion, must correspond to the sum of the infinitesimal spaces (each different from the other and always greater) covered with the infinite "increasing" *momenti di velocità*, in naturally accelerated motion, in the same infinite instants of time.

[†] The text originally stated "AD" when CD was likely intended.

Let us set aside – as we will examine it later[30] – the meaning that *momenta velocitatis* take on in this context, as well as the problems that arise from the ineffective geometric representation of space. For now, we are only interested in emphasizing Galileo's use of infinitesimal analysis techniques in some of the fundamental propositions of the *De motu naturaliter accelerato*: the very close **343**
relationship between speed (and its increasing *momenti*) and the passage of time is, in fact, emphasized in these demonstrations with the use of a very precise way of conceiving the infinite divisibility of geometric quantities and the composition of the continuum.

The use of this method of analysis appears even more evident in the *Scholium* to *Problema IX*:[31]

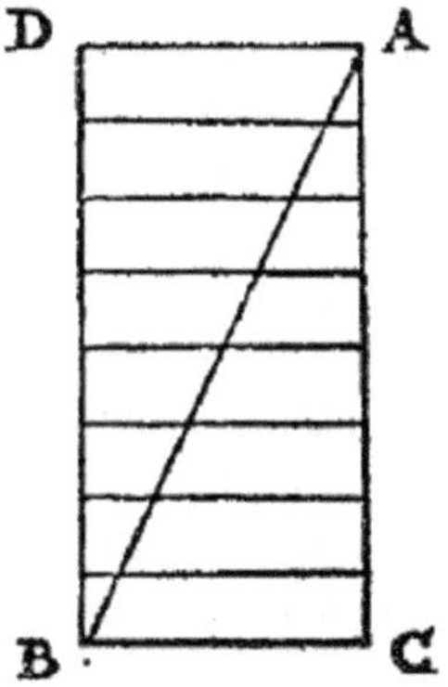

Figure 35

Resumentes enim triangulum ABC, nobis repraesentans in suis parallelis basi BC velocitatis gradus continue adauctos iuxta temporis incrementa, ex quibus, cum infinitae sint, veluti infinita sunt puncta in linea AC et instantia in quovis tempore, *exsurget* superficies ipsa trianguli; si intelligamus, motum per alterum tantum temporis continuari, sed non amplius motu accelerato, verum aequabili, iuxta maximum gradum velocitatis acquisitae, qui gradus repraesentatur per lineam BC; ex talibus gradibus *conflabitur* aggregatum consimile parallelogrammo ADBC, quod duplum est trianguli ABC: quare spatium quod cum gradibus consimilibus tempore eodem conficietur, duplum erit spatii peracti cum gradibus velocitatis a triangulis ABC repraesentatis[32]. [Resuming, then, the triangle ABC (see fig. 35), which represents

[30] Cf., *infra*, pp. 364 ff.
[31] "Dato spatio quovis tempore peracto ex quiete in perpendiculo, ex termino imo huius spatii planum inflectere, super quo post casum in perpendiculo tempore eodem conficiatur spatium cuilibet spatio dato aequale, quod tamen maius sit quam duplum, minus vero quam triplum spatii peracti in perpendiculo [Given a space completed at any time from rest along the perpendicular, position a plane at the lower end of this space, upon which, after falling along the perpendicular, in the same time, a space equal to any given space is completed, which, however, is greater than double but less than triple the space completed in the perpendicular]" (*G.G.*, VIII, pp. 240-241).
[32] *Ibid.*, pp. 242-243; the italics are ours. The same demonstration had already been proposed on the *Giornata Seconda* [Second Day] of the *Dialogo*. Regarding the relationship between the aggregates of parallels and the areas of the triangle and the rectangle, Galileo had, however, expressed himself with greater caution: "la qual infinità di linee, ci rappresenta in ultimo la superficie del triangolo AHD [of which the infinite lines ultimately represent to us the area of the triangle AHD]" (*G.G.*, VII, p. 255);

for us in its parallels to the base BC the continuously increasing degrees of speed according to the increments of time whereby, since they are infinite, just as there are infinite points on the line AC and infinite instants in any given time, the very area of the triangle *emerges*; if we consider the motion to continue for another equal duration, but no longer with accelerated motion, rather with uniform motion, according to the maximum degree of speed attained, which degree is represented by the line BC; an aggregate similar to the parallelogram ADBC will be *formed* from such degrees, which is double the triangle ABC: therefore, the space that will be traversed with uniform degrees of speed in the same time will be double the space traversed with the degrees of speed represented by the triangle ABC.]

Galileo has become more explicit. Until now, he had limited himself to saying that the comparison between two infinite parallel series could be considered fully equivalent to that between the areas in which these parallels are contained. Now, he declares with great ease that areas "arise" from the aggregation of infinite parallels. As Bonaventura Cavalieri will point out to him, to suppose that the comparison between aggregates of infinite parallels can be considered equivalent to that between the areas that contain them is very different from asserting that areas "consist" of the infinite parallels to their bases[33]. In doing so, one ends up admitting, in fact, that a line with length is produced by an infinite set of points that lack length. Or, if one
344 prefers, that the area (having length and width) is the result of infinite straight lines without width. Even in this case, we limit ourselves to emphasizing the problems of infinitesimal foundation, avoiding focus on the meaning of the *Scholium* for the definition of the functions of *velocitatis momentum*[34].

A further valuable indication of the use of infinitesimal techniques is provided by *Theorema III*, in which Galileo demonstrates that, in accelerated motion starting from rest, the times along the incline and along the perpendicular of equal elevation are in ratio to each other as the incline is to the perpendicular[35]. The demonstration of the theorem presupposes the postulate advanced at the beginning of the *De motu naturaliter accelerato*: the degrees of speed reached at equal vertical distances from rest, at any inclination, are equal[36]. Therefore

gradus in punctis I, E iidem erunt, nec non gradus in L et F. Quod si non hae tantum parallelae, sed ex punctis omnibus lineae AB usque ad lineam AC protractae intelligantur, momenta seu gradus velocitatum in terminis singularum parallelarum semper erunt inter se paria. Conficiuntur itaque spatia duo AC, AB

---

and, further on: "così il parallelogramma viene ad esser una massa ed aggregato di altrettanti gradi di velocità, ma ciascheduno eguale al massimo BC, la qual massa di velocità viene a esser doppia della massa delle velocità crescenti del triangolo, si come esso parallelogrammo è doppio del triangolo [thus the parallelogram comes to be a collection and aggregate of just as many degrees of speed, but each equal to the maximum BC, of which the collection of speeds comes to be double the collection of the increasing speeds of the triangle, just as this parallelogram is double the triangle]" (*ibid.*, p. 256).

33 Cf., *infra*, pp. 354 ff.

34 Cf. the next chapter, *infra*, pp. 370-371.

35 "Si super plano inclinato atque in perpendiculo quorum eadem sit altitudo, feratur ex quiete idem mobile, tempora lationum erunt inter se ut plani ipsius et perpendiculi longitudines [If a mobile is carried from rest along the inclined plane and the perpendicular, both having the same height, the times of travel will be to each other as the lengths of the plane itself and the perpendicular]" (*G.G.*, VIII, p. 215). Note that the strict dependence on the Postulate forces Galileo to limit the validity of the *Scholium* to the case of the descents of the same mobile.

36 *Ibid.*, p. 205.

iisdem gradibus velocitatis[37]. [the degrees at points I and E (see fig. 36) will be the same, as well as the degrees at L and F. And if not only these parallels, but also the lines drawn from all points of line AB up to line AC are considered, the momenta or degrees of speed at the ends of each parallel will always be equal to one another. Thus, the two spaces AC and AB are traversed with the same degrees of speed.]

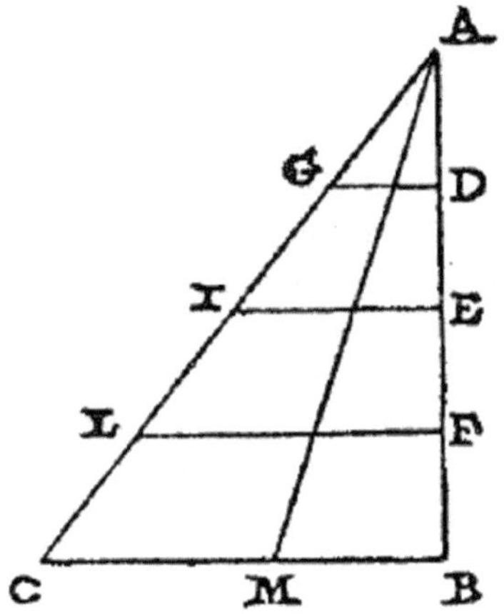

Figure 36

First of all, it will be convenient to note that the figure is only seemingly the same as the previous ones. The perpendicular and the inclined indicate, in fact, the spaces and, ultimately, also the times. Furthermore, it is obvious that the *momenti di velocità* depicted by the parallels to the base are acquired along AB and along AC at different times. It seems that Galileo – and this is quite interesting – clearly **345**
emphasizes the relationship between each point of AB and each point of the greater AC. In fact, the degrees of speed gained along AB are as many as along AC. On the other hand, the "sum" or "aggregate" of these two infinite series of points produces two different lengths, which correspond to the two planes and, therefore, to the times of fall[38]. The points marked on AC by the parallels LF, IE, GB, etc. (the *momenta velocitatis*), if not fewer in number than those on AB, should, therefore, each be proportionally greater than their corresponding ones. Otherwise, one would have to conclude that, with the same degrees of speed, one can cross two different spaces (two infinite sets of equal points) in the same time — which is against all logic, as well as against the very construction itself.

There is a close relationship between this theorem and the demonstration attempted in the Latin draft that was later dropped. It is worth noting that Galileo

---

[37] *Ibid.*, p. 216.

[38] "Si duo spatia conficiantur a mobili quod iisdem velocitatis gradibus feratur, quam rationem habent ipsa spatia, eamdem habent tempora lationum: ergo tempus lationis per AC ad tempus per AB est ut longitudo plani AC ad longitudinem perpendiculi AB: quod erat demonstrandum [If two spaces are traversed by a mobile with the same degrees of speed, the ratio of the spaces is the same as the ratio of the times of travel: therefore, the time of travel along AC (see fig. 36) to the time along AB is as the length of the plane AC to the length of the perpendicular AB: which was to be demonstrated]" (*ibid.*, p. 217).

shows that he is aware of the serious aporias that can arise from the application of infinitesimal analysis techniques to this case. In fact, he invites us to draw parallels to the base from all points of the line AB towards AC and concludes that the *momenta velocitatis*, "in terminis singularum parallelarum semper erunt inter se paria [at the ends of each of the parallels, will always be equal to one another]". In this way, he can avoid explicitly asserting a full one-to-one correspondence between the infinite points of AB and all the infinite points of AC.

The examples we have considered indicate, therefore, Galileo's conscious and constant recourse to a whole set of concepts and techniques of infinitesimal analysis in the demonstrations of the principal theorems of the *De motu naturaliter accelerato*. This practice is presented as completely natural, and no doubt is raised as to its validity. However, Galileo had previously reflected at length on the legitimacy of these concepts and techniques and had endeavored to provide some justification for them. The *Discorsi*, the *Dialogo* and other works preserve very important pages, which allow us to understand how Galileo had, at a certain point, oriented himself to resolve the aporias of the infinite, which – after Zeno's arguments – seemed to make a rigorous geometric-mathematical treatment of movement impossible.

Galileo's interest in the composition of the continuum dates back to his Paduan period. In the famous letter to Belisario Vinta, in which he indicated some of the works already underway, which he intended to complete in Florence, Galileo, in addition to the three books *De motu locali* and the two books *De sistemate seu constitutione universi*, declared that he had, among the "treatises" dedicated to natural questions, also a *De compositione continui*, of which, however, he did not
346 give any specifics[39]. On the other hand, in many fragments of the *Discorsi* examined in the previous chapter, it is not difficult to recognize the use of infinitesimal analysis techniques. It is not unfounded to suppose that Galileo's planned *De compositione continui* constituted a writing intended to justify the infinitesimal procedures employed in the demonstrations of the properties of naturally accelerated motion. The dates support this supposition since, in 1610, Galileo had certainly already been working for some time on the *de motu* theorems based on the correct principle that speed increases with time. No traces remain of the *De compositione continui* as a unified work. Galileo probably never wrote it. But it is not difficult to identify in Galileo's writings – especially in the *Giornata Prima* [First Day] of the *Discorsi* – abundant indications of the manner in which the Pisan scientist intended to approach the problem of infinite divisibility.

Abandoning an attempt at a comprehensive examination of Galileo's conceptions of the composition of the continuum, of the nature of the infinite and of its comprehensibility for the human mind, as well as of the evident and very important cosmological implications present in these reflections (the examination alone would require an entire book), we will focus our attention on those aspects of the discussion that best shed light on the effort made in the attempt to legitimize the procedures that we have emphasized in the Latin draft, in *Theorema I*, in *Theorema III* and in the *Scholium* to *Problema IX* of the *De motu naturaliter accelerato*. In essence – as we

[39] *G.G.*, X, p. 352.

have seen – the points that required a suitable foundation could be reduced to two: *a*) the possibility of proposing the comparison between two infinite series of linear magnitudes as a comparison no less legitimate than that between the equal areas that comprise them; and *b*) the assertion that the perpendicular and the inclined line (of a right-triangle) are both composed of an infinite number of points.

In some of the *Postille* to Rocco's *Esercitazioni Filosofiche*[40] and, more extensively, in a large group of pages in the *Giornata Prima* [First Day] of the *Discorsi*[41], Galileo devotes considerable space to the examination of traditional questions on the divisibility of the continuum, a true challenge for scientific and philosophical thought for many centuries. In the *Discorsi*, Galileo's considerations take their cue from the enunciation of a series of mechanical paradoxes that refer to various ways of conceiving the composition of the continuum and infinite divisibility. At a certain point, the attention of the interlocutors focuses on the mechanical *topos* of the so-called "Aristotle's wheel", since it is presented in the pseudo-Aristotelian *Questions of Mechanics*[42]. The paradox depends on the **347**
difficulty in understanding what causes the fact that (cf. figure 37), when the larger circumference completes a full rotation (corresponding to the linear development BF), the smaller circumference, even though it also completes only one rotation, seems to trace on the plane, through successive contacts, the line CE, which is significantly greater than its length.

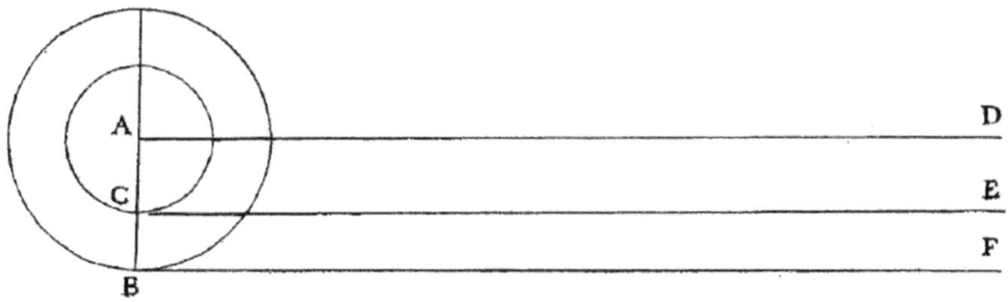

Figure 37

The solution to the paradox proposed by Salviati presupposes the choice of equivalence between a circumference and a polygon with infinite sides. The case of the two circumferences can thus be reduced to that of two concentric polygons. Now, by having the concentric polygons rotate on a plane – as Salviati observes – the larger one will successively touch with its sides the whole line BS (see fig. 38) while the sides of the smaller one will proceed along CV, so to speak, "in leaps". This finding suggests an explanation of the wheel paradox:

[40] *G.G.*, VII, pp. 682-683 and 744-750.
[41] *G.G.*, VIII, pp. 67 ff.
[42] On the paradox of the "wheel", cf. I. E. Drabkin, *Aristotle's Wheel: Notes on the history of this paradox* ("Osiris", IX, 1950, pp. 162-198). For Galileo's treatment of the *topos*, see the essay by P. Costabel, in *Galilée. Aspects de sa vie et de son œuvre*, P.U.F., Paris 1968, pp. 277-288. Cf. also Max Jammer, *Storia del concetto di spazio*, trad. it. di A. Pala, Feltrinelli, Milano 1966 (2nd ed.), pp. 63-64.

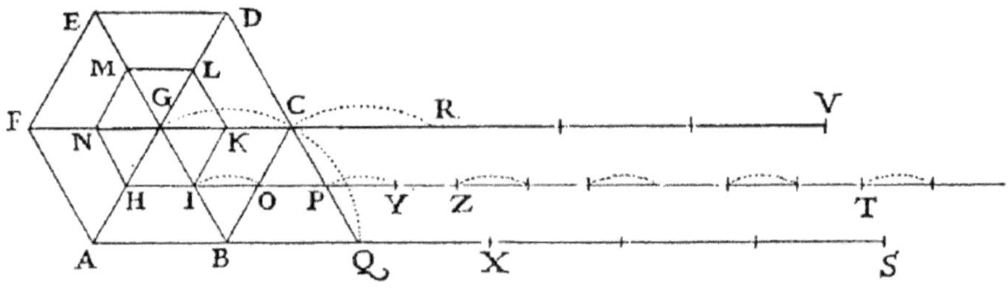

Figure 38

348 sì come ne i poligoni di cento mila lati alla linea passata e misurata dal perimetro del maggiore, cioè da i cento mila suoi lati continuamente distesi, è eguale la misurata da i cento mila lati del minore, ma con l'interposizione di cento mila spazii vacui traposti; così direi ne i cerchi (che son poligoni di lati infiniti) la linea passata da gl'infiniti lati del cerchio grande, continuamente disposti, esser pareggiata in lunghezza dalla linea passata da gl'infiniti lati del minore, ma da questi con l'interposizion d'altrettanti vacui tra essi[43]. [just as in polygons of one hundred thousand sides, the line passed and measured by the perimeter of the larger one, that is by its one hundred thousand sides continuously extended, is equal to the line measured by the one hundred thousand sides of the smaller one, but with the interposition of one hundred thousand empty spaces between them; so I would say that, in circles (which are polygons of infinite sides), the line passed by the infinite sides of the large circle, continuously disposed, is matched in length to the line passed by the infinite sides of the smaller one, but by these with the interposition of as many empty spaces between them.]

Immediately after proposing his solution, Salviati adds an important clarification:

e sì come i lati della circonferenza non sono quanti, ma bene infiniti, così gl'interposti vacui non son quanti ma infiniti: quelli, cioè, infiniti punti tutti pieni, e questi infiniti punti parte pieni e parte vacui[44]. [and just as the sides of the circumference are not a certain quantity, but rather infinite, so too the intervening voids are not a certain quantity but infinite: in the former, namely, the infinite points are all filled, and in the latter, the infinite points are partially filled and partially empty.]

The circumference is, therefore, composed of infinite "non quanti [non-quantifiable]" points. Salviati feels the need to distinguish the division into "quante [quantifiable]" parts from that into non-"quantifiable" parts.

E qui voglio che notiate, come risolvendo e dividendo una linea in parti quante, e per consequenza numerate, non è possibile disporle in una estensione maggiore di quella che occupavan mentre stavano continuate e congiunte senza l'interposizione d'altrettanti spazii vacui; ma imaginandola risoluta in parti non quante, cioè ne' suoi infiniti indivisibili, la possiamo concepire distratta in immenso senza l'interposizione di spazii quanti vacui, ma sì bene d'infiniti indivisibili vacui. E questo che si dice delle semplici linee, s'intenderà detto delle superficie e de' corpi solidi, considerandogli composti di infiniti atomi non quanti[45]. [And here I want you to note that, by resolving and dividing a line into quantifiable parts, and consequently numbered, it is not possible to arrange them in an extension greater than that which they occupied while they were continuous and joined without the interposition of an equal number of empty spaces; but imagining it resolved into non-quantifiable parts, that is, into its infinite indivisibles,

[43] *G.G.*, VIII, p. 71.
[44] *Ibid.*
[45] *Ibid.*, pp. 71-72.

we can conceive it immensely expanded without the interposition of quantifiable empty spaces, but indeed of infinite indivisible vacuous spaces. And what is said of simple lines, will be understood to be said of areas and solid bodies, considering them composed of infinite non-quantifiable atoms.]

In essence, Galileo states that, in terms of division into quantifiable parts, the two circumferences remain, in any case, different magnitudes. Conversely, in terms of division into non-quantifiable parts, the smallest, with the “interposition” of infinite non-quantifiable empty spaces, can “distrarsi [expand]” until it reaches the size of the largest and even surpasses it. Salviati emphasizes how this interpretation of the atomic composition of matter appears, among other things, particularly suitable for explaining the phenomenon of rarefaction and concentration[46]. The explanation
suggested by Salviati raises, as usual, objections from Simplicio, who primarily **349**
emphasizes his own doubts about the legitimacy of the procedures employed by the interlocutor:

quel comporre la linea di punti, il divisibile di indivisibili, il quanto di non quanti, mi paiono scogli assai duri da passargli[47]. [that composing the line of points, the divisible of indivisibles, the quantity of non-quantifiables, seem to me to be very difficult obstacles to overcome.]

Salviati promptly replies to the Aristotelian’s objections, observing that we are dealing with extremely delicate and extraordinarily subtle questions, since the infinite “è per sé solo da noi incomprensibile [is by itself alone incomprehensible to us]”[48]. Further complications arise from the fact that this is a matter of understanding the infinite and the indivisible together[49]. However, he does not despair at demonstrating that the continuum is infinitely divisible and, at the same time, that the continuum is composed of indivisibles. Since, however, truth is singular, this means “che il dire che il continuo costa di parti sempre divisibili, col dire che il continuo costa d’indivisibili, siano una medesima cosa [that saying that the continuum consists in always divisible parts, and saying that the continuum consists of indivisibles, are the same thing]”[50].

---

[46] “In questa guisa non repugna distrarsi v.g. un piccolo globetto d’oro in uno spazio grandissimo senza ammettere spazii quanti vacui, tutta volta però che ammettiamo l’oro esser composto di infiniti indivisibili [In this way, it is not contrary to expand, for example, a small gold pellet in a very large space without admitting any quantifiable empty spaces, all the while admitting that gold is composed of infinite indivisibles]” (*ibid.*); on this point, which refers to the question of Galileo’s “atomism”, see H. E. Le Grand, *Galileo’s matter theory*, in *New perspectives on Galileo* cit., pp. 197-208.

[47] *Ibid.*, p. 73.

[48] *Ibid.*, p. 76

[49] “Se vogliamo compor la linea di punti indivisibili, bisogna fargli infiniti; e così conviene apprender nel medesimo tempo l’infinito e l’indivisibile [If we want to compose a line of indivisible points, we must make them infinite; and so it is necessary to understand the infinite and the indivisible at the same time]” (*ibid.*, p. 77).

[50] “Io Sig. Rocco – aveva scritto Galileo nelle *Postille* alle *Esercitazioni Filosofiche* – essendo certo che il continuo costa di parti sempre divisibili, dico che è verissimo e necessario che la linea sia composta di punti, ed il continuo d’indivisibili; e cosa forse più inopinata vi aggiungo, cioè che, essendo il vero un solo, conviene che il dire che il continuo costa di parti sempre divisibili, col dire che il continuo costa d’indivisibili, siano una medesima cosa [I, Signor Rocco – Galileo had written in the *Postille* to the *Esercitazioni Filosofiche* – being certain that the continuum is made up of parts that are always

Salviati believes that he can favor one of the traditional arguments as particularly effective against the thesis of the composition of the continuum from indivisible, non-quantifiable parts. The argument consists in emphasizing how, even if the notion of an indivisible part is admitted, in other words, non-quantifiable (otherwise it would have parts, and would therefore be divisible), the quantity cannot arise from the sum of two or more indivisibles. In that case, of course, one would have to admit that even the union of any odd number of non-quantifiable, indivisible parts generates a quantity. And such quantity could then be divided into two parts, with the absurd consequence of the division of the indivisible[51]. Salviati believes that this objection can be satisfactorily answered by conceding that "né dieci, né cento, né mille [neither ten, nor a hundred, nor a thousand]" indivisibles can compose an indivisible magnitude, but by admitting that the quantity can be generated by an infinite number of non-quantifiable indivisibles[52].

Simplicio has an objection ready: given two unequal terminated lines, must we then admit that the larger one contains an infinite number of indivisible points more than the infinite points of the smaller one? Salviati recognizes the difficulty and
350 believes that it constitutes an eloquent proof of the limits of human reason, which tends to automatically transfer to the infinite attributes that we are accustomed to conferring "alle cose finite e terminate [to finite and terminated things]". Hence the error and the paradox, since it is evident that "questi attributi di maggioranza, minorità ed egualità non convenghino a gl'infiniti, de i quali non si può dire uno essere maggiore o minore o eguale all'altro [these attributes of greater, lesser and equality do not apply to the infinite, of which one cannot be said to be greater or lesser or equal to another]"[53].

Salviati proposes some considerations to convince Simplicio of the impossibility of transferring to infinite magnitudes the comparisons and proportionalities legitimately employed for finite magnitudes. In fact, he forces Simplicio to concede that the squares are fewer in number than the numbers themselves. Then, he induces the poor Aristotelian to admit that the squares are as many in number as their roots. Lastly, he concludes that there are as many roots of all squares as there are numbers, "poiché non vi è numero alcuno che non sia radice di qualche quadrato [since

---

divisible, say that it is very true and necessary that the line is composed of points, and the continuum of indivisibles; and what is perhaps more unexpected, I add to it, namely that, since the truth is singular, it is fitting that saying that the continuum is made up of parts that are always divisible, with saying that the continuum is made up of indivisibles, are the same thing]" (*G.G.*, VII, p. 745).

[51] *G.G.*, VIII, p. 77.

[52] *G.G.*, VIII, p. 76.

[53] *Ibid.*, p. 78. "Il conferire divisibilità – aveva obbiettato al Conte Rocco – è virtù della infinità, la quale è una materia lontanissima dall'esser capace di quelli attributi e condizioni alle quali soggiaciono i nostri numeri o grandezze comprese dal nostro intelletto: là non entra maggioranza, minoranza, né equalità, non vi ha luogo né il pari, né il dispari; ogni parte (se parte si può chiamare) dell'infinito è infinita… [Conferring divisibility – he had objected to Count Rocco – is a virtue of the infinite, which is a matter very far from being capable of those attributes and conditions to which our numbers or magnitudes understood by our intellect are subject: there applies neither greater, lesser, nor equality, nor is there a place for even, nor odd; every part (if it can be called a part) of the infinite is infinite…]" (*G.G.*, VII, p. 749).

there is no number that is not the root of some square]"[54]. The example serves to demonstrate the contradictions that inevitably arise when someone attempts to establish relations of magnitude between infinities. It is precisely by playing on these paradoxes that, at this point, Salviati replies to Simplicio's objection:

i punti nell'una esser quanti i numeri quadrati, in un'altra maggiore quanti tutti i numeri, in quella piccolina quanti sono i numeri cubi, non potrei io avergli dato sodisfazione col porne più in una che nell'altra, e pure in ciascheduna infiniti?[55] [the points in one being as many as the square numbers, in another larger one, greater than all the numbers, and in the smaller one, as many as the cubic numbers, could I not have satisfied him by placing more of them in one than in the other, and yet infinitely many in each?]

Outside the paradox, therefore, the difficulty raised by Simplicio must be answered by stating that in two unequal finite lines, the indivisible points are "né più, né manco, né altrettanti ma in ciascheduna infiniti [neither more, nor less, nor as many, but in each one, infinite]"[56]. Sagredo intervenes to clarify, with a similar method, that no proportion can be established between the infinite and the finite[57].

As for the possibility that infinite, non-quantifiable indivisibles can generate a quantifiable magnitude, Salviati believes that this is not only logically plausible, but even necessary. Especially necessary for those who admit that a magnitude (for example, a line) is divisible into parts that are always further divisible, that is, quantifiable. In fact:

una divisione e subdivisione che si possa proseguir perpetuamente suppone che le parti siano infinite, perché altramente la subdivisione sarebbe terminabile; e l'esser le parti infinite si tira in consequenza l'esser non quante, perché quanti infiniti fanno un'estensione infinita: e così abbiamo il continuo **351**
composto d'infiniti indivisibili[58]. [a division and subdivision that can be continued perpetually supposes that the parts are infinite, because otherwise the subdivision would be terminable; and the fact that the

---

[54] *G.G.*, VIII, p. 78.
[55] *Ibid.*, p. 79.
[56] *Ibid.*
[57] *Ibid.*, pp. 79-80.
[58] *Ibid.*, p. 80. "Se la divisione e suddivisione si ha da poter continuare sempre, bisogna necessariamente che la moltitudine delle parti sia tale che già mai non si possa superare; e sono, dunque, le parti infinite, altrimenti la divisione si finirebbe; e se sono infinite, bisogna che non siano quante, perché infiniti quanti compongono un quanto infinito e noi parliamo di quanti terminati: e però, gli altissimi ed ultimi, anzi primi componenti del continuo, sono indivisibili infiniti. Non vedete voi che il dire che il continuo costa di parti sempre divisibili, importa che, dividendo e suddividendo, non si arrivi mai a' primi componenti? I primi componenti, dunque, sono quelli che non sono più divisibili ed i non più divisibili sono gl'indivisibili [If division and subdivision are to be able to continue forever, it is necessary that the multitude of parts be such that it can never be exceeded; and therefore the parts are infinite, otherwise the division would end; and if they are infinite, it is necessary that they are non-quantifiable, because infinite quantifiables compose an infinite quantity and we are speaking of finite quantifiables: and therefore, the highest and ultimate, or rather first, components of the continuum are infinite indivisibles. Do you not see that saying that the continuum consists of parts that are always divisible, means that, by dividing and subdividing, one never arrives at the prime components? The prime components, therefore, are those that are no longer divisible and the no longer divisible are the indivisibles]" (*Postille* to the *Esercitazioni Filosofiche di Antonio Rocco*, *G.G.*, VII, p. 746).

parts are infinite implies that they are non-quantifiable, because infinite quantifiable parts make an infinite extension: and thus we have the continuum composed of infinite indivisibles.]

The very possibility of dividing a magnitude into ever more divisible parts rests, therefore, on the condition that its ultimate components are non-quantifiable. Simplicio begins to show some signs of uncertainty, and Salviati immediately returns to press him, asking him whether he believes that "le parti quante nel continuo [the quantifiable parts in the continuum]" are finite or infinite. The answer – as was to be expected – is purely Aristotelian: infinite in potentiality and finite in actuality. Simplicio's anguish is endless. He is, in fact, first urged to recognize that the "actual" division of an infinite magnitude does not produce parts which, when added together, are greater than the divided magnitude. Then, he must reluctantly acknowledge that, in the case of the division of a finite magnitude into infinite quantifiable parts, the sum of the "parts" would produce an infinite magnitude.

Leaving the unfortunate Aristotelian to ponder his own misfortunes, Salviati, urged by Sagredo, finally declares himself willing to provide the solution to the riddle. The quantifiable parts into which the "continuo terminato [terminated continuum]" can be divided are neither finite nor infinite. In fact,

parlando delle quantità discrete, parmi che tra le finite e l'infinite ci sia un terzo medio termine, che è il rispondere ad ogni segnato numero; sì che, domandato, nel presente proposito, se le parti quante nel continuo siano finite o infinite, la più congrua risposta sia il dire, non esser né finite né infinite, ma tante che rispondono ad ogni segnato numero: per il che fare è necessario che elle non siano comprese dentro a un limitato numero, perché non risponderebbono ad un maggiore; ma né anco è necessario che elle siano infinite, perché niuno assegnato numero è infinito: e così ad arbitrio del domandante una proposta linea gliela potremo assegnare segata in cento parti quante, e in mille e in cento mila, conforme a qual numero più gli piacerà; ma divisa in infinite, questo non già. Concedo dunque a i Signori filosofi che il continuo contiene quante parti quante piace loro, e gli ammetto che le contenga in atto o in potenza, a lor gusto e beneplacito; ma gli soggiungo poi, che nel modo che in una linea di dieci canne si contengono dieci linee d'una canna l'una, e quaranta d'un braccio l'una, e ottanta di mezzo braccio, etc., così contiene ella punti infiniti[59]. [speaking of discrete quantities, it seems to me that, between the finite and the infinite, there is a third middle term, which corresponds to every given number; so that, when asked, in the present case, whether the quantifiable parts in the continuum are finite or infinite, the most appropriate answer is to say that they are neither finite nor infinite, but so many as to correspond to every given number: for this to be the case, it is necessary that they are not contained within a limited number, because they would not correspond to a greater one; but neither is it necessary that they be infinite, because no assigned number is infinite: and so, at the discretion of the questioner, we can assign to him a proposed line divided into as many as one hundred quantifiable parts, and into a thousand, and into a hundred thousand, in accordance with whatever number he pleases; but divided into infinites, this is not the case. I therefore grant to the gentlemen philosophers that the continuum contains as many quantifiable parts as they like, and I admit to them that it contains them in act or in potentiality, according to their taste and pleasure; but I then add that, just as a line of ten rods contains ten lines of one rod each, and forty of one armspan each, and eighty of half an armspan, etc., so it contains infinite points.]

352 Let us pause here. The pages of the *Giornata Prima* present other notable testimonies of Galileo's reflections on the composition of the continuum and the nature of the infinite, which we will not dwell on. Nor will we dwell on the problem

[59] *G.G.*, VIII, p. 81.

of the evident analogy that Galileo's theory *de compositione continui* presents with the philosophy of Kalām, even though research aimed at verifying the possibility of such an influence and at identifying its possible channels appears quite inviting and appropriate[60]. In the passages we have examined, we find all the necessary elements to be able to return to the Latin draft and to the theorems of the *Giornata Terza* examined previously to verify whether there is any correspondence between the analysis techniques used therein and the reflections proposed here by Galileo.

The pages devoted to the composition of the continuum are neither easy nor clear. This explains some of the different "interpretations" that have been given. In essence, Galileo established two different propositions, which have not always been appropriately distinguished: the finite continuum is composed of infinite indivisible non-quantifiable parts; the continuum can be divided into an arbitrary (indeterminate, but not infinite) number of quantifiable parts[61].

Let us return to the figures of *Theorema I* (see fig. 34) and the *Scholium* of *Problema IX* (see fig. 35). In both cases, the perpendicular leg of the right-triangle **353**
represents time, while the *velocitatis momenta* are represented by the lines parallel to the base leg. The demonstration is based on the comparison between the area of the right-triangle and that of a rectangle of equal height (in the *Theorema*, with half the base, but in the *Scholium*, with equal base), whose lines parallel to the base represent the degrees of (constant) speed of a uniform motion. From the comparison between the areas, we obtain the equality of the spaces traversed in equal times, both in naturally accelerated motion and in uniform motion with a speed equal to half of the maximum *velocitatis momentum* reached in the accelerated motion (in

[60] As is known, the thinkers of Kalām (an Islamic theological school that flourished primarily in the 10th century, but was also influential thereafter) supported a peculiar atomistic conception, in which the central idea was that extension arises from the aggregation of indivisibles that lack extension. To tell the truth, according to the Kalām theorists, to create extension, one does not need infinite indivisibles, rather, two (for linear extension), four (for two-dimensional area), or eight (for three-dimensional extension) are sufficient. The polemical target of the Kalām was Aristotle's *Physics* and the refutation of the Stagirite's thesis that the ultimate elements of space and time are indivisible. No one has ever tried to establish how Galileo might have drawn upon these ideas. Nor is it known that he ever studied the philosophy of Kalām, of which a precise exposition had been provided in Maimonides's *Guide of the Perplexed*, translated into Latin by Buxtorf the Younger in the 17th century (although Maimonides is never cited by Galileo). The philosophy of Kalām certainly had a strong influence on Leibniz (cf. Foucher de Careil, *Leibniz, la philosophie juive et la Cabbale*, Parigi 1861 and M. Gueroult, *Leibniz. Dynamique et métaphysique*, Parigi 1967, 2nd ed.). According to Jammer, Galileo, in the pages of the *Discorsi* that we have been analyzing, would be "memore degli antichi insegnamenti del Kalām [mindful of the ancient teachings of the Kalām]" (*Storia del concetto di spazio*, trad. it. cit., p. 64). The Israeli scholar, however, does not provide any evidence for this claim. The concept of space in Kalām has been given considerable attention by H. A. Wolfson (see *The Philosophy of the Kalam*, Cambridge, Mass., and London 1976).

[61] I am referring to the important and often-cited unpublished doctoral dissertation (primarily Ch. IV) by Thomas Settle, who deserves credit for being the first to perceive the close connection between Galileo's general discussions dedicated to infinite divisibility and the composition of the continuum and the development of the concept of *momentum velocitatis*, as well as the demonstration techniques employed in the *De motu locali*. However, Settle, if I am not mistaken, does not properly distinguish the division into (indefinite) quantifiable parts from that into (infinite) non-quantifiable parts. And he believes that Galileo conceived the continuum as composed also of indefinite quantifiable parts (cf. *ibid.*, p. 186), whereas the Pisan scientist – as we have seen – asserts that the constitutive elements of the continuum are only the infinite non-quantifiable parts.

the *Scholium*, the accelerated motion covers a space equal to half of that covered by the uniform motion with a speed corresponding to the maximum degree of the accelerated motion – which is exactly the same). The perpendicular leg, representing time, is explicitly stated to be divisible into infinite instants, "veluti infinita sunt puncta in linea AC [as there are infinite points in the line AC]". Galileo, therefore, proposes that the representation of time with a line is perfectly adequate: "tal parte – aveva scritto nella *Giornata Prima* – è un instante d'un tempo quanto, qual'è un punto d'una linea, che ne contiene infiniti [such a part – he had written in the *Giornata Prima* – is an instant of a quantified time, just as a point is on a line, which contains infinite points]"[62].

He then adds that the representation of the *momenta velocitatis* with the parallels to the base of the triangle, continuously increasing according to the passage of time, can be considered equally legitimate and appropriate. The *momenta* are also infinite, just as the parallels to the base included in the right-triangle are infinite.

It will not escape notice that the legitimacy of the representation of the instants of time and of the *momenta velocitatis*, both infinite, respectively with a line and with an area is guaranteed precisely by the first of the two conclusions that we have summarized of Galileo's analysis of the continuum conducted in the *Giornata Prima* of the *Discorsi*: the ultimate components of all finite magnitudes are infinite and non-quantifiable, that is, indivisible. But finite magnitudes (such as, in our case, a given time, or a given degree of accelerated speed) are constituted by infinitely many indivisible components. Neither an instant of time, nor a point of a line, nor a *momentum velocitatis* constitute the time, the line, or the speed, "ma si bene infiniti [but rather infinitely many do]"[63]. The representation of time (composed of infinitely many instants) with a finite line, and of accelerated speed (composed of infinitely many *momenti*) with a finite area, rests on the assumption that non-quantifiable infinite points generate lines of finite quantity and that from infinite lines (non-quantifiable in width) a bounded area arises. It should be recalled that the Pisan scientist had described, not by chance, the formation of the area from the aggregation of *infiniti momenti* as an *exurgere* [emergence] and the effect of such aggregation as a *conflare* [formation][64].

354 Galileo could, therefore, consider the transposition of time and *momenta velocitatis* into figures to be effectively well-founded. Less justifiable, however, was the procedure by which, from the comparison of equal finite areas (as, in *Theorema I*, between the triangles IEF and GAI: see fig. 34), the Pisan scientist concluded the equality of the two infinite series of parallels (that is, the *momenta velocitatis*) contained in the two triangles. To tell the truth, the method must have seemed instinctively plausible to him: if areas resolve into the contained aggregate of infinite non-quantifiable parallels to the base, then two equal areas will be composed of two equal aggregates. Yet, in the *Giornata Prima*, Galileo had warned and distrusted comparing infinite series and attempting to establish proportions

[62] *G.G.*, VIII, p. 95.
[63] *Ibid.*, p. 77.
[64] *Ibid.*, p. 243.

of greater, lesser and equality among them. It is true that the case presented in those pages, which had pushed Galileo to deny any possibility of comparison, was different. It was, in fact, a matter of establishing whether the points contained in a given line were, so to speak, "more infinite" than the points contained in a smaller line. But the principle had been established, and Galileo knew he had no justification for proposing such a comparison. He, nevertheless, gambled on this instinctive plausibility. Anyone who had followed him up to that point, granting him all the undeclared assumptions he had relied on, would certainly not have wavered in the face of such minutiae. Who, after having admitted that the infinite parallels to the base "constitute" the area of the triangle, would have denied the foundation of the inverse procedure?

The important fact, which is worth pressing, is that the Pisan scientist was not only clearly aware of the difficulty, but had even discussed it with Bonaventura Cavalieri. It is known that the faithful Jesuate friar, author of the fundamental *Geometria indivisibilibus*, had come to inform the Master of his progress and the results obtained with the new method of indivisibles. The manuscript of the *Geometria* was in Galileo's hands well before the final drafting and publication of the work, which took place only in 1635[65]. In *Theorema II* of the second book of the *Geometria*, Cavalieri had demonstrated that "Tutte le linee di figure piane uguali sono uguali, e tutti i piani di solidi uguali sono uguali [All lines of equal plane figures are equal, and all planes of equal solids are equal]"[66]. This was exactly
the principle that Galileo needed to postulate in order to propose the comparison **355**
between infinite series of parallels to the base of equal figures. In a letter dated September 1634, unfortunately lost, Galileo had, however, curiously contested the legitimacy of the theorem with Cavalieri. This can be deduced from Cavalieri's response to the lost letter of the Pisan scientist.

> Ella dice che se tutte le linee di due superficie eguali sono eguali, diminuendole egualmente, l'ultime esinanizioni di esse dovriano esser eguali: il che poi non appare nell'esempio della scodella e del cono, restando in quella una circonferenza di cerchio, et in questo un punto, infinitamente minore di quella[67]. [You say that if all the lines of two equal areas are equal, by diminishing them equally, their final

---

[65] *Geometria indivisibilibus continuorum nova quadam ratione promota. Authore F. Bonaventura Cavalerio...*, Typis Clementis Ferronii, Bononiae 1635. For the events relating to the composition of *Geometria indivisibilibus* and for the anticipations that Cavalieri gave to Galileo, see the *Introduzione* of the *Geometria degli indivisibili di Bonaventura Cavalieri*. Edited by Lucio Lombardo Radice, U.T.E.T., Torino 1966 (with complete Italian translation). Regarding Cavalieri, despite the fact that many have indicated and continue to indicate his decisive importance, there is still no study based on rigorously scientific foundations. At a biographical level and for the recognition of the surviving works and manuscripts (unfortunately not many), we are, in fact, still forced to use the now almost centenary essay by Favaro (*Bonaventura Cavalieri e lo Studio di Bologna*, Fava e Garagnani, Bologna 1885); cf. also *D.S.B.*, III (1971), pp. 149-153 (E. Carruccio).

[66] *La Geometria degli indivisibili*, trad. it. cit., p. 208.

[67] *G.G.*, XVI, pp. 136-137. Cavalieri's letter is dated 2 October 1634. The Milanese mathematician admitted that he expected to provoke some objections and doubts, given that the subject was so arduous and delicate: "Ho però sentito con gusto ch'ella li habbi dato una scorsa, né li paia il mio metodo del tutto improbabile, benché ella dichi di havervi molte difficoltà; né me ne meraviglio, mentre par che io trapassi allo infinito, che porta seco tanti dubii quanti ella sa [I have, however, heard with pleasure that you have given it a glance, and that my method does not seem to you to be entirely improbable, although

remnants should be equal: which then does not appear in the example of the bowl and the cone, since in the former there remains a circumference of a circle, and in the latter a point, infinitely smaller than the former.]

The example of the "scodella [bowl]" is given in the same pages of the *Giornata Prima* of the *Discorsi* in which Galileo addresses the *de compositione continui* questions. The proof consists in constructing a figure by rotating a rectangle, in which a semicircle is inscribed whose center is joined to the ends of the base of the rectangle (see figure 39). From the rotation, a cylinder is formed containing a hemisphere and a cone. Once the hemisphere is removed, leaving the cone, a figure remains which Galileo imaginatively calls a "scodella [bowl]", due to its resemblance to the household utensil. The cone and the "bowl" are equivalent. By taking the plane GN, parallel to DE and moving it upwards, while always keeping it parallel to DE, the parts into which this plane divides the cone and the "bowl" remain always equal to each other. However, having reached the uppermost terminus, that
356 is, the position of AB, the plane GN cuts a circumference on the bowl, while on the cone it cuts only a point[68].

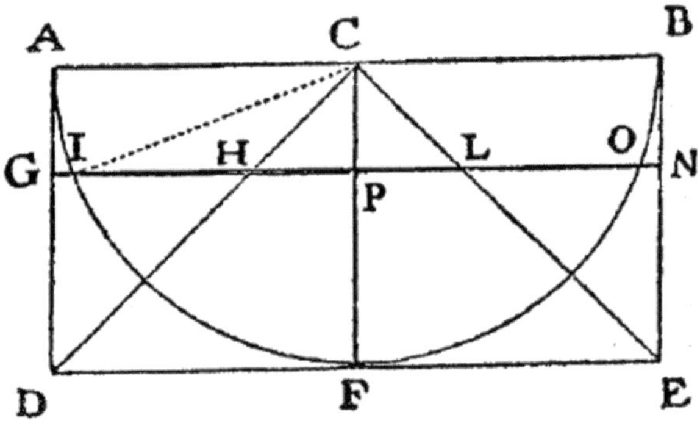

Figure 39

In essence, by proposing the bowl proof, Galileo had contested that from the equality of two areas one could pass to the equality of all the infinite lines that compose them (and, conversely, consider two equal areas constituted by two infinite series of equal lines). In the case of the "bowl" and the cone, in fact, the sections cut on the two solids by the plane parallel to the base remained equal in pairs until the last "esinanizione [remnant]". At this point, on one side, there remained a circumference of a circle, on the other, a point.

Bonaventura Cavalieri replied to the Master's "beautiful" objection. He argued that, in the "bowl" example, the indivisibles were "planes" and that both the

you say that there are many difficulties in it; nor am I surprised, since it seems that I am passing into the infinite, which brings with it as many doubts as you know]" (*ibid.*, p. 136).

[68] *G.G.*, VIII, pp. 74-75. Galileo took the demonstration of the "scodella [bowl]" from the *De centro gravitatis solidorum* cit. by Luca Valerio (*Liber II, Propositio XII*).

circumference and the point could be, in a certain sense, considered equal, "essendo noi arrivati al nullo piano tanto nel cono quanto nella scodella [since we have arrived at the null plane both in the cone and in the bowl]"[69]. Galileo accepted the Jesuate's justification. In fact, he included the example of the bowl in the *Discorsi*, maintaining that the circumference and the point at which the cylinder and the cone were ultimately divided were to be considered equal. He even drew from it an unusual argument to support the paradoxical statement "che un sol punto sia eguale a una linea [that a single point is equal to a line]"[70].

Naturally, beyond Cavalieri's reasons, it was practical needs that convinced Galileo to change his opinion. If he had maintained his criticism of Cavalieri's *Theorema II*, he would have been forced to give up the comparison between infinite aggregates of straight lines constituting two equal areas; in essence, the entire *De motu locali* would have fallen apart.

Although the Pisan scientist conceded to the Jesuate that the demonstration of the "bowl" was of no value against *Theorema II* of the *Geometria indivisibilibus*, the gap between Galileo's conception of the composition of the continuum and that of Cavalieri remained significant, and would never be bridged. In the Jesuate's first reaction upon reading the *Discorsi*, behind the lofty praises for the Master's **357**
superhuman ability, who, "con la scorta della geometria e con la tramontana del suo altissimo ingegno [with the guidance of geometry and the north star of his lofty genius]" was able to navigate successfully through "l'immenso oceano degli indivisibili, de' vacui, degli infiniti [the immense ocean of indivisibles, voids, infinities]" etc., there was evident conflict and even some bitterness for the misconduct committed by Galileo in reporting his opinions.

Io non ardii di dire che il continuo fosse composto di quelli indivisibili ma mostrai bene che fra continui non vi era altra proporzione che dalla congerie de gl'indivisibili (presi però equidistanti, se parliamo delle linee rette e delle superficie piane, particolari indivisibili da me considerati); il che mi metteva veramente in sospetto che quello che ha finalmente pronuntiato potesse esser vero. Se io havessi

---

[69] *G.G.*, XVI, p. 137. "In somma – aggiungeva – parmi che le ultime esinanitioni devano essere niente di quel genere che si diminuisse, non importando poi che differischino in altro genere [In summary – he added – it seems to me that the last remnant must be nothing of the kind that was diminished, it not mattering if they differ in another kind]". See also how Cavalieri set out, in another way, his objections to the "scodella [bowl]" argument in a later letter to Galileo dated 19 December 1634 (G.G., XVI, pp. 175-176).

[70] "E questo sarà – chiariva Galileo, introducendo la dimostrazione della "scodella" – col mostrarvi due superficie eguali ed insieme due corpi pur eguali e sopra le medesime dette superficie, come basi loro, collocati, andarsi continuamente ed egualmente, e queste e quelli, nel medesimo tempo diminuendo, restando sempre tra di loro eguali i loro residui, e finalmente andare, si le superficie come i solidi, a terminare le lor perpetue egualità precedenti, l'uno dei solidi con l'una delle superficie in una lunghissima linea, e l'altro solido con l'altra superficie in un sol punto, cioè, questi in un sol punto e quelli in infiniti [And this will be – as Galileo clarified, introducing the demonstration of the "bowl" – by showing you two equal areas and likewise two equal bodies placed upon these very same areas as their bases, continuously and equally diminishing, both the areas and the bodies, at the same time, with their remaining parts always equal to each other, and finally, both the areas and the solids reaching the termination of their preceding perpetual equalities, one of the solids with one of the areas in a very long line, and the other solid with the other area in a single point, that is, the latter in a single point and the former in infinitely many points]" (*G.G.*, VIII, p. 73).

havuto tanto ardire, l'havrei pregata a non tralasciarne questa confermatione, se non per la verità di essa conclusione, almeno acciò altri più attentamente havessero fatto riflessione a questa mia nuova maniera di misurare i continui[71]. [I did not dare to say that the continuum was composed of the indivisibles, but I showed well that among continua there was no other proportion than that of the aggregates of these indivisibles (but taken equidistantly, if we speak of straight lines and planar areas, particular indivisibles as considered by me); which really made me suspect that what you finally pronounced could be true. If I had had so much audacity, I would have asked you not to omit this confirmation, if not for the truth of this conclusion, at least so that others would reflect more carefully on this new way of measuring continua.]

Five years earlier, Cavalieri had stated with even greater clarity that he did not
conceive of a continuum as composed of indivisibles and that he limited himself to
showing that "i continui hanno la proportione degli aggregati di questi indivisibili
[continua have the proportion of the aggregates of these indivisibles]"[72]. On the
other hand, Galileo's infinitesimal conceptions had also appeared questionable
358 to other disciples and correspondents. Aproino[73] and de Ville[74] openly disclosed

[71] *G.G.*, XVIII, p. 67. It is worth recalling that Galileo had not cited Cavalieri's *Geometria* in the pages of the *Giornata Prima* of the *Discorsi*. Curiously, however, he cited the name of the Milanese mathematician in the same pages, not as the discoverer of the method of indivisibles, but as the author of the *Specchio Ustorio*: "Intorno a gli effetti de gli specchi d'Archimede mi rese credibile ogni miracolo che si legge in più scrittori, la lettura de i libri dell'istesso Archimede, già da me con infinito stupore letti e studiati; e se nulla di dubbio mi fosse restato, quello che ulteriormente ha dato in luce intorno allo specchio ustorio il P. Buonaventura Cavalieri e che io con ammirazione ho letto è bastato a cessarmi ogni difficoltà [Concerning the effects of Archimedes's mirrors, every miracle that is written about by many writers was made credible to me by the reading of the books of Archimedes himself, which I had already read and studied with infinite amazement; and although no doubt had remained in my mind, what Father Buonaventura Cavalieri has further revealed about the burning mirror, and which I have read with admiration, has been enough to put an end to all my difficulties]" (*G.G.*, VIII, pp. 86-87).

[72] *G.G.*, XVI, p. 138.

[73] [Paolo] Aproino (1586-1638), originally from Treviso, was a student of Galileo in Padua, with whom he worked primarily on experiments concerning the force of percussion. For this reason, in addition to remembering his early death, the Pisan scientist included him as an interlocutor in the dialogue of the so-called *Giornata Sesta* [Sixth Day] of the *Discorsi* (cf. *G.G.*, VIII, pp. 312-346). Aproino, as early as February 1635, had received from Micanzio the text of the *Giornata Prima* of the *Discorsi* and in a letter had communicated to Galileo the impression he had gathered, especially from reading the pages dedicated to the analysis of the nature of the infinite and the composition of the continuum. In essence, Aproino was dissatisfied with the "transito di comparatione che si fa dal poligono di moltissimi lati al circolo, imaginandolo di infiniti; perché, se ben in quantità si va prossimando alla misura, nel specie però della figura si va sempre più allontanando, ché il poligono di mille lati mi pare sì differente dal circolo che non è il triangolo, tanto quanto mille è più differente da l'uno che non è tre [transition of comparison that is made from the polygon of very many sides to the circle, imagining it to have infinite sides; because, although in quantity it is approaching the measure, in the specific case of the figure, it is always moving further away, because the polygon of a thousand sides seems to me as different from the circle as the triangle is, just as a thousand is more different from one than three is]" (*G.G.*, XVI, p. 219; letter of 3 March 1635, from Venice). Furthermore, Aproino also declared his doubts regarding the equality between two unequal circumferences or lines, in the name of the infinity of points present in both. According to this principle, in fact, "si potrebbe similmente demostrare, ogni linea essere eguale ad ogn'altra, anche irregulare, maggiore o minore che sii di lei [it could be similarly demonstrated that every line is equal to any other, even if irregular, whether larger or smaller than it]" (*ibid.*, p. 220). For Aproino, cf. *D.B.I.*, III (1961), pp. 646-648 (edited by M. Gliozzi).

[74] Micanzio had also shown the *Giornata Prima* of the *Discorsi* to Antonio de Ville, a military engineer and architect in the service of the Venetian Republic. De Ville, like Aproino, had expressed to Galileo,

more than one quandary. Descartes, a few years later, wrote to Mersenne that all of Galileo's reflections on the infinite appeared unsatisfactory[75]. Some of the criticisms he received must have seemed convincing to the Pisan scientist himself, but he could not easily dispense with his own speculations on this point. They were, in fact, one of the indispensable foundations for his own science *de motu*. In acknowledging its weakness, he would have been forced to renounce the *De motu locali* or to solemnly proclaim its character as a purely hypothetical construction devoid of any reference to the real movement of bodies. Which is something our

---

in a long letter, his doubts both regarding the treatise on resistances and regarding the speculations on infinity. As for the pages dedicated to infinity, de Ville was not satisfied with the asserted infinity of all square numbers (which would be as many as all the numbers), nor with the identification of unity-infinity made by Galileo. Nor had he been convinced by Galileo's speculation to justify the paradox of the "wheel". In this regard, the Venetian engineer had elaborated another, very interesting explanation: "Quanto a me, estimo che il moto del minore circolo si fa cossi: quando qualche parte del circolo magiore si move sopra il piano, si move anche parte proporzionale del minore, perché di tutti li punti assignabili del magiore si può tirare semidiametri al centro, le quali levaranno portioni proportionali delli circoli concentrici, et come sono li circoli cossi le portioni.... Donde se ne segue che il continuo è composto di parti divisibili, in infinito proportionalmente, di modo che le parti proportionali del minore sianno minori che le parti proportionali del magiore; et cossi le parti del minor circolo che caminano col tatto, benché sianno divisibili in infinito, saranno tutte alle parti del magiore in quella proportione che è il circolo minore al magiore [As for me, I estimate that the motion of the smaller circle is as follows: when some part of the larger circle moves on the plane, a proportional part of the smaller one also moves, because of all the assignable points of the larger one, one can draw semidiameters to the center, which will take proportional portions of the concentric circles, and as the circles are such, so are the portions.... Hence it follows that the continuum is composed of divisible parts, infinitely proportionally, so that the proportional parts of the smaller are smaller than the proportional parts of the larger; and thus the parts of the smaller circle that move in contact, although they are infinitely divisible, will all be to the parts of the larger in that proportion which the smaller circle is to the larger]" (*G.G.*, XVI, pp. 225-226; letter of 3 March 1635, from Venice). De Ville then contested Galileo's view that the reduction of the circle to a polygon with infinite sides was illicit (*ibid.*, p. 227) and declared that he did not understand "la forza ni la deduzione della dimostratione che un circolo divenghi eguale a un punto [the force or the deduction from the demonstration that a circle becomes equal to a point]" (p. 228). After the radical criticism leveled at Galileo's exposition on the infinite, de Ville felt the need, not inappropriate, to specify that his aim was not "di contrariare, ma di renderne più capace delle Sue propositioni, le quali, essendo alte et difficili et tanto sottilmente probate, non puonno essere capite che con difficoltà [to contradict, but to make myself more capable of understanding His propositions, which, being lofty and difficult and so subtly proven, can only be understood with difficulty]" (*ibid.*). However, de Ville was not actually convinced by Galileo's speculations. This can be inferred from a letter to Galileo from Micanzio, dated 10 March 1635, in which the famous Servite, referring to de Ville, asserted: "Egli non può assentire all'infinito et indivisibile [He cannot assent to the infinite and indivisible]" (*ibid.*, p. 229).

[75] Descartes's criticisms are contained in the long letter to Mersenne, written in Egmond-Binnen on 11 October 1638. Descartes contested the validity of Galileo's proposed solution to the "wheel" paradox and declared that he considered Galileo's proclamation of the equality between a line and a point to be a sophism. He concluded, lastly, with a general and final criticism: "Il manque en tout ce qu'il dit de l'infini, en ce que, nonobstant qu'il confesse que l'esprit humain, estant fini, n'est pas capable de le comprendre, il ne laisse pas d'en discourir tout de mesme que s'il le comprenoit [It is lacking in everything he says about the infinite, in that, despite admitting that the human mind, being finite, is not capable of understanding it, he still goes on to discuss it as if he understood it]" (*Oeuvres de Descartes*. Publiées par Ch. Adam et P. Tannery..., 12 voll., Parigi 1897-1917, *Correspondance II*, Lettera CXLVI, pp. 379-405; the passage quoted is at p. 383). On the Cartesian criticisms of Galileo's conceptions of the infinite, see the recent considerations of W. R. Shea, *Descartes as a critic of Galileo*, in *New perspectives on Galileo* cit., pp. 139-159.

scientist had absolutely no intention of doing.

359 In the examination to which we have previously subjected the demonstration techniques employed by Galileo in *Theorema III* of the *De motu naturaliter accelerato*, another difficulty (or foundational defect) emerged – as should be recalled – which is even more serious. *Theorema III* established that the times of descent along the incline and along the perpendicular of identical elevation are as the incline is to the perpendicular. Starting from the Postulate, Galileo obtained the demonstration by showing that the parallels to the base of the triangle mark points on the perpendicular and on the incline (at equal vertical distances from rest) in which the *momenta velocitatis* are identical (see fig. 36). Now, since "ex punctis omnibus [from all the points]" of the perpendicular one can draw parallels at the base towards the incline, Galileo concluded that along the two planes, the same degrees of speed occur: the times, therefore, will be as the planes[76].

The procedure adopted by the Pisan scientist seems to postulate that, on the incline and on the perpendicular, there is, so to speak, the same infinity of points; in essence, that the incline and the perpendicular are equal. How did Galileo think he could justify this procedure?

The *Giornata Prima* in the *Discorsi* preserves precise traces of Galileo's reflections even regarding this difficulty. It should be recalled that Simplicio, in fact, promptly replied to the theory of the composition of finite magnitudes of indivisibles expounded by Salviati, observing that, if this principle were accepted, it would no longer be possible to distinguish a larger line from a lesser one, since both are composed of infinite indivisibles[77]. The Aristotelian had raised precisely the issue that must be resolved in order to provide a minimal foundation for the techniques adopted in *Theorema III.*

A limited continuum – Galileo asserted – can be divided into as many parts as desired. Note that here the parallels – that is, the *momenta velocitatis* – have nothing to do with the generation of the triangular figure, because the focus is on the instantaneous degrees of speed, not on the total speed (the sum of the degrees). It is, therefore, not necessary to suppose that they are infinite, as in the previous figures. It will be sufficient that they are a number as large as desired, and, nevertheless, "indefinite", but not infinite.

It is beginning to appear clear how Galileo intended to found the procedure adopted in *Theorema III.* It is possible to divide AB and AC (see fig. 36) into an "indefinite" number of parts, always obtaining corresponding sections (the extremes of the parallels) that maintain the proportion between AB and AC. Galileo is operating here, therefore, not at the level of infinity (even though the treatment of
360 the theorem would seem to suggest it)[78], but rather in comparison between always proportional quantifiable parts[79], a comparison that he tried to legitimize, up to the

---

[76] *G.G.*, VIII, pp. 215-217.

[77] Cf. *supra*, pp. 344-345.

[78] Galileo speaks, in fact, of drawing the parallels to the base "ex punctis omnibus lineae AB, usque ad lineam CA [from all the points of the line AB, up to the line CA]" (*G.G.*, VIII, p. 216).

[79] This solution had been proposed to Galileo by de Ville in the letter cited above (cf. *supra*, note 74).

frontier of the "indefinite", in the *Giornata Prima* of the *Discorsi*.

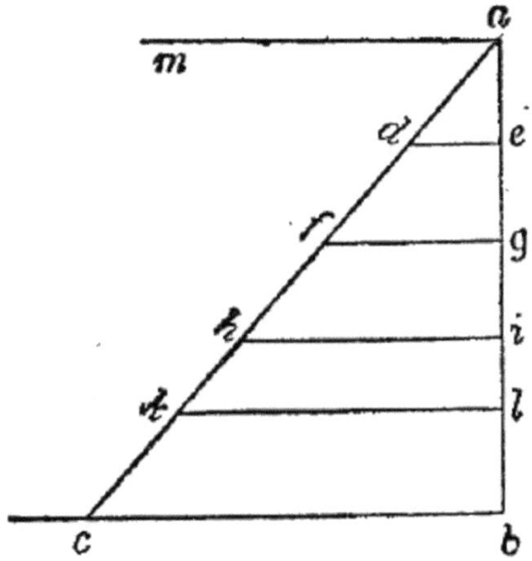

Figure 40

Any remaining doubt about this type of interpretation falls away when attention is focused on a Latin note that has already been examined previously from a different perspective[80]. This is a very early draft of *Theorema III*, given that Galileo still uses the erroneous supposition of the proportionality between acceleration and distance from rest (see fig. 40). The expressions that Galileo uses in the fragment to explain how he understands the correlation between the points of *ab* and those of *ac* leave no room for uncertainty. He states, in fact, that on the perpendicular and on the incline there are "quasi innumera quaedam spaciola [nearly innumerable small spaces]" (therefore not infinite!), equal in number to those that are marked on *ac* "per lineas innumeras parallelas [by innumerable parallel lines]", which start from *ab*. All the terms of each pair of corresponding *spaciola* [small spaces] are to each other as *ab* to *ac*: "intercepta enim spatia *ad*, *df*, *fh* etc. ad spatia *ae*, *eg*, *gi* etc. respondent singula singulis secundum rationem *ac* ad *ab* [for the intercepted spaces *ad*, *df*, *fh*, etc., correspond one-to-one to the spaces *ae*, *eg*, *gi*, etc., according to the ratio of *ac* to *ab*]"[81].

Thus, by avoiding reference to the infinite *momenta velocitatis*, as in *Theorema I*, and taking refuge in the concreteness of the quantifiable parts, Galileo was able to avoid Simplicio's objection, basing the procedure used in *Theorema III* on the principle of the indefiniteness of the quantifiable parts contained in a finite magnitude.

Nevertheless, the difficulty remained. And the objection raised by Simplicio **361**
continued to hang menacingly like the sword of Damocles. The Postulate establishes that the degrees of speed reached on the perpendicular and on the incline at equal

[80] Cf. *supra*, pp. 294 ff.
[81] *G.G.*, VIII, p. 388.

vertical distances from rest are equal. On the other hand, Galileo stated, with the utmost clarity, that speed increases constantly and without jumps: before reaching a certain degree, it passes through the infinite series of smaller degrees. This means (cf. fig. 40) that every point of the perpendicular indicates a degree of speed and that to every point of the incline there corresponds one and only one point on the perpendicular[†]. How can the paradox be resolved?

There remains evidence that Galileo, at some point, attempted to use the method of indivisible components to also compare magnitudes of the same kind, but of different size. This is demonstrated by another passage from the aforementioned letter from Cavalieri from 1634. The Jesuate replies to a series of observations made by Galileo on concentric circles and on the leg and hypotenuse of a rectangle cut by lines parallel to the base according to the method of indivisibles. From the answer, one can guess the concerns that tormented Galileo. Cavalieri specifies that, in order to avoid misunderstandings, he distinguished points or lines of "retto transito [direct transit]" from points and lines of "obliquo transito [oblique transit]", then using only the former and excluding the possibility of using them indiscriminately[82]. This precaution is indispensable, because a line is "cut off" for a certain portion by the perpendicular parallel and for increasingly larger portions as the parallels become oblique:

che poi tanti punti si causino da tutte le parallele, così nella perpendicolare come nell'obliqua, questo non lo negarò, come anco nelle circonferenze concentriche; ma che perciò dovesse dirsi tanto longa l'una come l'altra, mentre volessimo compor le linee di punti, dico che la differenza di questi transiti può cagionare questo, potendosi credere che detti punti siano forsi più diradati nell'obliqua che nella perpendiculare[83]. [that many points are caused by all the parallels, both in the perpendicular and in the oblique, this I will not deny, as also on concentric circumferences; but as for whether one should be said to be as long as the other, if we wish to compose the lines of points, I say that the difference of these transits may cause this, since it is possible to believe that said points are perhaps more sparse on the oblique than on the perpendicular.]

Galileo had thus proposed the case in the form of a paradox to his friend, whom

---

[†] The original text stated "...e che ad ogni punto dell'inclinata ne corrisponde uno ed uno soltanto nell'inclinata [...and that to every point of the incline there corresponds one and only one point on the *incline*]". It's believed that "...one and only one point on the *perpendicular*" was intended.

[82] "Quanto alle circonferenze de' centri concentrici, dico che per liberarmi da questi argomenti che si ponno fare, massime intorno alle linee rette o curve, segate da tutte le linee o da tutti i piani di varie figure, io ho distinto i punti di retto transito da quelli di obliquo transito, si come anco le linee di retto transito e di obliquo transito, non parendomi che si debbano cambiare quelli di retto transito con quelli di obliquo transito; e per misura de' continui ho assunto, per le linee i punti di retto transito, e per i piani le linee di retto transito [As for the circumferences of concentric centers, I say that to free myself from these arguments that can be made, especially concerning straight or curved lines, cut by all the lines or by all the planes of various figures, I have distinguished the points of direct transit from those of oblique transit, as well as the lines of direct transit and of oblique transit, not seeming to me that those of direct transit should be exchanged for those of oblique transit; and for the measure of the continua, I have assumed, for the lines, the points of direct transit, and for the planes, the lines of direct transit]" (*G.G.*, XVI, p. 137). For the definition of "direct" and "oblique transit", see Definition I of Book Two of the *Geometria degli indivisibili* (trad. it. cit., pp. 191-192).

[83] *G.G.*, XVI, p. 138.

he knew to be an expert in these problems, just as Simplicio would do again in the
Giornata Prima of the *Discorsi*. And Cavalieri had replied that, "mentre volessimo
compor le linee di punti [if we wish to compose the lines of points]" (Galileo's
point of view, which Cavalieri rejects in the *Geometria*), it could be granted that **362**
the hypotenuse and the leg are "cut off" in an equal number of points *by all the*
parallels, preserving the difference in length with the expedient of "direct transit"
on the perpendicular and of the "oblique" on the incline.

The Pisan scientist did not accept the suggestion or, at least, did not consider it explicitly proposable. In fact, he never referred to the direct or oblique transit. He preferred to reiterate that no comparison can be made between infinite numbers and, in the case of the perpendicular and the incline, he implied that it was a comparison between quantifiable parts that were not infinite, but indefinite.

In this way, with some uncertainty, some clever concealment, and, above all – by separating the theoretical foundation from the use of the principles stated in the *De motu locali* – he managed somehow to present as plausible and justified the demonstrative procedures and the representation of speed with geometric magnitudes, with its infinite *momenti*, of time and space, infinitely divisible. Ultimately, this was his goal, at least during the years of final drafting of the *Discorsi*.

At this point, our scientist could say he was at least partially satisfied. He had developed a theory of natural motion that accounted for both acceleration and its equality for all bodies. The use of demonstration techniques based on infinitesimal procedures seemed somehow justified and adequate to reflect the very pattern of natural reality. The second book of the *De motu locali* could finally be launched. After many years, Galileo thus decided to reveal the pages that were certainly the most important and most fruitful for subsequent developments that he had ever written.

Chapter VI

"MOMENTUM" IN THE "DE MOTU LOCALI"

1. The Momento of Accelerated Speed

In *Theorema I* of the *De motu naturaliter accelerato*, Galileo had established the 363
equality of distances traversed, for a given amount of time, by a naturally accelerated motion and by a uniform motion with a speed equal to the mean degree of the acceleration. To demonstrate the *Theorema*, Galileo resorted to the comparison between the rectangle and the triangle, constituted respectively by the infinite uniform (constant) *momenti della velocità* and of accelerated speed (progressively increasing in proportion to time). In the now familiar figure (see fig. 41), the space traversed with both motions was indicated by the line CD. By suggesting, not entirely legitimately – as we have seen – that the sum of the infinite *momenta velocitatis* (each different from the other) of the triangle is equal to that of the all-equal *momenta* of the rectangle, Galileo could conclude that, in equal times, uniform motion and naturally accelerated motion would have traversed equal spaces[1].

This conclusion presupposed a precise line of reasoning. To each *momento* of uniform speed there corresponds, in equal times, a constant traversed distance; to each *momento* of accelerated speed there corresponds, instead, in equal times, a different and constantly increasing traversed distance, as it goes from A towards B. Since the sums of the *momenti* of uniform speed and those of accelerated speed in equal times are equal (in the case of *Theorema I*, the uniform motion has a speed equal to the mean degree of the accelerated motion), the sums of the spaces traversed 364
in the same time with all the *momenti* of accelerated speed and with all the *momenti* of uniform speed will also be equal.

The accelerated *grado* or *momento di velocità* is thus clearly presented as the peak speed reached in natural motion at a given temporal distance from rest. This speed is the sum of all the equal increments of speed (also sometimes indicated by Galileo as *momentum*)[2] obtained starting from rest in equal times. Its specific characteristic 365

[1] *G.G.*, VIII, pp. 208-209.

[2] If I am not mistaken, no one has ever pointed out, as seems appropriate to do, that Galileo uses *momentum velocitatis* with a dual meaning in the *De motu naturaliter accelerato* theorems of the

P. Galluzzi, *Momento: Galilean Studies*, Archimedes 78,
https://doi.org/10.1007/978-3-032-35049-7_10

is its infinitesimal duration: a new addition of speed immediately intensifies the *momento.* At the same time, *momentum seu gradus velocitatis* appears to be closely connected to the idea of increasing space traversed in successive equal times from rest. On the one hand, therefore, the expression referred to the "maximam temporis atque motus affinitatem [maximum affinity of time and motion]"[3] (also suggested

---

Discorsi. On the one hand, in fact, the expression indicates the instantaneous speed reached after a certain time of falling. On the other hand, Galileo also uses it when referring to the increase in speed, corresponding to the succession of time instants, and always equal, instant after instant. The dual use is essentially illegitimate or, at the very least, ambiguous. The instantaneous speed is, in fact, the sum of equal instantaneous increments of speed, but in this sense, it should always be referred to in the plural (*momenta*). In contrast, Galileo often uses *momentum velocitatis in* b[†] to indicate instantaneous speed and unequivocally compares it to *gradus*, which he never uses to indicate instantaneous increment of speed. Let's give some examples to clarify the dual use of the term. The definition of naturally accelerated motion offers an example of the use of *momentum celeritatis* in the sense of an ever-equal addition of speed for each instant of time: "Motum aequabiliter, seu uniformiter acceleratum dico illum qui, a quiete recedens, temporibus aequalibus aequalia celeritatis momenta sibi superaddit [I speak of a uniformly or evenly accelerated motion as one which, departing from rest, adds equal momenta of swiftness to itself in equal time]" (*G.G.*, VIII, p. 198). See, for the other use, the conclusion of *Theorema I*: "Patet igitur, aequalia futura esse spatia tempore eodem a duobus mobilibus peracta, quorum unum motu ex quiete uniformiter accelerato moveatur, alterum vero motu aequabili iuxta momentum subduplum momenti maximi velocitatis accelerati motus [It is therefore evident that the spaces traversed in the same time will be equal for the two mobiles, one moving with uniformly accelerated motion from rest, and the other moving with uniform motion according to a momentum half the maximum momentum of speed of the accelerated motion]" (*ibid.*, p. 209). A careful review of the uses of the expression in the *Discorsi* allows us to establish that the most frequent usage is that of instantaneous speed. This is demonstrated by the fact that Galileo, in the geometric schema of the right-triangle, calls the parallels to the base of the triangle *momenta velocitatis.* Each different from the other, these do not indicate the equal increase in speed in equal times, but rather the different instantaneous speeds after a certain time of free fall (the most frequent use even in the *Giornata Quarta* is for *momentum* = instantaneous speed). The ambiguity arises from the fact that the relationship between non-uniform speed and time involves two distinct problems: 1) finding the measurement of that speed at a given instant, and 2) isolating the constant increase in speed for each instant. Galileo believed he could use *momentum velocitatis* to express both concepts, cleverly playing on the verbs that accompany the expression: *momentum*, as an instantaneous increase in speed, accumulates ("sibi superaddit [adds to itself]"), whereas *momentum*, as instantaneous speed, increases ("augentur [is augmented]") as time passes. It is worth noting that Clavelin attributed to Galileo's *momentum velocitatis* only the meaning of an infinitesimal increase in speed: "On ne forcera donc en aucune façon le texte de Galilée si l'on dit que le momentum velocitatis, dont dépend l'accélération de la vitesse, correspond déjà exactement à ce que la mécanique classique appellera le taux de variation de la vitesse. Ce qui prend corps dans cette définition du mouvement naturellement accéléré, ce n'est en fin de compte rien d'autre que la notion de différentielle de vitesse [We will therefore not in any way force Galileo's text if we say that the *momentum velocitatis*, on which the acceleration of speed depends, already corresponds exactly to what classical mechanics will call the rate of change of speed. What takes shape in this definition of naturally accelerated movement is ultimately nothing other than the notion of the differential of speed]" (*La philosophie naturelle de Galilée* cit., p. 303; but see also pp. 302 and 304). Even Thomas Settle did not perceive the dual meaning of *momentum velocitatis.* He – unlike Clavelin – emphasized that the meaning of the expression is that of instantaneous speed. (*Galilean science* cit., pp. 165-167). Lastly, it should be noted that, in one case, Galileo uses the expression *accelerationis gradus* with the meaning of instantaneous speed: "enim assumptum est, accelerationis gradus eadem ratione augeri, nempe secundum temporis incrementum, seu dicas secundum simplicem ac primam numerorum seriem [for it has been assumed that the degree of acceleration increases in the same way, namely according to the increase of time, or you could say according to the simple and first series of numbers]" (*G.G.*, VIII, p. 214).

[†] See the discussion on p. 378 in which *momentum velocitatis* is acquired at *b*.

[3] *Ibid.*, p. 197.

by the figure of *Theorema I*), on the other, it opened the way to a comparison between motions in terms of spaces traversed in equal times[4].

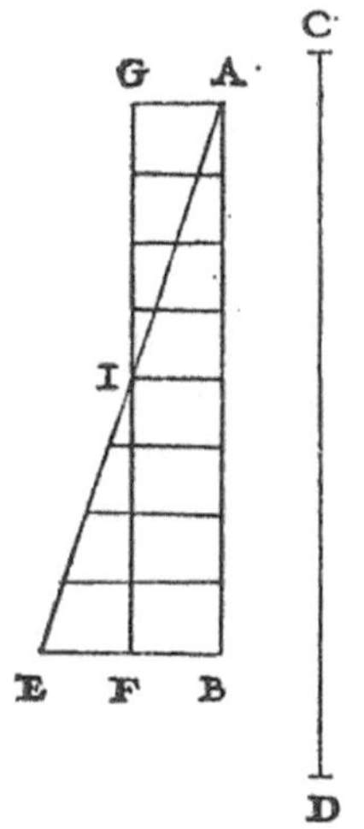

Figure 41

Without focusing on the attempt made by Galileo to justify the comparison between the infinite aggregates of parallels, as illustrated in the previous chapter, we will examine another supposition of considerable weight – also not explicitly presented by Galileo – which sheds considerable light on the operation carried out in *Theorema I* and on the meaning of *momentum velocitatis*. In essence, in the first of the *De motu naturaliter accelerato* theorems, Galileo asserts that speed not only can be divided into an infinite number of *momenta*, corresponding to the speeds reached at each successive instant of time from rest, but, above all, that such *momenta seu gradus* are to be understood not as accelerated speed, but rather as "uniform" speed.

[4] It is worth noting that Thomas Settle believed he perceived, in the use of *momentum velocitatis* in the *Discorsi*, an attempt to bridge considerations based on the relationship between time and speed with those concerning the distances traveled. According to Settle, the fact that in *Theorema I* of the *De motu naturaliter accelerato* (*G.G.*, VIII, pp. 208-209), *momentum* replaces *gradus*, used in the first part, only when Galileo is preparing to conclude that the distances traversed in a given time in natural motion starting from rest will be equal to those traversed with uniform motion, whose degree of speed is equal to half the maximum of the accelerated motion, indicates that *momentum* served a function that *gradus* could not perform: "to accomplish this general purpose of finding information about distances from information concerned initially only with time and speed, Galileo needed to invent or to find some means of transition. For him this role was filled by the notion of *momentum*" (*Galilean science* cit., p. 166). Although it is true that *momentum*, as we have already pointed out when examining the *Mecaniche*, often refers to spaces traversed in equal times, and although it is true that – as we will shortly see – also in the *De motu locali* the *momentum*-space traversed analogy is very strong, one cannot, however, forget that both uses of the term in the *De motu naturaliter accelerato* (instantaneous speed and constant increase in speed, in equal times) indicate a close connection with time, or rather with the instant of time.

Without this supposition, the comparison – indispensable for Galileo – between accelerated motion and uniform motion would have been impossible. How could he have, in fact, reduced a constantly accelerated motion to a uniform motion? Thus, in order to achieve his goal, Galileo suspends, so to speak, the acceleration at an infinitesimal level. However, he avoids begging the question[†]. And he prefers to
366 use it naturally – as if there were no doubts about its legitimacy. He also neglects to point out – as he never will – that the individual *momenta velocitatis* of natural motion are all uniform.

Why, then, these silences? In reality, it is not difficult to provide an answer. Although the supposition was undoubtedly indispensable for reducing the accelerated motion to an equivalent uniform motion, it was, however, incompatible with the very definition that Galileo had given of naturally accelerated motion: a motion in which, starting from rest, there is a constant increase in speed and, before reaching a certain *grado* or *momento di velocità*, it passes through all the infinite smaller degrees without ever stopping at one of them (it should be recalled that Galileo had responded with this same argument to Simplicio's objection on the infinite time necessary to surpass the infinite degrees)[5]. In such a motion, there is never a suspension of acceleration, nor is there ever, not even for an infinitesimal instant, a uniform speed. If this were to happen, then it would take an infinite amount of time to reach any *momento di velocità.*

On the other hand, Galileo cannot limit himself to saying that, by dividing the accelerated speed into an "indefinite" or arbitrarily large number of "instantaneous" *momenti* or *gradi*, they can be assumed to be uniform. He needs to attribute uniform speed to all the infinite *momenta velocitatis* of accelerated motion; otherwise, this "breaks" the possibility of comparing the triangle and the rectangle. He thus prefers once again to gloss over the difficulty, relying on the inattention or benevolence of the reader. Here, then, is a first conclusion: *momentum velocitatis* is predominantly an instantaneous speed; in very few cases, it indicates instead the elementary quantity of speed that the mobile accumulates instant after instant from the beginning of the natural motion.

There is another important observation to make. As the conclusion of *Theorema I* demonstrates, the *momenta velocitatis* have a very close relationship with the distance traversed. By dividing time into equal parts, they directly represent the space traversed. However, in *Theorema I*, the space indice is placed outside the

---

[†] The sentence here originally states, "Egli, però, evita di dichiarare questa fondamentale petizione di principio", which literally translates as, "He, however, avoids stating this fundamental petition-of-principle". Petition-of-principle is related to the informal logical fallacy of "begging the question" (*petitio principii*) such that an argument's premise assumes the truth of a conclusion).

[5] In response to Simplicio's objection, Salviati had responded by observing that this would happen: "quando il mobile andasse per qualche tempo trattenendosi in ciaschedun grado; ma egli vi passa solamente, senza dimorarvi oltre a un istante; e perché in ogni tempo quanto, ancor che piccolissimo, sono infiniti instanti, però son bastanti a rispondere a gl'infiniti gradi di velocità diminuita [when the mobile were to remain for some time at each degree; but it only passes through them, without remaining there for more than an instant; and because in any given time interval, however small, there are infinite instants, they are nevertheless sufficient to correspond to the infinite degrees of diminished speed]" (*G.G.*, VIII, pp. 200-201).

figure, where it remains extraneous, and Galileo does not attempt to emphasize its relation to time and to degrees of speed.

The same situation is repeated in *Theorema II*, in which, from the previous theorem and from *Theorema IV* of *De motu aequabili*[6], it can be deduced that, in naturally accelerated motion, the spaces traversed are as the square of the times spent. AB (see fig. 42) represents the *fluxus temporis* [flow of time], OD, PE **367**
the *gradus velocitatis* and the straight line HI, divided into equal parts, the line of motion, that is, the distance of fall. In this case, not having the need to compare the areas, Galileo even avoids "enclosing" the angle, thus canceling any relationship between the indices of time and speed and that of space, still essentially extraneous to the representation.

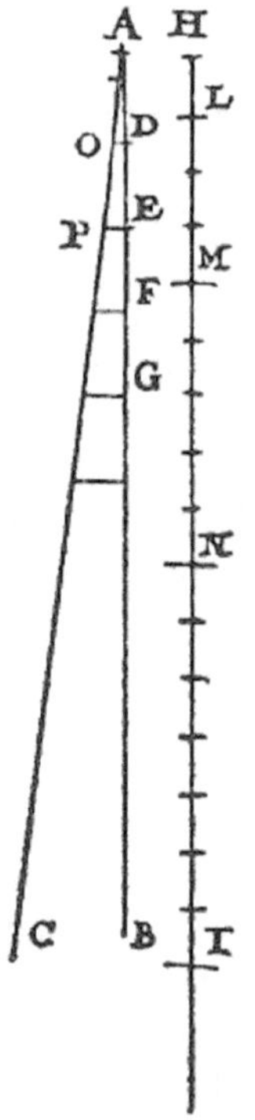

Figure 42

In the subsequent *Corollarium I*, Galileo translates the law established in *Theorema II* into an equivalent formulation: "dum igitur gradus velocitatis augentur iuxta seriem simplicem numerorum in temporibus aequalibus, spatia peracta iisdem

[6] "Si duo mobilia ferantur motu aequabili, inaequali tamen velocitate, spatia temporibus inaequalibus ab ipsis peracta habebunt rationem compositam ex ratione velocitatum et ex ratione temporum [If two bodies move with uniform motion, but with unequal speeds, the spaces traveled in unequal times by them will have a ratio composed of the ratio of the speeds and the ratio of the times]" (*ibid.*, p. 194).

temporibus incrementa suscipiunt iuxta seriem numerorum imparium ab unitate [thus, as the degrees of speed increase according to the simple series of numbers in equal times, the spaces traversed in those same times increase according to the series of odd numbers starting from unity]”[7]. The presentation of this new formulation of the proportions between speed and space, induces Sagredo to intervene to propose an even simpler way of establishing the proportions between *momenta velocitatis* and spaces traversed in equal times. Sagredo’s intervention brings the problem of the representation of space back into focus. Let’s observe figure 43. AO is, as usual, the time line, divided into three equal parts, AC, CI, IO. CB represents the maximum speed reached by the mobile as it descends with accelerated motion from rest from A for the time interval AC. This time, the space indice is completely absent, and so is the lateral line with which it was represented in the first two theorems. Yet, the entire reasoning aims to establish conclusions regarding the proportions between times and distances. In fact, it is a question of showing, using *Theorema I*, in what proportion the spaces traversed with uniform motion are equivalent to the natural accelerated motion in equal successive times from the beginning of the motion. The technique consists of applying the mean speed theorem to the triangle ABC, which represents the free descent from rest for the time interval AC. Since the maximum *momentum velocitatis* reached in this descent is BC, the natural motion can be considered fully equivalent to the uniform motion that occurs for the same time with a speed degree equal to EC (half of BC). The equivalence of the two motions means that equal distances will be traversed in equal times, which, however, Galileo avoids indicating in the figure. Then, the focus shifts to the triangle AFI, that is, to the accelerated motion for twice the time AI, repeating the reduction of the accelerated motion to the uniform motion. If the motion continued uniformly with
368 the degree of speed BC, in the second time interval CI, a space twice that traversed (according to *Theorema I*) in the first time interval AC would be traversed. The degree of speed at I would be represented by GI, equal to BC. But the motion is accelerated; thus, by doubling the time, the *momento della velocità* will also double (they are in fact directly proportional), which will therefore be equal to FI (double of GI). The “actual” acceleration in this second half is, therefore, FG. But we can substitute the accelerated motion with maximum degree FG with a uniform motion with a degree of speed equal to half the maximum, that is NG. We can thus see that, in the second time interval CI, the heavy body moved with a uniform speed triple that with which it traveled in the first time interval AC. “Convince lo spazio passato nel secondo tempo CI dovere esser triplo del passato nel primo tempo CA [It is convincing – as Sagredo concludes – that the space traversed in the second half CI should be triple that elapsed in the first half CA]”[8].

---

[7] *Ibid.*, p. 210.
[8] *Ibid.*, pp. 210-212.

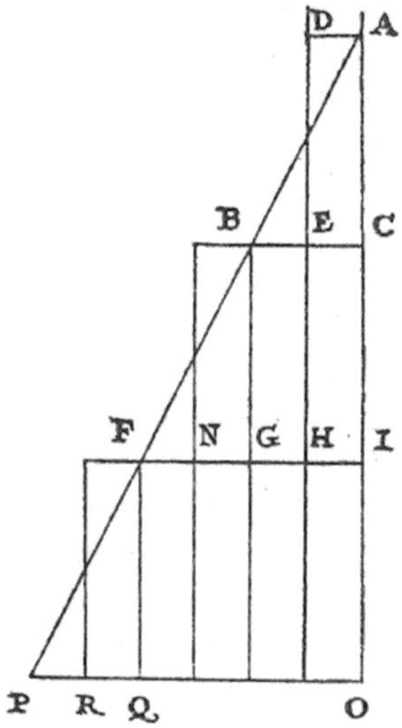

Figure 43

In the *Corollarium*, as in the first two theorems, Galileo reduced the progressively increasing speeds during the time of fall to a series of equal *momenti*, each of which is indicated in the figure by a line parallel to the base of the equal rectangles. He **369**
also stated that each of these uniform degrees, which operates for an instant of time (a point on the time line), corresponds to a (always equal) distance traveled. How else could he have gone from the triple mean degree NI to the triple distance traveled? Yet, Galileo avoids pointing out that the rectangles "represent" both the set of infinite *momenti di velocità* and the spaces traversed with such *momenti*.

It is evident that Galileo does not distinguish, on a "logical" level, between the sum of the infinite *momenti di velocità* (accelerated and uniform) and spaces traversed (always increasing or all equal) in equal times. However, he indicates space with a straight line external to the figure, or even avoids providing its indice. In the figure of *Theorema I* (see fig. 41), just like the one that accompanies Sagredo's proposal, the "visualization" of the relationships between space, on the one hand, and speed and time, on the other, could be obtained by identifying the indices of space in the area of the triangle and the rectangle, that is, by attributing the set of infinite *momenta velocitatis* to the spaces traversed. In the two figures (see figs. 41 and 43), in fact, for whatever sections are taken between the vertex and a parallel to the base (i.e., the uniform or accelerated *momenta velocitatis*), they are always related to each other as the square of the heights (i.e., of the times). Furthermore, the rectangle (i.e., the space traveled with uniform *momenta velocitatis*) is equal to the triangle (i.e., the space traveled with increasing *momenta velocitatis*, the last of which is twice the uniform one). In the last figure (see fig. 43), it was, finally, easy to show that the increase of spaces, in successive equal times, according to the series of odd numbers, with respect to the growth of speed according to the series of natural numbers, could be visualized by the triangle FAI as being triple that of ABC.

Galileo, in short, imagines distance as the "sum" of all the ever-increasing minimum distances covered by all the constantly increasing *momenta velocitatis* (or all the equal minimum distances covered by all the uniform *momenti* of mean speed), but he seems reluctant to indicate, in the figure, that the surfaces "generated" by the infinite *momenta velocitatis* give the proportions of the spaces traversed in equal times.

There are several explanations for this peculiar reticence. First of all, one can see Galileo's hesitation (more than somewhat perplexed by the use of procedures based on the composition of a finite magnitude from infinite non-quantifiable parts) in reducing the space traveled to the "sum" of the infinite degrees of speed corresponding to the infinite instants of time. In doing so, in fact, he would then have had to establish – and we know that he lacked the capacity to do so – the comparison between different spaces traversed with two different series of degrees of speed, but both infinite. Furthermore, he would have found the representation of the space of fall – a linear distance, according to common experience – with an area unconvincing. Without a doubt, this type of "visualization" must not have seemed perfectly adequate to him.

370 There exists, however, an even more convincing reason to explain Galileo's attitude and his hesitation to indicate in the figure space with the "sum" of the *momenta velocitatis*. Of course, it will be enough to consider that, in Theorems I and II, what must be demonstrated is precisely that the spaces traveled in equal times are equal both in accelerated motion and in uniform motion with a degree of speed = half of the maximum *momentum velocitatis*. Galileo was aware that, if he had explicitly declared that the sum of accelerated and uniform *momenta*, i.e., equal triangles and rectangles, "represent" the spaces traversed, he would have been rightly accused of "petere principium [begging the question]", that is, of advancing as a condition what was to be demonstrated. In the figure, in fact, the triangle and the rectangle are equal *by construction.* This is why Galileo carefully avoided pointing out that space could be represented by the area of the triangle and the rectangle. To further "conceal" the close relationship between the spaces traveled and the sum of the infinite *momenta*, he placed a straight line next to the figure, suggesting that it represented the indice of space[9].

But the relationship existed, and Galileo needed to declare it. Here, then – and this confirms the interpretation of a conscious concealment – having now reached a safe distance from *Theorema I* and in a context in which the coincidence between the aggregate of *momenta velocitatis* and the distance traveled was no longer the objective of the demonstration – Galileo candidly proposes it as the most evident and intuitive of truths. This is a text that we have already commented on, the *Scholium* to *Problema IX*, but it is worth rereading now from this perspective. The

---

[9] Thomas Settle was the first to present this quite convincing interpretation (*Galilean science* cit., p. 166, note 9). Settle rightly emphasizes that, strictly speaking, the figure of *Theorema I* presents the problem exclusively in terms of speed and time (*ibid.*, p. 165).

operation is made even more clever by the initial invitation to revisit the figure used in *Theorema I*, that is, the right-triangle ABC (see fig. 44),

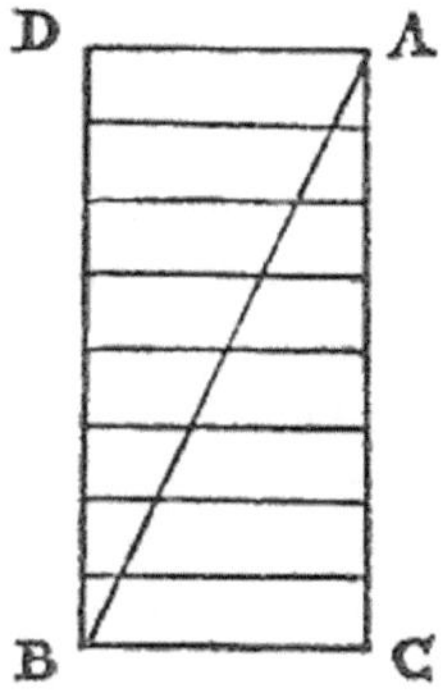

Figure 44

nobis repraesentans in suis parallelis basi BC velocitatis gradus continue adauctos iuxta temporis incrementa, ex quibus, cum infinitae sint, veluti infinita sunt puncta in linea AC et instantia in quovis tempore, exurget superficies ipsa trianguli; si intelligamus motum per alterum tantum temporis continuari, sed non amplius motu accelerato, verum aequabili, iuxta maximum gradum velocitatis acquisitae, qui gradus repraesentatur per lineam BC; ex talibus gradibus conflabitur aggregatum consimile parallelogrammo ADBC, quod duplum est trianguli ABC: quare spatium quod cum gradibus consimilibus tempore eodem conficietur, duplum erit spatii peracti cum gradibus velocitatis a triangulo
ABC repraesentatis. At in plano horizontali motus est aequabilis, cum nulla ibi sit causa accelerationis **371**
aut retardationis; ergo concluditur *spatium CD peractum tempore aequali temporis AC duplum esse spatii AC*: hoc enim motu ex quiete accelerato, iuxta parallelas trianguli conficitur; illud vero iuxta parallelas parallelogrammi, quae dum fuerint infinitae, duplae sunt ad parallelas infinitas trianguli[10]. [represents to us in its parallels to the base BC, degrees of speed continuously increasing according to the increments of time, from which, since they are infinite, just as there are infinite points on the line AC and instants at any time, arises the area itself of the triangle; if we understand motion to continue through just as much time again, but no longer with accelerated motion, rather with uniform motion, according to the maximum degree of speed acquired, which degree is represented by the line BC; from such degrees, there is formed an aggregate similar to the parallelogram ADBC, which is double the triangle ABC: thus, the space that will be completed with similar degrees in the same time will be double the space covered with the degrees of speed represented by the triangle ABC. But on the horizontal plane, motion is uniform since there is no cause of acceleration or retardation; therefore, it is concluded that the *space CD covered in the equal time of time AC is double the space AC*: for the latter motion is accelerated from rest, formed according to the parallels of the triangle; but the former motion, according to the parallels of the parallelogram, which, since they are infinite, are double the infinite parallels of the triangle.]

Each *momento di velocità* thus operates in an instant of time and can be considered a quantity of distance (constant or increasing) that is added to the previous ones. The analogy between *momenta velocitatis* and the spaces traversed with such *momenta* in equal times is very strong. *Momentum velocitatis* is, in a certain sense, the "link" between time and space[11]. Let us resume the comparison

[10] *G.G.*, VIII, pp. 242-243; the italics are ours.
[11] Cf. the considerations made in note 4.

established by Galileo in *Theorema I* between the uniform *momenti* of the rectangle and the accelerated *momenti* of the triangle: "Quod enim deficit momentorum in prima motus accelerati medietate (deficiunt enim momenta per parallelas[†] trianguli AGI repraesentata), reficitur a momentis per parallelas trianguli IEF repraesentatis [For what is lacking in the momenta in the first half of the accelerated motion (for the momenta represented by the parallels of triangle AGI) is restored by the momenta represented by the parallels of the triangle IEF (see fig. 41)]"[12]. Translated into spatial terms, this means that the space covered in the first part of the descent by the accelerated motion is equal to the excess space covered by the accelerated motion in the second part compared to the uniform motion with a speed equal to half the maximum reached in the accelerated motion.

The very close connection between *momentum velocitatis* and the distance traveled in equal times refers, by analogy, to the *momento della gravità*, which – as we have emphasized – Galileo always conceived and represented as a distance, or rather, as the tendency, the propensity, the inclination to complete a given path (with uniform speed) in a given time. The same could be said for the *momentum velocitatis* of the *De motu locali*. In fact, it appears to be nothing other than the tendency, the propensity, the inclination to move quickly (with motion that is still uniform), no longer constant but rather variable according to the temporal distance from rest.

The close analogy between *velocitatis momentum* and distance traveled also provides valuable indications for an evaluation of the relationship between *momentum* and *gradus*, terms that often compete in the *Discorsi*. In general, it seems, in fact, that Galileo entrusted *momentum* with the task of indicating speed as a function of distances traveled in successive equal fractions of time. One might even say that, in this sense, *momentum velocitatis* is indistinguishable from space. The Pisan scientist, however, still feels the need to conceive of speed as an autonomous quantity, distinguishing it from space and time: an "intensive" quantity, endowed with its own *status* and specific dimension. It was certainly the legacy of an old way
372 of conceiving speed, the scholastic one. This type of need was appropriately met by *gradus velocitatis*, which seemed to restore to speed an identity and an independence that, in Galileo's eyes, it risked losing when attributed to the relationship between space and time. The scale of "degrees" allowed us to compare two speeds and determine their greater or lesser "intensity".

This distinction between *gradus* and *momentum* essentially holds to a careful reading of the *De motu locali* theorems. On the other hand, it finds confirmation in the history of the terms[13]. If the relationship is not always so clearly specified, this is because speed, although still conceived in traditional terms, was almost always indicated by Galileo in the *Discorsi* as the distance traveled in an instant of time.

[†] "enim momenta per parallelas" was omited, now restored.

[12] *G.G.*, VIII, p. 209.

[13] *Gradus velocitatis* is a characteristic expression of scholasticism. It indicates the increase (*latitudo* [extent]) of speed, which is an autonomous reality, an independent form (cf. M. Clagett, *La scienza della meccanica*, trad. it. cit., pp. 240-241 and *passim*).

Grado also had to be adapted to express this reality.

2. The Momento or Impeto of Projectiles

The occurrences of *momento* in the *Discorsi* are not limited to the *Giornata Terza* [Third Day]. The use of the term is, in fact, also very frequent in the Third Book *De motu locali*, placed by Galileo in the *Giornata Quarta* [Fourth Day], and entitled *De motu proiectorum*. Galileo's reflections on projectiles date back to his youth and were closely connected from the beginning with his interests in natural motion[14]. The intertwining connection between *de proiectis* considerations and

[14] Galileo's interest in the motion of projectiles is already evident in the *De motu* (*G.G.*, I, pp. 307-315 and 337-340), in which the Pisan scientist tried to demonstrate the substantial equivalence of natural and violent motions. It must have developed further and become more precise in the second decade in Padua, when Galileo devoted himself continually to the study of movement. Writing – as we have seen – to Vinta on 7 May 1610, shortly before his return to Tuscany, Galileo mentioned, in fact, a treatise *De motu locali*, divided into three books. Surely the third was, as in fact happens in the *Discorsi*, the *De motibus proiectorum* (*G.G.*, X, pp. 351-352). The documents we have allow us to affirm that Galileo had discovered the parabolic trajectory of projectiles at the beginning of 1609. This can be deduced from a letter to Don Antonio de' Medici, dated 11 February 1609, in which the Pisan scientist mentions his studies "intorno al moto de i proietti [concerning the motion of projectiles]" (*ibid.*, p. 229). It should be recalled, however, that when Bonaventura Cavalieri, in 1632, included in the *Specchio Ustorio* the assertion of the parabolic nature of the trajectory described by projectiles, Galileo, resentful of the liberty taken by his pupil, who had not consulted him about the appropriateness of divulging the discovery, let it be understood to Cesare Marsili, on 11 September 1632, that he had reached this conviction about forty years earlier: "Io non posso nascondere a V.S. Ill.ma tale avviso essermi stato di poco gusto, nel vedere come di un mio studio di più di 40 anni, conferitone buona parte con larga confidenza al detto Padre [Cavalieri]*, mi deva ora essere levato le primizie, e sfiorata quella gloria che tanto avidamente desideravo e mi promettevo da si lunghe fatiche [I cannot hide from Your Most Illustrious Excellency the fact that such news has been of little pleasure to me, seeing how my studies of more than 40 years, a good part of which I entrusted with great confidence to the said Father [Cavalieri]*, must now be stripped of the first fruits, and have that glory, which I so avidly desired and promised myself from such long efforts, diminished]" (*G.G.*, XIV, pp. 386-387). Galileo's assertion would indicate that the discovery of the parabolic trajectory dates back to the years of the *De motu*, in which, however, there is no trace of in-depth investigations into the trajectories described by projectiles. Probably, in protesting, Galileo deliberately exaggerated the backdating. However, it has been demonstrated that, as early as 1601, Guidobaldo del Monte conducted an experiment that suggested the parabolic trajectory, and that Galileo witnessed the experiment. Furthermore, Guidobaldo's experiment is substantially identical to the one described by Galileo in the *Giornata Seconda* of the *Discorsi* (*G.G.*, VIII, p. 185). It therefore seems fair to conclude that, at the turn of the century, Galileo had already formed the conviction of the parabolic trajectory of projectiles. Galileo's presence at Guidobaldo's experiment is attested by a letter from Cavalieri, in which the Jesuate tried to calm Galileo's resentment, which arose from the passage in the *Specchio Ustorio*, by claiming as justification his belief that the discovery was now in the public domain: "e questo, molto me l'ha fatto credere il sentirla fatta tanto pubblica e per tanto tempo, che l'Oddi mi disse dieci anni sono, ch'ella ne haveva fatto qualche esperienza con il Sig. Guid'Obaldo del Monte [and this, I was very much convinced by hearing it made so public and for so long, that Oddi told me ten years ago, that you had had some experiments with Signor Guidobaldo del Monte]" (*G.G.*, XIV, p. 395; letter of 21 September 1632). It was Fredette who drew attention to Guidobaldo's experiment and the presence of Galileo (Unpublished dissertation cit., pp. 147-160). Naylor later revisited the topic, with a detailed examination of Guidobaldo's experiment and a rigorous comparison with the one described by Galileo in the *Discorsi* (*The evolution of an experiment: Guidobaldo del Monte and Galileo's Discorsi*

the study of natural motion and its laws is most clearly evident in the oft-cited
373 Ms. 72, many pages of which are, in fact, dedicated to this problem[15]. Lastly, it has been demonstrated that some of the *de motu naturaliter accelerato* theorems were suggested to Galileo precisely by the observation of the parabolic path of projectiles[16].

For this reason, a cursory examination of the uses of *momento* in the *Giornata Quarta* of the *Discorsi* cannot be clearly separated from the evaluations we have reached by analyzing the meaning assigned to the term in the *De motu naturaliter accelerato.* Galileo, in fact, intends to extend some of the conclusions reached in the treatment of accelerated motion to the analysis of this type of violent or, in any case, artificial motion, using substantially analogous techniques and concepts[17]. The gap that separates violent motions from natural ones is indeed more apparent than real. Nor are projectiles, in fact, exempt from the invariable law that compels every body, by virtue of its own heaviness, to descend with a constantly accelerated speed from rest. However, they also possess a tendency to move with uniform speed parallel to the plane of the horizon, a tendency impressed upon them by the projecting
374 agent. This "mixture" of the uniformly accelerated natural motion and the equal horizontal motion produces, as Galileo has no difficulty in demonstrating, parabolic trajectories whose amplitude varies according to the intensity of the uniform speed conferred on the projectile by the projector[18]. While it is carried parallel to the horizon by the uniform motion produced by the "violence", the projectile – like every body – descends with increasing speed, covering, in successive equal times, spaces which are to each other as the series of odd numbers.

At this point, after having carefully and precisely specified all the material inconveniences that prevent "real" motions from perfectly corresponding to mathematical laws[19], Galileo can attempt to subject the motion of projectiles to an analysis no less rigorously geometric than that developed for natural falls from rest. In this treatment, the role of *momento* will once again be central, which we will see very frequently coupled with *impeto.*

The fundamental questions are addressed in Theorems II and III. In *Theorema II*, it is shown that, in the case of two uniform motions, one vertical and the other horizontal, "impetus seu momentum lationis ex utroque motu compositae erit potentia aequalis ambobus momentis priorum motuum [the impetus or momentum of the motion composed of two motions will be a result equal to both momenta

---

demonstration of the parabolic trajectory, "Physis", XVI, 1974, fasc. 4, pp. 323-346).

[15] See the oft-cited work of Wisan, who repeatedly emphasizes the intertwining, in Ms. 72, of the *de motu naturaliter accelerato* theorems with those of the *de motu proiectorum.*

[16] See for example, folio 152*r* of Ms. 72 (*G.G.*, VIII, pp. 426-427, only partially published), in which Wisan recognized an attempt to derive the double distance rule (i.e., *Theorema I* of the *De motu naturaliter accelerato*) from the parabolic trajectory of the projectiles (*The new science of motion* cit., pp. 211-214). Favaro had included the fragment of folio 152*r* among the notes relating to the motion of projectiles.

[17] This was an old ambition of Galileo. It would be enough to recall the pages of the *De motu* on which we have previously focused (cf. *supra*, p. 182), in which natural and violent movements are traced back to the same principles.

[18] *G.G.*, VIII, pp. 269-274.

[19] *Ibid.*, pp. 273-279.

of the preceding motions]”. The demonstration clarifies the meaning of Galileo’s words:

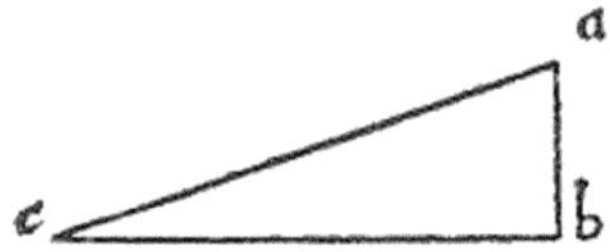

Figure 45

> Moveatur enim aliquod mobile aequabiliter duplici latione, et mutationi perpendiculari respondeat spatium *ab*, lationi vero horizontali eodem tempore confectae respondeat *bc*. Cum igitur per motus aequabiles conficiantur eodem tempore spatia *ab*, *bc*, erunt harum lationum momenta inter se ut ipsae *ab*, *bc*: mobile vero, quod secundum hasce duas mutationes movetur, describit diagonalem *ac*; erit momentum suae velocitatis ut *ac*. Verum *ac* potentia aequatur ipsis *ab*, *bc*; ergo momentum compositum ex utrisque momentis *ab*, *bc* est potentia tantum illis simul sumptis aequale: quod erat ostendendum[20]. [For a mobile moved uniformly by a double motion, and the space corresponding to the perpendicular motion is *ab* (see fig. 45), while the space corresponding to the horizontal motion, completed at the same time, is *bc*. Therefore, since equal spaces *ab*, *bc* are covered in equal time by uniform motions, the momenta of these motions will be in the same ratio as the spaces *ab*, *bc*: the mobile, which is moving according to these two motions, describes the diagonal *ac*; the momentum of its speed is proportional to *ac*. But *ac* in power (i.e., as the square) is equal to the combination of *ab*, *bc*; therefore, the momentum composed of both momenta of *ab*, *bc* is equal in power only to them taken together: which was to be demonstrated.]

If two uniform motions complete two different distances at the same time, it can be said that their *momenti* are as the distances. Here, *momento* indicates speed, which, by definition of uniform motions, is constant. This is why Galileo uses *momentum* in the singular instead of the plural that we are accustomed to. **375**
Whenever a time interval of a uniform motion is compared with an equivalent one of another motion of different intensity, the two *momenta velocitatis* will always be as the speeds of the two motions. These two motions, when combined, give rise to another uniform movement whose *momentum* corresponds to the square root of the sum of the squares of the component *momenti*.

This simple rule, therefore, teaches us how to combine two uniform motions. But the problem that Galileo must face is different: one of the two motions is, in fact, accelerated. The solution to this different question is entrusted to *Theorema III*, in which Galileo attempts to develop and justify a precise technique of representation suitable for the new case under consideration. Everything that is demonstrated in *Theorema III* has already been proven in the *De motu naturaliter accelerato*, but Galileo needs to modify the geometric schema and must propose a justification for it. Our scientist intends, in fact, to indicate the different magnitudes in the natural motion of fall from rest from *a* along *ab* (see fig. 46) using *ac* (= half of *dc*[†]) as

---

[20] *Ibid.*, p. 280.

[†] The text originally indicates “metà della *ab* [...half of *ab*]”; however, Galileo’s *Theorem III* of the Fourth Day does not indicate nor restrict itself to this relationship, but instead recites that *ac* is half the

denoting each the space of fall *ac*, the time to descend from *a* to *c*, and, lastly, as "mensura impetus seu momenti in puncto *c* ex descensu *ac* acquisiti [the measure of impetus or momentum at point *c* acquired from the descent of *ac*]"[21].

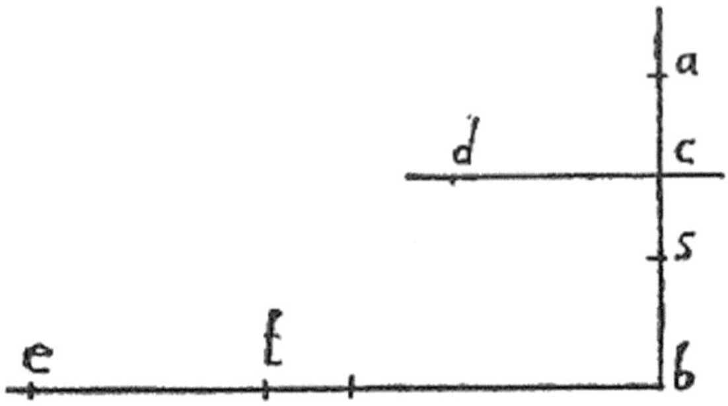

Figure 46

Starting from this new schema – quite different from the right-triangle used in the *De motu naturaliter accelerato* – it is now necessary to establish how to identify in the figure precise and adequate indices for the *momenta velocitatis* reached at *c* and *b*, as well as for the different times taken to reach them.

Developing this approach and, primarily, some of the main conclusions demonstrated in the *Giornata Terza*, Galileo has no difficulty in showing that:

1) The *impetus* at *c*, after the natural descent from rest, can be represented by *cd*, twice that of *ac* (according to *Theorema I* of the *De motu naturaliter accelerato*).

376 2) The time for the natural descent along *ab*, if *ac* is equal to *ac*‡, will be equal to *as*, the mean proportional between *ac*, *ab* (according to the second *Corollarium* of *Theorema II* of the *De motu naturaliter accelerato*).

3) The *impetus* at *b*, at the end of the descent from rest at *a*, is equal to *as*, since speed increases with time. If, therefore, the impeto and time for *ac* were *ac*, the impeto and time for *as* will both be equal to *as*.

4) Thus, the impeto at *b* is represented by *bl*, twice that of *as*. The *momenta celeritatis* at *c* and *b* are therefore to each other as *dc* to *bl*.

5) But *eb* (twice that of *ab*) : *bl* (twice that of *as*) = *ab* : *as*. Therefore *dc* : *bl* = *ac* : *as*.

Therefore, "ut momentum celeritatis in *c* ad momentum celeritatis in *b*, ita *ca* ad *as*, hoc est tempus per *ca* ad tempus per *ab* [as the momentum of swiftness at *c*

---

horizontal length *dc*, while *ba* is half the horizontal *be*, all the while maintaining that as is the mean proportional of *ac* and *ab*. For this reason, "...half of *ab*" is believed to be a typographical error and is now corrected to read *dc*.

[21] *Ibid*., pp. 281-283. Of *Theorema III*, a draft is preserved "della mano giovanile di Galileo [from the youthful hand of Galileo]", according to Favaro (*ibid*., p. 281). On the original purpose of the *Theorema*, cf. Wisan, *The new science of motion* cit., pp. 263-264.

‡ This statement alone appears tautological, but it's presumed to mean, "the time for the natural descent along ab, if *the time for the natural descent to traverse* ac is equal to ac, will be equal to...".

is to the momentum of swiftness at *b*, so is *ca* to *as*, that is, the time for *ca* to the time for *ab*]"[22]. Thus, it is demonstrated that the *momenta velocitatis* at different distances from rest are proportional to the times.

As we anticipated, Galileo did not demonstrate anything new. All these conclusions had been proven, and in a much simpler way, in the previous *Giornata.* But now, he is forced to use a representation very different from the very convenient and easy one of the triangle. And in this new figure, imposed on him by the need to analyze not a "simple" motion but a "mixed" motion, he needs to use the same line to indicate time, *momenta velocitatis* and distance. It is, therefore, his specific duty to show that the laws that regulate the relationships between these quantities in natural motion, demonstrated using the right-triangle schema, can also be demonstrated using this different technique of representation.

Once this first problem has been solved, another immediately arises, still concerning the conception of the geometric schema. How to insert uniform motion into the figure? Natural motion is, in fact, unique and always the same. It follows, in other words, a universal rule for all bodies: spaces increase as the square of the times. Conversely, uniform violent motion can be assigned infinitely many speeds. It is, therefore, necessary to find a way to adequately represent in the figure the various speeds of different intensities that can be assigned to uniform violent motion. As Galileo asserts, "necesse habemus definire aliquam communem mensuram iuxta quam utriusque motus velocitatem, impetum, seu momentum, dimetiri valeamus [we must define some common measure by which we can measure the speed, impetus, or momentum of both motions]"[23]. It is, in fact, essential to "mix" the *gradus velocitatis* of an accelerated motion, proportional to the time of fall from rest, with a single and determined degree of speed "ex illis innumeris [from those innumerables]", which can be assigned to a uniform motion.

The solution devised by Galileo consists in "producing" the determined intensity of uniform speed, using the natural motion of falling.

In the case of naturally accelerated motion along the parabola *ab* (see fig. 47),
the impeto at *b* is "determined" by the height *ac*, "unus enim atque idem est semper 377
impetus mobilis ex eadem altitudine cadentis [for the impetus of a mobile falling from the same height is always one and the same]", according to the supposition premised on the discussion of naturally accelerated motion. How can we "determine" the speed of uniform horizontal motion, in a way that we can "quasi digito monstrare [as if to point to or show with a finger]" in its exact measurement, managing "a reliquis segregare [to separate it from the rest]"? It will be enough to extend *ca*, for example up to *e* (*ae* is the *sublimitas* of the parabola), "ex qua si cadens ex quiete in *e* mente concipiam, patet impetum eius in termino *a* acquisitum, unum esse cum quo idem mobile, per horizontalem *ad* conversum, ferri concepero [from which, if I imagine a fall from rest at point *e*, it is clear that the impetus acquired at the endpoint *a* is the same as that with which I will have conceived the same mobile to be carried, in reverse, along the horizontal line *ad*]". And this determined

[22] *G.G.*, VIII, pp. 281-282.
[23] *Ibid.*, p. 282.

degree of uniform speed will be that and only that with which at the same time it will cover a distance double that of *ea*. The length of *ea* will thus indicate, in the figure, the intensity of the uniform motion (and it will be longer, the faster the violent motion)[24].

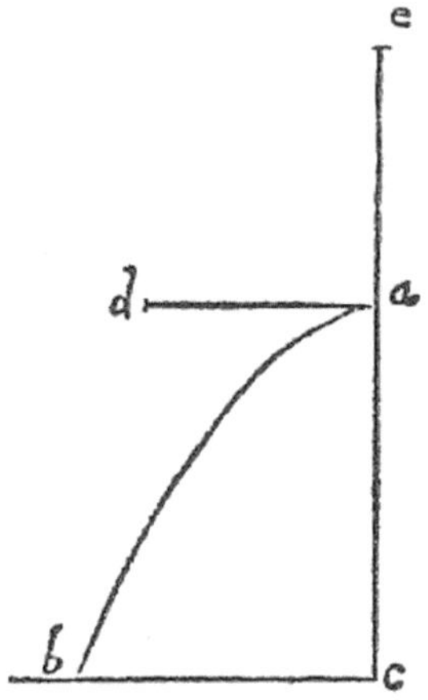

Figure 47

This is quite a peculiar fact. Galileo poses, with the status of a *Theorema*, a series of considerations relating to the construction of a geometric representation schema, different from those previously used, in which the different quantities taken into consideration are conveniently indicated[25].

378 The Pisan scientist is now finally able to propose a precise demonstration to determine the proportions of *impetus* or *momenta velocitatis* at the various points of the parabola, that is, of the trajectory described by the compound motion of the accelerated natural and the uniform violent. Let us take the semi-parabola *bec* (see fig. 48), whose width is *cd* and height *bd* (equal to half of *cd*). Let us assume that *cd* is equal to *ad* (making *ab* the "sublimity" of the parabola and *ib* the degree of speed reached by descending with natural motion from rest at *a* to *b*). We will have *bi* = *ba* = *bd*. We assume that *ab* is a measure of the space of fall, of the time of fall from *a* to *b*, and also of the *momentum velocitatis* acquired at *b*. This *momentum velocitatis*, converted at *b* into equal motion, will allow the mobile, in the same time *ab*, to cover a distance equal to *cd*, twice that of *ab*. On the other hand, falling freely from rest at *b*, the same mobile will obtain, at *d*, after the time *bd* (equal to the time *ab*), a *momento* or *impeto* equal to *bd*. Let us now see how it is possible to "integrate"

[24] *Ibid*., pp. 282-283.

[25] At this point, Galileo cleverly makes Sagredo propose again the "Platonic myth", already presented in the *Discorsi* (*G.G.*, VII, pp. 43-45). The "myth" offers, in fact, a suggestive example of the "production" of different uniform speeds, those of the planetary orbits, through the natural descent from the same point for different times. Thus, the technique proposed by Galileo to represent the motion of projectiles and the two component motions seemed to find confirmation in the "myth" of creation (*G.G.*, VIII, pp. 283-284).

the two *momenta*, at *b* and at *d*, by resolving them into that of the parabola at *c*, composed of both. If *ab* is a measure of the *momentum* acquired at *b*, then *bi* (equal to *ab*) can then indicate the equal *momento* acquired at *d*. Both *momenti*, at the instant when the mobile is located at *c*, are of uniform speed. By *Theorema II*, the "composite" *momento* of two uniform motions is just the square root of the sum of the squares of the component *momenti*. Therefore, since the *momenti* to be composed correspond to the legs of the right-triangle *abi*, the hypotenuse *ia* will be "quantitas seu mensura integri momenti, quo proiectum veniens per parabolam *bc* impetum facit in *c* [the quantity or measure of the integral momento, with which the projectile, coming along the parabola *bc*, makes its impetus at *c*]"[26].

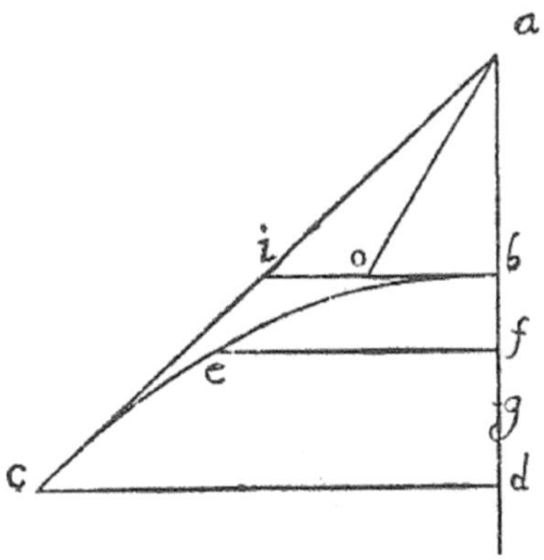

Figure 48

Theorema III allows us to establish the *impeto* or *momento della velocità* at 379
any other point of the parabola described by the "mixed" motion of the projectiles. Let *e*, for example, be the point of the parabola at which we intend to measure the "integral" *momento*. It is clear that the uniform *momento della velocità* is constant. It can, therefore, also in this case, be represented by *ab*. The *momento della velocità* of the natural motion at *e* will correspond, instead, to that reached at *f*, falling freely from rest at *b*. This *momento* will be measured by *fg*, the mean proportional between *bf*, *bd*. Let *ob* = *bg* be taken on *ib*, and join *o* to the vertex *a*. Again, according to *Theorema III* of *Giornata Quarta*, *oa* is, therefore, the *momento della parabola* at *e* (see fig. 48):

est enim *ab* determinatrix posita temporis et impetus in *b*, qui conversus in horizontali semper servatur idem; *bo* vero determinat impetum in *f* seu *e* per descensum ex quiete in *b* in altitudine *bf*; his autem *ab*, *bo* potentia aequipollet *ao*.[27] [for indeed *ab* is set as the determinant indicator of time and impetus at *b*, which, when reversed on the horizontal, always remains the same; but *bo* determines the impetus at *f* or

[26] *Ibid*., p. 285. In figure 48, the variation of the "composite" *momento* is indicated by the variation of the hypotenuse of the right-triangle *aib*, up to the point where it coincides with the perpendicular leg *ab*.
[27] *Ibid*.

e by descent from rest at b at the height bf; moreover ab, bo are equivalent in the resultant ao.]

It follows that the "integral" *momenti* or *impeti* of the projectiles are indicated by segments (*determinatrices* [determinants]). No longer parallel, just like the representing segments in the right-triangle, the *momenta velocitatis* of natural motion, the "determinants" of the "composite" *momento* of the projectiles are, however, also infinite. In fact, they correspond to the infinite segments drawn from the infinite points of ib to the vertex a. The triangle iab contains the infinite aggregate of such segments, which can be smaller and smaller infinitely than ai, the maximum "integral" *momento*, depending on whether a smaller projection time is taken (and we know that a given time can be infinitely divided). The extreme decrease of ai will occur at ab, that is, when the projection time is nullified.

Momentum velocitatis – as we had anticipated – also plays a central role in the motion of projectiles. The fundamental concept that allows the comparison, first, and the composition, then, of uniform and accelerated motion is, in fact, precisely that of *momento della velocità*, in the same sense, in essence, that we illustrated in the *Giornata Terza*. The operation of "integration" (Galileo will call the result of the component *momenti* "composto [composite]" or "integro [integral]" *momento*) of the two motions finds justification only at the level of infinitesimal *momenti* of speed. And this operation, to be possible, again requires the undeclared fundamental supposition that Galileo used throughout the *De motu naturaliter accelerato*: the *momenta* of natural motion are all *momenti* of uniform speed. Given this, there is no difficulty in integrating all the *momenti* or *impeti* of uniform speed with all the *momenti* or *impeti* of accelerated speed. At the level of the instant in time, both motions are, in fact, uniform.

We have previously emphasized how Galileo, with the notion of *momento di*
380 *velocità*, practically reduced accelerated natural motion to a series of very small equal times during which the motion is uniform, but, in each successive time, with greater speed. In our demonstration, the comparison appears to be made even more plausible through the distortion, so to speak, of the uniform motion. It is, in fact, introduced with its specific *grado* of speed as the consequence of an accelerated motion from which the cause of the acceleration is suddenly suppressed. Thus, the problem became even simpler. The summation concerned *momenti di velocità* of motions that were both naturally accelerated.

The construction that Galileo built to demonstrate his *de motu proiectorum* theorems is, therefore, the product of a new, profound foundational effort. The difficulties, uncertainties and concerns of the Pisan scientist faced with the task of comparing two speeds that are different not only in intensity but also in type (one accelerated, the other uniform) are clearly evident in the pages that follow the statement of *Problema I*. Galileo clearly explains the complexity of the problem he faced and the considerations that led him to adopt the geometric representation and demonstration techniques illustrated above. The greatest difficulty, in essence, consisted in "representing" with an analogous criterion (in such a way that they could first be compared, then "composed") two different speeds: one with degrees of constant intensity, the other, instead, accelerated according to the proportion of time from rest.

Galileo had compared (but not composed) two similar speeds in the *Giornata Terza* using the comparison between the right-triangle (for accelerated speed) and the rectangle (uniform speed). Only, in that case, he had made the uniform speed correspond either to the maximum degree of speed reached in the accelerated motion, or to half of that degree. Here, however – in addition to the fact that it is necessary to be able to "visualize" the effect of the composition of the two speeds – the degree of uniform speed cannot be subjected to such limitations. We need a representation schema in which it is possible to insert the indice of a uniform speed degree of any size[28]. For this, Galileo introduces the *sublimitas* of the parabola and generates the determined uniform motion from a natural descent of the duration necessary to acquire that degree, then deviating, at that point, the movement along a plane parallel to the horizon, so that the speed remains uniform[29].

But this is not the only reason that induced Galileo to present uniform motion as **381**
the consequence of a natural descent abruptly converted, at a certain point, into equal motion. Galileo must, in fact, find a common criterion of measurement for the two component motions, and this criterion must function at the level of their respective *momenta velocitatis*. And the stratagem of generating uniform motion through free fall provides an absolutely clear and precise one: time. In fact, Galileo obtained two *momenti di velocità* of two accelerated motions. The measure of these *momenti* will, therefore, appear to be the time of fall from rest. The "integral" *momento* will thus be the result of the composition of the times of the two accelerated motions. This is, in fact, what happens in the demonstration of *Problema I*. The two legs of the triangle *abi* represent the time it takes for the projectile to fall to reach the uniform speed and the time it takes for it to fall naturally from rest. One need only observe that the second time was a "real" time, while the first was the result of a complete abstraction, nor did it present any reference to the actual unfolding of the

[28] "Resta poi il trovar modo di determinare anco la quantità dell'impeto in un moto equabile in guisa tale che tutti coloro che circa di quello discorrino, si formino l'istesso concetto della grandezza e velocità sua, sì che uno non se lo figuri più veloce e un altro meno, onde poi nel congiugnere e mescolar questo da sé concepito equabile con lo statuito moto accelerato, da diversi uomini ne vengano formati diversi concetti di diverse grandezze d'impeti [It remains then – as Galileo had specified, after having clarified that the natural motion from rest of all heavy bodies is always the same – to find a way to also determine the quantity of the impeto in an equal motion in such a way that all those who discuss it will form the same concept of its magnitude and speed, so that one does not imagine it faster and another less, so that then in combining and mixing this conception as equal with the established accelerated motion, different concepts of different magnitudes of impeti are not formed by different people]" (*ibid.*, p. 286).

[29] "Per determinare e *rappresentare* cotal impeto e velocità particolare non ha trovato il nostro Autore altro mezo più accomodato, che 'l servirsi dell'impeto che va acquistando il mobile nel moto naturalmente accelerato: del quale qualsivoglia momento acquistato, convertito in moto equabile, ritien la sua velocità limitata precisamente, e tanta, che in altrettanto tempo quanto fu quello della scesa passa doppio spazio dell'altezza dalla quale è caduto [To determine and *represent* this particular impeto and speed, our Author has found no other means more suitable than to use the impeto that the mobile acquires in naturally accelerated motion: any momento acquired by it, converted into equal motion, retains its speed precisely limited, and such that in the same time as that of the descent, it passes double the distance of the height from which it fell]" (*ibid.*, pp. 286-287; the italics are ours).

motion of the projectiles[30].

After having emphasized – as we did in the previous paragraph – the analogy of the *momenta velocitatis* with the spaces traversed in equal times, we must now place emphasis on the close relationship between *momentum velocitatis* and the time of fall from rest. In these pages of *Giornata Quarta*, *momento* and *impeto* are, in fact, all but indistinguishable from time. What Galileo composes are "times", and the "integral" or "composite" *momento* ends up being a function of the "real" time of projection and the "artificial" time necessary to reach the uniform speed of the projectile.

This entails a different technique for measuring speed compared to the one predominantly used in the *Giornata Terza*: the *grado* or *momento* or *impeto di velocità* is not a function of the space traveled in equal times, but rather a direct expression of time. To achieve this result, it was necessary to reduce uniform motion to a naturally accelerated motion. In uniform motion, in fact, it is impossible to express the relationships between the different speeds simply in terms of time.

On the contrary, in the *De motu naturaliter accelerato*, where the problem was
382 to establish the relationships between the speeds of natural motions from rest, in equal successive times, in terms of distances traveled, Galileo had been forced to reduce the infinite increasing *momenta velocitatis* to equivalent uniform speeds (*Theorema I*). Only in uniform motions, in fact, when equal times are taken, does it appear indisputable and evident that the speeds (the set of infinite *momenti* of uniform speed) are as the distances traversed.

It does not seem legitimate, therefore, to link Galileo's *momentum velocitatis* inextricably to either space or time. The shift that the expression presents towards these two poles clearly derives from the different contexts in which Galileo used it. *Momentum* had, from the beginning, a twofold value. In fact, it constituted a kind of link between two ways of considering speed: as a function of distance and of time.

Throughout *Giornata Quarta* we have been observing a very high frequency of *impeto-impetus* and a continuous relationship of the term with *momentum velocitatis*. *Impetus* – as is worth emphasizing – was completely absent from the *de motu naturaliter accelerato* theorems. Both of these phenomena appear easily justifiable. The high frequency of *impeto* finds, in fact, a logical explanation in the object of Galileo's attention in the *Giornata Quarta*: the motion of projectiles, which produce

[30] The decisive function played by time after the reduction of equal motion to an equivalent natural acceleration is clearly indicated by Galileo: "il che non si può determinare senza la considerazione del tempo decorso dal principio della composizione de i 2 moti, la qual considerazione di tempo non si richiede nella composizione de i moti equabili, le velocità ed impeti de i quali son sempre i medesimi; ma qui, dove entra nella mistione un moto che, cominciando dalla somma tardità, va crescendo la velocità conforme alla continuazion del tempo, è necessario che la quantità del tempo ci manifesti la quantità del grado di velocità nell'assegnato punto [which cannot be determined without considering the time elapsed from the beginning of the composition of the two motions, which consideration of time is not required in the composition of equal motions, the speeds and impeti of which are always the same; but here, where a motion enters into the mixture that, starting from the greatest slowness, continues to increase in speed in accordance with the continuation of time, it is necessary that the quantity of time manifests to us the quantity of the degree of speed at the assigned point]" (*ibid.*, p. 289).

blows, percussion, violent impacts on the obstacles that stand in their way. Even though he conducted a purely geometric-mathematical discussion, this background was influential. Galileo was therefore prompted to use terminology appropriate to the topic being discussed. And *impeto*, in this sense, is already a "classical" term.

On the other hand, the use of *momentum* and *gradus* must be maintained in order to emphasize that, in the treatment of the motion of projectiles, one proceeds with the same rigor as in the previous exposition and to legitimize the use of conclusions demonstrated in the *Giornata Terza.* Thus Galileo uses the three terms interchangeably, and we find *impeto* even in the plural, with exactly the same meaning that we have recognized as customary for *momento* and *grado* in the *De motu naturaliter accelerato*. However, *momento* never occurs when the effect of speed is explicitly emphasized. In these cases, *impeto* appears on its own, always conceived as proportional to the *momento* or *grado di velocità.* Galileo, for example, speaks of "impeto e percossa in *e* [the impeto and percussion at *e*]" which "sarà maggiore della percossa in *b* [will be greater than the percussion at *b*]"[31]. Or he alludes to "questi impeti, colpi o vogliam dir percosse di tali proietti [these impeti, blows or we might say percussions of such projectiles]"[32]. The proportional relationship between *momentum* or *gradus* and *impetus* (the impact produced by such speed) appears explicit in other passages: "*ia* erit ergo quantitas seu mensura integri momenti, quo proiectum veniens per parabolam *bc* impetum facit in *c* [*ia* will therefore be the quantity or measure of the integral momento, with which the projectile, coming through the parabola *bc*, makes its impetus at *c*]"[33]. In essence, the "integral" *impeto*, like *momento*, is a function of the times of the two component **383**
motions[34].

After the clarifications of recent Galilean historiography, it is almost unnecessary to specify that Galileo's *impeto* has very little to do with the *impetus* of the Parisian physicists or with the dynamic conceptions of Buridan[35]. The analogy is, in fact, only apparent. Galileo's *impeto* is almost always – but not always – the physical effect of speed[36]. It is as a consequence of the progressive increase in speed that the

[31] *Ibid.*, p. 290.
[32] *Ibid.*, p. 291.
[33] *Ibid.*, p. 285.
[34] For a different evaluation of the relationship between *impeto* and *momento*, cf. the essay by Altieri Biagi (*Galileo e la terminologia* cit., pp. 55-59). It should be noted, however, that Altieri Biagi essentially limits herself to comparing the two terms in their "mechanical" meaning (*impeto* and *momento della gravità*), neglecting to express a judgment on the relations between the two terms in relation to the use made of them by Galileo in the *Giornata Quarta* of the *Discorsi.*
[35] Cf. *supra*, pp. 178-179 (note 94).
[36] Note, in fact, that in a few cases, Galileo resorts to *impetus*, in the *Giornata Quarta*, assigning it the value of the cause of motion and speed. This occurs in the statement of *Propositio VIII*: "Amplitudines parabolarum a proiectis *eodem impetu explosis factarum*… [The amplitudes of the parabolas made by projectiles *thrown with the same impetus*…]" (*ibid.*, p. 297; the italics are ours). This use is found again, twice, in Sagredo's speech, which precedes the presentation of the *tabula* of shots: "Questa vedrò io molto volentieri, mentre che per essa potrò venir in cognizione della differenza de gl'*impeti* e delle forze che si ricercano per cacciar il proietto nella medesima lontananza con tiri che chiamano di volata... se altri volesse alla elevazione di 3 o 4 gradi, o di 87 o di 88, far cadere la palla dove fu cacciata alla elevazione di 45 (dove si è mostrato ricercarsi l'impeto minimo), credo si ricercherebbe un eccesso

impeto increases; the *impeto* does not react upon the speed, causing it to intensify. *Impeto* is not, namely, a dynamic force, but rather a simple effect of movement, an effect to be understood in relation to the resistances opposing the movement. There are also different *gradi* or *momenti* of *impeto*, each of which is directly proportional to the corresponding *momento* or *grado di velocità*. We have seen the expression “grado dell’impeto [degree of impeto]” in the letter to Sarpi. In the *Giornata Quarta* of the *Discorsi*, Galileo prefers to express himself in terms of *quantitas impetus*[37], “quantità dell’impeto [quantity of impeto]”[38].

* * *

It now seems appropriate to attempt a general assessment of the meaning presented by the different values that we have seen *momento* take on in Galileo’s writings illustrated thus far.

384 It should be recalled that the term entered Galileo’s lexicon in the context of the mechanical investigations carried out in Padua during the last decade of the 16th century, with the specific meaning of inclination produced by weight (*momento della gravità*). We have also endeavored to show that, especially in the second decade in Padua, Galileo attempted to employ the “static” concept of *momento* as a decisive element in the construction of a new science *de motu naturaliter accelerato*. From this attempt we have, first, seen the expression *momentum velocitatis* emerge as a direct derivation of *momentum gravitatis*. Then, having become aware of the inability to develop, starting from heaviness as the cause of motion, a justification for the acceleration that is equal for all bodies, Galileo progressively conferred greater autonomy to *momentum velocitatis*, which ultimately came to be configured as an instantaneous speed, that is, as a uniform speed of infinitesimal duration (or, also, the space traversed with such a speed during such an infinitesimal duration). Thus conceived in this way, and made proportional to time, *momenta velocitatis* became the cornerstone element of the science *de motu*. Galileo had entrusted, namely, this concept with the task of making the acceleration of the natural motion of falling rationally comprehensible.

In the history of the Galilean *momento*, the “critical” phase is therefore constituted by the abandonment of the attempt at a mechanical foundation of movement. From that point on, the development of the concept of *momentum velocitatis* follows a completely autonomous path. Galileo deliberately suppresses

---

immenso di forza [I will see this very willingly, as through it I can come to know the difference in the *impeti* and forces required to project the projectile to the same distance with shots that they call “volata [volleys]”... if someone wanted an elevation of 3 or 4 degrees, or 87 or 88, to make the ball fall where it was projected at an elevation of 45 (where it has been shown that the minimum *impeto* is sought), I believe an immense excess of force would be required]” (*ibid.*, p. 306; the italics are ours). This observation indicates, against easy generalizations, how complicated a careful lexical investigation is for authors who experience significant innovation, such as Galileo, and how some traditional meanings of technical terms tend to survive, almost unconsciously, even for authors who have contributed to dismantling the theories of which they were an expression.

[37] *G.G.*, VIII, p. 289.

[38] *Ibid.*, p. 285.

any reference to heaviness (and its *momenti*) and entrusts the growth of the *momenta velocitatis*, equal for all mobiles starting from rest, to a natural law, established, ultimately, by God.

Now, in the Latin pages of the *Giornata Terza* of the *Discorsi*, *momentum*, in addition to appearing strictly connected to speed and time, has come to take on an increasingly evident infinitesimal connotation. The term has come to mean, in fact, "atom", "point", "minimum quantity" (of time – corresponding to the duration of the natural fall of a given "uniform" speed, immediately supplanted by another greater one – and, at the same time, of space – corresponding to the distance, also infinitesimal, traversed with such uniform speed in the atom of time).

As the analysis of the traditional uses of *momento* carried out in Part One of our work has shown, alongside the value of *momentum* = "inclination", there was another, no less widespread, corresponding to "point", "atom", infinitesimal element[39]. The context in which we have verified the use of the term in this sense is that – as very well known and certainly long pondered by Galileo – of the sixth book of Aristotle's *Physics* and of the main commentaries on those fundamental pages. In this context, the discussion is centered on the composition of the continuum, the
nature of infinity, and the infinite divisibility of space, motion and time. The Latin **385**
translators and commentators of these pages had generally used *momentum motus* to indicate the indivisibles of movement. The *momenta* of motion corresponded to the points of the line and to the instants of time. On the question of whether motion is composed of *momenta*, we have seen a great debate taking place, during which different proposals were put forward. Themistius, Simplicius and other famous commentators had insisted on these themes. But even widely circulated texts such as the commentaries on the *Physics* by Toledo and the Conimbricenses had reported extensively on the debate, in which – as it is worth remembering – the motion under analysis was uniform, not accelerated, and the *momentum* indicated an infinitesimal of constant motion.

As far as the investigation carried out in Part One allows us to conclude, the two most common meanings of *momento* (the "mechanical" one of "inclination" and that of "small quantity", in general, applied – as we have seen – to motion, to time, but also to weight)[40] are clearly distinct in the tradition, even if not necessarily irreducible. In the historical aspects of the Latin term reconstructed in Part One, it is not easy to find grounds for a convincing reunification, nor to maintain that the two values had a common origin. Moreover, in Greek, the two concepts were expressed with distinct and completely independent terms (*ῥοπή* and *κίνημα*). Galileo himself – as we have made evident – possesses and uses *momento* in the *Mecaniche*, at least once, with the meaning of minimum quantity[41], without realizing that in that context it can generate several misunderstandings, lending itself to being confused with the abundantly used technical value of "inclination" produced by weight.

---

[39] Cf. Ch. IV of Part One.
[40] Cf. Ch. II of Part One.
[41] Cf. *supra*, pp. 202-203 and 209.

Nevertheless, it is difficult to avoid the impression that the Pisan scientist, at a certain point, made a conscious shift from one meaning to another. This must have happened as a consequence of the profound "crisis" brought about by the abandonment of the mechanical approach in the science of movement. Having eliminated all relation between speed and heaviness and its *momenti*, why did Galileo not discard the *momentum velocitatis*, which was the most evident indice of such a foundational effort?

It must have been then, while preparing to give a rigorously geometric-mathematical treatment of natural motion, whose speed continually increases in proportion to time, that Galileo discovered that *momentum* could still be of use to him. He could, in fact, have liquidated it and resorted to *gradus*, or used new terms. Instead, he kept *momentum* and even made it a central term.

Of course, the meaning assigned to the term had profoundly changed. The
386 "dynamic" *momento* had no relation to the heaviness of the mobile. It was measured only in terms of time from rest. It became the infinitesimal element of uniform speed whose sum composed the accelerated speed. Moreover, each *momentum velocitatis*, constituting an infinitesimal of natural motion, corresponded perfectly to one of the infinite instants composing the time of fall. Lastly, it could also be "conceived" as the minimum amount of space traversed in such an infinitesimal amount of time.

In this precise context, the term was recovering, by curious fate, meanings different from the "mechanical" ones, but no less documented by tradition. And Galileo, by reproposing it in this new guise, could have a twofold advantage: 1) to use a "technical" term, already extensively tested in the "classical" discussions on the composition of continuous quantities (space, motion, time), a term that immediately recalled the analogy with the instant of time, to which Galileo was continually comparing it; 2) to keep open, albeit in an ambiguous form, the possibility of reproposing the mechanical foundation of movement, that is, the possibility – in the event that he had succeeded in solving the problems of general foundation – of presenting *momentum velocitatis* as a derivation of *momentum gravitatis*. Ultimately, Galileo never abandoned the belief that the cause of natural motion was to be found in heaviness, and although he did not openly display this opinion in the *Discorsi*, it was only because – as we have seen – he was unable to convincingly deduce from it the equality of all natural motions from rest.

If the observations we have been making have – as it seems to us – a foundation, the history of Galileo's *momento* appears quite significant and notably instructive as compared to the brief analyses or lexical mechanics of keen and high-level authors, who are also quite adept at muddying the waters and circumventing difficulties with clever writing, just like Galileo. We are, in fact, in the presence of a lexical "case" of enormous complexity. Initially, our author uses the term with the two primary meanings documented by tradition, showing that he considers them clearly distinct and irreducible. Later, he coins the expression *momentum velocitatis* modeled on that of *momento della gravità* (i.e., from *momento* = "inclination"). In a third phase, he finds that he has no sufficient reason to support the derivation of *momentum velocitatis* from that of heaviness. He, therefore, disconnects them but does not suppress the expression. Instead, he maintains it, assigning it, however, the different meaning of infinitesimal of *uniform* speed. In doing so, *momentum* recovers the other value (*momento* = infinitesimal quantity) with which Galileo uses it abundantly in the *De motu locali*.

All this happens without the slightest warning being given to the unfortunate reader. Indeed, not even the most attentive and perceptive of readers would have been able to grasp the successive stages of this complex operation – just recall the example of Mersenne, who showed that he did not understand the meaning of *momentum velocitatis*[42]. On the basis of Galileo's printed works alone, it is, **387**
indeed, almost impossible to reconstruct the long-attempted mechanical approach and, therefore, the original genesis of *momentum velocitatis* modeled after that of heaviness. Without the *De motu*, the *Mecaniche* and, especially, the so-called *Frammenti*, the two *momenti* of the *Discorsi* appear to be distinctly separate and independent entities, nor is it easy to suspect that they were at a certain point closely connected.

Before declaring this peculiar affair concluded, we should make one last point. The cycle of use of *momento* is not, in fact, completed. And the last page of this story brings non-negligible confirmation to the division that we have proposed. *Momento della velocità*, in its infinitesimal dimension and very close relationship with time, will ultimately offer, through a peculiar reversal, the model for a radical revision of the *momento della gravità*. With this, we reach the force of percussion and the conclusion of our work.

[42] Throughout the abbreviated version of the *Discorsi*, Mersenne will systematically use *degrez de vitesse* [degree of speed] to render *momentum velocitatis*. See, for example, the definition of naturally accelerated motion ("...temporibus aequalibus aequalia celeritatis momenta sibi superaddit [at equal times, it adds equal momenta of swiftness to itself]": *G.G.*, VIII, p. 198), in which the French translator rendered by passing from *momento* to *grado*: "Le mouvement naturel qui augmente sa vitesse uniformement et esgalment, est celuy qui depuis son repos jusques à la fin de sa cheute acquiert des degrez esgaux de vitesse en des temps esgaux [The natural movement which increases its speed uniformly and equally is that which, from its rest until the end of its fall, acquires equal degrees of speed in equal times]" (*Les nouvelles pensées de Galilée*, ed. crit. cit., I, p. 104).

CHAPTER VII

THE MOMENTO OF PERCUSSION

Galileo gave attention to the force of percussion on several occasions during 389
his very long career as a scientist. The first interests related to the problem of percussion appear in the *Mecaniche*. In the concluding pages of the Paduan treatise, he presented some brief considerations "della forza della percossa [on the force of percussion]" and its "effetto veramente meraviglioso [truly marvelous effect]"[1]. In reality, these were just some methodological suggestions on how to approach the problem. In concluding, Galileo promised to respond to the doubts and difficulties that such indications would inevitably have raised "con poca fatica [with little effort]" in the "problemi meccanici, che in fine di questo discorso si aggiungeranno [mechanical problems, which will be added at the end of this discourse]"[2]. The

[1] *G.G.*, II, pp. 188-190.

[2] *Ibid.*, p. 190. The *Problemi*, which Galileo refers to here for the first time, constituted a project to which he thought of dedicating himself on various occasions. Probably following the model of the pseudo-Aristotelian *Questions of Mechanics*, the Pisan scientist, in essence, intended to add an interpretation of some practical phenomena to the *Mecaniche*, using the same principles illustrated in the Paduan treatise. Galileo mentioned the *Problemi* project on several occasions. In fact, in the famous letter to Vinta of 7 May 1610, referring to the *Mecaniche*, he specified that the treatise was divided into three books "due attenenti alle demostrazioni de i principii et fondamenti, et uno de i problemi [two concerning the demonstrations of the principles and foundations, and one concerning the problems]" (*G.G.*, X, p. 352). Then, no news until 1637, when, having finished writing the *Discorsi*, the Pisan scientist returned to work on the works to be completed. On 24 April and 7 November of that year, he communicated to Diodati his intention to make "al netto... una parte dei problemi parte fisici e parte matematici [a fair copy of... a portion of the problems, partly physical and partly mathematical]" (*G.G.*, XVII, pp. 63 and 213). In another letter to Diodati, dated 23 January 1638, Galileo informed his friend that he had "buon numero di problemi e questioni spezzate, tutte, al mio consueto, nuove e con nuove dimostrazioni confermate [a good number of problems and miscellaneous questions, all, as usual, new and with new confirmed demonstrations]" (*ibid.*, p. 262). However, Galileo did not complete his project. Only a few fragments of the *Problemi* written by Galileo at various periods of his life have reached us. Instead, we have preserved ten *Problemi o questioni spezzate* which, according to Viviani's testimony, were drawn up by Vincenzo Galileo "su le soluzioni spiegategli dal proprio padre già cieco, in alcuni giorni ne' quali avanti il mio stanziare in Arcetri [cioè prima del 1639]*, egli andava colà a visitarlo [on the solutions explained to him by his father, who was already blind, on certain days in which, before my settling in Arcetri [i.e., before 1639], he went there to visit him]" (cf. The *Avvertimento* of Favaro in the edition of the *Problemi* and of the fragments presumably relating to this Galilean writing in *G.G.*, VIII, pp. 561-570). In what we have of the work on the *Problemi* (*ibid.*, pp. 571-642) we do not find, however, any specific attention to questions relating in any way to the force of percussion.

P. Galluzzi, *Momento: Galilean Studies*, Archimedes 78,
https://doi.org/10.1007/978-3-032-35049-7_11

390 essence of Galileo's methodological observations consisted in declaring with the utmost determination that the effect of percussion, although at first glance it may appear very different from those produced by mechanical instruments, can be reduced to a principle that derives "non da altro fonte, che da quello stesso onde scaturiscono le ragioni d'altri effetti meccanici [from no other source than the same one from which the reasons for other mechanical effects arise]"[3]. That is,

> che la forza, la resistenza ed il spazio per il quale si fa il moto, si vanno alternamente con tal proporzione seguendo, con legge tale rispondendo, che resistenza eguale alla forza sarà da essa forza mossa per egual spazio e con egual velocità di quella che essa si muova. Parimente, forza che sia la metà meno di una resistenza potrà muoverla purché si muova essa con doppia velocità, o vogliam dire per distanza il doppio maggiore di quella che passerà la resistenza mossa[4]. [that the force, the resistance and the space through which the motion occurs, alternate in such a proportion, responding to such a law, that a resistance equal to the force will be moved by that force through an equal space and with the same speed as that with which it moves. Likewise, a force which is half as great as a resistance will be able to move it, provided that it moves with double the speed, or we may say, over a distance twice as great as that which the resistance moved will traverse.]

Once percussion is brought back into the family of mechanical instruments, the "meraviglia dell'effetto, il quale non esce punto da i termini delle naturali constituzioni e di quanto s'è detto [wonder of the effect, which does not go beyond the limits of the natural constitutions and of what has been said]" will cease[5].

Apart from these brief hints, the *Mecaniche* offers no concrete indication of how Galileo intended to proceed. We find evidence of the effort to develop some experiments to demonstrate the infinite force of percussion in a brief mention in the letter to Guidobaldo dated 29 November 1602[6]. Then, essentially, we have to wait

---

[3] *G.G.*, II, p. 188.

[4] *Ibid.*, p. 189.

[5] *Ibid.*

[6] "Al Sig.r Francesco mi farà grazia rendere un baciamano, dicendogli che con un poco d'ozio gli scriverò una esperienza che già mi venne in fantasia per misurare il momento della percossa [I will do Signor Francesco the favor of kissing his hand, telling him that with a little leisure I will write him an experiment that has occurred to me for measuring the momento of percussion]" (*G.G.*, X, p. 100). The experiment to which Galileo refers here is, in all probability, one of the two reported by Torricelli in the *Terza* of the *Lezioni Accademiche* and attributed to "quel famosissimo Vecchio [that very famous Old Man]", when he was "in Padua". The first of the experiments consisted of taking a bow and placing it parallel to the horizon. On the bowstring, at its central point, he hung a given weight with a string an arm's length long. He lifted it up, then let it fall freely, and observed how "incurvasse e si tirasse dietro la forza dell'arco [the bowstring would bend and draw along with it the force of the bow]". Next, he applied different heavy bodies to the bowstring until he obtained the same bend in the string. The weight of the heavy body required to produce that bend would give the measure of the force of natural percussion. In this way, Galileo tried to demonstrate the infinite force of percussion. The second experiment, which consisted of striking a lead sphere placed on an anvil with a hammer, also served to confirm its infiniteness. The sphere, of course, becomes dented. To produce the same dent with a dead weight, it was necessary to apply a certain weight on top of the sphere. But – as Galileo observed – while a second blow of the hammer, identical to the first, causes a new dent, the repeated application of the same weight does not produce any new effect on the lead sphere. To match the effect of subsequent blows with a dead weight, it is therefore necessary to continually increase the weight. "E questo succederà sempre con progresso sino in infinito.... Segno manifesto che la forza della percossa sia infinita [And this will always happen with progress towards infinity.... A clear sign that the force of percussion is infinite]" (E. Torricelli,

until 1634 to find, in the correspondence, some explicit reference to this theme[7]. **391**
Between 1634 and 1638, in addition to the final draft of the *Discorsi*, Galileo also worked on the *Giornata* that he planned to dedicate to percussion. Due to his own delays and the impatience of the publishers, he was finally forced to give it up[8]. Death overtook him in 1642 with the work still unfinished, despite the assiduous assistance of Torricelli, who in his last months was writing for the now-blind Master the "added" *Giornata* on the theory of proportions and the notes on percussion.

Fortunately, we still have a fairly elaborate dialogue draft of the *Giornata* on percussion, which Galileo must have dictated to Marco Ambrogetti around 1638, and which he planned to include in a later edition of the *Discorsi*[9]. This is an unfinished

---

Opere scelte cit., pp. 572-574).

[7] Clear evidence of Galileo's interest in the study of percussion can be found in his correspondence from 1634 to 1640. Niccolò Aggiunti, on 1 February 1634, told the Master that he had greatly appreciated his proposal regarding percussion, and, above all, the part relating to the affirmation of the infiniteness of the effect that it produces: "qualunque anco lieve percossa haverà forza infinita, perché proposto qualsivoglia grandissimo peso, potremo trovare un tal resistente nel quale una leggier percossa opererà non meno che la premente gravezza del proposto grandissimo peso [even any light percussion will have infinite force, because if we propose any great weight, we will be able to find such a resistant body in which a light percussion will operate no less than the pressing heaviness of the proposed great weight]" (*G.G.*, XVI, p. 31). Dino Peri's letter to Galileo, four years later, allows us to establish that the Pisan scientist, despite his terrible state of health, had "tuttavia internatosi nella profondissima speculatione della percossa e ne aveva acquistato la quasi intera sodisfazione [nevertheless delved into the very deep speculation of percussion and had acquired almost complete satisfaction from it]" (*G.G.*, XVII, p. 328; letter of 21 April 1638). On 1 August 1639, writing to Baliani (who was familiar with the pages on percussion that concluded the *Mecaniche*, pages that Galileo no longer remembered having written), Galileo expressed his intention to add, in a subsequent edition of the *Discorsi*, "altre speculationcelle et in particolare quella attenente alla forza della percossa, nell'investigatione della quale ho consumate molte centinaia e migliaia di hore, e finalmente ridottala ad assai facile esplicatione, siché altri in manco di mez'hora di tempo potrà restarne capace [other little speculations, and in particular that concerning the force of percussion, in the investigation of which I have spent many hundreds and thousands of hours, and finally reduced it to a very easy explanation, so that others will be able to understand it in less than half an hour]" (*G.G.*, XVIII, p. 78).

[8] Cf. the *Avvertimento* of Favaro (*G.G.*, VIII, pp. 26 ff.). While writing the *Discorsi*, Galileo must have hoped to be able to include the *Giornata* dedicated to percussion, of which he gave an explicit preview in the *Giornata Quarta*, which he then either forgot or was unable to remove: "*Salv.* spedita che avremo la lettura di questo trattato de i proietti, gli spiegherò tutte quelle fantasie o voglian dire stravaganze, che de i discorsi dell'Accademico mi son rimaste in memoria [*Salv.* As soon as we have read this treatise on projectiles, I will explain to you all those fantasies or, I mean, extravagances that have remained in my memory from the Academician's discourses]" (*ibid.*, p. 293).

[9] That this draft, which remained unknown even to Torricelli, was probably dictated by Galileo to Ambrogetti (whose copy is now lost), is attested by a statement by Viviani: "Ultimo Congresso del signor Galileo intorno alla forza della percossa, datomi a copiare dal Signor Vincenzio Galileo, dopo la morte del Padre. Questo non è stampato, ma l'originale si tróva appresso agli eredi di detto Vincenzio, e non mi sovviene se sia di mano del medesimo signor Galileo, oppure di Marco Ambrogetti, come piuttosto io mi credo… [Last Congress of Signor Galileo on the force of percussion, given to me to copy by Signor Vincenzio Galileo, after the death of his Father. This is not printed, but the original is in the possession of the heirs of the said Vincenzio, and I do not remember whether it is by the hand of the same Signor Galileo, or by Marco Ambrogetti, as I rather believe…]" (*Quinto Libro degli Elementi di Euclide, ovvero Scienza Universale delle Proporzioni spiegata colla Dottrina del Galileo...*, Alla Condotta, in Firenze 1674, p. 103). See the *Avvertimento* to the *Discorsi* cit. by Favaro (p. 27) and the *Discorsi* edited by Carugo-Geymonat cit., pp. 849-851. The draft presumably dictated to Ambrogetti was published for the first time in the Florentine edition of Galileo's Works (Tartini and Franchi, Florence 1718, II,

392 work, not without irreconcilable inconsistencies, but capable, however, of indicating with sufficient precision the reflections and the method with which Galileo intended to conduct the analysis of natural percussion (that is, the one produced by a heavy body after a free descent of any duration). For our investigation, these are very important pages, as they preserve for us the last attempt of our scientist to bring about a substantial reunification between statics and dynamics.

In the fragment of the dialogue, Simplicio is absent. His place is taken by the Venetian Paolo Aproino, a friend and pupil of Galileo in Padua and his correspondent during the Florentine years. Aproino had died in 1638, and it is extremely probable that his inclusion among the interlocutors represents a way of fittingly honoring his memory and merit[10].

The discussion on natural percussion in the "added" Giornata is introduced by a remark by Aproino, who has the task of recalling the nature and conclusions of the Academician's early studies on this theme (i.e., the *Mecaniche*). He recalls a complex experiment conceived by the Academician to measure the force of natural percussion in terms of weight. The aim of the attempt was, obviously, to reduce percussion to the same principles that govern the functioning of mechanical instruments: in essence, it was a question of reducing a purely dynamic action to an equivalent static one.

The experiment is extremely complicated and does not allow for easy and exact measurements[11]. Salviati, therefore, intervenes to propose a more appropriate technique to achieve the same goal. He invites us to imagine a *berta* (pile driver) weighing 100 pounds, capable of driving a pole 4 finger-widths into the ground when dropped from a height of four arm-lengths. Assuming that the same pole is driven in by four finger-widths with a one thousand pound weight ("peso morto [dead weight]") which acts "colla sola gravità, senza moto precedente [with heaviness

---

pp. 693-710) with the following title: "*Giornata Sesta del Galileo, della forza della percossa, da aggiugnersi a i Discorsi e alle Dimostrazioni mattematiche intorno alle due nuove scienze appartenenti alle meccaniche ed a i movimenti locali* [*Sixth Day of Galileo, on the force of percussion, to be added to the Discourses and Mathematical Demonstrations on the two new sciences belonging to mechanics and local motions*]".

[10] Regarding Aproino, cf. *D.B.I.*, III (1961), pp. 646-648 (edited by M. Gliozzi).

[11] In essence, the experiment consists of hanging a bucket full of water, with a hole in the bottom, from the end of a balance beam, to which a second empty bucket is connected with a rope and placed exactly underneath the first. At the other end of the balance, a heavy body hangs that can counterbalance the two buckets. Once the hole in the first bucket was opened, water begins to fall into the second one below. As it falls, it would have produced a percussion whose force would have been measured by the weight that had to be added to the counterweight to balance the bucket system after the hole had been opened. Unfortunately, as Aproino admits, the experiment produced results different from those predicted. However, it served to establish that the *momento della percossa* of the water "equivale al momento ed al peso di quella quantità d'acqua cadente che si trova sospesa in aria tra le due acque delle due secchie, superiore ed inferiore, la qual quantità d'acqua non gravita punto né contro alla secchia superiore, né contro all'inferiore [is equivalent to the momento and weight of that quantity of falling water which is suspended in the air between the two bodies of water of the two buckets, upper and lower, which quantity of water does not exert heaviness at all against either the upper bucket or the lower one]" (*G.G.*, VIII, p. 324). The difficulty of the experiment lies in the far from easy measurement of the falling water, as Salviati honestly acknowledges (*ibid.*, p. 325).

alone, without previous motion]"[12], Salviati asks his interlocutors "se possiamo per la detta sperienza asserire che l'aggravio del peso morto farà sempre il medesimo effetto sopra una resistenza che fa il peso di cento libbre cadente dall'altezza di 393 quattro braccia [can we, from the aforementioned experiment, assert that the increase of the dead weight will always have the same effect on a resistance produced by a weight of one hundred pounds falling from a height of four arm-lengths]"[13]. After an initial affirmative response from both interlocutors, Sagredo reconsiders and recognizes the deception. The dead weight of 1000 pounds, which equals the effect of the pile driver's first blow, is not sufficient to equal the effect of the second blow, with which the pile driver drives the pole another two finger-widths into the ground: how can it be admitted, as Sagredo acknowledges, "che l'avernelo tolto solamente e poi rimessoglielo sopra torni a cacciarlo due altre dita? [that simply having removed it and then putting it back again would drive it two more finger-widths?]"[14].

Thus, it seems that this path inevitably leads to a paradoxical conclusion. Each subsequent blow of the pile driver, despite being equal to the previous ones, continues to drive the pole into the ground, although to an ever lesser extent, while the dead weight, to produce the same effect, must be significantly increased each time. On the one hand, we have the blow which is always the same; on the other, the dead weight, which varies. Which of these dead weights – asks Sagredo – "doveremo noi prendere per ferma e certa misura della forza del colpo, che pur, quanto a se stesso, è sempre il medesimo? [should we take as a fixed and certain measure of the force of the blow, which, in itself, is always the same?]"[15]. Faced with these paradoxes, as Salviati observes, it seems that one cannot avoid admitting that "la forza della percossa essere infinita o vogliamo dire indeterminata o indeterminabile, e farsi ora minore ed ora maggiore, secondo che ella viene applicata ad una maggiore o minore resistenza [the force of percussion is infinite or, we might say, indeterminate or indeterminable, and becomes now smaller and now larger, depending on whether it is applied to a greater or lesser resistance]"[16]. In any case, it does not appear that it can be measured by means of dead weight. In order to match the effect of the successive blows, indeed, a dead weight is needed that is "maggiore e maggiore in immenso [greater and greater in immensity]"[17]. To absolutely oppose the force of percussion, the resistance should not yield even by an imperceptible space, that is, it should be infinite.

Despite this *impasse*, Salviati does not despair of reducing the operation of percussion to the same principles that govern the functioning of the other machines[18].

---

[12] *Ibid.*, p. 325.
[13] *Ibid.*, p. 326.
[14] *Ibid.*, p. 327.
[15] *Ibid.*, p. 328.
[16] *Ibid.*
[17] *Ibid.*
[18] It seems – as Galileo asserts – "che non bene si comprenda come l'operazione della percossa, che sembra infinita, non debba procedere per mezzi diversi da quelli di altre macchine, che con pochissima forza superano resistenze immense: tuttavia, io non dispero di poter esplicare come in questa ancora si procede nella medesima maniera [that it is not well understood how the operation of percussion, which

To do this, it will first be necessary to recall the general principle of all mechanical operations:

> Generalmente dunque diciamo il momento del men grave pareggiare il momento del più grave, quando la velocità del minore alla velocità del maggiore abbia l'istessa proporzione che la gravità del maggiore a quella del minore; al quale ogni poco vantaggio che si conceda, supera l'equilibrio e s'introduce il moto[19]. [Generally, then, we say that the momento of the less heavy body equals the momento of the heavier body when the speed of the lesser has the same proportion to the speed of the greater as the heaviness of the greater has to that of the lesser; to which any granted slight advantage overcomes the equilibrium and introduces motion.]

394 This is the principle whose centrality we emphasized when illustrating the *Mecaniche*: speed adds *momento* to weight. This principle, according to Salviati, makes it possible to demonstrate that the operation appears infinite not only in percussion, "ma tale si mostra ella in qualsivoglia altro meccanico ordigno [but reveals itself to be such in any other mechanical device as well]"[20]. In fact, if we imagine a weight applied to one end of the steelyard as large as we wish, it will always be counterbalanced by another, however small we wish, "se noi faremo che lo spazio per lo quale scenderà sia cento o mille e più volte maggiore dello spazio della salita dell'altro, cioè se la velocità di quello sia cento e mille volte maggiore di questo [if we ensure the space through which it descends a hundred or a thousand or more times greater than the space of the ascent of the other, that is, if the speed of the former is a hundred or a thousand times greater than that of the other]"[21].

To verify whether the effect of the natural percussion can be attributed to those of other machines, it is necessary, however, to abandon the model of the pile driver that hits the pole. In this model, in fact, despite appearances, each of the blows is different, with the resistance always yielding differently[22]. Instead, it is necessary that the resistance be invariable, so as to provide a univocal measure of percussion. It is, therefore, necessary to resort to a model in which the resistance offers a constant obstacle to the percussion.

Salviati, therefore, invites us to consider a pulley mounted with a very long rope, from one end of which is suspended a body of given weight, which rests on the ground. At the other end of the rope, a smaller weight is held, "lasciandolo da qualche altezza cadere ed osservando quello che segue nell'altro gran solido nel sentir la strappata dell'altro peso cadente, la quale strappata sarà ad esso gran peso come un colpo che lo voglia cacciare in sù [letting it fall from some height and observing what happens to the other large solid when it feels the jerk of the

---

seems infinite, should not proceed by means different from those of other machines, which with very little force overcome immense resistances: nevertheless, I do not despair of being able to explain how even in this case it proceeds in the same manner]" (*ibid.*, p. 329).

[19] *Ibid.*, p. 330.

[20] *Ibid.*

[21] *Ibid.*

[22] As a result of the first blow, the pole sinks four finger-widths deep, and this yielding causes the speed of the pile driver to decrease. The second blow, however, causes the pole to yield by only 2 finger-widths. This means that the subtraction of speed will be less, and therefore the strike will be more effective. (*ibid.*, pp. 331-332).

other falling weight, of which the jerk will be to that large weight like a blow that wants to push it upwards]"[23]. According to the mechanical principle mentioned above, any weight, however small it may be, "prevalerà ad un altro quanto si voglia maggiore [will prevail over another however large it may be]" when the speed of the smaller "abbia maggior proporzione alla velocità del maggiore che non ha la gravità del maggiore alla gravità del minore [has a greater proportion to the speed of the larger than the heaviness of the larger has to the heaviness of the smaller]"[24]. According to Galileo, this situation of ascendancy always occurs in the case of the pulley, no matter how small the weight of the heavy body dropped. Its speed, in fact, "supera d'infinito intervallo [exceeds by an infinite interval]" that of the other weight, "la quale è nulla [which is none]". Between none and any given quantity, there is an infinite distance. The small weight, falling freely from any height, will lift the heavy body resting on the ground for a certain distance. But according to what proportions?

To verify this and, therefore, establish conclusively whether the effect of **395**
percussion is equal to that of other mechanical instruments[25], Salviati proposes an "experiment" consisting of hanging two heavy bodies from a string and letting one fall from the height of one arm-length. The "experiment" is "imagined", not actually performed, and the results are deduced from the general principles set out above[26]. What Salviati "inclina a credere [inclines to believe]" would happen following the "strappata [jerk]" and the reasons on which he based his prediction are quite interesting:

sebbene il peso cadente è un aggregato di gravità e di velocità, l'operazione della gravità nel sollevar l'altro è nulla, avendo a sé opposta e renitente altrettanta gravità dell'altro peso, il quale è manifesto che mosso non sarebbe senza l'aggiunta all'altro di qualche piccola gravità: l'operazion dunque per la quale il peso cadente dee sollevar l'altro, è tutta della velocità, la quale altro che velocità non può conferire; né potendo conferirne altra che quella che egli ha, e non avendo altra che quella che, partendosi dalla quiete, ha guadagnata nello spazio della scesa di un braccio, per altrettanto spazio e con altrettanta velocità spignerà l'altro all'insu, confermandosi con quello che in varie esperienze si può riconoscere, che è che il grave cadente, partendosi dalla quiete, si trova in ogni sito aver tant'impeto, che basta per ridur se stesso alla medesima altezza[27]. [although the falling weight is an aggregate of heaviness and speed, the operation of heaviness in lifting the other is null, since it has opposed and resisted the same amount of heaviness in the other weight, which clearly would not be moved without the addition of some small heaviness to the other: the operation, therefore, by which the falling weight must lift the other, is entirely

---

[23] *Ibid.*, p. 333.

[24] *Ibid.*

[25] "Seguita ora che cerchiamo d'investigare quanto sia per essere lo spazio al quale la ricevuta percossa lo solleverà, e se forse questo risponda a quello delli altri strumenti meccanici... [Let us now proceed to investigate how much space the received percussion will raise it, and whether perhaps this corresponds to that of other mechanical instruments...]" (*ibid.*).

[26] The whole experimental discourse in question (ibid., pp. 333-335) is "hypothetical": "Figuriamoci di fare la prima esperienza col fare cadere... [Imagine making the first experiment by dropping...]" (*ibid.*). Furthermore, no reference is made to the actual behavior of the two heavy bodies hanging at the end of the rope. However, the strength of the "reasons" advanced by Salviati overcome the reticence of Aproino and Sagredo, despite the absence of a true experimental verification. This is another example of the complexity of Galileo's position regarding the "experiments".

[27] *Ibid.*, p. 335.

a matter of speed, which cannot confer anything other than speed; nor being able to confer any other than that which it has, and having no other than that which, departing from rest, it has gained in the space of the descent of one arm-length, for it will push the other upwards in the same space and with the same speed, confirming itself with what can be recognized in various experiments, which is that the falling heavy body, departing from rest, is found in every point to have such impeto as suffices to bring itself back to the same height.]

According to Salviati's analysis, heaviness remains constant during free fall, and its effect consists of perfectly matching the resistance of the counterweight being lifted[28]. The lifting operation is, therefore, the effect of speed alone. But what is its measure? The answer is simple. The weight falls freely; therefore, it will acquire, at every instant, a new increment of speed:

sicché, nel caso nostro, il grado massimo di velocità del grave, mentre liberamente scende, è quel che si trova avere nel punto che egli comincia a sollevare il suo compagno; ed è manifesto che tal grado di
396 velocità non si andrà più augumentando, essendo tolta la cagione dello augumento che era la gravità propria di esso grave descendente, la quale non opera più, essendo tolta la sua propensione di scendere dalla repugnanza del salire di altrettanto peso del suo compagno. Si conserverà, dunque, il detto grado massimo di velocità, ed il moto, di accelerato si convertirà in equabile: quale poi sia per essere la futura velocità, è manifesto dalle cose dimostrate e vedute ne' passati giorni, cioè che la velocità futura sarà tale che in altrettanto tempo quanto fu quello della scesa, si passerà doppio spazio di quello della caduta[29]. [so that, in our case, the maximum degree of speed of the heavy body, while it freely descends, is that which it has at the point where it begins to lift its companion; and it is clear that this degree of speed will no longer increase, the cause of the increase having been taken away, which was the heaviness of the descending heavy body, which no longer operates, its propensity to descend having been taken away by the resistance of the ascent of a matching weight of its companion. Therefore, the said maximum degree of speed will be maintained, and the motion will be converted from accelerated to uniform: what then the future speed will be is clear from the things demonstrated and seen in the past days, that is, that the future speed will be such that in the same time as that of the descent, twice the space of that of the fall will be passed.]

Heaviness is the "cagione dell'augumento [cause of the increase]" of speed, that is, of acceleration. This is really interesting. Until now, Galileo has carefully avoided stating that acceleration depends on heaviness[30]. Moreover, Galileo clearly distinguishes heaviness and speed, analyzing their effects separately. He also applies the laws of naturally accelerated motion to speed, suggesting, with utmost clarity, that in falling, heaviness remains the same, it does not increase. Lastly, he compares the case of the pulley, at the ends of which are suspended heavy bodies of the same weight, to that of a heavy body placed on a plane parallel to the horizon: in both cases, there is a neutralization of the effects of heaviness, that is, a substantial indifference to motion and rest. That is, Galileo asserts a sort of inertial principle of percussion: after the percussion, the weight that was dropped will continue to move indefinitely with uniform motion, with the maximum degree of speed reached in

[28] Galileo always refused to accept the Aristotelian thesis of the intensification of heaviness in the *deorsum* [downward] motion as a consequence of the intensification of the "desire" for the natural place, as the mobile approaches it.
[29] *Ibid.*, p. 337.
[30] Cf. *supra*, Part Two, Ch. IV.

free descent[31].

The explanation appears ingenious to Sagredo, who, however, admits to still being quite perplexed about many other aspects of percussion, which do not seem to be attributable to the model of the pulley with equal bodies suspended at the ends. What would happen if the dropped weight were less than the counterweight? The separate analysis of heaviness and speed could not, in fact, be performed in such a case. What role would the laws of naturally accelerated motion and the principle of conservation of motion then have? The problems, as can be seen, were neither few nor simple. And Galileo, preparing to answer Sagredo's doubts, returns to propose as indisputable principles some conclusions established "nel trattato del moto [in the treatise on motion]": the equality of the degrees of speed at the foot of the incline and of the perpendicular of equal elevation, the ratio between the times of descent along the incline and the perpendicular equal to that between the perpendicular **397**
and the incline, and, lastly, the laws of "mechanical" *momenti*[32]. Then the writing abruptly ends, leaving us with no possibility of guessing how the Pisan scientist intended to develop his considerations.

We still preserve, in dialogical form, another fragment, which is prefaced with the title *Proposizione*, of which its relationship with the text illustrated so far is unknown. It is evident, however, that the *Proposizione* records an attempt to explain the mechanism of percussion and its infiniteness that is substantially different from that of the previous passage[33]. In fact, Galileo shows himself to be directed towards pursuing a rigorously "static" justification of the phenomenon, setting aside any consideration relating to naturally accelerated motion. The *Proposizione* opens with the statement to be proved:

[31] "Ed è manifesto che tal grado di velocità non si andrà più augumentando, essendo tolta la cagione dello augumento, che era la gravità propria di esso grave descendente, la quale non opera più, essendo tolta la propensione di scendere dalla repugnanza del salire di altrettanto peso del suo compagno. Si conserverà dunque il detto grado massimo di velocità, ed il moto di accelerato si convertirà in equabile: quale poi sia per essere la futura velocità è manifesto dalle cose dimostrate e vedute ne' passati giorni, cioè che la velocità futura sarà tale che in altrettanto tempo quanto fu quello della scesa, si passerà doppio spazio di quello della caduta [And it is evident that such a degree of speed will no longer increase, the cause of the increase having been removed, which was the own heaviness of the descending weight, which no longer operates, the tendency to descend having been removed by the resistance of the ascent of the matching weight of its companion. The said maximum degree of speed will therefore be preserved, and the motion, from accelerated, will turn into uniform: what the future speed will be is evident from the things demonstrated and observed in the past days, that is, that the future speed will be such that in the same amount of time as that of the descent, twice the distance of the fall will be passed]" (*ibid.*, p. 337).

[32] *Ibid.*, p. 338.

[33] The change of register was evident even to the first editors (Florence 1718) of the fragment on percussion, who noted: "Avverta il Lettore che il discorso seguente non bene sarà connesso colle cose supposte di sopra, perché l'Autore ha avuto pensiero di distenderlo diversamente da quello che aveva in animo quando si notarono le sopraddette conclusioni [The Reader should be warned that the following discourse will not be well connected with the things supposed above, because the Author has had the thought of developing it differently from what he had in mind when the aforementioned conclusions were noted]" (*Opere di Galileo* cit., II, p. 702).

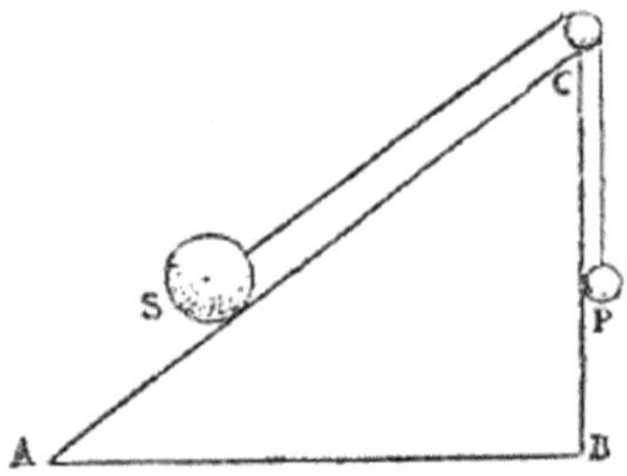

Figure 49

Se l'effetto che fa una percossa del medesimo peso, e cadente dalla medesima altezza, caccierà un resistente sempre eguale per qualche spazio, e che per fare un simile effetto ci bisogni una determinata quantità di peso morto, che senza percossa prema, dico che quando il medesimo percuziente sopra un altro resistente maggiore, con tal percossa, lo caccerà, v.g., per la metà dello spazio che fu cacciato l'altro, per far questa seconda cacciata non basta la pressura del detto peso morto, ma ve ne vuole altro il doppio più grave; e così in tutte le altre proporzioni, quanto una cacciata fatta dal medesimo percuziente è più breve, tanto, per l'opposito, con proporzione contraria vi si ricerca, per far l'istesso, gravità maggiore di peso morto premente[34]. [If the effect produced by percussion of the same weight, falling from the same height, will always drive an equal resister for some space, and that to produce such an effect a determined quantity of dead weight is required, which presses without percussion, I say that when the same percussive body on another greater resister, with such percussion, drives it, for example, for half the space that the other was driven, to make this second drive, the pressure of the said dead weight is not sufficient, but another heavy body twice as heavy is required; and so in all other proportions, as a drive made by the same percussive object is shorter, so much, conversely, in inverse proportion, a greater heaviness of pressing dead weight is required to do the same.]

Now, the Pisan scientist intends to show that the model of the perpendicular and the incline of equal elevation, on which two bodies are connected by a string that
398 passes through the vertex C of the triangle ACB (see figure 49), whose weights are to each other as the lengths of the planes on which they rest, is perfectly equivalent to that of the pile driver of a given weight that strikes a pole by descending freely from a given height and drives it into the ground for a given space. To establish this equivalence in a concrete way, Galileo shows by what reasoning it is possible to reduce the case of a ten-pound pile driver, which, by falling freely from a height of four arm-lengths, drives the pole into the ground by four finger-widths, to the tried and tested "static" model of the inclined plane. Let us first assume that, to obtain the same effect, it is necessary to apply to the pole the pressure of a dead weight ten times greater (one hundred pounds). Now, we know that the weight of ten pounds "dovendo calare a perpendicolo [having to fall perpendicularly]" is sufficient to lift (to tell the truth, merely to balance) a weight of one hundred pounds on a plane whose length is ten times the elevation. The "impeto" acquired by the ten pounds descending along the perpendicular will be sufficient to lift the one hundred pounds along the incline for a distance equal to that of its descent. But the displacement along the incline of the greater weight corresponds to a "vertical"

[34] *G.G.*, VIII, p. 340.

displacement equal to one-tenth of that of the lesser weight. At this point, Galileo can cleverly "close" the reasoning, emphasizing the validity of the analogy between percussion and mechanical instruments. The force, in fact, capable of lifting one hundred pounds

> è eguale alla forza colla quale il medesimo peso delle cento libbre calca in giù, e questa era la potente a cacciare il palo, postavi sopra e premendo: ecco dunque esplicato come la caduta di dieci libbre di peso è potente a cacciare una resistenza equivalente a quella che ha il peso di cento libbre per essere sollevato, ma la cacciata non sarà più che per la decima parte della scesa del percuziente[35]. [is equal to the force with which the same weight of one hundred pounds presses down, and this was the force capable of pushing down the pole, placed on top and pressing: thus, here is explained how the fall of ten pounds of weight is capable of pushing down a resistance equivalent to that which the weight of one hundred pounds has to be lifted, but the drive will not be more than a tenth part of the descent of the striker.]

The model of the inclined plane and the static principles on which it depends (especially the *momento della gravità* and its laws) serve to adequately explain the apparent increase in intensity of the successive percussions and to confirm the statement of the *Proposizione*[36]. The increase in intensity of successive percussions leads to the affirmation of the infiniteness of the percussion:

> E così, moltiplicando la resistenza in infinito, sempre la medesima percossa la potrà superare, ma col cacciare il resistente sempre per minore e minore spazio con alterna proporzione: onde pare che noi ragionevolmente possiamo asserire la percossa essere infinita[37]. [And so, by multiplying the resistance infinitely, the same percussion will always be able to overcome it, but by always pushing the resister away for a smaller and smaller space with alternating proportions: hence, it seems that we can reasonably assert that the percussion is infinite.]

On the other hand, the "forza del premente senza percossa [force of pressing without percussion]" also seems to be infinite. In fact, when it overcomes the resistance of the pole, "lo caccerà non per quello spazio solo che lo averà cacciato la percossa, ma seguiterà di cacciarlo in infinito [it will not only drive it for the space that the percussion will have driven it, but will continue to drive it infinitely]"[38]. **399**

Let us take stock of Galileo's complex reasoning. In essence, he started from the intention, already expressed in the draft dialogue, to demonstrate that natural percussion is always measurable in terms of "dead" weight, that is, it can be reduced to the static principles that regulate the *momenti* of heavy bodies along planes of

---

[35] *Ibid.*, p. 341.

[36] "E se noi ponemo la resistenza del palo esser raddoppiata o triplicata, sicché vi bisogni per superarla la pressura di dugento o trecento libbre di peso morto, replicando simil discorso troveremo l'impeto delle dieci libbre cadenti a perpendicolo esser potente a cacciare, siccome la prima, la seconda e la terza volta il palo, e come nella prima la decima parte, così nella seconda volta la ventesima, e nella terza la trentesima parte della sua scesa [And if we suppose that the resistance of the pole is doubled or tripled, so that to overcome it, the pressure of two hundred or three hundred pounds of dead weight is required, by repeating a similar argument, we will find that the impeto of the ten pounds falling perpendicularly is powerful enough to drive the pole the first, second, and third time, and as in the first time it is the one-tenth part, so in the second time it is the one-twentieth, and in the third time it is the one-thirtieth of its descent]" (*ibid.*).

[37] *Ibid.*

[38] *Ibid.*

different inclinations but equally elevated. Then he showed that the same natural percussion (i.e., produced by the same heavy body falling freely for an equal distance) varies according to the different yielding of the resistance. And it is greater the less the resistance yields. It tends to become infinite as the resistance tends to oppose without any yielding. The infiniteness of natural percussion automatically implies the affirmation of the infiniteness of pressure. Each successive strike of the pile driver on the pole corresponds to a greater and greater pressing weight, capable of determining the same sinking of the resistance. The dead weights needed to match the effect of successive percussions are, in the end, related to each other as the sinkings of the pole.

It is not difficult, taking these conclusions into account, to recognize that the *Proposizione*, intended to reduce natural percussion to the field of mechanical operations and to interpret it in the light of static principles, has obtained decidedly modest results. First of all, it provides no indication of the causes of the enormous force aroused by natural percussion. Secondly, it is strictly linked to the comparison between successive natural percussions of the same heavy body, which strikes while falling from the same height onto the same pole. Again, Galileo cannot establish anything precise about the relationship between the natural percussion of different heavy bodies, falling from different heights onto continuously variable resistances.

The complete mastery of the force of percussion could not ignore the clarification of these fundamental aspects, nor could one avoid taking a stance on the problem of the "mechanism" that made percussion infinite. The *Proposizione* thus ends in a new failure. Reducing natural percussion to a case of statics sheds little or no light on its nature. In the end, even between the model of natural percussion and that of pressure, to which Galileo intends to reduce it, an irreconcilable incongruity has appeared: natural percussion drives the pole through an ever smaller space, while the equivalent dead weight drives it through an infinite space[39]. Sagredo himself
400 must admit that there is a long way to go: percussion can "essere creata in tante e tante maniere ed applicata a tante varietà di resistenze [be created in so many ways and applied to so many varieties of resistance]" which will have to be explained "almeno alcune, l'intelligenza delle quali potrebbe aprirci la mente all'intelligenza di tutte [at least some of which, the understanding of which could open our minds to the understanding of all]"[40].

It is not within the scope of our work to highlight the errors, paralogisms, and inconsistencies present in the surviving Galilean pages dedicated to percussion.

[39] "Moltiplicando la resistenza in infinito sempre la medesima percossa la potrà superare, ma col cacciare il resistente sempre per minore e minore spazio con alterna proporzione: onde pare che noi ragionevolmente possiamo asserire la forza della percossa essere infinita: imperocché quando ella supera la resistenza del palo, lo caccerà non per quello spazio solo che lo averà cacciato la percossa, ma seguiterà di cacciarlo in infinito [By multiplying the resistance to infinity, the same percussion will always be able to overcome it, but by always driving the resister through a smaller and smaller space with alternating proportions: hence, it seems that we can reasonably assert that the force of percussion is infinite: because when it overcomes the resistance of the pole, it will not only drive it through the space that the percussion will have driven it, but will continue to drive it infinitely]" (*ibid.*).

[40] *Ibid.*

What we are interested in pointing out, however, is that, up till now, we have essentially seen the emergence of two types of approach. In the first dialogued part, the role of the accelerated speed acquired by the heavy object appeared central. The force of percussion has been attributed, in fact, to the effect of the naturally accelerated speed, independent of that of heaviness. In the *Proposizione*, every reference to accelerated speed is instead suppressed. The measurement of the force of percussion is attempted in terms of vertical displacement of resistance, nor is any reference to fall times encountered. Moreover, the accelerated motion with which the heavy body strikes the pole, falling freely from a given height, is considered equivalent to the uniform motions that occur in the sliding along the perpendicular and the incline of two connected heavy bodies, without acknowledging the difficulty. The approach appears rigorously static in the *Proposizione*, while, in the dialogue fragment that precedes it, we see the intention to attempt to explain the infiniteness of percussion by using the principles of the science of motion. The *Proposizione* is closer to the purpose, declared in the *Mecaniche*, of reducing percussion to the operations of mechanical instruments; but it ends with the recognition of an only apparent analogy and of the substantial non-mechanical nature of percussion. Even in the dialogued fragment, there is no lack of references to the *Mecaniche*, especially the use of the principle of "virtual speeds" to explain the equality of the *momenti* of two different weights applied at different distances from the fulcrum of the balance. The mechanical principle, of course, is valid when comparing given quantities of weight and speed. In the dialogued fragment, Galileo claims to use it to demonstrate that the force of a natural percussion is infinite, based on the impossibility of comparing a given speed – that, in the example of the pulley, of the lesser heavy body that descends freely – and the absence of speed of the weight to be lifted. The infiniteness of the force of natural percussion would, namely, be a consequence of the infinite magnitude that any speed has with respect to rest. The "mechanical" principle, as interpreted, seemed to justify that a heavy body, however small, left to fall freely, would always lift, by a certain distance, the other heavy
body, however large, hanging at rest at the end of the rope of the pulley. In essence, **401**
the *momento* of its speed (infinite compared to that of the resistance) justified the infiniteness of the *momento della percossa*.

One gets the sense that, in these notes, our scientist is once again intent on finding a unified solution (at least in terms of the principles used in the analysis) for phenomena such as pressure and natural percussion, which belong to substantially irreducible fields. Natural percussion is, in fact, closely connected to natural motion, of which it represents an aspect, and requires, in order to be correctly interpreted, the analysis of the dynamic principles, as well as the concept of "mass". The pressure of dead weight is, on the contrary, a pure case of statics, the interpretation of which presupposes the use of the principles that govern the functioning of mechanical instruments, such as the steelyard or the equilibrium of two connected heavy bodies persisting on the perpendicular and on the incline of identical elevation.

Again, we find Galileo at an *impasse*: the *momento della velocità* cannot be measured in terms of the *momento della gravità*; that is, natural percussion cannot be traced back to the pressure of the dead weight. From Galileo's mechanics and its principles, the laws of natural motion cannot be deduced, nor can the phenomenon of natural percussion be conveniently explained. The awareness of the new failure, encountered in yet another and ultimate attempt at reunification, is clearly evident from the fact that the *Proposizione* stops precisely at the moment in which the nature of percussion appears irreducible to that of mechanical instruments.

Fortunately, we have another note, certainly written after the dialogued fragment and the *Proposizione*. It is an important document because it explicitly records the growing awareness of the impossibility of realizing the project of mechanical treatment of percussion declared at the end of the *Mecaniche* many years earlier. Galileo speaks in the first person and refers to the question that, in the fragment, Sagredo addresses to Salviati, regarding the possibility of reducing the operation of percussion to that of the other instruments[41]. The fact that percussion overcomes any resistance but for a finite interval, while the pressing body moves a finite resistance for an infinite interval, makes Galileo doubt

> che il quesito del Sig. Sagredo sia inesplicabile, come quello che cerchi di agguagliar cose non proporzionabili che tali credo io che sieno l'azioni della percossa e quella della pressione[42]. [that Signor Sagredo's question is inexplicable, as it tries to equate things that are not proportional, such as I believe the actions of percussion and that of the pressure are.]

402 It is, therefore, useless to continue along the path taken in the first dialogued fragment and in the *Proposizione*. It is necessary to change register, placing the attention on percussion, without attempting to attribute it to the operations of mechanical instruments. Of the new approach that Galileo had to attempt after having reached the conviction that unification was impossible, we have retained a beautiful passage of definitive nature.

> Il momento di un grave nell'atto della percossa altro non è che un composto ed aggregato di infiniti momenti, ciascuno di essi eguale al solo momento, o interno e naturale di se medesimo (che è quello della propria gravità assoluta, che eternamente egli esercita posando sopra qualunque resistente), o estrinseco e violento, quale è quello della forza movente. Tali momenti nel tempo della mossa del grave si vanno accumulando di instante in instante con eguale additamento[43] e conservando in esso nel modo appunto che si va accrescendo la velocità di un grave cadente; ché siccome negli infiniti instanti di un tempo, benché minimo, si va sempre passando da un grave per nuovi ed eguali gradi di velocità,

[41] In the dialogued passage, Sagredo had invited Salviati to clarify whether, in the effect of percussion, the art "defraudi la natura, cosa che nella prima apparenza par che facciano altri strumenti meccanici, alzandosi gravissimi pesi con poca forza in virtù della leva, della vite, della taglia ed altri [defrauds nature, something that at first glance other mechanical instruments seem to do, lifting very heavy weights with little force by virtue of the lever, the screw, the pulley and others]" (*ibid.*, pp. 328-329). Sagredo, in essence, showed that he did not believe that percussion could be traced back to other mechanical instruments. The reference to the dialogued passage obviously proves that this fragment is later. The fragment is preserved in a copy by Viviani, who declares that he took it "da un foglio originale del Galileo, di sua mano, tra le cose della percossa [from an original sheet of Galileo's, by his own hand, among the things related to percussion]" (cf. *ibid.*, p. 343, critical apparatus).

[42] *Ibid.*

[43] Note the close analogy of this definition with that of naturally accelerated motion on which it is clearly modeled (cf. *ibid.*, p. 205).

con ritener sempre gli acquistati nel tempo percorso, così, anche nel mobile, si vanno conservando di instante in instante e componendosi quei momenti, o naturali o violenti, conferitigli o dalla natura o dall'arte[44]. [The momento of a heavy body in the act of percussion is nothing other than a composite and aggregate of infinite momenti, each of which is equal to the sole momento, either internal and natural to itself (which is that of its own absolute heaviness, which it eternally exerts when resting on any resisting object), or extrinsic and violent, such as that of the motive force. Such momenti, during the time of the motion of the heavy body, are accumulated from instant to instant with equal addition and are conserved in it in the same way that the speed of a falling heavy body is increased; for just as in the infinite instants of a time, however minimal, a heavy body always passes through new and equal degrees of speed, always retaining those acquired in the elapsed time, so too, in the mobile, those momenti, whether natural or violent, conferred upon it by nature or by art, are preserved from instant to instant and are composed together.]

The scene and the method have changed profoundly. The infiniteness of natural percussion[45] does not depend on the stillness of the object being struck, which comparatively is "infinite" for any speed of the striker. Nor does it rest on the **403**
infinite divisibility of the yielding space of the resistance. It is conferred, instead, by the naturally accelerated motion of falling. It is an entirely new perspective compared to those attempted in the dialogued fragment and in the *Proposizione*. The *momenti* mentioned in the note are, of course, those *della gravità*, and they increase according to the same mechanism that governs the growth of the *momenti della velocità* in natural fall: that is, according to the passage of time. We are facing a decisive step. The infinitesimal analysis, which has accounted for the relationship between the growth of *momenta velocitatis* and the passage of time, is now also proposed to explain the increase in the *momento della gravità*, which occurs in free descent. What increases, Galileo seems to suggest, is not heaviness (as Aristotle claimed), but its *momenti*.

This is, however, a particularly delicate passage that should be analyzed with the utmost care. It should be noted, first of all, that no causal relationship or dependence

[44] *Ibid.*, p. 344. This fragment is preserved in a modern copy in Ms. Galil. 84 (c. 37*r*), preceded by a foreword that attributes its authorship to Galileo: "Roba copiata da un esemplare del Galileo che si trova in mano del Sig. Vincenzio suo figliolo, di mano di questo, e tutto appresso del Sig. Cosimo [Copied things from a manuscript by Galileo, which is in the possession of Signor Vincenzio, his son, by his own hand, and all afterward by Signor Cosimo]". This fragment was also published in the Florentine edition of Galileo's *Opere*. Drake has raised some doubts about the attribution of this important note to the Pisan scientist, which he considers rather characteristic of Torricelli's opinions on the subject of percussion (*Galileo, Two new sciences* cit., p. 304, note 20). Drake's suspicions are based primarily on the indiscriminate use of *momento* in the passage, both in a "static" and a "dynamic" sense. In reality, there is no evidence that allows us to deny Galileo the authorship of these lines and the note prefacing the modern copy (with references to the names of Vincenzio and Cosimo Galilei) actually suggests that whoever prepared it had an older document in front of him. The same Florentine editors, who perhaps consulted the original, had no suspicions. As for the use of *momento*, there is no particular ambiguity recorded. Throughout the entire passage, the term is always used in the "static" sense of *momento della gravità*. The alleged ambiguity perceived by Drake depends on the way, in my opinion, largely unsatisfactory, in which the North American scholar has classified the different uses of the term in Galileo's work (cf. *id. ibid.*, *Glossary*, pp. XXXV-XXXVI; see also what we have observed in this regard *supra*, p. 156). If, as it seems fair to admit, this passage was preserved among Galileo's papers inherited by his son Vincenzio (that is, alongside the dialogued fragment and the *Proposizione*, it is likely that it was even unknown to Torricelli).

[45] And also of the violent one, which is explicitly alluded to in the above-mentioned passage.

is explicitly suggested between the growth of the *momenti della velocità* and the growth of the *momenti della gravità*. The two phenomena unfold in parallel and independently, according to the passage of time, without their relationship being adequately specified. Indeed, to further complicate the reader's task, Galileo avoids using the term *momento della velocità* and uses only the term *grado*. It seems clear that this must be attributed, without any possibility of equivocation, to Galileo's desire to avoid suggesting that the increase in speed depends on the accumulation of *momenti della gravità*. And the reason is not difficult to guess. How could he, in fact, justify the equivalence of all the falling speeds? And, in the case that he had succeeded with some stroke of genius, how could he have justified the difference in the percussions?

If the impossibility of adequately answering these problems led Galileo to propose the accumulation of the *momenti della gravità* as the cause of the continuous growth of the *momenti della velocità*, this implies that it would be a serious error to recognize in this fragment a return by the Pisan scientist to the positions represented by the theory of *impetus*. The analogy is only apparent. Galileo does not propose, in fact, any connection between the increase in the *momenti della gravità* and the growth of speed (Torricelli will do so later – as we will see shortly – by developing this type of approach). Instead, he limits himself to indicating what he believes is the mechanism that, through the increase in the *momenti della gravità* in free descent, explains the infinite force of the natural percussion. The acceleration that characterizes free fall is mentioned in the note only because of the close analogy it presents with the rate of accumulation of the *momenti della gravità*.

Each of the infinite *momenti della gravità* that aggregate during the descent is equivalent to the "momento interno e naturale di se medesimo [internal and natural momento of itself]", that is, to the weight of the heavy body (absolute heaviness).
404 Which weight is exerted "eternally" on any resistance. This very brief reference allows us to explain how Galileo conceived the accumulation of the *momenti della gravità*. A careful examination of the *Lezioni Accademiche* contributes significantly to understanding the meaning of his reflections on this point. In them, Torricelli, who was well acquainted with Galileo's speculations on percussion, intended to explain them, attempting to develop them in a systematic treatment[46]. The reference to Torricelli and to the pages of the *Lezioni* is explicitly suggested by the reference, in our fragment, to the "eternal" exertion of heaviness. In the *Lezione Seconda*, the scientist from Faenza invites us to consider the case of a weight of one hundred pounds placed on a marble table capable of supporting up to a thousand pounds. The weight of one hundred pounds would, therefore, have had no effect.

Non si nega che il momento di tal grave non sia cento libre, come realmente egli è, e che moltiplicato non possa rompere la tavola; anzi si afferma che con questo momento di libre cento gravita egli non solamente adesso, ma graviterà sempre uniformemente sopra il piano a lui sottoposto, in tal maniera però

[46] Three of Torricelli's *Lezioni Accademiche* (first published in 1715) are devoted to percussion (II, III and IV). On Torricelli's theory of percussion and its relations with Galilean conceptions, see the essay by S. Moscovici, *Torricelli's* Lezioni Accademiche *and Galileo's theory of percussion*, in *Galileo man of science* cit., pp. 432-448.

che in ciascuno istante del tempo che continuamente scorre, egli va facendo la sua violenza solamente di cento libre per volta alla tavola di marmo.... Dall'altra parte poi il marmo sottoposto in ciascuno istante del tempo che corre, va continuamente corrispondendo al grave premente con un momento di resistenza non come cento, ma come mille....

Se noi pigliassimo dieci palle eguali ad essa tutte insieme; ovvero se noi potessimo racchiudere in una sola tutta la virtù e tutta l'attività delle dette dieci palle, averessimo una forza di mille libre unite insieme; e sarebbe apunto tale che posandola sopra quel marmo... esso marmo resterebbe rotto[47].

[It is not denied that the momento of such a heavy body is one hundred pounds, as it really is, and that, if multiplied, it cannot break the table; rather it is affirmed that with this momento of one hundred pounds, it is not only heavy now, but will always be uniformly heavy upon the plane beneath it, in such a manner, however, that in each instant of time that continually passes, it goes on doing this violence only one hundred pounds at a time to the marble table.... On the other hand, then, the marble subjected in each instant of time that passes, continually responds to the pressing heavy body with a momento of resistance not as one hundred, but as one thousand....

If we were to take ten balls equal to it all together; or if we could enclose in a single one all the virtue and all the activity of the said ten balls, we would have a force of a thousand pounds united together; and it would be precisely such that placing it on that marble... that marble would be broken.]

Thus, for every instantaneous push of the heaviness of the body, there corresponds – in the case of the heavy body placed on a marble surface – an equally instantaneous counter-push of the resistance, which, if it is greater, will always prevail, supporting the heavy body. What will happen then – Torricelli asks – in the case that the heavy body, if instead of resting on a plane whose resistance is greater than its own heaviness, it falls freely into the void and, finally, violently hits an obstacle?

La definizione medesima che il Galileo adduce del moto naturalmente accelerato basta per isvelare questo arcano della natura, intorno alla forza della percossa. Aprasi la scaturigine della gravità. Sollevisi la palla grave in alto, in maniera tale che possa poi, quando ella ricaderà all'ingiù, dimorare per l'aria dieci istanti di tempo, e per conseguenza generare dieci di quei suoi momenti. Io dico che detti momenti 405
si conserveranno, e si aggregheranno insieme. Ciò è manifesto per l'esperienza continua de i gravi cadenti, e del moto accelerato; vedendosi che i gravi dopo le cadute hanno maggior forza, che non avevano quiescenti[48]. [The same definition that Galileo adduces of naturally accelerated motion is sufficient to reveal this mystery of nature, concerning the force of percussion. Let the spring of heaviness be revealed. Let the heavy ball be raised aloft, in such a way that it can then, when it falls back down, remain in the air for ten instants of time, and consequently generate ten of its momenti. I say that these momenti will be conserved, and will aggregate together. This is evident from the continuous experiment of falling heavy bodies, and of accelerated motion; it is seen that heavy bodies, after falling, have greater force than they had when quiescent.]

The heavy body that "eternally" exerts its *momento* on the plane that supports it, when left to fall freely, records a continuous accumulation of the *momenti* generated in the infinite instants of time. "La gravità ne i corpi naturali è una fontana dalla quale continuamente scaturiscono momenti [The heaviness in natural bodies – as Torricelli will say with effective and beautiful imagery – is a fountain from which

[47] *Opere scelte* cit., pp. 556-557.
[48] *Ibid.*, p. 559.

momenti continually spring forth]"[49].

Torricelli's *Lezioni*, therefore, serve admirably to explain the "eternal" load that Galileo refers to in the above-mentioned passage. Unlike Torricelli, who did not hesitate to state with the utmost clarity that he considered the acceleration of natural motion to be a direct consequence of the accumulation of *momenti della gravità* generated at each instant[50], Galileo – as we have already emphasized – avoids establishing any relationship of causal dependence between the two phenomena. The absence of precise indications on this point is not to be considered accidental. It responds, instead, to a precise design, which involves the exclusion of heaviness and its *momenti* from the causes of the acceleration of natural motion. One might be tempted to say that Galileo, far from proposing the "fountain of heaviness" to explain the increase in speed, resorts to the "costituzione naturale [natural constitution]" (already previously invoked to justify the equality of the degrees of speed impressed on all bodies starting from rest) to account for the accumulation of the *momenti della gravità*.

Galileo's hesitation to take the step that Torricelli boldly took can be explained by the awareness that our scientist had of not having a theory capable of convincingly
406 explaining the causal role of heaviness. I have already stated previously that I am certain that Galileo never abandoned the conviction that the accelerated speed of fall depends on the action of heaviness. But he never succeeded, starting from heaviness, in explaining acceleration, and, above all, in justifying the equality of all naturally accelerated descents from rest. He therefore fell back on the "constitution" of nature and on the divine law which had *ab origine* predisposed every heavy body to descend from rest according to the same degrees of speed. After the failure of the mechanical approaches, Galileo eventually thought of using this interpretative model to explain the infinite force of percussion, which he presented, in fact, as a natural motion with acceleration of heaviness. Thus – even though Galileo does

---

[49] *Ibid.*, p. 558.

[50] "La velocità ne i gravi cadenti altro non è che un non so che di posteriore e propriamente un effetto causato da i momenti intrinseci del corpo che discende: ma i momenti intrinseci sono un certo che precedente, e sono la vera e l'unica causa della maggiore e minore velocità e possono stare e sussistere da se stessi, senza l'aiuto o compagnia di velocità alcuna. Ciò si vede ne i gravi applicati alla libra con diverse lontananze, ovvero posti sopra piani diversamente inclinati, dove hanno i diversi momenti in atto, ma le diverse velocità solo in potenza. Ma la velocità per se stessa non può già sussistere senza i momenti interni [The speed of falling bodies is nothing but a certain latter thing and properly an effect caused by the intrinsic momenti of the descending body: but the intrinsic momenti are something prior, and are the true and only cause of the greater and lesser speed and can exist and subsist by themselves, without the help or company of any speed. This is seen in the bodies applied to the scales at different distances, or placed on differently inclined planes, where they have the different momenti in action, but the different speeds only in potential. But the speed by itself cannot subsist without the internal momenti]" (*ibid.*, p. 572). Galileo never went as far as to assert with such energy the direct proportionality between *momento-velocità* (except in the *De motu*, where however he did not yet possess the term *momento*). His hesitation (as we have already illustrated earlier) was due to foundation difficulties. Torricelli, on the other hand, tenaciously defended the proportionality between *momento-velocità*, as asserted in his *De motu*, even from the rather effective objections raised by Mersenne and Roberval, developing a theory that justified the equality of the increase in speed from rest for all heavy bodies in natural motion (cf. what we observed *supra*, Part Two, ch. IV, pp. 323-326).

not state it explicitly – the accumulation of the *momenti della gravità* generated at each successive instant is not the cause, but rather, in a certain way (that is, on a logical and analogical level), the consequence of the growth, according to time, of the *momenta velocitatis*.

Made proportional to the passage of time, which becomes, just as in natural motion, the characteristic element, the *momenti della percossa* could not but be infinite, like the *momenta velocitatis*. It should be noted that Galileo speaks of infinite *momenti* which, by accumulating, constitute the *momento della percossa*, which, however, is not called "infinite". "La forza della percossa è di infinito momento, tuttavolta che ella si applichi in un momento ed in uno instante dal grave percuziente sopra materia non cedente [The force of percussion – as he will clarify in another note – is of infinite momento, whenever it is applied in a momento and in one instant by the heavy percussive body on non-yielding material]"[51]. But he will hasten to specify that this condition never occurs, thus placing the infiniteness of percussion as a "limit". A definition of the "infinite" *momento della percossa*, included in the same group of fragmentary notes, helps us to understand what Galileo really meant by this expression:

> La forza della percossa è di infinito momento, perché non vi è resistenza, benché grandissima, che non venga superata da forza di percossa minimissima[52]. [The force of percussion is of infinite momento, because there is no resistance, although very great, that cannot be overcome by the very slightest force of percussion.]

The resistance he refers to, "benché grandissima [although very great]", is not infinite, and overcoming it does not require infinite force.

It is evident that we are faced with the same type of "infinitesimal" considerations previously proposed by Galileo to justify the *momenti della velocità*. In natural percussion, as in free fall, the infinite *momenti* generated at each instant form, by accumulation, a finite magnitude (a certain *momento*, corresponding to a certain percussion). It was certainly by thinking of the analyses carried out in the *Discorsi* that Galileo thought he could resolve the aporias arising from the percussion/natural acceleration analogy. In terms of quantity, rather than being "infinite", in fact, the *momenti della percossa* – like those *della velocità* – can be properly said to **407**
be "indefinite". Any span of time cannot be divided into an infinite number of quantifiable parts, but rather into any large number of quantifiable parts, in each of which the acquisition of a *momento della gravità* will occur. The sum of all the *momenti* acquired will, therefore, give a finite magnitude. And the longer the time of fall, the greater this finite magnitude (i.e., the percussion) will be. In this way, Galileo could delude himself into escaping the objection of the infiniteness of all

[51] *G.G.*, VIII, p. 345. For this note, also published in the Florentine edition of the *Opere* of Galileo (1718), the foreword given in note 44 applies.
[52] *Ibid.*

percussions and, therefore, of their substantial equivalence[53].

The importance of Galileo's notes on the nature of percussion that we have been illustrating is truly remarkable. In fact, they allow us to close the last page of our work on *momento* with a note that is not at all insignificant. The *momento* cycle of the use in Galileo's work now appears complete. The Pisan scientist has long sought by every means to base the science of movement on a model that can be traced back to static principles and, above all, to the concept and expression of *momento della gravità* – from which he consciously derived the concept and expression of *momentum velocitatis*. He then had to acknowledge the impossibility of realizing such a project, ending up separating *momentum velocitatis* from that of heaviness. The difficulties encountered in proposing a suitable explanation of percussion ultimately led him to outline a new approach, which completely overturned the initial attitude. Such an approach consists in taking *momenta velocitatis* as it appears conceived in the *De motu locali* and its rate of increase (equal for all bodies in natural motion from rest according to the inflexible statute of nature, that is, of God) as a model for interpreting the increase of *momenta gravitatis*, from which the extraordinary effect of the natural percussion depends. *Momenti della gravità* are thus also made proportional to time (no longer to the vertical displacement from rest, as in the *Mecaniche* or in several Paduan fragments).

For the Galilean *momento*, it is an unpredictable outcome. In fact, the
408 "mechanical" *momento*, which had long tried to impose its own characteristic physiognomy on the "dynamic" one, ended up being subjugated by it. The project of deriving dynamics from static principles, pursued by Galileo with great tenacity for many years, suddenly seems to undergo a radical reversal of trend. The laws of *momenti della gravità* do not serve to account for the acceleration of natural motion. Rather, it is the laws of natural motion that offer – when understood correctly – a model capable of making the "mechanical" *momento* and its variations fully intelligible.

This surprising outcome effectively measures – as it seems to us – the length and complexity of the path taken by Galileo, in the dynamic field, from the starting

[53] In the *Lezioni Accademiche* dedicated to percussion, Torricelli had very honestly recognized the difficulty inherent in comparing two infinite series of *momenti* as a consequence of the infinite divisibility of time. And he had premised that in order to properly compare the momenti of percussion produced by the same heavy body falling from different heights, it was necessary to refer to the "al foro del meraviglioso Fra Bonaventura, appresso il quale non solo non è assurdo che un infinito sia maggiore di un altro, ma è necessario. Che tutte le linee d'un parallelogrammo maggiore a tutte le linee d'un parallelogrammo minore abbiano la medesima proporzione che il parallelogrammo al parallelogrammo, se ben sono infinite; e che tutti i cerchi d'un cilindro minore, siano come il cilindro al cilindro, se bene sono infiniti, appresso di lui sono verità che vanno fra gl'ingressi della sua dottrina [forum of the marvelous Brother Bonaventura, according to whom not only is it not absurd that one infinite is greater than another, but it is necessary. That all the lines of a greater parallelogram have the same proportion to all the lines of a lesser parallelogram as the parallelogram to the parallelogram, even though they are infinite; and that all the circles of a lesser cylinder are as the cylinder to the cylinder, even though they are infinite, are truths which are among the entries of his doctrine]" (*Opere scelte* cit., p. 570).

positions, substantially attributable to traditional approaches, up to the threshold of the modern conception of movement, to the assertion of which he contributed more than anyone else. And the story of *momento* in the work of the Pisan scientist – a story that is neither simple nor linear, but twisted and full of problematic aspects – reflects with precise fidelity the meaning of this immense attempt, highlights the issues faced, the obstacles encountered, sometimes insurmountable, as well as the different solutions attempted.

Thus, an investigation into the Galilean "lexicon" – or rather, substantially, into a single term, although extremely significant – has been transformed, almost automatically, into a general essay on the genesis of the Galilean conception of movement, on its successive developments, on the theoretical elaboration, the "foundation", which accompanied it. If this was possible, it is because *momento* has proven to be, beyond the initial expectations, a crucial term.

APPENDIX[†]

[†] Aside from page numbers, the following brackets and asterisks are according to the original text. The exceptions to this are the translated titles indicated within brackets in the *Table of References*, which are provided by the Translator.

P. Galluzzi, *Momento: Galilean Studies*, Archimedes 78,
https://doi.org/10.1007/978-3-032-35049-7

411 The examples with context and the Index of occurrences cover the entirety of occurrences of *momento*/*momentum* in the 21 volumes of the Edizione Nazionale of the *Opere di Galileo Galilei*, including the critical apparatus. The occurrences of *momento* have been distributed into six sections, while those of *momentum* are divided into five sections. Within some sections, examples are distributed into subsections. In both the sections and the subsections, the contexts are arranged in chronological order. The chronological order is also respected in the index of *Other places*, which in turn appears divided into sections. Since the dating of the composition is highly controversial for many works and impossible for some (for example, the corpus of *Frammenti attenenti ai Discorsi*), we have applied the following criteria in the arrangement. First of all, we have assumed the year of publication as the date of writing for works printed during Galileo's lifetime. The collections of fragments and early drafts of a work are placed immediately before the final draft. For works not published by Galileo and of uncertain date, we relied on the prevailing opinion among scholars of the Pisan scientist, or on new elements that emerged from our research. We have considered as "opere" the letters-treatise that Favaro, in the Edizione Nazionale, preferred to isolate from the rest of Galileo's correspondence. In regard to the contexts and places of the *Lettere*, since these are documents for which we have absolutely certain dates of composition, we have deemed it appropriate to present them at the end of the individual sections and subsections, arranging them in chronological order.

The references that follow the contexts, as well as those that make up the index of *Other places*, contain the abbreviation of the work (cf. the Table of References), the volume of the Edizione Nazionale in which it is published (Roman numeral), and the page and the line (Arabic numerals) in which *momento*/*momentum* occur. The Arabic number in parentheses, which sometimes follows the reference to the line, indicates how many times our term appears there. The asterisk placed after the work's abbreviation * indicates that the context in question is recorded in the critical apparatus of the Edizione Nazionale. Lastly, for the *Lettere*, we thought it useful to add the date (in parenthesis) to the usual reference system.

TABLE OF REFERENCES

De motu	=	De motu antiquiora. [Older writings on motion.] 413
Tasso	=	Considerazioni al Tasso. [Considerations on Tasso.]
Architett.	=	Breve instruzione all'Architettura Militare. [Brief instruction on Military Architecture.]
Mecan.	=	Le Mecaniche. [On Mechanics.]
Cosmogr.	=	Trattato della sfera ovvero cosmografia. [Treatise on the sphere or cosmography.]
Mazzoni	=	Lettera a Jacopo Mazzoni. [Letter to Jacopo Mazzoni.]
Sider.	=	Sidereus Nuncius. [Starry Messenger.]
Tratt.	=	Diversi fragmenti attenenti al Trattato delle cose che stanno in su l'acqua. [Various fragments pertaining to the Treatise on Things that stand in water.]
Discorso	=	Discorso intorno alle cose che stanno in su l'acqua e che in quella si muovono. [Discourse on things that stand in water and move in it.]
Incogn.	=	Postille e frammenti della risposta alle *Considerazioni* di Accademico Incognito. [Marginal notes and fragments of the reply to the *Considerations* of the Unknown Academician.]
Nozzol.	=	Lettera a Tolomeo Nozzolini. [Letter to Tolomeo Nozzolini.]
Risposta	=	Risposta alle opposizioni di L. delle Colombe e di V. di Grazia contro al Trattato delle cose che stanno su l'acqua e che in quella si muovono. [Reply to the objections of L. delle Colombe and V. di Grazia against the Treatise on Things that Stand on Water and Move in It.]
Istoria	=	Istoria e dimostrazioni intorno alle macchie solari e loro accidenti comprese in tre lettere scritte a Marco Welseri. [History and demonstrations concerning sunspots and their phenomena, included in three letters written to Mark Welser.]
Dini	=	Lettere a Mons. Piero Dini. [Letters to Monsignor Piero Dini.]
Cristina	=	Lettera a Madama Cristina di Lorena, Granduchessa di Toscana. [Letter to Madame Cristina of Lorraine, Grand Duchess of Tuscany.]
Copern.	=	Considerazioni circa l'opinione copernicana. [Considerations on the Copernican opinion.]
Flusso	=	Discorso del flusso e reflusso del mare. [Discourse on the flow and reflux of the Sea.]

P. Galluzzi, *Momento: Galilean Studies*, Archimedes 78,
https://doi.org/10.1007/978-3-032-35049-7

Longit.	=	Relazione generale del nuovo trovato di G.G. in proposito del prendere in ogni tempo e luogo la longitudine. [General report on the new discovery of G.G. concerning the taking of longitude at any time and place.]
Sagg.	=	Il Saggiatore. [The Assayer.]
Ingoli	=	Lettera a Francesco Ingoli in risposta alla *Disputatio de situ et quiete Terrae.* [Letter to Francesco Ingoli in response to the *Disputation on the position and rest of the Earth.*]
Cavallo	=	Scritture concernenti il quesito in proposito della stima di un cavallo. [Writings concerning the question about the estimation of a horse.]
Idraul.	=	Scritture attenenti all'idraulica. [Writings concerning hydraulics.]
Accel.	=	De motu accelerato. [On accelerated motion.]
Dial.	=	Dialogo sopra i due massimi sistemi del mondo. [Dialogue Concerning the Two Chief World Systems.]
Rocco	=	Postille alle *Esercitazioni Filosofiche* di A. Rocco. [Marginal notes on the *Philosophical Exercises* of A. Rocco.]
Framm.	=	Frammenti attenenti ai Discorsi e dimostrazioni matematiche. [Fragments pertaining to the Discourses and mathematical demonstrations.]
Discorsi	=	Discorsi e dimostrazioni matematiche intorno a due nuove scienze. [Discourses and mathematical demonstrations concerning two new sciences.]
Percossa	=	Della forza della percossa. Principio di Giornata aggiunta. [On the force of percussion. Principle of the added Day.]
Scritture	=	Scritture e frammenti di data incerta. [Writings and fragments of uncertain date.]
Lettere	=	Lettere di Galileo. [Letters of Galileo.]

MOMENTO

1. Istante, punto di tempo: 415

L'altre aeree impressioni trascorrono in minimi m o m e n t i di tempo non pur la faccia del Sole, ma spazi molto maggiori [*Istoria*, V, 128,13]. Le stelle erranti e fisse di m o m e n t o in m o m e n t o ci si mostrerebbono [*Sagg.*, VI, 318,7,8]. (SIMP.) Voi, Sig. Sagredo, questa volta vi sete dimostrato dissimile a voi medesimo, che solete in un m o m e n t o penetrar tutte le cose [*Dial.*, VII, 238,30]. (SALV.) L'accelerazione si fa continuamente di m o m e n t o in m o m e n t o , e non intercisamente di parte quanta di tempo in parte quanta [*Dial.*, VII, 255,19(2)]. (SALV.) se noi figureremo il tempo diviso in m o m e n t i , cioè in minime particole eguali [*Discorsi*, VIII, 148,27]. Per non si potere in un solo passaggio così bene osservare il m o m e n t o stesso sul principio del moto [*Lettere* (29/XI/602), X, 99,51]. Tali congiunzioni e separazioni vengono osservabili senza errore di mezzo minuto d'ora, mediante la velocità del lor moto e 'l piccolissimo m o m e n t o che media tra 'l toccare e 'l non toccare [*Lettere* (15/VIII/636), XVI, 464,67]. Finito il discostamento, immediatamente, senza interporre m o m e n t o di quiete, l'ho vista tornar in dietro [*Lettere* (30/I/638), XVII, 271,88].

1. Instant, point in time:

The other aerial impressions pass in short m o m e n t i of time not only across the face of the Sun, but over much greater spaces [*Istoria*, V, 128,13]. The wandering and fixed stars would show themselves to us from m o m e n t o to m o m e n t o [*Sagg.*, VI, 318,7,8]. (SIMP.) You, Signor Sagredo, this time have shown yourself to be unlike yourself, who are accustomed to penetrate all things in a m o m e n t o [*Dial.*, VII, 238,30]. (SALV.) Acceleration occurs continuously from m o m e n t o to m o m e n t o , and not intermittently from one part of time to another [*Dial.*, VII, 255,19(2)]. (SALV.) if we imagine time divided into m o m e n t i , that is, into equal minimal particles [*Discorsi*, VIII, 148,27]. For it is not possible to observe the m o m e n t o itself at the beginning of the motion so well in a single pass [*Lettere* (29/XI/602), X, 99,51]. Such conjunctions and separations are observable without an error of half a minute of an hour, by means of the speed of their motion and the very small m o m e n t o that mediates between touching and not touching [*Lettere* (15/VIII/636), XVI, 464,67]. Once the deviation was over, immediately, without interposing a m o m e n t o of rest, I saw it go back [*Lettere* (30/I/638), XVII, 271,88].

Other places:

Cosmogr., II, 216,16; II, 217,24; II, 220,30; II, 221,3; II, 240,17. *Risposta*, IV, 747,11. *Istoria*, V, 208,24. *Flusso*, V, 385,33. *Longit.*, V, 423,10; V, 424,26; V, 424,34. *Sagg.*, VI, 286,14. *Dial.*, VII, 54,20; VII, 87,26; VII, 194,35; VII, 195,3; VII, 203,7; VII, 249,34; VII, 255,4; VII, 300,27; VII, 338,9; VII, 358,4; VII, 377,11(2); VII, 381,7(2); VII, 442,5(2); VII, 455,37. *Rocco*, VII, 723,32; VII, 724,4; VII, 729,28; VII, 733,28; VII, 737,26. *Discorsi*, VIII, 59,21; VIII, 75,20; VIII, 110,4; VIII, 128,30; VIII, 129,3; VIII, 130,15; VIII, 143,4; VIII, 144,18; VIII, 148,12; VIII, 148,18; VIII, 148,29,31; VIII, 148,34-149,1,2; VIII, 150,6; VIII, 224,6; VIII, 225,1; VIII, 276,35; VIII, 344,23. *Lettere* (29/XI/602), X, 98,28; (29/XI/602), X, 99,79; (12/II/611), XI, 46,4; (21/V/611), XI, 112,232; (15/VI/613), XI, 525,40;

P. Galluzzi, *Momento: Galilean Studies*, Archimedes 78,
https://doi.org/10.1007/978-3-032-35049-7

(19/X/622), XIII, 98,21; (5/VII/631), XIV, 281,9; (15/VIII/636), XVI, 464,67; (5/VI/637), XVII, 91,101; (24/I/640), XVII, 154,42.

2. *Fig.* Importanza, peso, autorevolezza:

Mi par che noi assai frequentemente diciamo «Questo è negozio ben grave, ma l'altro è di poco m o m e n t o » e «Noi consideriamo le cose leggiere e trapassiamo quelle che son di m o m e n t o »: metafore, stimere'io, tolte dalla meccanica [*Discorso*, IV, 20,21]. Conosce di non poter produr cosa che sia di m o m e n t o contro all'insuperabil verità [*Risposta*, IV, 549,17]. Quando per altro ella fusse
416 piena d'errori o di cose di poco m o m e n t o [*Dini*, V, 300,7]. Esaminare i m o m e n t i delle ragioni portate dall'una e dall'altra parte [*Dial.*, VII, 396,14]. Non senza m o m e n t i di ragioni [*Lettere* (1/IX/611), XI, 199,732].

2. *Fig.* Importance, weight, authority:

It seems to me that we very often say "This is a very serious matter, but the other is of little m o m e n t o " and "We consider the trivial things and pass over those which are of m o m e n t o ": metaphors, I would think, taken from mechanics [*Discorso*, IV, 20,21]. He knows that he cannot produce anything of m o m e n t o against the insurmountable truth [*Risposta*, IV, 549,17]. When, moreover, it was full of errors or things of little m o m e n t o [*Dini*, V, 300,7]. Examine the m o m e n t i of the reasons given by both sides [*Dial.*, VII, 396,14]. Not without m o m e n t i of reasons [*Lettere* (1/IX/611), XI, 199,732].

Other places:

Tasso, IX, 110,14. *Architett.*, II, 38,15. *Mazzoni*, II, 199,5. *Risposta*, IV, 680,19. *Istoria*, V, 125,20. *Cristina*, V, 322,22. *Copern.*, V, 369,7. *Sagg.*, VI, 234,17; VI, 234,21; VI, 253,31; VI, 277,14; VI, 296,8; VI, 306,7; VI, 314,17; VI, 362,2. *Ingoli*, VI, 533,1; *Cavallo*, VI, 588,4. *Idraul.*, VI, 629,11. *Dial.*, VII, 169,32; VII, 193,23; VII, 198,5; VII, 344,11; VII, 367,27; VII, 399,5. *Rocco*, VII, 721,30; VII, 724,5. *Discorsi*, VIII, 121,14; VIII, 213,30; VIII, 309,6. *Lettere* (8/I/588), X, 23,27; (II/609), X, 232,7; (21/V/611), XI, 116,353; (1/IX/611), XI, 202,832; (26/XII/615), XII, 214,4; (8/I/616), XII, 223,37; (16/I/616), XII, 225,2; (VI/617), XII, 322,35; (9/X/623), XIII, 135,13; (12/IV/625), XIII, 264,20; (7/IV/629), XIV, 32,59; (13/X/632), XIV, 409,121; (3/IX/639), XVII, 98,18.

3. Piccola quantità, in generale:

È già manifesto come l'uno [peso] contrapeserà l'altro e, per consequenza, come, se a uno di essi fosse aggiunto un minimo m o m e n t o di gravità si moverebbe al basso inalzando l'altro [*Mecan.*, II, 163,30]. (SALV.) e quale e quanta sia la forza che da numero immenso di debolissimi m o m e n t i insieme congiunti risulta, porgacene evidentissimo argomento il veder noi un peso di milioni di libbre, sostenuto da canapi grossissimi, cedere e finalmente lasciarsi vincere e sollevare dall'assalto de gl'innumerabili atomi di acqua [*Discorsi*, VIII, 67,11]. (SALV.) È manifesto che, posti i sostegni H, I sotto l'estremità D, F, ogni m o m e n t o che si aggiunga di forza o di peso in E, quivi si farà la rottura [*Discorsi*, VIII, 173,23]. Adunque il fil di ferro sosterrà 1200 braccia, meno un m o m e n t o della sua propria lunghezza [*Lettere* (6/VIII/630), XIV, 129,53].

3. Small amount, in general:

It is already clear how one [weight] will counterbalance the other and, consequently, how, if a minimum m o m e n t o of heaviness were added to one of them, it would move downwards, raising the other [*Mecan.*, II, 163,30]. (SALV.) and what and how great is the force that results from an immense number of very weak momenti joined together, let this furnish us the most evident proof in which we see a weight of millions of pounds, supported by very thick ropes, yield and finally allow itself to be overcome and lifted by the assault of the innumerable atoms of water [*Discorsi*, VIII, 67,11]. (SALV.) It is evident that, placing the supports H, I under the extremity D, F, every momento that force or weight is added to E, the break will occur there [*Discorsi*, VIII, 173,23]. Therefore, the wire will support 1200 arm-lengths, minus one momento of its own length [*Lettere* (6/VIII/630), XIV, 129,53].

Other places:
Mecan., II, 168,17. *Lettere* (18/XII/628), XIII, 463,10.

4. Inclinazione naturale, propensione, capacità (proporzionale al peso):

Ma se nella medesima circonferenza fusse applicata forza animata, la quale avesse momento di far impeto per tutti i versi, potria far l'effetto [*Mecan.*, II, 168,36]. Momento denota non solamente quella forza che ha un corpo al muovere un altro, ma anco quella abilità naturale che hanno i mobili a essere mossi [*Risposta*, IV, 706,32].

4. Natural inclination, propensity, ability (proportional to weight):

But if on the same circumference an animated force were applied, which had the momento to exert an impeto in all directions, it could produce the effect [*Mecan.*, II, 168,36]. Momento denotes not only the force that a body has in moving another, but also the natural ability that mobiles have to be moved [*Risposta*, IV, 706,32].

Other places:
Risposta, IV, 706,30; IV, 707,1. *Discorsi*, VIII, 109,28; VIII, 125,16.

5. Inclinazione determinata, oltre che dal peso, dalla distanza dal fulcro, o dalla velocità del moto:

Momento della gravità

Definizioni generali: Momento è la propensione di andare al basso cagionata non tanto dalla gravità del mobile, quanto dalla disposizione che abbino tra di loro diversi corpi gravi, mediante il **417**
qual momento si vedrà molte volte un corpo contrapesare un altro di maggior gravità [*Mecan.*, II, 159,17,19]. È dunque il momento quell'impeto di andare al basso, composto di gravità, posizione e di altro, dal che possa essere tal propensione cagionata [*Mecan.*, II, 159,26]. Momento appresso i meccanici significa quella virtù, quella forza, quella efficacia, con la quale il motor muove e 'l mobile resiste; la qual virtù depende non solo dalla semplice gravità, ma dalla velocità del moto, dalle diverse inclinazioni degli spazi sopra i quali si fa il moto [*Discorso*, IV, 68,12]. Dal vulgare vostro non solamente non è usata questa voce momento , ma né anco niun'altra comunissima in tutte le matematiche [*Incogn.*, IV, 158,33]. Essendo voi... assai ignudo delle scienze matematiche, delle quali quella parte che considera i momenti della gravità e della velocità de' corpi, che si chiama meccanica, è membro assai nobile e principale [*Rocco*, VII, 726,21]. (SALV.) Devo metter in considerazione come queste forze, resistenze, momenti , figure etc., si posson considerar in astratto e separate dalla materia, ed anco in concreto e congiunte con la materia [*Discorsi*, VIII, 154,15]. (SALV.) È necessario che noi convenghiamo in por distinzione tra queste due maniere di considerare, chiamando un prendere assolutamente quello quando intenderemo lo strumento preso in astratto, cioè

separato dalla gravità della propria materia; ma congiungendo con le figure semplici ed assolute la materia, con la gravità ancora, nominaremo le figure congiunte con la materia momento o forza composta [*Discorsi*, VIII, 155,8].

Variabile secondo la distanza dal fulcro: Nella statera si vede un picciolo contrapeso alzare un peso grandissimo non per eccesso di gravità, ma sì bene per la lontananza dal punto donde viene sostenuta la statera, la quale, congiunta con la gravità del minor peso, gli accresce momento ed impeto di andare al basso, col quale può eccedere il momento dell'altro maggior grave [*Mecan.*, II, 159,25,26]. Centro della gravità si diffinisce essere in ogni corpo grave quel punto, intorno al quale consistono parti di eguali momenti [*Mecan.*, II, 159,30]. I momenti di pesi diseguali vengono pareggiati dall'essere sospesi contrariamente in distanze che abbino la medesima proporzione [*Mecan.*, II, 163,19]. Il momento del peso C non sarà più eguale al momento del peso A, per esser diminuita la distanza del punto F dalla linea della direzione che dal sostegno B, secondo la BI, va al centro della Terra [*Mecan.*, II, 181,26,27]. Quanto la distanza KB è diminuita dalla distanza BA, tanto il momento del peso F è scemato dal momento del peso A. E così parimente, inchinando più il peso, come saria secondo la linea BL, il suo momento verrà scemando e sarà come se pendesse dalla distanza BM, secondo la linea ML [*Mecan.*, II, 181,33(2)-182,1]. (*Salv.*) Conviene ora che cominciamo a investigare secondo qual proporzione vadia crescendo il momento della propria gravità, in relazione alla propria resistenza all'essere spezzato in un prisma o cilindro, mentre, stando parallelo all'orizonte, si va allungando; il qual momento trovo andar crescendo in duplicata proporzione di quella dell'allungamento [*Discorsi*, VIII, 158,35-159,2]. (SALV.) I prismi e cilindri crescere il lor momento sopra le proprie resistenze secondo i quadrati delle lunghezze loro (mantenendo però sempre la medesima grossezza) [*Discorsi*, VIII, 162,25]. (SALV.) De i cilindri e prismi simili i momenti composti, cioè risultanti dalle lor gravità e dalle loro lunghezze, che sono come leve, hanno tra di loro la proporzione sesquialtera di quella che hanno le resistenze delle medesime lor basi [*Discorsi*, VIII, 163,30]. (SAGR.) Io benissimo avevo, compreso la facultà della leva, e come, crescendo o scemando la sua lunghezza cresceva o calava il momento della forza e della resistenza [*Discorsi*, VIII, 175,22].

Variabile secondo la diversa inclinazione dei piani: Nell'inclinare a basso per la circonferenza CFLI il peso posto nell'estremità della linea BC, viene a scemarsi il suo momento ed impeto d'andare a basso di mano in mano più, per esser sostenuto più e più dalle linee BF, BL [*Mecan.*, II, 182,6]. Venendo al basso il grave dal punto C per la circonferenza CFLI, nel primo punto C il suo momento di discendere sia totale ed integro: perché non viene in parte alcuna sostenuto dalla circonferenza, e non è, in esso primo punto C, in disposizione a moto diverso da quello, che libero
418 farebbe nella perpendicolare e contingente DCE [*Mecan.*, II, 182,19]. Il momento integro ed assoluto che ha il mobile nella perpendicolare all'orizonte a quello che ha sopra il piano inclinato HF, avere la medesima proporzione che la linea HF alla linea FK [*Mecan.*, II, 183,1]. Molto adunque importa il considerare per quali linee si facciano i moti e massime ne i gravi inanimati: dei quali i momenti hanno il loro total vigore e la intiera resistenza nella linea perpendicolare all'orizonte; e nell'altre, transversalmente elevate o inchinate, servono solamente quel più o meno vigore, impeto, o resistenza, secondo che più o meno le dette inchinazioni s'avvicinano alla perpendicolar elevazione [*Mecan.*, II, 186,11]. (SALV.) Considerisi... che i momenti o le velocità d'un istesso mobile son diverse sopra diverse inclinazioni di piani e che la massima è per la linea perpendicolarmente sopra l'orizonte elevata e che per l'altre inclinate si diminuisce tal velocità, secondo che quelle più dal perpendicolo si discostano, cioè più obliquamente s'inclinano; onde l'impeto, il talento, l'energia, o vogliamo dire il momento, del descendere vien diminuito nel mobile dal piano soggetto [*Discorsi*, VIII, 215,12,16]. (SALV.) Sopra l'orizontale, che qui s'intende per una superficie egualmente lontana dal medesimo centro e perciò affatto priva d'inclinazione, nullo sarà l'impeto o momento di detto mobile [*Discorsi*, VIII, 215,36]. (SALV.) E perché siamo convenuti che di un mobile tanto sia l'impeto, l'energia, il momento o la propensione al moto, quanta è la forza o resistenza minima che basta a fermarlo e s'è concluso che il grave H è bastante a proibire il moto al grave G, adunque il minor peso H, che nella perpendicolare FC esercita il momento totale, sarà la precisa misura del momento parziale che il maggior peso G esercita per il piano inclinato FA [*Discorsi*, VIII, 217,22,25(2)]. (SAGR.) Parmi che facilmente si possa dedurre argumentando *ex aequali* con la proporzione perturbata, che i

m o m e n t i dell'istesso mobile per piani diversamente inclinati... che abbino l'istessa elevazione, son fra loro in reciproca proporzione de' medesimi piani [*Discorsi*, VIII, 217,36].

Variabile secondo le diverse velocità: Possiamo venire in cognizione come la velocità del moto sia potente ad accrescere m o m e n t o nel mobile, secondo quella medesima proporzione con la quale essa velocità di moto viene augumentata [*Mecan.*, II, 164,24]. Pesi assolutamente eguali, mossi con eguali velocità, sono di forze e di m o m e n t i eguali nel loro operare [*Discorso*, IV, 68,11]. Il secondo principio è che il m o m e n t o e la forza della gravità venga accresciuto dalla velocità del moto; sì che pesi assolutamente eguali, ma congiunti con velocità diseguali, sieno di forza, m o m e n t o e virtù diseguale e, più potente il più veloce, secondo la proporzione della velocità sua alla velocità dell'altro [*Discorso*, IV, 68,32,34]. Onde noi ancora possiamo prendere per verissimo assunto che pesi assolutamente diseguali, alternatamente si contrappesano e si rendono di m o m e n t i eguali, ogni volta che le loro gravità con proporzione contraria rispondono alle velocità de' lor moti [*Discorso*, IV, 69,24]. Una grandissima trave che, v.g., pesi 1000 libbre, potrà essere alzata e sostenuta da acqua che non ne pesi 50; e questo avverrà quando il m o m e n t o dell'acqua venga compensato dalla velocità del suo moto [*Discorso*, IV, 71,6]. Il m o m e n t o composto della poca gravità assoluta dell'acqua ENSF e della grandissima velocità nello abbassarsi, pareggia la forza e 'l m o m e n t o che risulta dalla composizione dell'immensa gravità dell'acqua ABCD con la grandissima tardità nell'abbassarsi [*Discorso*, IV, 77,15,17]. (SAGR.) Ma credete voi che la velocità ristori per l'appunto la gravità? che cioè tanto sia il m o m e n t o e la forza di un mobile, v.g. di quattro libbre di peso, quanto quella di un di cento, qualunque volta quello avesse cento gradi di velocità e questo quattro gradi solamente? [*Dial.*, VII, 241,18].

Momento della percossa

(SIMPL.) Che il grave descendente *vires acquirat eundo*, crescendo la velocità a ragion dello spazio, e che il m o m e n t o dell'istesso percuziente sia doppio venendo da doppia altezza, mi paiono proporzioni da concedersi senza repugnanza e controversia [*Discorsi*, VIII, 203,26]. (SALV.) Imperò che, essendo quello che percuote il medesimo, non può determinarsi la differenza e m o m e n t o **419**
delle percosse se non dalla differenza della velocità [*Discorsi*, VIII, 204,6,8]. (SALV.) Il m o m e n t o di un grave nell'atto della percossa altro non è che un composto ed aggregato d'infiniti m o m e n t i , ciascuno di essi eguale al solo m o m e n t o , o interno e naturale di se medesimo... o estrinseco e violento, quale è quello della forza movente. Tali m o m e n t i nel tempo della mossa del grave si vanno accumulando di instante in instante con eguale additamento e conservando in esso nel modo appunto che si va accrescendo la velocità di un grave cadente; ché siccome negl'infiniti instanti di un tempo, benché minimo, si va sempre passando da un grave per nuovi ed eguali gradi di velocità, con ritener sempre gli acquistati nel tempo percorso, così anche nel mobile si vanno conservando di instante in instante e componendosi quei m o m e n t i , o naturali o violenti, conferitigli o dalla natura o dall'arte [*Percossa*, VIII, 344,8,9,10,13,20]. (SALV.) La forza della percossa è di infinito m o m e n t o , tutta volta che ella si applichi in un momento ed in uno instante dal grave percuziente sopra materia non cedente [*Percossa*, VIII, 344,22]. (SALV.) Se dunque si fa in tempo la cedenza nel luogo della percossa, in tempo ancora si farà l'applicazione di quei m o m e n t i acquistati nel moto dal percuziente; il quale tempo è bastante ad estinguere ed a smorzare in parte quell'aggregato de' sopraddetti m o m e n t i [*Percossa*, VIII, 345,3,5]. (SALV.) La forza della percossa è di infinito m o m e n t o , perché non vi è resistenza, che non venga superata da forza di percossa minimissima [*Percossa*, VIII, 345,24].

5. Inclination determined, in addition to the weight, by the distance from the fulcrum, or from the speed of motion:

Momento della gravità

General definitions: Momento is the propensity to go downwards caused not so much by the heaviness of the mobile, as by the disposition that different heavy bodies have among themselves, through which momento one will often see a body counter-weigh another of greater heaviness [*Mecan.*, II, 159,17,19]. Momento, therefore, is that impeto to go downwards, composed of heaviness, position and other things, from which such a propensity may be caused [*Mecan.*, II, 159,26]. Momento among mechanics means that power, that force, that efficacy, with which the motor moves and the mobile resists; which power depends not only on simple heaviness, but on the speed of the motion, on the different inclinations of the spaces over which the motion occurs [*Discorso*, IV, 68,12]. In your vernacular not only is this word momento not used, but neither is any other very common word in all mathematics [*Incogn.*, IV, 158,33]. You being... quite ignorant of the mathematical sciences, of which that part which considers the momenti of heaviness and the speed of bodies, which is called mechanics, is a very noble and principal member [*Rocco*, VII, 726,21]. (SALV.) I must consider how these forces, resistances, momenti, figures, etc. can be considered in the abstract and separate from matter, and also in the concrete and connected with matter [*Discorsi*, VIII, 154,15]. (SALV.) It is necessary that we agree to distinguish between these two ways of considering, calling an absolute taking that in which we intend the instrument be taken in the abstract, that is, separated from the heaviness of its own matter; but when joining the matter with the simple and absolute figures, with heaviness as well, we will call the figures joined with the matter momento or composite force [*Discorsi*, VIII, 155,8].

Variable according to the distance from the fulcrum: In the steelyard we see a small counterweight lifting a very large weight not because of excess heaviness, but rather because of the distance from the point where the steelyard is supported, which, combined with the heaviness of the smaller weight, increases its momento and impeto to go downwards, with which it can exceed the momento of the other greater heavy body [*Mecan.*, II, 159,25,26]. The center of heaviness is defined to be, in every heavy body, that point around which parts of equal momenti are arranged [*Mecan.*, II, 159,30]. The momenti of unequal weights are equalized by being suspended oppositely at distances having the same proportion [*Mecan.*, II, 163,19]. The momento of weight C will no longer be equal to the momento of weight A, because the distance of point F from the line of direction that from support B, according to BI, towards the center of the Earth, has decreased [*Mecan.*, II, 181,26,27]. As much as the distance KB is diminished by the distance BA, so much the momento of weight F is diminished by the momento of weight A. And so likewise, by inclinding the weight more, as it would be along the line BL, its momento will be diminished and it will be as if it were hanging from the distance BM, along the line ML [*Mecan.*, II, 181,33(2)-182,1]. (*Salv.*) It is now fitting that we begin to investigate in what proportion the momento of its own heaviness increases, in relation to its resistance to being broken into a prism or cylinder, while, standing parallel to the horizon, it is elongated; which momento I find to increase in double the proportion of that of the elongation [*Discorsi*, VIII, 158,35-159,2]. (SALV.) Prisms and cylinders increase their momento above their own resistances according to the square of their lengths (always, however, maintaining the same thickness) [*Discorsi*, VIII, 162,25]. (SALV.) Of similar cylinders and prisms, the composite momenti, that is, resulting from their heavinesses and their lengths, which are like levers, have among themselves the sesquialteral proportion of that which the resistances of their same bases have [*Discorsi*, VIII, 163,30]. (SAGR.) I had very well understood the faculty of the lever, and how, by increasing or decreasing its length, the momento of the force and of the resistance increased or decreased [*Discorsi*, VIII, 175,22].

Variable according to the different inclination of the planes: In inclining downward along the circumference CFLI, the weight placed at the extremity of the line BC comes to be diminished little by little in its momento and impeto to go downward, by being sustained more and more by the lines BF, BL [*Mecan.*, II, 182,6]. Coming down from the point C along the circumference CFLI, the m o m e n t o of descent of the heavy body at the first point C is total and complete: because it is not supported in any part by the circumference, and is not, at the first point C, disposed to a motion other than that which it would freely make in the perpendicular and contingent DCE [*Mecan.*, II, 182,19]. The integral and absolute m o m e n t o that the mobile has on the perpendicular to the horizon to that which it has on the inclined plane HF, has the same proportion that the line HF has to the line FK [*Mecan.*, II, 183,1]. It is therefore very important to consider along which lines the motions are made, especially in inanimate heavy bodies: the m o m e n t i of which have their total force and their entire resistance on the line perpendicular to the horizon; and in the other lines, transversally elevated or inclined, they serve only with more or less force, impeto, or resistance, according as the said inclinations more or less approach the perpendicular elevation [*Mecan.*, II, 186,11]. (SALV.) Consider... that the m o m e n t o or speeds of the same mobile is different on different inclinations of planes and that the maximum is for the line perpendicularly above the horizon and that for the other inclined lines this speed is diminished, according as they depart further from the perpendicular, that is, more obliquely inclined; hence the impeto, the talent, the energy, or we may say the m o m e n t o , of the descent is diminished in the mobile by the subject plane [*Discorsi*, VIII, 215,12,16]. (SALV.) On the horizontal, which here is understood as a surface equally distant from the same center and therefore completely devoid of inclination, the impeto or m o m e n t o of said mobile will be zero [*Discorsi*, VIII, 215,36]. (SALV.) And since we have agreed that the impeto, energy, m o m e n t o or propensity to motion of a mobile is as great as the minimum force or resistance that is sufficient to stop it, and we have concluded that the heavy body H is sufficient to prohibit the motion of the heavy body G, therefore the lesser weight H, which in the perpendicular FC exerts the total m o m e n t o , will be the precise measure of the partial m o m e n t o that the greater weight G exerts along the inclined plane FA [*Discorsi*, VIII, 217,22,25(2)]. (SAGR.) It seems to me that one can easily deduce by arguing *ex aequali* with the perturbed proportion, that the m o m e n t i of the same mobile along differently inclined planes... which have the same elevation, are among themselves in reciprocal proportion to the same planes [*Discorsi*, VIII, 217,36].

Variable according to different speeds: We can come to understand how the speed of motion is capable of increasing the m o m e n t o in the mobile, according to the same proportion with which the speed of motion is increased [*Mecan.*, II, 164,24]. Absolutely equal weights, moved with equal speeds, have equal forces and m o m e n t i in their operation [*Discorso*, IV, 68,11]. The second principle is that the m o m e n t o and force of heaviness are increased by the speed of motion; so that weights absolutely equal, but joined with unequal speeds, are of unequal force, m o m e n t o and power, and the faster the more powerful, according to the proportion of its speed to the speed of the other [*Discorso*, IV, 68,32,34]. Hence we can also take it as a very true assumption that absolutely unequal weights alternately counterbalance each other and become of equal m o m e n t i , whenever their heavinesses correspond in contrary proportion to the speeds of their motions [*Discorso*, IV, 69,24]. A very large beam, for example, weighing 1000 pounds, can be lifted and supported by water that does not even weigh 50 pounds; and this will occur when the m o m e n t o of the water is compensated by the speed of its motion [*Discorso*, IV, 71,6]. The m o m e n t o composed of the small absolute heaviness of the water ENSF and the very great speed in lowering, balances the force and the m o m e n t o that results from the composition of the immense heaviness of the water ABCD with the very great slowness in lowering [*Discorso*, IV, 77,15,17]. (SAGR.) But do you believe that speed exactly compensates for heaviness? That is, that the m o m e n t o and force of a mobile, say, weighing four pounds, are as great as those of one weighing one hundred, provided that the former had one hundred degrees of speed and the latter only four degrees? [*Dial.*, VII, 241,18].

M o m e n t o d e l l a p e r c o s s a

(SIMPL.) That the heavy descending body acquires forces in its motion, the speed increasing in proportion to the space, and that the m o m e n t o of the same percussive body is double, coming from double height, seem to me to be proportions that can be granted without repugnance or controversy

[*Discorsi*, VIII, 203,26]. (SALV.) Because, since that which strikes is the same, the difference and momento of the percussions cannot be determined except by the difference in speed [*Discorsi*, VIII, 204,6,8]. (SALV.) The momento of a heavy body in the act of percussion is nothing but a composite and aggregate of infinite momenti, each of them equal to the sole momento, either internal and natural to itself... or extrinsic and violent, such as that of the moving force. Such momenti in the time of the motion of the heavy body are accumulated from instant to instant with equal addition and are conserved in it in the same way that the speed of a falling heavy body is increased; for just as in the infinite instants of a time, however minimal, a heavy body is always passing through new and equal degrees of speed, always retaining those acquired in the time elapsed, so too in the mobile those momenti, whether natural or violent, are conserved from instant to instant and are composed together, conferred on it either by nature or by craft [*Percossa*, VIII, 344,8,9,10,13,20]. (SALV.) The force of percussion is of infinite momento, whenever it is applied in one moment and in one instant by the heavy percussive body on unyielding matter [*Percossa*, VIII, 344,22]. (SALV.) If therefore the yielding occurs in time at the place of the percussion, the application of those momenti acquired in the motion by the striker will also occur in time; which time is sufficient to extinguish and partly dampen that aggregate of the aforementioned momenti [*Percossa*, VIII, 345,3,5]. (SALV.) The force of percussion is of infinite momento, because there is no resistance that cannot be overcome by the very slightest force of percussion [*Percossa*, VIII, 345,24].

Other places:

Mecan., II, 160,16; II, 161,28; II, 162,17; II, 165,8; II, 168,27; II, 169,11; II, 170,6; II, 171,32; II, 171,33; II, 172,1; II, 172,4; II, 172,6; II, 172,11; II, 174,32; II, 175,20; II, 175,21(2); II, 175,29; II, 175,30; II, 175,31; II, 175,32; II, 177,1; II, 177,25; II, 178,2; II, 178,3; II, 178,26; II, 181,19; II, 181,31; II, 182,32; II, 182,34; II, 183,1; II, 183,10; II, 183,12. *Tratt.*, IV, 55,23. *Discorso*, IV, 68,18; IV, 68,31; IV, 70,4,5,6(2); IV, 70,8; IV, 70,9; IV, 70,12; IV, 74,6; IV, 74,7; IV, 75,33; IV, 75,34; IV, 75,35; IV, 75,37; IV, 76,1; IV, 77,34; IV, 78,17; IV, 78,35; IV, 78,36. *Incogn.*, IV, 158,33; IV, 183,17,18; IV, 184,11. *Nozzol.*, IV, 304,13; IV, 304,14; IV, 308,24(2); IV, 309,30. *Risposta*, IV, 314,35,36; IV, 457,37; IV, 553,10; IV, 581,23; IV, 581,32; IV, 592,31; IV, 599,31; *IV, 667,5; *IV, 667,13; IV, 673,5; IV, 705,15; IV, 705,24; IV, 705,32; IV, 706,30; IV, 706,32; IV, 707,1; IV, 707,6. *Dial.*, VII, 240,35; VII, 279,29. *Rocco*, VII, 726,26; VII, 727,12. *Framm.*, VIII, 438,12; VIII, 438,14,15; VIII, 573,16; VIII, 573,32; VIII, 578,3; VIII, 578,14; VIII, 581,32; VIII, 581,34; VIII, 582,6; VIII, 582,8; VIII, 583,38; VIII, 584,6; VIII, 584,16. *Discorsi*, VIII, 119,16; VIII, 128,10; VIII, 154,30(2); VIII, 154,32; VIII, 155,18; VIII, 155,32; VIII, 155,33; VIII, 156,2; VIII, 156,7; VIII, 156,8; VIII, 156,10; VIII, 156,31; VIII, 156,32; VIII, 157,28; VIII, 157,29; VIII, 157,35; VIII, 158,1; VIII, 158,5; VIII, 159,11; VIII, 159,25; VIII, 159,29; VIII, 159,31; VIII, 163,35; VIII, 163,36; VIII, 164,2; VIII, 164,12(2); VIII, 164,18; VIII, 164,24; VIII, 165,7; VIII, 166,17; VIII, 167,15; VIII, 167,16; VIII, 167,17; VIII, 167,19; VIII, 167,20; VIII, 167,21; VIII, 167,22; VIII, 167,23; VIII, 167,26(2); VIII, 168,16; VIII, 168,17; VIII, 168,19(2); VIII, 168,21(2); VIII, 168,22; VIII, 168,25; VIII, 168,27; VIII, 168,28; VIII, 168,31(2); VIII, 168,33(2); VIII, 169,1; VIII, 169,2; VIII, 172,14; VIII, 172,16; VIII, 172,19; VIII, 172,20; VIII, 172,23(2); VIII, 172,25(3); VIII, 172,26; VIII, 172,27; VIII, 172,31; VIII, 174,30; VIII, 174,31; VIII, 175,13; VIII, 175,32; *VIII, 176,5; *VIII, 176,6; *VIII, 176,8; *VIII, 176,9; *VIII, 176,12; VIII, 178,24; VIII, 207,5; VIII, 207,7; VIII, 207,8; VIII, 207,11; VIII, 207,12; VIII, 207,14; VIII, 207,15; VIII, 207,26; VIII, 216,20; VIII, 216,21; VIII, 217,14; VIII, 217,26; VIII, 217,29; VIII, 291,2; VIII, 330,6. *Percossa*, VIII, 324,11; VIII, 324,15; VIII, 324,32; VIII, 324,33. *Lettere* (29/IX/602), X, 100,91; (27/IX/633), XV, 284,26; (27/IX/633), XV, 284,28; (27/IX/633), XV, 284,30; (1/VIII/639), XVIII, 77,70.

420 6. Grado di velocità (o di tardità):

(SALV.) l'accelerazione si fa continuamente di momento in momento e non intercisamente di parte quanta di tempo in parte quanta, essendo posto il termine A come momento minimo di velocità, cioè come stato di quiete e come primo instante del tempo susseguente AD, è manifesto che avanti l'acquisto del grado di velocità DH, fatto nel tempo AD, si è passato per altri infiniti gradi minori e minori, guadagnati ne gli infiniti instanti che sono nel tempo DA [*Dial.*, VII, 255,21]. (SALV.) Dico per tanto che un corpo grave ha da natura intrinseco principio di muoversi verso 'l comun centro de i gravi, cioè del nostro globo terrestre, con movimento continuamente accelerato, ed accelerato sempre egualmente, cioè che in tempi eguali si fanno aggiunte eguali di nuovi m o m e n t i e gradi di velocità [*Discorsi*, VIII, 118,31]. (SALV.) Per ora basta al nostro Autore che noi intendiamo che egli ci vuole investigare e dimostrare alcune passioni di un moto accelerato (qualunque si sia la causa della sua accelerazione) talmente, che i m o m e n t i della sua velocità vadano accrescendosi, dopo la sua partita dalla quiete, con quella semplicissima proporzione con la quale cresce la continuazione del tempo [*Discorsi*, VIII, 202,30]. (SAGR.) ma perché il mobile scende con velocità accresciuta uniformemente in tutti i tempi eguali, aggiugnerà al grado CB nel seguente tempo CI, quei m o m e n t i medesimi di velocità crescente secondo le parallele del triangolo BFG, eguale al triangolo ABC [*Discorsi*, VIII, 211,15]. (SALV.) Per determinare e rappresentare cotal impeto e velocità particolare, non ha trovato il nostro Autore altro mezo più accomodato, che 'l servirsi dell'impeto che va acquistando il mobile nel moto naturalmente accelerato: del quale qualsivoglia m o m e n t o acquistato, convertito in moto equabile, ritien la sua velocità limitata precisamente, e tanta, che in altrettanto tempo quanto fu quello della scesa passa doppio spazio dell'altezza dalla quale è caduto [*Discorsi*, VIII, 286,37].

6. Degree of speed (or slowness):

(SALV.) The acceleration occurs continuously from moment to moment and not intermittently from portion of time to portion of time, since the terminus A is taken as the minimum m o m e n t o of speed, that is, as the state of rest and as the first instant of the time subsequent AD, it is clear that before the acquisition of the degree of speed DH, achieved in the time AD, it has passed through other infinitely smaller and smaller degrees, gained in the infinite instants that are in the time DA [*Dial.*, VII, 255,21]. (SALV.) I say therefore that a heavy body has by nature the intrinsic principle of moving towards the common center of heavy bodies, that is, of our terrestrial globe, with a continuously accelerated motion, and always equally accelerated, that is, in equal times, equal additions of new m o m e n t i and degrees of speed are made [*Discorsi*, VIII, 118,31]. (SALV.) For now, it is enough for our Author that we understand that he wants to investigate and demonstrate to us certain properties of an accelerated motion (whatever be the cause of its acceleration) in such a way that the m o m e n t i of its speed continue to increase, after its departure from rest, with that very simple proportion with which the continuation of time increases [*Discorsi*, VIII, 202,30]. (SAGR.) but because the mobile descends with speed uniformly increasing in all equal times, it will add to the degree CB in the following time CI, those same m o m e n t i of increasing speed according to the parallels of the triangle BFG, equal to the triangle ABC [*Discorsi*, VIII, 211,15]. (SALV.) To determine and represent such a particular impeto and speed, our Author has found no other means more suitable than to use the impeto that the mobile acquires in naturally accelerated motion: any m o m e n t o of which, when converted into uniform motion, retains its speed precisely limited, and such that in the same time as that of the descent, it passes double the distance of the height from which it fell [*Discorsi*, VIII, 286,37].

Other places:

Framm., VIII, 425,19,20; VIII, 425,22(2). *Discorsi*, VIII, 218,27. *Lettere* (VI/617), XIII, 325,139.

FREQUENCIES of *momento*:

sec.	1	74
»	2	48
»	3	6
»	4	6
»	5	255
»	6	11
	Total	400

MOMENTUM

1. Istante, punto di tempo: **421**

Si ex alta turri duos lapides... eodem momento proiciantur [*De motu*, I, 263,19]. Sit tale tempus, cuius extrema momenta sint *cd*... ergo virtus in momento *c* est aequalis virtuti in momento *d* [*De motu*, I, 327,13,16,17]. Nec in tempore... poterunt assignari duo momenta in quibus aequalis fuerit virtus impellens [*De motu*, I,390,28].

1. Instant, point in time:

If from a high tower two stones... are thrown at the same momento [*De motu*, I, 263,19]. Let there be such a time, whose extreme momenta are *cd*... therefore the force at momento *c* is equal to the force at momento *d* [*De motu*, I, 327,13,16,17]. Nor in time... could two momenta be assigned in which the impelling force was equal [*De motu*, I,390,28].

Other places:

De motu, I, 391,8; I, 391,11; I, 391,14; I, 391,15.

2. *Fig.* Importanza, peso, autorevolezza:

Scopulum enim magni momenti indetectum reliquit [*De motu*, I, 320,6]. Ex his colligi potest, nullius esse momenti Aristotelis argumentum [*De motu*, I, 346,8]. Parvi momenti existimandum minime videtur [*Sider.*, III, 60,6].

2. *Fig.* Importance, weight, authority:

For he left a rock of great momenti undetected [*De motu*, I, 320,6]. From these things it can be gathered that Aristotle's argument is of no momento [*De motu*, I, 346,8]. It hardly seems right to consider it of little momento [*Sider.*, III, 60,6].

Other places:

De motu, I, 283,13; I, 313,22; I, 335,2; I, 359,5; I, 372,31; I,376,31; I,377,36; *Lettere* (8/II/610), X, 467,6.

3. Inclinazione naturale proporzionale al peso (o alla leggerezza):

Themistius... scribit: Cum vacuum itaque cedat aequaliter, sed neque cedat quidem... ita fit ut

P. Galluzzi, *Momento: Galilean Studies*, Archimedes 78,
https://doi.org/10.1007/978-3-032-35049-7

differentiae gravium et levium, idest rerum m o m e n t a , tollantur [*De motu*, I, 294,21].

3. Natural inclination proportional to weight (or lightness):

Themistius writes: Since the void, then, yields equally, but does not yield at all... it comes about that the differences of heavy and light things, that is, the m o m e n t a , are removed.

Other places:
De motu, I, 412,6.

4. Inclinazione determinata, oltre che dal peso, dalla distanza dal fulcro, o dalla velocità del moto:

Ponderum in libra suspensorum m o m e n t a habent rationem compositam ex ratione ipsorum
422 ponderum et ex ratione distantiarum [*Framm.*, VIII, 367,15]. m o m e n t a gravitatis eiusdem mobilis super plano inclinato et in perpendiculo permutatim respondent longitudini et elevationi eiusdem plani [*Framm.*, VIII, 376,1]. Ponatur igitur, augeri vel imminui motus velocitatem secundum proportionem qua augentur vel minuuntur gravitatis m o m e n t a ; et cum constet, eiusdem mobilis m o m e n t a gravitatis super plano *db* ad m o m e n t a super plano *da* esse ut longitudo *da* ad longitudinem *db*, idcirco velocitatem per *db* ad velocitatem per *da* esse ut *ad* ad *db* [*Framm.*, VIII, 379,3,4]. Velocitates mobilium quae inaequali m o m e n t o incipiunt motum, sunt semper inter se in eadem proportione ac si aequabili motu progrederentur [*Framm.*, VIII, 386,12]. Considera, m o m e n t u m in singulis circumferentiae quadrantis punctis imminui pro ratione accessus puncti perpendicularis... ad centrum [*Framm.*, VIII, 417,10]. Constat ex elementis mechanicis, m o m e n t u m ponderis super plano secundum lineam ABC elevato ad m o m e n t u m suum totale esse ut BE ad BA, eiusdem ponderis m o m e n t u m super elevatione AD ad totale suum m o m e n t u m esse ut DF ad DA [*Discorsi*, VIII, 222,2,3,5,6].

4. Inclination determined, other than the weight, by the distance from the fulcrum, or by the speed of motion:

Of weights suspended on the scales, the m o m e n t a have a compounded ratio from the ratio of the weights themselves and from the ratio of the distances [*Framm.*, VIII, 367,15]. The m o m e n t a of heaviness of the same mobile on an inclined plane and on the perpendicular correspond reciprocally to the length and the elevation of the same plane [*Framm.*, VIII, 376,1]. Let it therefore be assumed that the speed of motion increases or decreases according to the proportion by which the m o m e n t a of heaviness increase or decrease; and since it is established that the m o m e n t a of heaviness of the same mobile on plane *db* to the m o m e n t a on plane *da* are as the length of *da* to the length of *db*, therefore the speed along *db* to the speed along *da* is as *da* to *db* [*Framm.*, VIII, 379,3,4]. The speeds of mobiles that begin motion with unequal m o m e n t o are always in the same proportion to one another as if they were proceeding with uniform motion [*Framm.*, VIII, 386,12]. Consider that the m o m e n t u m at each point of the quadrant of the circumference diminishes in proportion to the approach of the perpendicular point... to the center [*Framm.*, VIII, 417,10]. It is established from the elements of mechanics that the m o m e n t u m of a weight on a plane elevated along the line ABC to its total m o m e n t u m is as BE to BA, and that the m o m e n t u m of the same weight on the elevation AD to its total m o m e n t u m is as DF to DA [*Discorsi*, VIII, 222,2,3,5,6].

Other places:
Framm., VIII, 367,17; VIII,367,22; VIII, 367,23; VIII, 367,28(2); VIII, 367,29(3); VIII, 367,30; VIII, 367,32; VIII, 368,1; VIII, 368,4(2); VIII, 368,5; VIII, 368,8(2); VIII, 368,10(2); VIII, 368,15; VIII,

368,16; VIII, 376,7; VIII, 376,8; VIII, 376,12; VIII, 376,16; VIII, 376,17; VIII, 376,20; VIII, 376,21; VIII, 376,22; VIII, 376,23; VIII, 377,1(2); VIII, 378,3; VIII, 378,5; VIII, 378,6; VIII, 378,8; VIII, 378,9; VIII, 386,5; VIII, 386,6; VIII, 386,9; VIII, 386,10; VIII, 386,15; VIII, 386,16; VIII, 417,16(2). *Discorsi*, VIII, 228,8,10.

5. Grado di velocità (o di tardità):

Cumque tarditas in infinitum augeri, aut velocitas imminui, possit, fateri oportebit, mobile aliquando tam immensum obtinuisse tarditatis m o m e n t u m , ut, cum eo latum, vel integri anni curriculo ne spatium transversi digiti pertransisset [*Accel.*, II, 263,15]. In motu ex quiete eadem ratione intenditur velocitatis m o m e n t u m et tempus ipsius motus [*Framm.*, VIII, 281,12]. M o m e n t a velocitatum cadentis ex sublimi sunt inter se ut radices distantiarum peractarum, nempe in subduplicata ratione illarum [*Framm.*, VIII, 380,13]. Si in linea naturalis descensus a principio lationis sumantur duae distantiae inaequales, m o m e n t a velocitatis cum quibus mobile permeat illas distantias sunt inter se in duplicata proportione ipsarum distantiarum [*Framm.*, VIII, 380,16]. In naturali descensu velocitatis m o m e n t a eadem semper reperiri in punctis aequaliter ab orizonte distantibus iuxta perpendiculares distantias [*Framm.*, VIII, 388,4]. Et quia velocitas semper intenditur pro ratione elongationis a termino *a*, constat, in latione *ab* tot esse velocitatis gradus seu m o m e n t a diversa, quot sunt in eadem linea *ab* puncta magis ac magis a termino *a* distantia [*Framm.*, VIII, 388,16]. Determinetur ergo impetus in singulis punctis parabolae *bec* ex potentia m o m e n t i acquisiti per descensum *ab*, quod semper servatur idem et determinat impetum orizontalen, et ex potentia alterius m o m e n t i acquisiti in descensu perpendiculari [*Framm.*, VIII, 427,4,6]. Si mobile lationem suam continuaret iuxta gradum seu m o m e n t u m velocitatis in prima temporis particula acquisitae, motumque suum deinceps aequabiliter cum tali gradu extenderet, latio haec duplo esset tardior ea, quam iuxta gradum velocitatis in duabus temporis particulis acquisitae obtineret [*Discorsi*, VIII, 198,10]. Motum aequalibiter, seu uniformiter, acceleratum dico illum, qui, a quiete recedens, temporibus aequalibus aequalia celeritatis m o m e n t a sibi superaddit [*Discorsi*, VIII, 198,17]. Apparet totidem velocitatis m o m e n t a absumpta esse in motu accelerato iuxta crescentes parallelas trianguli AEB, ac in motu aequabili iuxta parallelas parallelogrammi GB: quod enim m o m e n t o r u m deficit in prima motus accelerati medietate... reficitur a m o m e n t i s per parallelas trianguli **423**
IEF repraesentatis [*Discorsi*, VIII, 209,1,3,5]. Aequalia futura esse spatia tempore eodem a duobus mobilibus peracta, quorum unum motu ex quiete uniformiter accelerato moveatur, alterum vero motu aequabili, iuxta m o m e n t u m subduplum m o m e n t i maximi velocitatis accelerati motus [*Discorsi*, VIII, 209,8,9]. Superveniente vero casu per *bd* parabola *bc* designatur, cuius impetus in termino *c* est compositus ex aequabili transversali, cuius m o m e n t u m est ut *ab*, et ex altero m o m e n t o acquisito in descensu *bd* in termino *d* seu *c*; quae m o m e n t a aequalia sunt [*Discorsi*, VIII, 285,6,7,8]. Impetus seu m o m e n t u m cuiuslibet semiparabolae aequatur m o m e n t o naturaliter cadentis in perpendiculari ad horizontem, quae tanta sit quanta est composita ex sublimitate cum altitudine semiparabolae [*Discorsi*, VIII, 299,6].

5. Degree of speed (or slowness):

And since slowness can increase infinitely, or speed diminish, it must be admitted that a mobile at some point possessed so immense a m o m e n t u m of slowness, that, having been moved with it, would not have passed even the space of a finger's breadth in the course of a whole year [*Accel.*, II, 263,15]. In motion from rest, the m o m e n t u m of speed and the time of the motion are intensified in the same ratio [*Framm.*, VIII, 281,12]. The m o m e n t a of the speeds of a body falling from a height are among themselves as the roots of the distances traversed, that is, in the subduplicate ratio of those distances [*Framm.*, VIII, 380,13]. If on the line of natural descent from the beginning of the motion two unequal distances are taken, the m o m e n t a of speed with which the mobile traverses those distances are among themselves in the duplicate ratio of those distances [*Framm.*, VIII, 380,16].

In natural descent, the m o m e n t a of speed are always found to be the same at points equally distant from the horizon according to perpendicular distances [*Framm.*, VIII, 388,4]. And because speed is always increased according to the ratio of the elongation from the point *a*, it is evident that, in the motion *ab*, there are as many degrees of speed, or different m o m e n t a , as there are points in the same line *ab* more and more distant from the point *a* [*Framm.*, VIII, 388,16]. Let the impetus therefore be determined at each point of the parabola *bec* from the strength of the m o m e n t i acquired by the descent *ab*, which is always preserved the same and determines the horizontal impetus, and from the strength of another m o m e n t i acquired in the perpendicular descent [*Framm.*, VIII, 427,4,6]. If the mobile were to continue its motion according to the degree or m o m e n t u m of speed acquired in the first particle of time, and would then extend its motion uniformly with that degree, this motion would be twice as slow as it would be if it were to proceed according to the degree of speed acquired in two particles of time [*Discorsi*, VIII, 198,10]. I speak of a motion that is equally or uniformly accelerated, which, departing from rest, adds to itself equal m o m e n t a of speed in equal periods of time [*Discorsi*, VIII, 198,17]. It is apparent that the same number of m o m e n t a of speed are consumed in the accelerated motion along the increasing parallels of triangle AEB, and in the uniform motion along the parallels of parallelogram GB: for what is lacking in the first half of the accelerated motion's m o m e n t o r u m ... is replenished by the m o m e n t i s represented by the parallels of triangle IEF [*Discorsi*, VIII, 209,1,3,5]. The spaces traversed in the same time by two mobiles will be equal, one of which moves with uniformly accelerated motion from rest, and the other with uniform motion, according to a m o m e n t u m that is half the m o m e n t u m of the maximum speed of the accelerated motion [*Discorsi*, VIII, 209,8,9]. Coming upon the certain case of the parabola *bc* that is defined through *bd*, whose impetus at point *c* is composed of a uniform transverse motion, whose m o m e n t u m is as *ab*, and of another m o m e n t o acquired in the descent *bd* at point *d* or *c*; these m o m e n t a are equal [*Discorsi*, VIII, 285,6,7,8]. The impetus or m o m e n t u m of any semi-parabola is equal to the m o m e n t o of a body naturally falling perpendicularly to the horizon, which is as large as the composite of the height and the altitude of the semi-parabola [*Discorsi*, VIII, 299,6].

Other places:

Accel., II, 262,29; II, 266,23. *Framm.*, VIII, 375,19; VIII, 380,21; VIII, 380,22; VIII, 388,11; VIII, 388,15; VIII, 428,26(2); VIII, 429,23; VIII, 429,26; VIII, 430,2. *Discorsi*, VIII, 209,4; VIII, 216,11; VIII, 263,4; VIII, 280,3,4; VIII, 280,9; VIII, 280,11; VIII, 280,12; VIII, 280,13; VIII, 281,15; VIII, 282,6; VIII, 282,7(2); VIII, 282,8; VIII, 282,9(2); VIII, 282,12(2); VIII, 282,14; VIII, 282,22; VIII, 284,32; VIII, 285,13; VIII, 285,15; VIII, 285,19; VIII, 295,4; VIII, 299,11; VIII, 299,15; VIII, 299,16; VIII, 299,16; VIII, 299,23.

FREQUENCIES of *momentum*:	sec. 1	9
	» 2	11
	» 3	2
	» 4	57
	» 5	82
	Total	141

INDEX

P. Galluzzi, *Momento: Galilean Studies*, Archimedes 78,
https://doi.org/10.1007/978-3-032-35049-7

INDEX OF NAMES[†]

[†] The number refers to the pages; italics indicate that the name appears in the notes. Numbers linked by a hyphen indicate a succession of pages in which the author is at the center of the discussion. The reference to the page does not exclude that the name also appears in the notes.

P. Galluzzi, *Momento: Galilean Studies*, Archimedes 78,
https://doi.org/10.1007/978-3-032-35049-7

MIX
Papier aus verantwortungsvollen Quellen
Paper from responsible sources
FSC® C105338

If you have any concerns about our products,
you can contact us on
ProductSafety@springernature.com

In case Publisher is established outside the EU,
the EU authorized representative is:
Springer Nature Customer Service Center GmbH
Europaplatz 3, 69115 Heidelberg, Germany

Printed by Libri Plureos GmbH
in Hamburg, Germany